SEAWEEDS OF
THE SOUTHEASTERN
UNITED STATES

SEAWEEDS OF THE SOUTHEASTERN UNITED STATES

CAPE HATTERAS TO CAPE CANAVERAL

Craig W. Schneider and Richard B. Searles

Trinity College, Hartford, and Duke University, Durham

Illustrations by Julia S. Child

Duke University Press   *Durham and London*

© 1991 Duke University Press

All rights reserved.

Printed in the United States of America

on acid-free paper ∞

Library of Congress Cataloging-in-Publication Data

Schneider, Craig W.

Seaweeds of the southeastern United States : Cape

Hatteras to Cape Canaveral / Craig W. Schneider,

Richard B. Searles ; illustrations by Julia S. Child.

Includes bibliographical references and index.

ISBN 0-8223-1101-1

1. Marine algae—South Atlantic States—

Identification.

I. Searles, Richard B.   II. Title.

QK571.5.S68S36   1991

589.4′5—dc20 90-42415 CIP

Dedicated to Ginny and Georgie

CONTENTS

PREFACE

The purpose of this book is to provide a manual for the identification of the seaweeds along the southeastern Atlantic coast of the United States. It is intended as a field guide and laboratory manual for professional and amateur biologists with an interest in the identification of marine plants. The emphasis is therefore on keys, descriptions, and illustrations. Many of the larger, more distinctive seaweeds will be recognizable simply from the illustrations; more difficult taxa will require study of the keys and descriptions. Background and practical information are included in the introductory sections.

The genesis of this book was a joint effort by the authors to investigate the seaweed flora of the deep offshore waters of the Carolinas in the early 1970s. One of the joys of our research was the investigation of the offshore "reefs" and shipwrecks by SCUBA diving. The opportunity to collect while swimming over rock outcroppings never seen before by phycologists or, in most cases, any divers was a unique opportunity that allowed us to explore a special wilderness at our doorstep. At the beginning of our collaboration the only comprehensive report of the seaweeds for the Carolinas was Hoyt's 1920 publication on the marine algae of Beaufort, N.C. During the course of our studies Donald Kapraun published his accounts of the nearshore flora of the Carolinas, but the richness of the offshore waters indicated to us that a comprehensive account of both inshore and offshore seaweeds was needed. At the same time, we began to collect plants in Georgia and realized that with a few additions we could extend the geographic coverage of our book throughout the natural biogeographic region defined by Cape Hatteras and Cape Canaveral. Included in the flora are over 300 species of green, golden brown, brown, and red seaweeds. Faced with the scope of this project and our own limitations of time, energy, and expertise, we chose not to incorporate the Cyanobacteria and vascular flowering plants that grow in the marine and estuarine waters of the region.

The coastal waters of the southeastern United States are a valuable resource, both regionally and nationally. In order to safeguard that resource, we need to know the biological components of the region and understand their interactions. The seaweeds are one of the critical elements in the biological communities in the estuaries and ocean waters of the region, and we hope that this book will make people more aware of these plants, make the plants easier to identify, and stimulate additional studies of them for years to come.

ACKNOWLEDGMENTS

In our preparation of this book and the study of the seaweeds described here we have had the assistance of numerous friends and colleagues, and, at the risk of leaving some out, we thank a few of them: Chuck Amsler, David Ballantine, Dick Barber, Steve Blair, John Brauner, Dan Duerr, Sherry Epperley, Chris van den Hoek, Bill Kirby-Smith, Max Hommersand, Eric Houston, Bill Johansen, Donald Kapraun, Geoff Leister, Paulette Peckol, Joe Ramus, Richard Reading, Steve Ross, Tony Scheer, Susan Vlamynck, James Willis, and Reid Wiseman. Special thanks are due to Paul Silva and Mike Wynne, who, in addition to other kinds of help, each read the rough draft of the manuscript and offered many crucial, critical, and constructive suggestions. Our editor was amazed at the detailed and careful work that they did. Errors of commission and omission have surely crept in since their reading, and for these we accept all responsibility. At the penultimate moment, Geoff Leister came forward to assist with the conversion of the manuscript to the software required for publication.

We are also grateful to the curators and keepers of the herbaria that have made collections available to us. These include the Agardh Herbarium, Lund, Sweden; Allan Hancock Foundation; Botanical Museum, University of Copenhagen; Farlow Herbarium, Harvard University; Muséum National d'Histoire Naturelle, Paris; Harbor Branch Foundation; New York Botanical Garden; Smithsonian Institution; University of Georgia; University of Michigan; University of North Carolina, Chapel Hill; University of North Carolina, Wilmington; and the University of Puerto Rico, Mayagüez.

During the course of the studies leading to the writing of this book we have been supported by grants from the National Science Foundation, the National Oceanographic and Atmospheric Administration, the state of Georgia, and the state of North Carolina. The Duke University Marine Laboratory has been an important base for much of our fieldwork, and without its logistical support our task would have been much more difficult. Our home institutions—Trinity College and Duke University—have been very generous with funds for both research and illustrating the book. Most of the illustrations are the work of a gifted scientific illustrator, Julia S. Child, whom we were fortunate to involve in the project. A few of the illustrations are our own or those of former students at Trinity College, Wendy A. Pillsbury, Richard Reading, Margaret Soltysik, and Shelby Tupper. Charles D. Amsler contributed his expertise on the local species of the Ectocarpaceae by preparing the taxonomic treatment of that family.

The study of the seaweeds in this region has occupied a combined total of over forty years of the authors' lives. During this time our wives, Ginny and Georgie, have provided both the emotional and practical support needed, and we want them to know that it has been appreciated and crucial to the completion of this work.

INTRODUCTION

The coast of the southeastern United States appears barren of seaweeds, but in fact it supports a great diversity of them. In an early account of the seaweeds of the region Harvey (1852) noted that the comparable coast of Europe had more kinds of seaweeds. He attributed the lower diversity here to the prevalence of sandy shores, as contrasted with the rocky coast of Europe. With hundreds of miles of open, sandy beaches, the coast appeared to Harvey as it appears to others: a region mostly devoid of suitable habitats for seaweeds. Investigations going back over 100 years have shown, however, that the waters between Cape Hatteras and Cape Canaveral have a rich and interesting flora of more than 300 seaweed species, even though on some of its shores, particularly the open beaches, seaweeds are few in number. The increase in our knowledge of the flora is reflected graphically in figure 1, which indicates an initial flurry of activity in the mid-nineteenth century, a period of inactivity, and then a steady increase in knowledge through most of the present century, beginning with W. D. Hoyt's notable publication in 1920.

The diversity of seaweeds reflects to some degree the diversity of habitats in the region. Because of the system of barrier islands that fringe the coast and enclose the shallow-water sounds, there are actually many more miles of shore than the distance between the two capes defining the region suggests. Natural rocky shores are almost nonexistent, but in many places jetties, seawalls, groins, bridge supports, and other semipermanent hard structures have been erected that may support algal growth (Hay and Sutherland 1988). Furthermore, large expanses of sheltered, intertidal, and shallow subtidal wetlands lie protected inside barrier islands. Seaweeds are often plentiful in such habitats; they are anchored to shells, worm tubes, or other animal remains, and they grow on or tangled among the marsh grasses and subtidal flowering plants that cover extensive areas in the sounds. These shallow sounds produce large quantities of some seaweeds—enough *Gracilaria*, for instance, to support an agar industry during World War II. In addition, there are areas on the continental shelf offshore where sedimentary rocks emerge through the mantle of sandy sediment to provide a habitat within reach of sunlight that supports a substantial growth of seaweeds.

The seaweeds in the region are of great importance to their marine communities because they provide much of the three-dimensional structure. Along with the vascular plants of the estuaries and the phytoplankton, the seaweeds are the source of the primary production which fuels the abundant populations of herbivores, carnivores, omnivores, detritivores, commensals, and parasites in these waters.

The geographic extent of the flora covered in this work corresponds with the warm temperate waters of the southeastern United States (Searles 1984a). The greatest diversity of seaweeds in this region occurs in North Carolina between

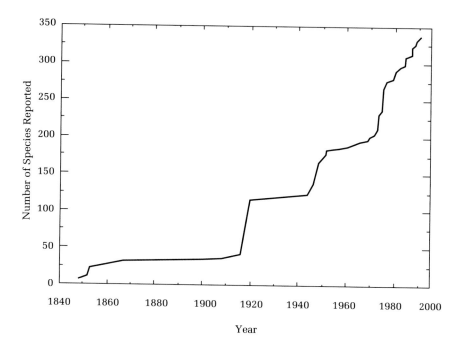

Figure I. Changes in the number of species reported in the region since the 1840s.

Cape Lookout and Cape Fear. A few additional species appear in the waters south of Cape Fear, but no marked change in the flora is observed until the more tropical waters and seaweeds of southern Florida are reached; conversely, north of Cape Hatteras the characteristic species have centers of distribution in cool, northern waters.

Harvey (1852, 1853, 1858), in his three-part treatise on the seaweeds of North America, listed 28 species from the southeastern United States. By the early part of this century, Hoyt (1920) was able to list 123 species and varieties (excluding Cyanobacteria), mostly from North Carolina. At present, 337 taxa are known, and it seems an appropriate time to bring together the increased knowledge of seaweeds into a single volume.

The marine multicellular representatives of four divisions of eukaryotic algae are included in this flora: the Chlorophyta, the Phaeophyta, the Chrysophyta, and the Rhodophyta. Prokaryotic algae (Cyanobacteria or Cyanophyta) have not been included. For identification of marine Cyanobacteria we suggest reference to Humm and Wicks (1980). For identification of the marine and brackish-water flowering plants we suggest Godfrey and Wooten (1979, 1981).

**SEAWEEDS OF
THE SOUTHEASTERN
UNITED STATES**

HISTORY OF PHYCOLOGY IN THE REGION

Study of the seaweed flora of the mid-Atlantic region began in the mid-1800s when local naturalists sent specimens to experts, most of whom were in Europe. Some of these experts also visited the region, including the noted phycologist William H. Harvey from Trinity College, Ireland. In addition, information came from reports by people responsible for general surveys of the total flora. The situation began to change in the early 1900s as local colleges began to develop into major universities; phycologists were added to their faculties and they gradually developed the diversity of expertise in marine botany we now enjoy.

The first records of seaweeds in the region were published by Jacob W. Bailey (1848, 1851), a member of the faculty at West Point. He reported on plants sent to him from Charleston, South Carolina, by Lewis R. Gibbes and plants he himself collected on a trip to South Carolina and Florida. Bailey corresponded with several of the phycologists in Europe, with Harvey in particular.

Harvey was an oddity among the phycologists of his era, for he actually traveled abroad, personally making many of the collections on which he reported. Most of his colleagues, in contrast, remained in Europe and only processed plants collected and sent to them by resident overseas naturalists or those who were accompanying the voyages of exploration and commerce that were characteristic of the times. In 1850 Harvey traveled along the coast collecting seaweeds between Halifax, Nova Scotia, and Key West, Florida. In January, and again in March, he visited Charleston, South Carolina. There he collected seaweeds with Lewis Gibbes, a local collector who had previously sent specimens to him in Dublin. Harvey also received Charleston specimens from H. W. Ravenel and specimens from North Carolina from C. Congdon. The results of his observations on these plants were published in 1852, 1853, and 1858 as his *Nereis Boreali-Americana; or Contributions to a History of the Marine Algae of North America* under the auspices of the Smithsonian Institution. In this three-part tome he listed twenty-eight species from the region: two from North Carolina, twenty-four from South Carolina, and two not specifically from the region but from all parts of the American coast. One of the South Carolina plants was described as the new species *Grateloupia gibbesii* Harvey, the epithet honoring its collector.

The Reverend Mr. M. A. Curtis (1867) was the first local botanist to publish on the seaweeds of the region; he listed thirty-six species of red, brown, and green seaweeds as well as four blue-green algae in his catalogue of indigenous and naturalized plants of North Carolina. Some of the specimens remain from this early collection, and those that have not been found are readily assignable to common members of the flora. Sixteen of his collections were new records for the region.

J. Cosmo Melvill (1875) published a short account of collections he made in the Charleston area and in Key West, Florida, but within the region these included only three species not previously listed by Harvey.

Duncan S. Johnson of Johns Hopkins University visited the U.S. Fish Commission Laboratory in Beaufort, North Carolina, in June 1899. Based on observations he made there, he published (1900) an account of the algae of the adjacent sounds, listing fifteen species of seaweeds.

The single most illuminating account of the seaweeds in the region in those early days came from a professor of biology at Washington and Lee University, W. D. Hoyt. He collected plants primarily in the region surrounding Beaufort, North Carolina, in the years 1903–1909, but he also visited sites from Okracoke, North Carolina, through South Carolina to Tybee, Georgia. His description of 133 species, varieties, and forms of seaweeds and Cyanobacteria (1920) was more comprehensive than previous accounts for the region and indicated a diverse and interesting flora. Hoyt described the seasonality of North Carolina seaweeds and the overlap of tropical and temperate species that occurs just south of Cape Hatteras. In addition to working along the shore, Hoyt made two dredge collections from offshore in deep water; these were a preview of the rich collecting that would follow at offshore locations in later years.

A further contribution by Hoyt was his study of the reproduction and life history of local populations of *Dictyota* (Hoyt 1907, 1927). With cultured plants he demonstrated the alternation of their generations and observed a monthly periodicity in the formation of gametes in which eggs and sperm are released at the time of the full moon at flood tides. This pattern contrasted with the twice-monthly release of gametes in what was considered to be the same species, *D. dichotoma*, in European waters. Hoyt therefore recognized a new variety, *menstrualis*, which has only recently (Schnetter et al. 1987) been elevated as a separate species, in part on the basis of the differences noted by Hoyt. In addition to that variety, Hoyt described alone (1920) or together with M. A. Howe of the New York Botanic Gardens (1916) ten new species of algae based on plants from the region, most of them small and obscure epiphytic or endophytic plants.

In 1904 James J. Wolfe, a biologist who had studied algae for his Ph.D. at Harvard, was hired on the staff at Trinity College in Durham, North Carolina (later to become Duke University). A native South Carolinian, he was the first resident specialist on the algae in the region. At Beaufort, North Carolina, he followed up Hoyt's study of *Dictyota* with similar studies of the local species of *Padina*, *P. gymnospora* (as *P. variegata*). He demonstrated its alternation of generations and the genetic determination of male and female sexuality while disproving the suggestion that eggs developed parthenogenetically (1918). He died at an early age in 1920, leaving unfinished a monograph on the marine diatoms of the region.

Wolfe was followed at Duke University by the first chairman of its new Department of Botany, Hugo L. Blomquist, an eclectic systematist who published extensively on bryophytes and flowering plants as well as seaweeds. He authored a series of papers, mostly on ectocarpalean algae (Blomquist 1954, 1955, 1958a, 1958b), but started with an initial account (Blomquist and Pyron 1943) of seaweeds which drifted up on the beach near Beaufort, North Carolina, following a major hurricane in 1940; that account was followed by a listing of additions to the local flora of Beaufort (Blomquist and Humm 1946).

One of Blomquist's students, Louis G. Williams, made reciprocal transplant studies of local *Codium* "species," demonstrating that plants reported as *C. tomentosum* were ecophenic variants of *C. decorticatum* (1948b). Williams's greatest contribution (1948a, 1949) was a detailed study of the seasonal changes in the flora at the Cape Lookout jetty in North Carolina. These observations on seasonality expanded those made by Hoyt at nearby Beaufort. In an investigation of the algae on the "Black Rocks" off the North and South Carolina coast, Williams (1951) also published the first report of seaweeds in the region collected by divers.

During this same general time T. A. and Anne Stephenson were conducting the North American part of their worldwide descriptive investigation of intertidal communities. In 1952 they published an account of intertidal plants and animals of Charleston, South Carolina, and Beaufort, North Carolina, complementing Williams's intertidal studies.

Harold J. Humm, another student of Blomquist's at Duke University, later became part of the faculty. Humm demonstrated a variety of interests. During the 1940s he became involved in the wartime effort to find substitutes for the agar products which had previously been imported from Japan. He and others published a series of papers describing agar, local seaweed resources, and the suitability and characteristics of extracts from *Hypnea musciformis* and *Gracilaria verrucosa* (as *G. confervoides*; Humm 1942, 1944, 1951; Humm and Wolf 1946; Micara 1946; Causey et al. 1946; DeLoach et al. 1946a, 1946b). During the war a viable agar industry was centered in Beaufort, North Carolina, using locally collected *Gracilaria* as well as plants brought in from elsewhere along the East Coast.

In 1952 Humm described the flora on the intertidal rocks at Marineland, Florida, the only account of seaweeds from the coast of Florida north of Cape Canaveral since the visit of Bailey to Saint Augustine more than 100 years earlier. Several papers followed, written by or with students about the seaweeds of North Carolina (Aziz and Humm 1962; Earle and Humm 1964; Humm and Cerame-Vivas 1964; Aziz 1967).

William Randolph Taylor published two landmark books on the floras of the western Atlantic Ocean which in part covered the waters of the southeastern United States. The first, published in 1937 (revised in 1957), covered the seaweeds of the northeastern United States, and the second, published in 1960, concerned the seaweeds of the tropical and subtropical coasts from Cape Hatteras to Brazil. The mid-Atlantic coast of the United States lies at the edges of these two regions, and Taylor's books were therefore the greatest stimulus and aid to study of the local seaweeds since the publication of Hoyt's 1920 report.

In the early 1960s the University of North Carolina at Chapel Hill added Max H. Hommersand to its staff, and he trained a number of students who contributed to the knowledge of the seaweeds of North Carolina and the region. Donald W. Ott investigated the marine representatives of the xanthophycean genus *Vaucheria* in North Carolina (Ott and Hommersand 1974); Carol Aregood (1975) studied the red alga *Nitophyllum medium* and placed it in a new genus, *Calonitophyllum*; Gayle I. Hansen (1977a, 1977b) completed a morphological and cultural study of another red algal genus, *Cirrulicarpus*, and in particular the local species *carolinensis*; Charles F. Rhyne (1973) investigated the biology of the local species of the green alga, *Ulva*, discovering that most plants belonged to two species not previously recognized as part of the flora; Joy F. Morrill (1976) included local Rhodomelaceae in her study of the dorsiventral members of that group; Joseph P. Richardson resolved questions about the ecology of the seasonal seaweeds *Dictyota menstrualis* (as *D. dichotoma*; 1979), *Dasya baillouviana* (1981), and *Bryopsis plumosa* (1982), which disappear during parts of the year; and Paul W. Gabrielson included regional members of the Solieriaceae in his extensive study of species in that family (Gabrielson 1983; Gabrielson and Hommersand 1982a, 1982b).

Humm moved on to the University of South Florida, and Richard B. Searles came to Duke University and trained several students who studied the local flora. Two of these—D. Reid Wiseman and James Fiore—began their studies with Humm. Wiseman's studies included a survey of the algae from South Carolina (1966, 1978); Fiore (1969, 1977) investigated the life histories of several brown algae, naming the new genus *Hummia* on the basis of some of those studies (1975). Nancy J. Alexander (1970) investigated the genus *Enteromorpha* on the jetty at Fort Macon; John F. Brauner (1975) made a seasonal investigation of the epiphytic algae on seagrasses in the Beaufort, North Carolina, area; and, somewhat later, Mitsu M. Suyemoto (1980) studied the crustose coralline red algae in the offshore waters of Onslow Bay.

Working with Searles, initially as a student, Craig W. Schneider began an investigation of the seaweeds in offshore waters, continuing the work started by Hoyt. Schneider, Searles, and their students combined to publish a series of papers expanding and refining the knowledge of the flora (Reading and Schneider 1986; Schneider 1974, 1975a, 1975b, 1975c, 1980, 1984, 1988, 1989; Schneider and Eiseman 1979; Schneider and Searles 1973, 1975, 1976; Searles 1972, 1984b,

1987; Searles, Hommersand, and Amsler 1984; Searles and Leister 1980; Searles and Lewis 1982; Searles and Schneider 1990; Wiseman and Schneider 1976). Schneider also published monographic studies of the genera *Audouinella* (1983) and *Peyssonnelia* (Schneider and Reading 1987) from the region and described two new genera of red algae, *Searlesia* and *Calliclavula*. Searles described the endophytic brown algal genus *Onslowia* and the red algal genus *Nwynea*. Between them they described twenty-five new species of green, brown, and red seaweeds and with others expanded the known flora from the offshore waters of the region by ninety-nine species. In the process they provided the data for an analysis of the biogeographic relationships of the flora within the region and between neighboring regions and for the productivity of the offshore waters (Schneider 1975d, 1976; Schneider and Searles 1979; Searles and Schneider 1980; Searles 1984a).

Another Searles student, Paulette Peckol, conducted descriptive and experimental ecological studies of the deep-water algal communities on a rock ledge off Cape Lookout (Peckol 1982; Peckol and Searles 1983, 1984). She also made comparative physiological studies of deep-water and shallow-water seaweeds (Peckol 1983; Peckol and Ramus 1985).

In Georgia, Russell L. Chapman (1971, 1973) published accounts of the marine algae of that coast, and Stephen M. Blair and Margaret O. Hall (1981) described a collection of plants from offshore South Carolina and Georgia waters. Searles, stimulated by the establishment of a small rocky area off the coast of Georgia as the Gray's Reef National Marine Sanctuary, published several papers (Searles 1981, 1983, 1987; Searles and Ballantine 1986) and a field guide (1988) describing the offshore flora and several new species. Joseph Richardson, having joined the faculty at Savannah State University, published two papers (Richardson 1986, 1987) on the seaweeds of the near-shore waters of Georgia.

With the appointment of Donald F. Kapraun to the faculty of the University of North Carolina at Wilmington in the mid-1970s there was for the first time a marine phycologist who was a full-time resident at the coast. Kapraun turned much of his attention to the local flora and produced a series of papers clarifying the taxonomic concepts in difficult genera such as *Polysiphonia*. He used controlled culture conditions to study the physiological, cytological, and morphological characteristics of the plants (1977a, 1977b, 1978a, 1978b, 1978c, 1979; Kapraun and Freshwater 1987; Kapraun and Luster 1980; Kapraun and Martin 1987). He also used data from his cultural studies, which related temperature and light conditions to growth and reproduction, in a floristic analysis of the region (1980b). Kapraun and Zechman (1982) made a study of the phenology and vertical distribution of the seaweeds on the jetty at Masonboro Inlet. Kapraun's observations of the seaweeds along the coast and in the sounds of North Carolina were brought together in a two-volume work (Kapraun 1980a, 1984) describing that flora, the first such work since the pioneering publication of Hoyt. One of Kapraun's students, Charles D. Amsler, investigated the ectocarpalean algae of the region (Amsler 1983, 1984a, 1984b, 1985; Amsler and Kapraun 1985), comparing plants in culture with field-collected plants. He has contributed the section on the Ectocarpaceae in this book.

Work on the local seaweeds continues as a new generation of phycologists brings new techniques and insights to the study of seaweeds. Taxonomic studies are giving way to, or are being supplemented by, ecological, physiological, genetic, and cytological investigations using the wealth of species available. These investigations will ultimately lead to a clearer taxonomic and floristic understanding of the seaweeds of the region.

THE GEOLOGICAL ENVIRONMENT

The region between Cape Hatteras and Cape Canaveral (figure 2) is almost exclusively a region of sandy shores. The nearest natural rocky coastlines to the north are in Long Island Sound. The sandy shore extends south from there, interrupted occasionally by capes and inlets, to southern Florida, where rocky outcroppings reappear as the limestone remains of coral formations. The most prominent cape on the mid-Atlantic coast of the United States is Cape Hatteras, which marks the northern boundary of the region covered in this flora. The only natural rocky shores within the region are very small outcroppings of sedimentary rock at Marineland, Florida, and Fort Fisher, just north of Cape Fear, North Carolina.

The mainland is drained by rivers that originate either in the mountains, the piedmont foothills, or within the coastal plain. The rivers deposit their sediments in the sounds behind the barrier islands. The sands of the outer coast beaches are not recent deposits; they are relict sediments incorporated into the beaches and islands as the shoreline moved inland.

Sea level retreated beyond the edge of the continental shelf more than 20,000 years ago; 18,000 years ago it began to rise and move inland; in the last 5,000 years sea level has risen only a few meters, but the rate at which it has been rising has been increasing in the present century and is currently reflected in high rates of erosion and island migration along many parts of the coast.

The characteristic features of the coast are the barrier islands which front the shallow-water sounds. At the northern end of the region the sounds stretch almost unimpeded up the coastal plain where they form the broad, shallow, highly productive waters of the Pamlico Sound and Neuse River estuary. Along much of the coast to the south the sounds push up against relict barrier islands and are long and narrow. Along the Georgia coast the new barrier islands have fused with old barrier islands to form the relatively broad Georgia Sea Islands. In Florida, like the Carolinas, the sounds are again long and narrow. In North Carolina and northeastern Florida, where tidal ranges are small, the barrier islands are long and inlets are infrequent. In the middle of the region (South Carolina and Georgia) tidal ranges double, increasing the volume of water that enters and exits the sounds twice each day. As a result, inlets tend to occur more frequently, making the islands shorter.

The continental shelf off the barrier islands varies in width. At Cape Hatteras it is about 35 km wide and forms Diamond Shoals. The continental shelf becomes progressively wider to the south until, off Georgia, it is 135 km wide; it then narrows along the Florida coast to 55 km off Cape Canaveral. On the continental shelf, sedimentary rocks are generally covered by a thin layer of shell and sand or muddy sediment unsuitable for anchoring most seaweeds. Rocks emerge through the sediments in some areas across the width of the shelf, and there is an almost continuous band of exposed rock along the outer margin of the continen-

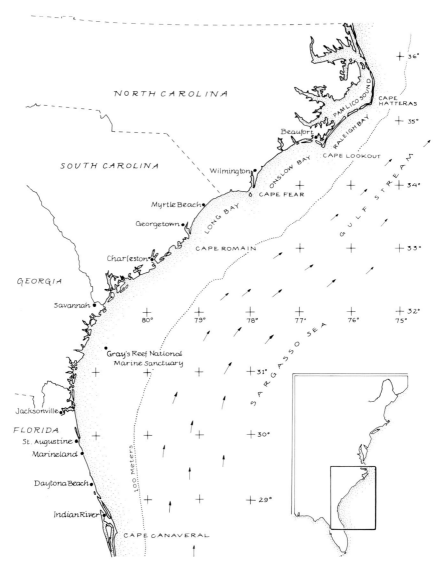

Figure 2. Coastline of the southeastern United States from Cape Hatteras to Cape Canaveral. For additional localities not shown, see the list of collection sites on page 509.

tal shelf. When these outcroppings lie within photic depths, they may support abundant seaweed populations. Such a situation occurs in Onslow Bay, North Carolina, into which no major sediment-carrying rivers empty. In Georgia, one of the outcrop areas with its associated wealth of plants and animals has been designated as the Gray's Reef National Marine Sanctuary.

THE HYDROGRAPHIC ENVIRONMENT

CURRENTS, TIDES, AND WAVES

The Florida Current, or Gulf Stream, originates in the Florida Straits from waters of the Gulf of Mexico. As it flows northward it is joined by the Antilles Current, which moves northward outside the Bahama Islands. The Florida Current usually lies offshore and parallel to the edge of the continental shelf as it moves along the coast to Cape Hatteras; there it leaves the coast and moves out to sea to become the Gulf Stream proper. The position of the Florida Current changes seasonally; in the winter it is driven offshore by the dominant northerly wind. When this happens, the cold waters of the Virginia Coastal Current may move into the northern part of the region, slipping around Cape Hatteras and along the North Carolina coast (Gray and Cerame-Vivas 1963; Stefansson et al. 1971). In summer the Florida Current moves inshore and brings warm, clear water to the continental shelf. This allows more sunlight to reach the bottom, which during the winter may have been below the compensation depth for seaweed photosynthesis. Inshore currents south of Cape Hatteras are variable. South-flowing geostrophic currents are periodically interrupted by inshore movement of the Florida Current, causing northward flow.

Tidal currents can be very strong, particularly adjacent to and inside the inlets through which the sounds empty and fill. Tidal ranges vary from a little more than 1 m near Cape Hatteras and Cape Canaveral, where the continental shelf is narrow, to over 2 m along the Georgia coast. The average height of waves is 1.7 m at Cape Lookout, North Carolina, and 0.8 m at Jacksonville, Florida.

LIGHT

Within the sounds, where tidal currents suspend large quantities of sediment and phytoplankton abundance can be high, light penetration is limited and seaweeds do not generally grow more than 1–2(–4) m below the low tide line. Water clarity can be much greater across the continental shelf, with seaweeds growing as deep as 90 m at the edge of the shelf (Schneider 1976). Water clarity can change markedly, however, in relatively short periods of time. In fall and winter, clarity decreases as storm waves suspend sediment particles and phytoplankton exhibits increased growth. In spring and summer, intrusions of clear "Gulf Stream" water allow increased light penetration, which is correlated with a seasonal increase in the biomass of seaweeds on the continental shelf (Schneider and Searles 1979). The episodic nature of the changes in light availability on the continental shelf are shown in the data of Peckol and Searles (1984) for a mid-shelf station off Cape Lookout, North Carolina.

TEMPERATURE

There are three different temperature transitions of importance within the region. From north to south there is the abrupt temperature change off Cape Hatteras, where the warm waters of the Gulf Stream meet Virginia coastal water (Gray and Cerame-Vivas 1963). A north-south temperature transition of less magnitude occurs at Cape Canaveral (Parr 1933).

There are also transitions in temperature from offshore to inshore and from season to season. In the sounds of North Carolina water temperatures can exceed 30°C in summer and drop to 0°C in winter or spring. In summer, bottom temperatures offshore are in the mid to upper 20s on the inner and middle shelf and down to 10°C at the edge of the shelf at depths where light becomes limiting for algal growth. In winter, bottom temperatures on the middle shelf remain in the middle teens; temperatures decrease toward the edge of the shelf and toward shore (Stefansson and Atkinson 1967). Seasonal temperature changes may therefore exceed 30°C in the sounds but vary only 15°C to 18°C on the inner and mid-shelf and 8°C to 9°C at the shelf edge. The central Florida coast also has high summer temperatures similar to those in the Carolinas (30°C), but these drop only to 14°C in the winter (Mook 1980).

WATER CHEMISTRY

Salinity ranges in ocean waters do not vary enough from 36 percent to affect seaweed growth. In the estuaries there is a full range of salinities from near oceanic values at the inlets to fresh water near the heads of the estuaries (Hoyt 1920; Kapraun and Zechman 1982; Litaker et al. 1987). The abundance of dissolved nutrients of importance to seaweed growth can vary drastically, particularly in the estuaries where there are large changes in the course of a day (Litaker et al. 1987; Ramus and Venable 1987). In offshore waters the changes recorded are tied more closely to the seasons, with relatively high nutrient concentrations in winter contrasted with low summer levels. Ramus and Venable (1987) have demonstrated marked differences in the ability of local seaweed species to utilize short-term, episodic increases in available nutrients. They found that ephemeral weedy plants such as *Ulva* can rapidly take up available nutrients; more persistent long-lived plants such as *Codium* are relatively unaffected by pulses of nutrients in the environment.

BIOGEOGRAPHY AND ECOLOGY

CHARACTERISTICS AND LIMITS OF THE REGION

Hoyt (1920) listed a large percentage of species in the Carolinas that were at their northern or southern limits of distribution. The Stephensons (1952) studied intertidal communities at Marineland, Florida; Charleston, South Carolina; and Beaufort, North Carolina. On the basis of the species of plants and animals present and the average monthly temperatures at these locations they characterized the region as warm temperate. They identified four elements within the flora: (1) eurythermic tropical species, (2) eurythermic cold-water species, (3) eurythermic cosmopolitan species, and (4) mid-Atlantic species with distributions centered in the region.

Humm (1969), instead of considering the region a distinct entity, viewed it as a long transition zone between a Caribbean-centered tropical flora and a North Atlantic–centered cool/cold temperate flora. He said that elements of each flora extended different distances into or across the region, but he recognized no species with a distribution centered in the region.

Van den Hoek (1975) agreed with the Stephensons in recognizing the region as warm temperate and was more specific and quantitative in describing its relationships with other floras. Searles (1984a) discussed the boundaries of the region. The northern boundary at Cape Hatteras is relatively distinct: north of the cape the flora is characterized by the gradual disappearance from north to south of northern species that reach their southern limit of distribution but are not replaced by southern species. Humm (1979) listed only 141 species, exclusive of the blue-green algae, in the Virginia flora. South of Cape Hatteras there is a marked increase in the number of species. Of the 338 taxa reported in the region, almost half (49 percent) are warm-water species at their northern limit of distribution, and most of these extend all the way to the waters of North Carolina. A smaller percentage (13 percent) are northern species that spill south around Cape Hatteras, mostly ending their distribution in North Carolina, with a small number extending on into South Carolina, Georgia, or northern Florida.

The southern boundary of the region is less distinct. Searles (1984a) reported that the waters just to the south of Cape Canaveral have a distinctly more tropical composition, with 42 percent of the species reported there being at their northern limit of distribution. In the less diverse flora north of Cape Canaveral at Marineland only 9 percent of the species are at their northern limit of distribution. The pattern of distribution is complicated by the presence of a number of temperate-water species in the northeastern Gulf of Mexico that are evident in winter (Earle 1969). Taylor (1955) characterized the flora in the northern and northeastern part of the gulf as an impoverished Caribbean flora rather than a temperate flora.

INSHORE AND OFFSHORE FLORAS

The differences between the flora of inshore shallow waters and that of the offshore continental shelf in the Carolinas have been emphasized by Searles and Schneider (1980). The offshore flora is dominated by tropical species, 37 percent of which are at their northern limit of distribution and only 1 percent at their southern limit. The shallow-water flora, on the other hand, is more evenly distributed between northern (10 percent) and southern (21 percent) elements. In both inshore and offshore floras most species have ranges extending north and south beyond the region.

Approximately one-third of the flora (36 percent) is restricted to shallow water, and one-third (38 percent) to deep water, with the remaining 26 percent represented in both environments.

Peckol and Ramus (1985) compared the photosynthetic abilities of different populations of *Sargassum filipendula*, a dominant species in both the shallow subtidal zone on jetties and in deep offshore waters, and demonstrated physiological differences between the populations suggestive of genetic divergence. Other species, such as *Dasya baillouviana*, show morphological and color variations between shallow-water and deep-water populations.

ENDEMIC AND DISJUNCT SPECIES

Van den Hoek (1975) recognized only five endemic species for the region. Searles (1984a) listed thirty-one species that are known only from the region or have distributions centered in the region but spill across the southern boundary at Cape Canaveral. The present twenty-five species restricted to the region (table 1, p. 510) constitute 7 percent of the total flora. Table 1 lists twelve other species whose ranges also extend farther into Florida or to Bermuda. An additional fifteen species with disjunct distributions are listed in table 2 (p. 510); these species are restricted in North America to the Cape Hatteras–Cape Canaveral region.

SEASONALITY

The observations on the seasonality of seaweeds made by Hoyt (1920) have been augmented in North Carolina by Williams (1948a, 1949), Richardson (1978), and Kapraun and Zechman (1982) for the intertidal and near-shore subtidal communities; by Schneider (1976), Peckol (1982), and Peckol and Searles (1983) for deep-water communities in the Carolinas; and by Searles (1987) for Georgia's offshore waters.

Seasonal changes in the shallow-water flora of southern North Carolina were described by Kapraun and Zechman (1982). A winter-spring flora disappears in May–June and is replaced by a summer flora that reaches a peak in species number in July–August and then slowly declines in the fall. These seasonal species are superimposed on a large group of species that are present throughout the year. Kapraun and his students have experimentally investigated the responses of a variety of plants from shallow water to combinations of light and temperature (Kapraun 1977b, 1978a, 1978b, 1979; Amsler 1984b, 1985; Amsler and Kapraun 1985), relating the results to the seasonal abundance of the species. Richardson (1979, 1981, 1982) demonstrated the ability of species of *Dasya*, *Dictyota*, and *Bryopsis* to survive as protonemalike or holdfast stages prior to their seasonal appearance as macroscopic plants on the jetties in North Carolina.

There is no distinct winter flora in the offshore waters. There are only a few perennial plants which persist through the winter and begin to renew growth in the spring. Beginning in the spring and extending through the summer, there is a gradual increase in the number of species and the biomass of plants; a peak in both parameters occurs in July–August (Schneider 1976, Schneider and Searles 1979; Searles 1987). In North Carolina this summer flora disappears gradually during the fall; at Gray's Reef in Georgia it disappears precipitously in late August or September (Searles 1987). A number of species (e.g., *Dasya baillouviana* and *Grinnellia americana*) that are restricted to the winter-spring flora in shallow waters are members of the summer flora offshore.

PHYSIOLOGICAL ECOLOGY

In addition to the studies listed above related to seasonality, several investigators have studied a variety of plants from the shallow-water sounds and from offshore waters in attempts to understand the photosynthetic, nutrient uptake, and growth characteristics of the species (Coutinho and Zingmark 1987; Duke, Lapointe, and Ramus 1986; Peckol 1983; Ramus and Rosenberg 1980; Ramus and Venable 1987; Rosenberg and Ramus 1982a, 1982b, 1984). Local species of seaweeds have also been used in refutation of Engelmann's theory of complementary chromatic adaptation as an explanation for the distribution of seaweeds along the gradients of light quality associated with increasing depth (Ramus 1983; Ramus and van der Meer 1983).

BIOLOGICAL INTERACTIONS

In their studies of the intertidal, Kapraun and Zechman (1982) observed seasonal changes in the vertical distribution of some species that they attributed to competition and narrow tolerance for some seasonally changing physical factors. Hay and Sutherland (1988) suggested that the major factors controlling the composition of the algal communities are desiccation during low tide and grazing by fish during high tide. Although the herbivorous parrotfish and surgeonfish of the tropics are absent or not important in the region, there are omnivorous fishes (e.g., the sheepshead, *Archosargus probatocephalus*) that prefer some intertidal algae and invertebrates (Ogburn 1984). Hay (1986) noted that genera such as *Ulva* and *Enteromorpha*, which are palatable to spottail pinfish (*Diplodus holbrooki*), tend to occur subtidally in the winter when these fish are rare or absent from the jetties. In the warm months, when pinfish are common around the jetties, these algae are mostly restricted to the refuge of the intertidal.

Richardson (1978) studied some aspects of subtidal colonization in the presence and absence of herbivores on a jetty in North Carolina where he excluded grazers with cages. His studies indicated that in the absence of large grazers (fish and sea urchins), hard surfaces were dominated by tube-building polychaete worms, bivalve molluscs, and serpulid worms, whereas seaweeds decreased in abundance. Peckol and Searles (1983) made similar studies of the seaweeds and invertebrates on a rock outcrop offshore on the continental shelf. Their studies suggested that the interactions controlling community structure are complex; there is competition for space when grazers are excluded, and the composition of the community that develops in part reflects the season in which space becomes available. In the absence of large grazers, the dominant organisms in that community appeared to be barnacles rather than algae; in the presence of those grazers, seaweeds were dominant and barnacles were rare. In both studies there were no experimental controls for the smaller grazers such as amphipod and polychaete crustaceans, which are capable of consuming large quantities of larger organisms (Hay et al. 1987, 1988).

Competition between species of seaweeds results in decreased rates of growth in understory plants associated with larger plants such as *Sargassum filipendula* (Hay 1986). Chemical defenses against herbivory have been demonstrated in a number of seaweeds, and Hay (1986) presented data from North Carolina suggesting that the common sea urchin *Arbacia punctulata* preferentially selected or avoided some species of seaweeds on a coastal jetty and specifically demonstrated a low preference for the brown alga *Dictyota menstrualis* (as *D. dichotoma*; Hay et al. 1987).

COLLECTION, PRESERVATION, AND MICROSCOPIC EXAMINATION OF SEAWEED SPECIMENS

COLLECTION

Along the coastline, seaweeds grow on hard surfaces such as seawalls, jetties, pilings, and the rare rock outcroppings. A diverse assemblage of seaweeds also grows epiphytically on the subtidal and intertidal plants of the estuaries. Species of genera such as *Vaucheria*, *Bachelotia*, *Caloglossa*, and *Bostrychia* grow on stable mud or peat deposits in sheltered habitats. Many drift plants are found cast up on the beaches and tangled among the stems of the marsh plants, particularly after storm activity.

Most seaweeds can be collected by wading or, in the subtidal of seawalls and jetties, by surface diving with mask and swimfins. Because of low water clarity near shore, the seaweeds there do not grow below the depths accessible by easy surface diving. To ensure that the whole plant is removed, the collector should use a putty knife or similar instrument to scrape plants from the points of attachment. In rare instances of tightly appressed calcareous algae, a chisel and hammer are necessary. Because hard surface habitats, and therefore seaweeds, are limited along this coast, collectors should be especially careful to remove only the minimum amount of plant material required for study.

The two practical methods for obtaining seaweeds in offshore waters are SCUBA diving and dredging. Dredging demands a sturdy geological rock dredge capable of breaking off pieces of rock or scraping plants and animals from smooth rock, thus requiring a large vessel with a winch and steel cable. Because some of the plants live in the cracks between rocks and are missed by dredges, and because some seaweeds are soft and delicate, SCUBA diving is the preferred method of sampling in all but the deepest locations.

In offshore waters macroscopic seaweeds grow attached to exposed rock along the edges of low ledges and on areas of bare rock on the flats behind the ledges, or more typically emerge through a light covering of shell and sand. Occasionally they grow on large pieces of shell. The latter are often found on the lower levels in front of the ledges where smaller sand particles appear to have winnowed away, perhaps due to the swirling of the currents as they pass over the ledges. Crustose fleshy and coralline red algae may adhere tightly to the rock, and a chisel and hammer may be required to break off pieces; they may also be found on shells and coral fragments.

The only special collecting item required when diving is a mesh bag to hold the specimens. It is useful to have a large bag for the big specimens and one or more small fine-mesh bags into which one can separate the smaller plants. In addition to obvious large algae, one should also collect sessile marine invertebrates (bryozoans, hydrozoans, gorgonian corals, molluscs) that have small algae growing on them.

PRESERVATION

Living specimens can be returned to the lab in coolers at ambient temperatures as long as they are not crowded. Specimens can be killed and the tissues fixed with a variety of reagents; the specifics depend on the use intended. For routine purposes specimens are killed and fixed in formalin-seawater in a ratio of 1:10. This is a noxious reagent, and when opening collections in the lab you are advised to wear gloves and work in a ventilated chemical hood. Specimens in formalin-seawater should be stored in the dark, where they will retain most of their color for long periods of time. After the initial killing and fixing, the plants can be transferred to 1:20 formalin-seawater for storage.

MICROSCOPY

Identification of specimens often requires microscopic examination. Small plants can be placed on slides under a coverslip and examined directly. Larger specimens may require sectioning. Coarse, sturdy plants such as *Gracilaria blodgettii* can be sectioned freehand with a single-edged razor blade. Small or membranous plants can be sandwiched between two glass slides and thin pieces cut off with a razor blade drawn against the end of the top slide as it is moved to expose the plant material. The secret in these hand-sectioning techniques is to cut a large number of sections; usually only a few out of two or three dozen are thin enough to be useful, but only one or two of these are necessary to see the diagnostic characteristics of the species.

Sections can be cut more accurately with a freezing microtome. The specimens are oriented and frozen on a mechanical stage in a few drops of water, or water and a solute such as gum arabic, and then cut with a microtome knife or razor blade as the block of tissue is moved into the path of the blade in small increments by the microtome. The sections are transferred to slides using a wet watercolor paintbrush.

Material can be viewed unstained using bright-field optics. The image may be enhanced with phase optics, differential interference-contrast (Nomarski) optics, or, most commonly, by stains. A frequently used stain is aniline blue (cotton blue). One of its advantages is that it is water soluble and the specimens do not have to be transferred to a nonaqueous medium. The stain is prepared as a 1 percent solution in water. Pieces of plant may be stained on a slide for a minute or more, depending on the bulk of the tissue. Excess stain is then removed using a pipette and the dye is fixed in the cells with a weak acid solution (1% HCl).

Slides may be made permanent by mounting them in a 1:1 corn syrup:water solution, covering with a coverslip, and then periodically replacing the water lost by evaporation with additional corn syrup solution until an equilibrium is established. The process is accelerated by placing the slide mounts on a warming table. The stock corn syrup solution is dilute enough to allow a growth of fungi, so a few drops of phenol per 100 cc of solution are recommended as a fungicide. After the slides have lost all excess water, the coverslip should be ringed with fingernail polish.

TAXONOMIC TREATMENT

LIST OF SPECIES

The following list of species shows the sequence in which the taxa are presented. Natural keys are provided in the text for all taxa, and there is an artificial key to the genera at the end of the taxonomy section.

The accepted name is given for each species, accompanied by the basionym if the species is based on an earlier name. These are followed by the species descriptions. The descriptions are intended to be complete enough to differentiate each species from others in the region. Because we have not necessarily observed all of the variation that has been described for each species, the descriptions of locally collected plants have been augmented with information from the literature. Where we know that the local populations differ significantly from the species as it has been described, the differences are noted at the end of each species treatment together with any problems of classification or historical notes of interest.

Previous records of each species from the region are listed as literature citations, first under the accepted name, followed by synonyms and misapplied names. Misapplied names are followed by the term "*sensu*" prior to the references using those names.

Habitat and seasonal data are provided for each species. The distribution of each species is listed beginning with the range from north to south in the western Atlantic. If a species occurs in other regions, these are listed sequentially starting in the northeastern Atlantic and continuing south along the coasts of Europe, the Mediterranean, West Africa, southern Africa, the Indian Ocean, Australia, the tropical Pacific islands, the western Pacific, and ending with the distribution north to south along the western shores of the Americas. For widely distributed species the listings are intended to indicate only the general range; we do not attempt to indicate distributional limits with any precision. Distributions have been determined from many sources, but chief among these are South and Tittley (1986), Taylor (1960), Oliviera F. (1977), Lawson and John (1982), Seagrief (1984), Scagel et al. (1986), Abbott and Hollenberg (1976), and Dawson's series of papers on the Marine Algae of Pacific Mexico.

CHLOROPHYTA

Ulvophyceae
Ulotrichales
  Ulotrichaceae
    Gomontia polyrhiza (Lagerheim) Bornet et Flahault
    Monostroma oxyspermum (Kützing) Doty
    Ulothrix flacca (Dillwyn) Thuret

Ulvales
  Ulvellaceae
    Entocladia viridis Reinke
    Phaeophila dendroides (P. Crouan et H. Crouan) Batters
    Pringsheimiella scutata (Reinke) Marchewianka
    Ulvella lens P. Crouan et H. Crouan
  Ulvaceae
    Blidingia marginata (J. Agardh) P. Dangeard
    Blidingia minima (Kützing) Kylin
    Enteromorpha clathrata (Roth) Greville
    Enteromorpha compressa (Linnaeus) Nees
    Enteromorpha flexuosa subsp. *flexuosa* J. Agardh
    Enteromorpha flexuosa subsp. *paradoxa* (C. Agardh) Bliding
    Enteromorpha intestinalis (Linnaeus) Nees
    Enteromorpha linza (Linnaeus) J. Agardh
    Enteromorpha prolifera (O. F. Müller) J. Agardh
    Enteromorpha ramulosa (J. E. Smith) Carmichael
    Enteromorpha torta (Mertens) Reinbold
    Pseudendoclonium submarinum Wille
    Ulva curvata (Kützing) De Toni
    Ulva fasciata Delile
    Ulva rigida C. Agardh
    Ulva rotundata Bliding
Cladophorales
  Cladophoraceae
    Chaetomorpha aerea (Dillwyn) Kützing
    Chaetomorpha brachygona Harvey
    Chaetomorpha crassa (C. Agardh) Kützing
    Chaetomorpha gracilis Kützing
    Chaetomorpha linum (O. F. Müller) Kützing
    Chaetomorpha minima Collins et Hervey
    Cladophora albida (Nees) Kützing
    Cladophora dalmatica Kützing
    Cladophora hutchinsiae (Dillwyn) Kützing
    Cladophora laetevirens (Dillwyn) Kützing
    Cladophora liniformis Kützing
    Cladophora montagneana Kützing
    Cladophora pellucidoidea van den Hoek
    Cladophora prolifera (Roth) Kützing
    Cladophora pseudobainesii van den Hoek et Searles

Cladophora ruchingeri (C. Agardh) Kützing
Cladophora sericea (Hudson) Kützing
Cladophora vadorum (Areschoug) Kützing
Cladophora vagabunda (Linnaeus) van den Hoek
Rhizoclonium riparium (Roth) Harvey
Valoniaceae
Valonia utricularis (Roth) C. Agardh
Anadyomenaceae
Anadyomene saldanhae Joly et Oliveira Filho
Microdictyon boergesenii Setchell
Boodleaceae
Struvea pulcherrima (J. E. Gray) Murray et Boodle
Chaetosiphonaceae
Blastophysa rhizopus Reinke
Caulerpales
Codiaceae
Codium carolinianum Searles
Codium decorticatum (Woodward) Howe
Codium fragile subsp. *tomentosoides* (van Goor) Silva
Codium isthmocladum Vickers
Codium taylorii Silva
Udoteaceae
Avrainvillea longicaulis (Kützing) Murray et Boodle
Boodleopsis pusilla (Collins) Taylor, Joly et Bernatowicz
Udotea cyathiformis Decaisne
Udotea flabellum (Ellis et Solander) Lamouroux
Caulerpaceae
Caulerpa mexicana Kützing
Caulerpa prolifera (Forsskål) Lamouroux
Caulerpa racemosa var. *laetevirens* (Montagne) Weber-van Bosse
Bryopsidaceae
Bryopsis pennata Lamouroux
Bryopsis plumosa (Hudson) C. Agardh
Derbesia marina (Lyngbye) Solier
Derbesia turbinata Howe et Hoyt
Ostreobiaceae
Ostreobium quekettii Bornet et Flahault

CHRYSOPHYTA

Xanthophyceae
Vaucheriales
  Vaucheriaceae
    Vaucheria acrandra Ott et Hommersand
    Vaucheria adela Ott et Hommersand
    Vaucheria arcassonensis Dangeard
    Vaucheria coronata Nordstedt
    Vaucheria erythrospora Christensen
    Vaucheria litorea C. Agardh
    Vaucheria longicaulis Hoppaugh
    Vaucheria minuta Blum et Conover
    Vaucheria nasuta Taylor et Bernatowicz
    Vaucheria velutina C. Agardh

PHAEOPHYTA

Phaeophyceae
Ectocarpales
  Ectocarpaceae
    Acinetospora crinita (Harvey) Kornmann
    Bachelotia antillarum (Grunow) Gerloff
    Botrytella micromora Bory
    Ectocarpus elachistaeformis Heydrich
    Ectocarpus fasciculatus Harvey
    Ectocarpus siliculosus (Dillwyn) Lyngbye
    Herponema solitarium (Sauvageau) Hamel
    Hincksia granulosa (J. E. Smith) Silva
    Hincksia irregularis (Kützing) Amsler
    Hincksia mitchelliae (Harvey) Silva
    Hincksia onslowensis (Amsler et Kapraun) Silva
    Hincksia ovata (Kjellman) Silva
    Phaeostroma pusillum Howe et Hoyt
    Streblonema invisibile Hoyt
    Streblonema oligosporum Strömfelt
  Ralfsiaceae
    Pseudolithoderma extensum (P. Crouan et H. Crouan) Lund
Chordariales
  Myrionemataceae
    Hecatonema floridanum (W. R. Taylor) W. R. Taylor

Hecatonema foecundum (Strömfelt) Loiseaux
Hecatonema maculans (Collins) Sauvageau
Myrionema magnusii (Sauvageau) Loiseaux
Myrionema strangulans Greville
Leathesiaceae
Leathesia difformis (Linnaeus) Areschoug
Myriactula stellulata (Harvey) Levring
Chordariaceae
Cladosiphon occidentalis Kylin
Spermatochnaceae
Nemacystus howei (W. R. Taylor) Kylin
Stilophora rhizodes (Turner) J. Agardh
Dictyosiphonales
Striariaceae
Hummia onusta (Kützing) Fiore
Striaria attenuata (C. Agardh) Greville
Punctariaceae
Asperococcus fistulosus (Hudson) Hooker
Punctaria latifolia Greville
Punctaria tenuissima (C. Agardh) Greville
Scytosiphonales
Scytosiphonaceae
Colpomenia sinuosa (Roth) Derbès et Solier
Petalonia fascia (O. F. Müller) Kuntze
Rosenvingea orientalis (J. Agardh) Børgesen
Scytosiphon lomentaria (Lyngbye) Link
Scytosiphon lomentaria var. *complanatus* Rosenvinge
Sphacelariales
Choristocarpaceae
Onslowia endophytica Searles
Sphacelariaceae
Sphacelaria rigidula Kützing
Sphacelaria tribuloides Meneghini
Dictyotales
Dictyotaceae
Dictyopteris delicatula Lamouroux
Dictyopteris hoytii Taylor
Dictyopteris membranacea (Stackhouse) Batters
Dictyota cervicornis Kützing
Dictyota ciliolata Kützing

Dictyota menstrualis (Hoyt) Schnetter, Hörnig et Weber-Peukert
Dictyota pulchella Hörnig et Schnetter
Lobophora variegata (Lamouroux) Womersley
Padina gymnospora (Kützing) Sonder
Padina profunda Earle
Spatoglossum schroederi (C. Agardh) Kützing
Zonaria tournefortii (Lamouroux) Montagne
Sporochnales
  Sporochnaceae
    Sporochnus pedunculatus (Hudson) C. Agardh
Desmarestiales
  Arthrocladiaceae
    Arthrocladia villosa (Hudson) Duby
Fucales
  Fucaceae
    Ascophyllum nodosum (Linnaeus) Le Jolis
    Fucus vesiculosus Linnaeus
  Sargassaceae
    Sargassum filipendula C. Agardh
    Sargassum filipendula var. montagnei (Bailey) Collins et Hervey
    Sargassum fluitans Børgesen
    Sargassum natans (Linnaeus) Gaillon

RHODOPHYTA

Rhodophyceae
  Bangiophycidae
Porphyridiales
  Porphyridiaceae
    Chroodactylon ornatum (C. Agardh) Basson
    Stylonema alsidii (Zanardini) Drew
Compsopogonales
  Erythropeltidaceae
    Erythrocladia irregularis Rosenvinge
    Erythrocladia endophloea Howe
    Erythrotrichia carnea (Dillwyn) J. Agardh
    Erythrotrichia vexillaris (Montagne) Hamel
    Porphyropsis coccinea (Areschoug) Rosenvinge
    Sahlingia subintegra (Rosenvinge) Kornmann
Bangiales

Bangiaceae
Bangia atropurpurea (Roth) C. Agardh
Porphyra carolinensis Coll et Cox
Porphyra leucosticta Thuret
Porphyra rosengurttii Coll et Cox
Florideophycidae
Acrochaetiales
Acrochaetiaceae
Audouinella affinis (Howe et Hoyt) C. W. Schneider
Audouinella bispora (Børgesen) Garbary
Audouinella botryocarpa (Harvey) Woelkerling
Audouinella corymbifera (Thuret) Dixon
Audouinella dasyae (Collins) Woelkerling
Audouinella daviesii (Dillwyn) Woelkerling
Audouinella densa (Drew) Garbary
Audouinella hallandica (Kylin) Woelkerling
Audouinella hoytii (Collins) C. W. Schneider
Audouinella hypneae (Børgesen) Lawson et John
Audouinella infestans (Howe et Hoyt) Dixon
Audouinella microscopica (Nägeli) Woelkerling
Audouinella ophioglossa C. W. Schneider
Audouinella saviana (Meneghini) Woelkerling
Audouinella secundata (Lyngbye) Dixon
Nemaliales
Helminthocladiaceae
Helminthocladia andersonii Searles et Lewis
Galaxauraceae
Galaxaura obtusata (Ellis et Solander) Lamouroux
Scinaia complanata (Collins) Cotton
Gelidiales
Gelidiaceae
Gelidium americanum (W. R. Taylor) Santelices
Gelidium pusillum (Stackhouse) Le Jolis
Bonnemaisoniales
Bonnemaisoniaceae
Asparagopsis taxiformis (Delile) Trevisan
Naccariaceae
Naccaria corymbosa J. Agardh
Corallinales
Corallinaceae

(Articulated Corallines)
  Amphiroa beauvoisii Lamouroux
  Corallina officinalis Linnaeus
  Haliptilon cubense (Kützing) Garbary et Johansen
  Jania adhaerens Lamouroux
  Jania capillacea Harvey
  Jania rubens (Linnaeus) Lamouroux
(Crustose Corallines)
  Fosliella farinosa (Lamouroux) Howe
  Leptophytum sp.
  Lithophyllum intermedium (Foslie) Foslie
  Lithophyllum subtenellum (Foslie) Foslie
  Lithothamnion occidentale (Foslie) Foslie
  Mesophyllum floridanum (Foslie) Adey
  Neogoniolithon accretum (Foslie et Howe) Setchell et Mason
  Neogoniolithon caribaeum (Foslie) Adey
  Phymatolithon tenuissimum (Foslie) Adey
  Pneophyllum lejolisii (Rosanoff) Chamberlain
  Pneophyllum sp.
  Titanoderma pustulatum (Lamouroux) Nägeli
Gigartinales
Peyssonneliaceae
  Peyssonnelia atlantica C. W. Schneider et Reading
  Peyssonnelia conchicola Piccone et Grunow
  Peyssonnelia inamoena Pilger
  Peyssonnelia simulans Weber-van Bosse
  Peyssonnelia stoechas Boudouresque et Denizot
Dumontiaceae
  Dudresnaya crassa Howe
  Dudresnaya georgiana Searles
  Dudresnaya puertoricensis Searles et Ballantine
Gymnophloeaceae
  Predaea feldmannii Børgesen
  Predaea masonii (Setchell et Gardner) De Toni fil.
Sebdeniaceae
  Sebdenia flabellata (J. Agardh) Parkinson
Halymeniaceae
  Cryptonemia luxurians (C. Agardh) J. Agardh
  Grateloupia cunefolia J. Agardh
  Grateloupia filicina (Lamouroux) C. Agardh

Grateloupia gibbesii Harvey
Halymenia bermudensis Collins et Howe
Halymenia floresia (Clemente y Rubio) C. Agardh
Halymenia floridana J. Agardh
Halymenia hancockii W. R. Taylor
Halymenia trigona (Clemente y Rubio) C. Agardh
Kallymeniaceae
  Cirrulicarpus carolinensis Hansen
  Kallymenia westii Ganesan
Plocamiaceae
  Plocamium brasiliense (Greville) Howe et W. R. Taylor
Sarcodiaceae
  Trematocarpus papenfussii Searles
Cystocloniaceae
  Craspedocarpus humilis C. W. Schneider
Solieriaceae
  Agardhiella ramosissima (Harvey) Kylin
  Agardhiella subulata (C. Agardh) Kraft et Wynne
  Eucheuma isiforme var. *denudatum* Cheney
  Meristiella gelidium (J. Agardh) Cheney et Gabrielson
  Meristotheca floridana Kylin
  Sarcodiotheca divaricata W. R. Taylor
  Solieria filiformis (Kützing) Gabrielson
Hypneaceae
  Hypnea cervicornis J. Agardh
  Hypnea musciformis (Wulfen) Lamouroux
  Hypnea valentiae (Turner) Montagne
  Hypnea volubilis Searles
Phyllophoraceae
  Gymnogongrus griffithsiae (Turner) Martius
  Petroglossum undulatum C. W. Schneider
Gigartinaceae
  Gigartina acicularis (Wulfen) Lamouroux
Petrocelidaceae
  Mastocarpus stellatus (Stackhouse) Guiry
Gracilariales
  Gracilariaceae
    Gracilaria blodgettii Harvey
    Gracilaria curtissiae J. Agardh
    Gracilaria cylindrica Børgesen

Gracilaria mammillaris (Montagne) Howe
Gracilaria tikvahiae McLachlan
Gracilaria verrucosa (Hudson) Papenfuss
Gracilariopsis lemaneiformis (Bory) Dawson, Acleto et Foldvik
Rhodymeniales
  Rhodymeniaceae
    Agardhinula browneae (J. Agardh) De Toni
    Botryocladia occidentalis (Børgesen) Kylin
    Botryocladia pyriformis (Børgesen) Kylin
    Botryocladia wynnei Ballantine
    Chrysymenia agardhii Harvey
    Chrysymenia enteromorpha Harvey
    Glioderma atlanticum Searles
    Glioderma blomquistii Searles
    Glioderma rubrisporum Searles
    Halichrysis peltata (W. R. Taylor) P. Huvé et H. Huvé
    Leptofauchea brasiliensis Joly
    Rhodymenia divaricata Dawson
    Rhodymenia pseudopalmata (Lamouroux) Silva
  Champiaceae
    Champia parvula (C. Agardh) Harvey
    Champia parvula var. *prostrata* L. Williams
  Lomentariaceae
    Lomentaria baileyana (Harvey) Farlow
    Lomentaria orcadensis (Harvey) W. R. Taylor
Ceramiales
  Ceramiaceae
    Anotrichium tenue (C. Agardh) Nägeli
    Antithamnion cruciatum (C. Agardh) Nägeli
    Antithamnionella atlantica (Oliveira F.) C. W. Schneider
    Antithamnionella elegans (Berthold) Price et John
    Antithamnionella flagellata (Børgesen) Abbott
    Calliclavula trifurcata C. W. Schneider
    Callithamniella silvae Searles
    Callithamniella tingitana (Bornet) Feldmann-Mazoyer
    Callithamnion cordatum Børgesen
    Callithamnion pseudobyssoides P. Crouan et H. Crouan
    Centroceras clavulatum (C. Agardh) Montagne
    Ceramium byssoideum Harvey
    Ceramium diaphanum (Lightfoot) Roth

Ceramium fastigiatum (Roth) Harvey
Ceramium fastigiatum f. flaccidum H. Petersen
Ceramium floridanum J. Agardh
Ceramium leptozonum Howe
Ceramium rubrum (Hudson) C. Agardh
Compsothamnion thuyoides (Smith) Schmitz
Griffithsia globulifera Kützing
Lejolisia exposita C. W. Schneider et Searles
Nwynea grandispora Searles
Pleonosporium boergesenii (Joly) R. Norris
Pleonosporium flexuosum (C. Agardh) De Toni
Ptilothamnion occidentale Searles
Rhododictyon bermudense W. R. Taylor
Spyridia clavata Kützing
Spyridia hypnoides (Bory) Papenfuss
Delesseriaceae
Acrosorium venulosum (Zanardini) Kylin
Apoglossum ruscifolium (Turner) J. Agardh
Branchioglossum minutum C. W. Schneider
Branchioglossum prostratum C. W. Schneider
Calloseris halliae J. Agardh
Caloglossa leprieurii (Montagne) J. Agardh
Calonitophyllum medium (Hoyt) Aregood
Grinnellia americana (C. Agardh) Harvey
Haraldia lenormandii (Derbès et Solier) J. Feldmann
Hypoglossum hypoglossoides (Stackhouse) Collins et Hervey
Hypoglossum tenuifolium (Harvey) J. Agardh
Myriogramme distromatica Boudouresque
Searlesia subtropica (C. W. Schneider) C. W. Schneider et Eiseman
Dasyaceae
Dasya baillouviana (Gmelin) Montagne
Dasya ocellata (Grateloup) Harvey
Dasya rigidula (Kützing) Ardissone
Dasya spinuligera Collins et Hervey
Dasysiphonia concinna C. W. Schneider
Dasysiphonia doliiformis C. W. Schneider
Heterosiphonia crispella var. laxa (Børgesen) Wynne
Rhodomelaceae
Acanthophora spicifera (Vahl) Børgesen
Bostrychia radicans (Montagne) Montagne

Bryocladia cuspidata (J. Agardh) De Toni
Bryocladia thyrsigera (J. Agardh) Schmitz
Bryothamnion seaforthii (Turner) Kützing
Chondria atropurpurea Harvey
Chondria baileyana (Montagne) Harvey
Chondria curvilineata Collins et Hervey
Chondria dasyphylla (Woodward) C. Agardh
Chondria floridana (Collins) Howe
Chondria littoralis Harvey
Chondria polyrhiza Collins et Hervey
Chondria tenuissima (Goodenough et Woodward) C. Agardh
Dipterosiphonia reversa C. W. Schneider
Herposiphonia delicatula Hollenberg
Herposiphonia tenella (C. Agardh) Nägeli
Laurencia corallopsis (Montagne) Howe
Laurencia pinnatifida (Gmelin) Lamouroux
Laurencia poiteaui (Lamouroux) Howe
Micropeuce mucronata (Harvey) Kylin
Polysiphonia atlantica Kapraun et J. Norris
Polysiphonia binneyi Harvey
Polysiphonia breviarticulata (C. Agardh) Zanardini
Polysiphonia denudata (Dillwyn) Harvey
Polysiphonia ferulacea J. Agardh
Polysiphonia flaccidissima Hollenberg
Polysiphonia gorgoniae Harvey
Polysiphonia harveyi Bailey
Polysiphonia harveyi var. *olneyi* Harvey
Polysiphonia havanensis Montagne
Polysiphonia howei Hollenberg
Polysiphonia nigrescens (Hudson) Greville
Polysiphonia pseudovillum Hollenberg
Polysiphonia scopulorum var. *villum* (J. Agardh) Hollenberg
Polysiphonia sphaerocarpa Børgesen
Polysiphonia subtilissima Montagne
Polysiphonia tepida Hollenberg
Polysiphonia urceolata (Dillwyn) Greville
Pterosiphonia pennata (C. Agardh) Falkenberg
Wrightiella tumanowiczii (Harvey) Schmitz

CHLOROPHYTA

Plants of marine and fresh waters; unicellular, colonial, multicellular, or acellular; cells uninucleate or multinucleate; acellular plants include large, macroscopic individuals formed from interwoven filaments; pigmented with chlorophylls *a* and *b*, alpha, beta, and gamma carotene, and the xanthophylls lutein, violaxanthin, zeaxanthin, antheraxanthin, and neoxanthin; plastids delimited by two membranes, thylakoids in stacks of two to five, often with one or more pyrenoids; cell walls present except in some unicells and contain some combination of cellulose, hydroxyproline glycosides, xylan, mannan, or calcium carbonate; storage compounds alpha-linked glucose (starch); motile cells commonly with two or four flagella, but uniflagellate and multiflagellate cells also characterize particular taxa; flagella naked, or less frequently with organic scales or hairs; sexual life histories either haplontic, diplontic, or diplohaplontic, the latter isomorphic or heteromorphic; asexual reproduction by fragmentation, asexual spores, or parthenogenesis.

The intertidal is often locally dominated by green algae, particularly members of the Ulotrichales (*Enteromorpha*, *Monostroma*, and *Ulva*), Cladophorales (*Chaetomorpha* and *Cladophora*), and Caulerpales (*Bryopsis* and *Codium*). In subtidal waters the Caulerpales are also conspicuously represented by species of *Avrainvillea*, *Caulerpa*, *Codium*, and *Udotea*.

ULVOPHYCEAE

Plants filamentous, parenchymatous, or siphonous; cell division with a persistent, closed, interzonal centric spindle, cytokinesis centripetal with neither phycoplast nor phragmoplast; motile cells scaly or naked, flagellar bases with 180-degree rotational symmetry and counterclockwise absolute orientation; life histories haplontic, diplohaplontic, and diplontic; predominantly marine.

The ordinal classification is based on O'Kelly and Floyd (1984).

Key to the orders

1. Plants acellular, quadriflagellate zoospores absent, cell walls containing mannan or xylan . . . . Bryopsidales
1. Plants cellular, quadriflagellate zoospores present, walls containing mannan or cellulose 2
2. Motile cells with scales, terminal cap of flagellar base simple and overlapping, sporangial exit not papillate . . . . Ulotrichales
2. Motile cells lacking scales, terminal cap of flagellar base absent or bilobed, sporangial exit papillate 3
3. Cells uninucleate, terminal cap of flagellar base bilobed, sporangial exit papilla without plug . . . . Ulvales

3. Cells coenocytic, terminal cap of flagellar base lacking, sporangial exit papilla with plug . . . . Cladophorales

ULOTRICHALES

Plants composed of uniseriate branched or unbranched filaments, or monostromatic blades; branched filaments may form pseudoparenchyma; cells uninucleate or multinucleate; motile cells scaly; gametes biflagellate, zoospores quadriflagellate; flagellar bases with simple, overlapping terminal caps; life history mostly an alternation of a multicellular gametophyte and a unicellular sporophyte, or more rarely with isomorphic phases.

A single family is included in the order here.

Ulotrichaceae

Key to the genera

1. Plants with blades . . . . Monostroma
1. Plants without blades 2
2. Branched, growing in shell or wood . . . . Gomontia
2. Unbranched, not growing in shell or wood . . . . Ulothrix

Gomontia Bornet et Flahault 1888

Plants growing in shell or wood; cells near surface irregular, those deeper in the substratum narrower and of more uniform diameter; asexual reproduction by zoospores.

Gomontia polyrhiza (Lagerheim) Bornet et Flahault 1888, p. 164.

Codiolum polyrhizum Lagerheim 1885, p. 22.

Filaments irregularly branched; cells 4–10 μm diameter, 2–6 times as long (20–50 μm); sporangia 30–40(–150) μm diameter, 150–250 μm long.

Hoyt 1920; Williams 1948a; Taylor 1957, 1960.

Found boring in shells, spring and summer.

Distribution: Canadian Arctic to Virginia, North Carolina, Bermuda, southern Florida, Caribbean, USSR to Portugal, Indian Ocean, Japan, Alaska to California.

Kornmann (1962) found *Codiolum*-like shell-boring stages similar to *Gomontia polyrhiza* in the life history of *Monostroma grevillei* (Thuret) Wittrock, an alga subsequently separated by Gayral (1964) into the genus *Ulvopsis*. Wilkinson and Burrows (1972) studied *Gomontia polyrhiza*-like algae in the British Isles and concluded that six different taxa, including *Eugomontia sacculata* Kornmann, *Entocladia perforans* (Huber) Levring, and *Coldiolum* stages in the

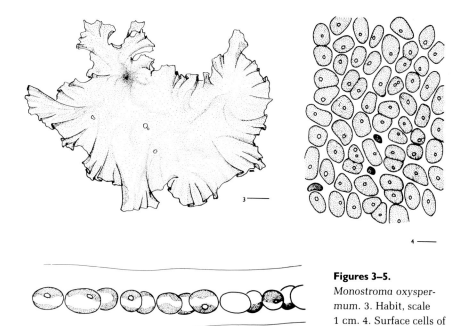

Figures 3–5.
Monostroma oxyspermum. 3. Habit, scale 1 cm. 4. Surface cells of blade, scale 10 μm. 5. Cross section of blade, scale 10 μm.

life histories of *Codiolum polyrhizum* Lagerheim and *Monostroma grevillei* (Thuret) Wittrock, were confused under the designation *Gomontia polyrhiza*. Hoyt's report of this alga from North Carolina indicated plants with filaments 4–8 μm diameter, sporangia 30–40 μm diameter, zoospores of two kinds (10–12 μm and 5–6 μm diameter, or 5 μm long and 3.5 μm diameter), and aplanospores 4 μm diameter. The status of this alga will be unclear until the local representatives are isolated and studied in culture.

Monostroma Thuret 1854
Plants monostromatic blades, developing from a discoid or filamentous germling.

Monostroma oxyspermum (Kützing) Doty 1947, p. 12.
    Ulva oxysperma Kützing 1843, p. 296.
    Figures 3–5
    Plants light green, flat or ruffled sheets, to 10(–60) cm tall; attached basally by rhizoids or free floating; cells irregularly arranged or in groups of two or four, in surface view 7–18(–26) μm diameter, rounded angular to round or oval, walls thick; blade 20–40(–60) μm thick, cells in section rounded rectangular with lumens 14–21 μm tall; sexual reproduction unknown.
    Chapman 1971. As *Ulvaria oxysperma*, Kapraun 1984; Richardson 1986, 1987.

On mud in marshes and on shells and seawalls in sounds, winter and spring.

Distribution: Canadian Arctic to Virginia, North Carolina, South Carolina, Georgia, Bermuda, southern Florida, Brazil, USSR to Portugal, Alaska to California.

The genus was placed in the Ulotrichales on the basis of flagellar ultrastructural characteristics (O'Kelly and Floyd 1984), but the generic placement of *oxyspermum* remains uncertain. This species reproduces only asexually and, based on studies of populations in different parts of the world, there is disagreement about the method of spore release (Gayral 1965; Tatewaki 1969; Golden and Garbary 1984) which makes the generic placement in *Monostroma* rather than *Ulvaria* arbitrary.

Ulothrix Kützing 1833
Plants filamentous, unbranched, usually attached by elongate basal cells; cells cylindrical, length equal to width or shorter, uninucleate; plastids single parietal bands completely or incompletely circling the cells, with one to several pyrenoids.

Ulothrix flacca (Dillwyn) Thuret in Le Jolis 1863, p. 56.

Conferva flacca Dillwyn 1809, p. 53, pl. 49.

Figure 6

Filaments 10–25 µm diameter, 0.25–0.75 diameters long, plastids incomplete bands covering entire length of inner walls with one to three pyrenoids; sporangial cells swollen to 50–60(–80) µm.

Blomquist and Humm 1946; Williams 1948b; Chapman 1971; Kapraun and Zechman 1982; Kapraun 1984.

Epiphytic and on intertidal rocks in winter and spring, rare in summer.

Distribution: Canadian Arctic to Virginia, North Carolina, South Carolina, Georgia, USSR to Portugal, Azores, Japan.

Although some species assigned to *Ulothrix* have symmetrical, chlorophycean flagellar bases and phycoplast-type cell division and are more appropriately placed in the Chlorophyceae, Sluiman et al. (1983) indicated that *Ulothrix flacca* is probably a member of the Ulvophyceae and should be retained in the

Figure 6. *Ulothrix flacca*, portion of filament, scale 5 µm.

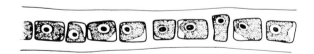

genus *Ulothrix*. Perrot (1972) studied "*Ulothrix flacca*" in France and demonstrated that collections from Roscoff consisted of two species, one with isogametes and isomorphic phases, the other with anisogametes and heteromorphic phases.

Local specimens have a single pyrenoid in each plastid.

ULVALES

Plants with branched, uniseriate filaments, sometimes united in pseudoparenchymatous discs, or plants forming single- or two-layered blades or tubes; cells typically uninucleate; bases of all flagella with a bilobed terminal cap and proximal sheath consisting of two equal subunits; exit pore of gametangia and sporangia papillate, producing a more or less rounded pore after motile cell release; life history diplohaplontic with isomorphic phases.

Key to the families

1. Plants filamentous and in some cases pseudoparenchymatous; rhizoplasts lacking in motile cells .... Ulvellaceae
1. Plants parenchymatous or pseudoparenchymatous; rhizoplasts present in motile cells .... Ulvaceae

Ulvellaceae

Plants filamentous, uniseriate, branched, heterotrichous or creeping, plant or base of plant sometimes discoid and erect filaments pseudoparenchymatous; colorless hairs (setae) sometimes present; motile cells without rhizoplasts.

Generic concepts in this group are not clear (Nielsen 1972; South 1974; Yarish 1975). The disposition of *Phaeophila* was not indicated by O'Kelly and Floyd (1984), but it is included here with the genera with which it is traditionally placed.

Key to the genera

1. Plants discoid 2
1. Plants not discoid 3
2. Cells of disc with aseptate setae .... Pringsheimiella
2. Cells of disc without aseptate setae .... Ulvella
3. Pyrenoids one, or rarely more than one per cell, cleavage of sporangia sequential... Entocladia
3. Pyrenoids more than one per cell, cleavage of sporangia simultaneous......
   .... Phaeophila

Entocladia Reinke 1879

Plants epiphytic or endophytic, creeping on surface or penetrating outer walls of host; filamentous, irregularly branched, with or without free erect filaments, but sometimes becoming pseudoparenchymatous in center of radiating filaments; plastids single, parietal, with one or more pyrenoids; cells uninucleate, with or without aseptate setae; zoospores quadriflagellate, division of sporangium sequential, spore release explosive; gametes biflagellate, isogamous.

Entocladia viridis Reinke 1879, p. 476.

Figures 7 and 8

Filaments much branched, without central pseudoparenchymatous region; cells 3.5–7(–10) µm diameter, 1–4(–6) diameters long, cylindrical, swollen, or ir-

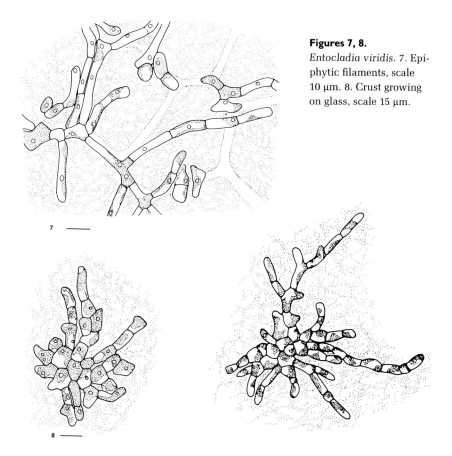

Figures 7, 8.
Entocladia viridis. 7. Epiphytic filaments, scale 10 µm. 8. Crust growing on glass, scale 15 µm.

7 ⎯⎯

8 ⎯⎯

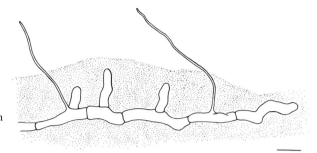

Figure 9. *Phaeophila dendroides,* growing in shell, scale 25 μm.

regular; plastids with usually one, though sometimes up to four, pyrenoids; with or without septate setae.

Williams 1948a, 1951; Taylor 1960. As *Endoderma viride,* Hoyt 1920; Humm 1952.

Growing in the cell walls of *Cladophora, Calonitophyllum, Hypoglossum,* and *Rhodymenia,* summer.

Distribution: Canadian Arctic to Virginia, North Carolina, northeastern Florida, Caribbean, Brazil, Norway to Portugal; elsewhere, widespread in temperate and tropical seas.

O'Kelly and Yarish (1980) used the differences in sporangial ontogeny given here in the generic descriptions to distinguish *Entocladia* from *Phaeophila.* They also (1981) distinguished *Entocladia* from *Acrochaete* as circumscribed by Nielsen (1979) by the absence of the erect filaments that characterize the latter genus.

Phaeophila Hauck 1876

Plants epi-endophytic or boring in carbonate shell; cells of branched, uniseriate filaments bearing one to three, long, aseptate hairs; hair cytoplasm not separated from cell by wall, hair bases not swollen; plastids lobed, parietal, with several pyrenoids; zoospores quadriflagellate, sporangial mother cells multinucleate, cleavage simultaneous.

Phaeophila dendroides (P. Crouan et H. Crouan) Batters 1902, p. 13.
    Ochlochaete dendroides P. Crouan et H. Crouan 1852, no. 346.
    Figure 9

Plants frequently and widely branched; cells 9–40 μm diameter; 15–50(–80) μm long, generally cylindrical, partly of irregular diameters; setae occasional to frequent, one to three per cell, their bases often spirally twisted; zoosporangia intercalary or terminal on short branches, often swollen, 16–40 μm diameter, 30–85 μm long.

Aziz and Humm 1962.

On *Dictyota menstrualis,* summer.

Distribution: Newfoundland to Virginia, North Carolina, Bermuda, southern Florida, Caribbean, Brazil, Norway to Portugal; elsewhere, widespread in temperate and tropical seas.

Pringsheimiella v. Höhnel 1920

Plants discoid, single layered, the filaments laterally closely united, cells in center of disc somewhat taller than those toward margins, which are radially elongate; growth marginal; young cells sometimes bear colorless hairs; plastids parietal, platelike, with single pyrenoids; zoospores and gametes quadriflagellate.

Pringsheimiella scutata (Reinke) Höhnel ex Marchewianka 1924, p. 42.

Pringsheimia scutata Reinke 1889, p. 33, pl. 25.

Plants discoid, 0.1–0.2 mm diameter, cells variable in size and shape, marginal cells horizontally elongate, laterally branched; central cells vertically elongate, to 12 μm diameter; zoosporangia oval to subpyriform, 15–22 μm diameter, 28–38 μm high, zoospores 15 μm diameter, gametes 4 μm diameter.

Aziz and Humm 1962.

On *Sargassum filipendula* and *Cladophora* spp., June and October.

Distribution: Canadian Arctic to Connecticut, North Carolina, Bermuda, Caribbean, Brazil, Norway to Spain, southern Africa, Japan, Washington.

Ulvella H. Crouan et P. Crouan 1859

Plants epiphytic, endophytic, or on rock; discoid, disc cells compressed and disc more than one cell thick in center, more clearly filamentous toward margins, and some marginal cells showing precocious branching (Y-shaped cells); without hairs or setae; cells uni- to multinucleate; plastids parietal, with or without pyrenoids; reproduction by biflagellate zoospores; zoosporangia formed by central cells of disc; sexual reproduction unclear.

Ulvella lens H. Crouan et P. Crouan 1859, p. 288, pl. 22, figs 25–28.

Figure 10

Discs 1–5 mm diameter, margins single layered, centers two to three layered; central cells 5–10(–20) μm diameter, irregularly polygonal, marginal cells elongate, 3–4(–15) μm diameter by 15–30 μm long, some peripheral cells forked; multinucleate; central cells forming four, eight, or sixteen zoospores.

Hoyt 1920; Williams 1948a, 1949, 1951; Humm 1952; Taylor 1960; Amsler and Searles 1981.

On shells, hydroids, and *Codium* and other algae, summer.

Distribution: Canadian Arctic, Massachusetts, North Carolina, South Carolina, northeastern Florida, southern Florida, Caribbean, Norway to Spain, Azores.

Kapraun (1984) pointed out that *Pseudulvella* Wille differs from *Ulvella* in having quadriflagellate zoospores. Since the zoospores of the local plants have not been observed, the generic identity of the local plants is not certain.

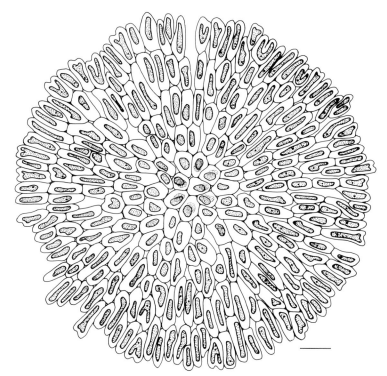

Figure 10. *Ulvella lens*, scale 10 μm.

Ulvaceae

Plants pseudoparenchymatous discs, single or two layered blades, or tubular; motile cells with rhizoplasts.

Key to the genera

1. Plants filamentous, pseudoparenchymatous, sometimes discoid, but not erect or tubular .... Pseudendoclonium
1. Plants not pseudoparenchymatous; erect and tubular or membranous 2
2. Plants tubular or if partly membranous the base of the blade tubular 3
2. Plants membranous throughout................................... Ulva
3. Plastids stellate, erect plant originating from a disc without rhizoids ....
.... Blidingia
3. Plastids not stellate, erect plant attached by rhizoids........ Enteromorpha

Blidingia Kylin 1947

First cell of germling receiving all of zoospore cytoplasm; germling closely branched, discoid, becoming two layered; mature plant erect, tubular, *Enteromorpha*-like; plastids one per cell, stellate, parietal, with one pyrenoid; asexual reproduction by quadriflagellate zoospores; sexual reproduction unknown.

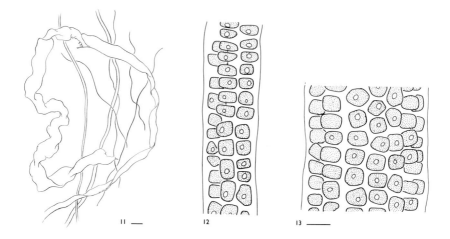

Figures 11–13. *Blidingia marginata.* 11. Young, slender, tubular filaments, and older, irregular, collapsed filaments, scale 25 μm. 12, 13. Surface cells of young filaments, scale 10 μm.

Key to the species

1. Plants simple, or with little proliferous branching, cells in longitudinal series, especially along blade margins, blades 12–100 μm thick . . . . *B. marginata*
1. Plant branched or unbranched, cells not in linear series, wall of tubular branches 8–10 μm thick or thicker . . . . *B. minima*

Blidingia marginata (J. Agardh) P. Dangeard 1958, p. 347.

　Enteromorpha marginata J. Agardh 1842, p. 16.

　Figures 11–13

　Plants simple or rarely with a few marginal proliferations, cylindrical when young, becoming flattened with age, 2–5 cm high; to 15–20 cells, 100–200 μm wide; 12–100 μm thick; cells to 10 μm long, sometimes in longitudinal series, particularly in young branches; attached by basal discs; blades narrow toward bases and tips, otherwise of constant diameter, sometimes slightly inflated; long, narrow threads forming tangled masses.

　Chapman 1971; Kapraun 1984; Richardson 1987.

　In salt marshes, winter–spring.

　Distribution: Canadian Arctic to Virginia, North Carolina, Georgia, Bermuda, ?Caribbean, Norway to Portugal, Australia, British Columbia to California.

　Parke and Burrows (1976) suggested that *Blidingia minima* and *B. marginata* may be conspecific.

Blidingia minima (Nägeli ex Kützing) Kylin 1947, p. 8.

　Enteromorpha minima Nägeli ex Kützing 1849 p. 482.

　Figure 14

　Plants deep green or yellowish, simple or branched, 1–24 cm tall, bases of

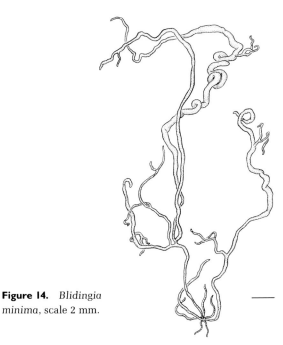

Figure 14. *Blidingia minima*, scale 2 mm.

fronds slender, one or more attached to basal discs; blades expanded above bases, linear, sometimes torulose and more or less tubular, 1–2(–5) mm broad; cells angular, (3–)5–7(–10) μm diameter, in no apparent order, almost cuboidal in section; walls thin; stellate plastids covering the exposed face of the cells.

Stephenson and Stephenson 1952; Kapraun 1984; Richardson 1987. As *Enteromorpha minima*, Williams 1948a; Taylor 1960.

On intertidal jetties and seawalls in sounds and on outer coast, winter–spring.

Distribution: Canadian Arctic to Virginia, North Carolina, South Carolina, Georgia, Bermuda, Iceland, British Isles to Portugal, southern Africa, Australia, Japan, Alaska to Pacific Mexico, Chile.

Enteromorpha Link 1820, nom. cons.

Plants in part hollow and tubular, walls of tubes one cell thick, lower part of plant always hollow, upper parts tubular or becoming filamentous in many species, flattened and bladelike in others; simple or branched, attached by non-septate rhizoidal outgrowth of cells, sometimes free floating; cells uninucleate, plastids parietal, platelike, cup shaped, or tubular, with one or more pyrenoids; asexual reproduction by quadriflagellate zoospores or fragmentation; sexual reproduction by isogamous or anisogamous, biflagellate gametes.

Key to the species and subspecies

1. Plants filiform, unbranched, consisting of four to ten cell rows, 30–60 μm in diameter; central cavity 10–15 μm diameter . . . . *E. torta*
1. Plants filiform or not, branched or unbranched; central cavity greater than 15 μm diameter 2
2. Pyrenoids one to two in majority of cells 3

2. Pyrenoids two to nine in majority of cells 8
3. Cells in middle and apical regions of main axis and broad branches unordered or arranged in well-ordered groups or rows among unordered groups; chloroplasts mostly tilted toward the apically oriented anticlinal cell wall and hence covering only the apical part of the outer cell wall 4
3. Cells in middle and apical regions of main axis and broad branches arranged in longitudinal and often transverse rows, or in large well-ordered groups; chloroplasts mostly covering the outer cell wall 6
4. Plants branched, cells in middle and apical regions in well-ordered groups among unordered groups .... *E. compressa* (in part)
4. Plants unbranched when full grown, or with few microscopic branches near the base; cells unordered or in well-ordered groups or rows among unordered groups 5
5. Axes to 0.5 cm broad; cells in middle and apical regions in well-ordered groups or undulating cell rows among unordered groups ....
.... *E. compressa* (in part)
5. Axes to 4(–10) cm broad; cells in middle and apical regions unordered; small well-ordered groups or undulating cell rows may occur ..... *E. intestinalis*
6. Plants unbranched, or seldom with one or two broad branches on the upper basal part of the main frond; more than 0.5 cm broad .... *E. linza*
6. Plants with many branches, or if unbranched the axis at most 0.5 cm broad
.... 7
7. Unbranched or with few branches similar to main axis; cells in the upper basal region may be rounded and conspicuously narrower and longer; pyrenoids one to two per cell in middle region, up to five in basal region ......
.... *E. flexuosa* subsp. *flexuosa*
7. Branched with a mixture of micro- and macroscopic branches; cells in the upper basal region irregularly polygonal with four to six rounded corners, or rounded; conspicuously narrower and longer-celled cell rows absent in basal region; pyrenoids one per cell throughout, rarely two .... *E. prolifera*
8. Pyrenoids five or less per cell 9
8. Pyrenoids more than five per cell, up to ten 10
9. Microscopic branches uniseriate at tip .... *E. flexuosa* subsp. *paradoxa*
9. Microscopic branches with broad base and multiseriate except for single terminal cell .... *E. ramulosa* (in part)
10. Short branches spinelike, with broad bases and mostly multiseriate except for terminal cell .... *E. ramulosa* (in part)
10. Short branches filiform, uniseriate tip generally more than five cells long
.... *E. clathrata*

Enteromorpha clathrata (Roth) Greville 1830, p. 181.

Conferva clathrata Roth 1806, p. 175.

Plants strap shaped to filiform, attached or free floating, to 40 cm long, 0.5–5(–10) mm wide; branches to several orders distributed along main axes, or concentrated toward bases of axes, or main axes not distinguishable from branches; apices of axes obtuse; young branches with uniseriate tips of five cells or more; cells in middle and apical regions rectangular, quadrangular, or irregularly polygonal, with mainly equal divisions, to 10 μm wide, 30 μm long; arranged in longitudinal rows, especially in narrower branches; cells in the lower basal region large, 35–45(–50) μm in largest dimension, irregularly polygonal to rounded; in upper basal regions sometimes becoming rectangular and forming undulating cell rows; plastids covering at most only part of the outer cell wall or tilted toward any anticlinal wall; pyrenoids two to six in middle and apical cells, up to twelve in basal cells.

Stephenson and Stephenson 1952; Aziz and Humm 1962; Alexander 1970; Brauner 1975; Wiseman 1978; Kapraun 1984; Richardson 1986, 1987.

Intertidal on jetties and seawalls and subtidal on *Zostera marina*, present in all seasons, abundant winter–spring.

Distribution: Canadian Arctic to Virginia, North Carolina, Georgia, Brazil, Iceland to Portugal; elsewhere, widespread in temperate and tropical seas.

Kapraun (1984) noted the plastids are netlike and have two to four pyrenoids in North Carolina plants.

Enteromorpha compressa (Linnaeus) Nees 1820, index [2].

Ulva compressa Linnaeus 1753, p. 1163.

Figures 15 and 16

Plants mostly branched, sometimes simple, to 35 cm tall; branches concentrated in the lower parts of the primary axis; large branches and main axes compressed or loosely adnate with hollow margins, main axes to 0.5(–1) cm broad, widening toward the top, attenuated at the apices; branches to 0.3 cm broad and often as long or longer than primary axes; cells in middle and apical regions irregularly polygonal to rectangular with four to six rounded to angular corners, to 15 μm wide, 18 μm long, rectangular cells forming islands of well-ordered longitudinal or transverse rows, cell divisions mostly unequal; cells in basal regions rounded polygonal to round, unordered; rhizoidal cells darkly pigmented and larger than vegetative cells; plastids parietal, often strongly tilted toward apical regions; pyrenoids one per cell, rarely two.

Bailey 1848; Harvey 1858; Curtis 1867; Melvill 1875; Alexander 1970; Kapraun 1980, 1984; Richardson 1986, 1987.

Subtidal and intertidal on jetties, summer–winter.

Distribution: Canadian Arctic to Virginia, North Carolina, South Carolina, Georgia, Bermuda, southern Florida to Uruguay, USSR to Portugal; elsewhere, widespread in temperate and tropical seas.

Enteromorpha flexuosa (Wulfen) J. Agardh 1883, p. 126.
  Conferva flexuosa Wulfen 1803, p. 11.

E. flexuosa subsp. *flexuosa* J. Agardh 1883, p. 126.
  Figures 17 and 18
  Plants to 25 cm tall, slender, filiform to strap shaped, hollow or the two layers

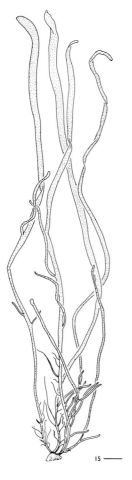

Figures 15, 16. *Enteromorpha compressa.* 15. Small plant, scale 1 mm. 16. Surface cells of large axis, scale 10 μm.

16 ———

15 ———

compressed in broad parts and adnate with hollow margins; plants not proliferous, sparsely branched or commonly simple; cells in middle and apical regions rectangular, quadrangular, or irregularly polygonal with or without rounded corners; cell divisions equal, forming longitudinal and short transverse rows unless cell order broken by oblique divisions; cells in upper basal region more rounded; cells in lower basal region also rounded, often elongate, and vegetative cells there rare or absent; plastids forming a hollow cylinder; pyrenoids one to two in middle and upper cells, to five in basal cells.

Hoyt 1920; Williams 1948a; Humm 1952; Taylor 1960; Alexander 1970; Chapman 1971; Schneider 1976; Kapraun 1984. As *E. lingulata*, Blomquist and Humm 1946; Williams 1948a, 1949; Taylor 1960; Earle and Humm 1964; Chapman 1971, Richardson 1987.

On jetties, in sounds and estuaries, winter–summer, and from 38 to 45 m in Onslow Bay, June.

Distribution: Quebec to Virginia, North Carolina, Georgia, northeastern Florida, Bermuda, southern Florida, Gulf of Mexico, Caribbean to Uruguay, Norway to Portugal; elsewhere, widespread in temperate and tropical seas.

E. flexuosa subsp. *paradoxa* (C. Agardh) Bliding 1963, p. 79, figs 42–45.
  Conferva paradoxa C. Agardh 1817, p. XXII.
  Figure 19
  Plants to 8 cm long, slender, profusely branched, branchlets uniseriate; cells rectangular to polygonal, to 25 μm wide and 25–50 μm long, in longitudinal and transverse rows throughout; plastids dentate discs with up to five pyrenoids.

Kapraun 1984. As *E. erecta*, Williams 1948a, 1949; Taylor 1960. As *E. plumosa*, Blomquist and Humm 1946; Williams 1948a, 1949; Humm 1952.

On exposed jetties and in sounds, in intertidal, year-round.

Distribution: Canadian Arctic to Virginia, North Carolina, northeastern Florida, Norway to Portugal, Central Pacific islands.

Kapraun (1984) reported local plants with cells 10 μm wide, 25 μm long, with plastids forming hollow cylinders with three to five pyrenoids.

Enteromorpha intestinalis (Linnaeus) Nees 1820, index [2].
  Ulva intestinalis Linnaeus 1753, p. 1163.
  Figure 20
  Plants simple, rarely with proliferations from the stipes, 30(–100) cm long, tubular and inflated, or compressed with the two layers weakly adnate and with hollow margins, to 4(–10) cm broad; cells in middle and apical regions irregularly polygonal with four to six rounded to angular corners, (7–)9–15 μm diameter, divisions mostly unequal, not in linear rows except in narrow branches, but

sometimes in short, curved rows; cells in basal region larger, to 18(–22) μm long, round to polygonal, not in linear rows; plastids parietal, usually strongly tilted toward apical sides of cells; one, rarely two, pyrenoids per cell.

Harvey 1858; Curtis 1867; Hoyt 1920; Williams 1948a, 1949; Stephenson and Stephenson 1952; Taylor 1957, 1960; Earle and Humm 1964; Alexander 1970; Brauner 1975; Schneider 1976; Kapraun 1984; Richardson 1986, 1987.

Intertidal on jetties and seawalls, drifting in sounds and estuaries, year-round, but primarily in spring; offshore from 28 to 50 m in Onslow Bay and Raleigh Bay, June–July.

Distribution: Canadian Arctic to Virginia, North Carolina, USSR to Portugal, Azores; elsewhere, widespread in temperate and tropical seas.

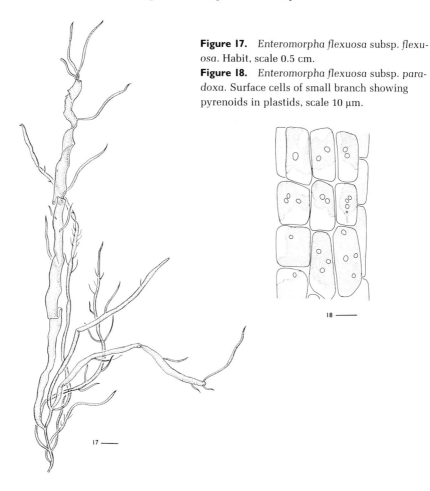

Figure 17. *Enteromorpha flexuosa* subsp. *flexuosa*. Habit, scale 0.5 cm.
Figure 18. *Enteromorpha flexuosa* subsp. *paradoxa*. Surface cells of small branch showing pyrenoids in plastids, scale 10 μm.

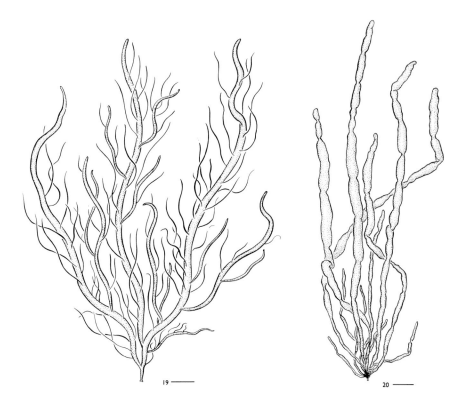

Figure 19. *Enteromorpha flexuosa* subsp. *paradoxa*, scale 2 mm.
Figure 20. *Enteromorpha intestinalis*, scale 1 cm.

Enteromorpha linza (Linnaeus) J. Agardh 1883, p. 134.

Ulva linza Linnaeus 1753, p. 1633.

Figure 21

Plants unbranched, or rarely with one or two broad branches from upper basal region; blades strap shaped, oblong, oblanceolate, or irregular; flat, the two layers separate only along the lower margins in the stipitate base; 50(–175) cm long, 10(–45) cm broad, (25–)35–60(–80) µm thick, the margins undulating to ruffled; cells in middle and apical regions quadrangular to rectangular, but often irregularly polygonal with mostly equal divisions, 10–16(–25) µm diameter; regions of ordered cells in longitudinal and transverse cell rows interspersed with regions with less order; cells in lower basal regions with rounded rhizoidal cells interspersed with vegetative cells in undulating cell rows; cells in upper basal region quadrangular to elongate, or irregularly polygonal with four to six rounded corners, or elliptical to round, cells unordered or in undulating cell rows; plastids covering outer cell walls or tilted toward any anticlinal wall; pyrenoids one to two per cell.

Hoyt 1920; Williams 1948a, 1949; Stephenson and Stephenson 1952; Taylor 1960; Earle and Humm 1964; Alexander 1970; Chapman 1971; Kapraun 1984; Richardson 1987.

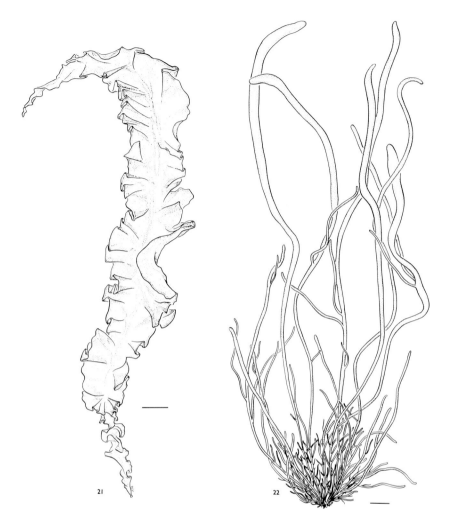

Figure 21. *Enteromorpha linza*, scale 2 cm.

Figures 22–24. *Enteromorpha prolifera*. 22. Habit, scale 1 mm. 23. Surface cells of young slender branch, scale 10 μm. 24. Surface cells of older branch, scale 10 μm.

On jetties and seawalls in the midtidal, most common winter–spring, rare in summer.

Distribution: Canadian Arctic to Virginia, North Carolina, South Carolina, Georgia, USSR to Portugal; elsewhere, widespread in temperate and tropical seas.

Enteromorpha prolifera (O. F. Müller) J. Agardh 1883, p. 129, pl. 4, figs 103–104.
    Ulva prolifera O. F. Müller 1778, p. 7, pl. 763, fig. 1.
    Figures 22–24
Plants with distinct primary axes, often intensely proliferous, to 60 cm tall

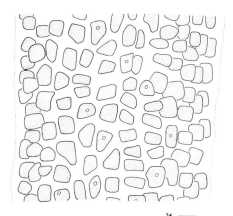

23 ———

24 ———

or larger, 1.5 cm wide, in brackish water becoming threadlike; small to middle-sized cells arranged in longitudinal and sometimes in transverse series in narrow parts of plants, 9–18(–27) μm in longest dimension; cells of macroscopic branches in surface view rectangular, often square; young branches in part multiseriate; pyrenoids one per cell, rarely two.

Hoyt 1920; Stephenson and Stephenson 1952; Alexander 1970; Brauner 1975; Kapraun 1984; Richardson 1986, 1987.

In sheltered and exposed intertidal and shallow subtidal locations, year-round, abundant in spring.

Distribution: Labrador to Virginia, North Carolina, South Carolina, Georgia, USSR to Portugal; elsewhere, widely distributed in temperate and tropical seas.

Hoyt (1920) described plants several meters long and 2 cm in diameter. Alexander (1970) reported plants 10–20 cm long with fronds 1–5 mm wide and branches either concentrated at the base of the plant or continuous along the primary axis. Alexander also reported a few plants corresponding to subspecies *radiata* J. Ag., which is recognized by some as a separate species. The subspecies is based on characteristics which are either not significantly different from the species or are not consistently present (e.g., rhizoidlike cells at the bases of slender branches, or trabeculate ingrowths from the inner cell walls into the central cavity). The large cells of some specimens (to 27 μm diameter) correspond to the cell size of the subspecies *gullmariensis* Bliding.

Enteromorpha ramulosa (J. E. Smith) Carmichael in Hooker 1833, p. 315.

Ulva ramulosa J. E. Smith 1810, pl. 2137.

Plant richly branched, epiphytic or in mats on stones; branches produce alternating long and short, spinelike branchlets, the latter with broad bases and normally one very small tip cell; cells in older parts of plants polygonal or rounded, nearly unordered, 30 μm wide, 40 μm long; in younger parts of plants square to rectangular, in fairly distinct longitudinal series, 15 μm wide, 20 μm long, some forms with smaller, unordered cells; plastids against outer or anticlinal walls; pyrenoids two to ten.

Kapraun 1984; Richardson 1986, 1987.

Coastal jetties, winter–spring.

Distribution: Labrador, North Carolina, Georgia, USSR to Portugal, Azores, Mediterranean, Central Pacific islands.

Kapraun (1984) reported plants from North Carolina with cells 10 by 25 μm and disc-shaped plastids that did not cover the entire face of the cell and had two to three pyrenoids. Scagel (1966) placed this species in synonymy with *Enteromorpha crinata* (Roth) J. Agardh, but his plants had smaller cells and usually a single pyrenoid per cell.

Enteromorpha torta (Mertens) Reinbold 1893b, p. 201.

Conferva torta Mertens in Jürgens 1822, Dec. 13.

Filaments cylindrical, hollow, 25–50 μm diameter, without branches; cells in three to twelve very distinct longitudinal rows, sometimes obliquely spiraled around the long axis of the frond; central cavity 10–15 μm diameter; square cells 12–14 μm on a side, rectangular cells 13(–28) by 11(–16) μm, the longer dimension usually longitudinal; plastids parietal, completely covering outer walls in young regions, sometimes tilted toward anticlinal wall in older cells; pyrenoids one, rarely two or three.

Kapraun 1984.

Rare, in southern North Carolina salt marshes, winter.

Distribution: Quebec, Maine, New Hampshire, Virginia, North Carolina, USSR to Portugal, Alaska to California.

Plants reported by Kapraun (1984) had cells to 8 by 14 μm.

Questionable Record

Enteromorpha lingulata J. Agardh 1883, p. 143.

Chapman's (1971) Georgia specimens have not been located.

Pseudendoclonium Wille 1901

Plants crustose; irregularly branched prostrate filaments forming short, erect filaments, pseudoparenchymatous cells, and short, single-celled or rarely multi-cellular rhizoids; plastids single in each cell, parietal and with a single pyrenoid; colorless hairs or setae lacking; asexual reproduction by akinetes or zoospores; zoospores bi- or quadriflagellate, with pyrenoid, with or without eyespots; sexual reproduction unknown.

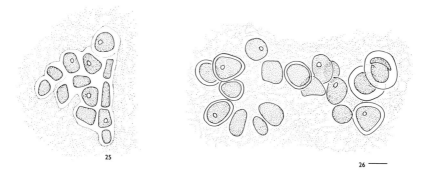

Figures 25, 26. *Pseudendoclonium submarinum*, growing on shell, scale 10 μm.

Pseudendoclonium submarinum Wille 1901, p. 29, pl. 3, figs 101–132.
    Figures 25 and 26
    Plants pseudoparenchmatous, crustose, with highly branched prostrate and erect filaments; primary filaments fifteen to twenty cells long, erect filaments three to seven cells long, young cells 6–9 μm diameter, 7–14 μm long; older cells short and much broader, almost spherical, central cells compressed and angular; terminal cells tapered toward tips, to 15 μm long; rhizoidal cells sometimes present; zoospores formed by any cell, ovoid, 3.6–4.5 μm wide, 4.8–7.2 μm long; flagella 1.5 times as long as cell.
    As *Protoderma marinum*, Williams 1948a; Humm 1952; Aziz and Humm 1962.
    On shells of oysters, barnacles, and crabs, in the intertidal, spring.
    Distribution: Canadian Arctic to Virginia, North Carolina, northeastern Florida, Caribbean, USSR to Portugal, southern Africa.
    The identity of the algae reported as *Pseudendoclonium submarinum* is arguable. Aleem and Schulz (1952) placed the species in synonymy with *Protoderma marinum* Reinke, creating the new combination *Pseudendoclonium marinum* (Reinke) Aleem et Schulz, but Yarish (1975) kept them as separate taxa because of uncertainties about the characteristics of *Protoderma marinum*. As examples of the continuing problems with these organisms, Wille (1901) reported that the species lacked any eyespots in the zoospore, but Yarish reported them present. Reinke and Yarish both reported quadriflagellate zoospores, but South (1974) reported biflagellate zoospores.

Ulva Linnaeus 1753, nom. cons.
Plants developing from filamentous germlings to form linear or expanded, foliose, two-layered blades attached by nonseptate, rhizoidal outgrowths of cells near the point of attachment; cells of blades mostly uninucleate, rhizoid-producing cells multinucleate; chloroplasts single, cup shaped or platelike, usually at outer faces of cells, with one or more pyrenoids; sporophytic reproduction by quadriflagellate zoospores; gametophytic reproduction by isogamous or anisogamous, biflagellate gametes; life history diplohaplontic, isomorphic.

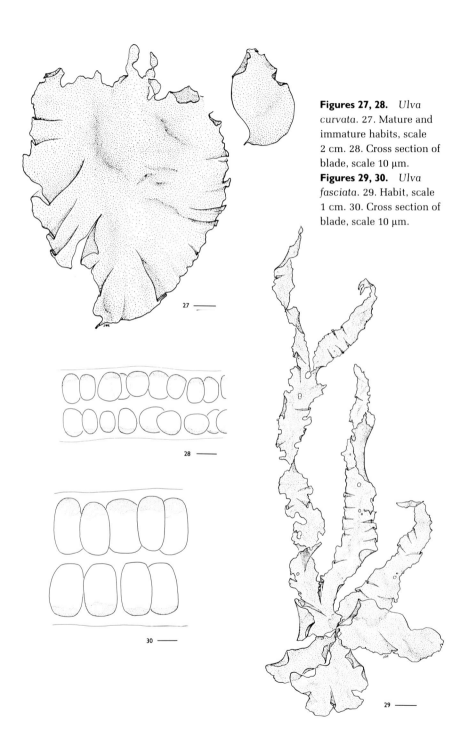

Figures 27, 28. *Ulva curvata.* 27. Mature and immature habits, scale 2 cm. 28. Cross section of blade, scale 10 μm. **Figures 29, 30.** *Ulva fasciata.* 29. Habit, scale 1 cm. 30. Cross section of blade, scale 10 μm.

Key to the species

1. Plants basally divided into several long, narrow strips .... *U. fasciata*
1. Plants lobed or marginally divided, but essentially consisting of a single blade
   2
2. Pyrenoids single; blade asymmetric and curved from base .... *U. curvata*
2. Pyrenoids two to three; blades symmetrical at first, rounded or lobed 3
3. Blades irregularly lobed, with microscopic marginal teeth; cells in cross sections of younger parts rounded; cells in older parts taller than wide; a plant of exposed coastal habitats .... *U. rigida*
3. Blades initially rounded, becoming lobed, lacking marginal teeth; cells in cross section rounded to wider than tall; a plant of hyposaline creeks and estuaries .... *U. rotundata*

Ulva curvata (Kützing) De Toni 1889, p. 116.
   Phycoseris curvata Kützing 1845, p. 245.
   Figures 27 and 28
   Plants 10–35(–100) cm long, blade and stipe asymmetrical, margins of blade entire; cells in surface view rectangular to polygonal, in rows or unordered, 11–18(–20) μm wide, 7–11(–15) μm long; blades 33–40 μm thick in upper parts, 75 μm thick toward the base, and 85 μm thick in rhizoidal region; chloroplasts typically lying against lateral walls of cells; pyrenoids one to two.
   Rhyne 1973; Kapraun 1984; Richardson 1987.
   In marshes and sounds on mud, shell, and seawalls, winter–spring.
   Distribution: Virginia, North Carolina, South Carolina, Georgia, Baltic Sea to Portugal.

Ulva fasciata Delile 1813, p. 297, pl. 58, fig. 5.
   Figures 29 and 30
   Plants to 30(–150) cm long, stipitate, the base cuneate, expanded above the base and divided irregularly and palmately into linear segments; margins entire, flat or ruffled; blade segments 0.5–5.0 cm bands divided irregularly and palmately into linear segments; margins entire, flat or ruffled; blade segments 0.5–5.0 cm broad, to 100 μm thick; height of cells in cross section of blades often double their width, particularly in middle of blade, in surface view 8–10 μm wide, 14–16 μm long; one to two pyrenoids per cell.
   Hoyt 1920; Williams 1948a; Humm 1952; Chapman 1871; Kapraun 1984; Richardson 1987.
   Common in lower intertidal on jetties along the coast.
   Distribution: North Carolina, Georgia, northeastern Florida, Bermuda, Carib-

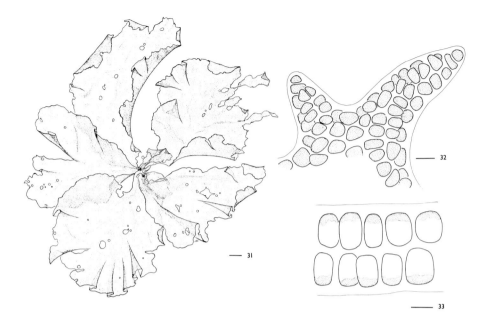

Figures 31–33. *Ulva rigida.* 31. Habit, scale 1 cm. 32. Detail of two teeth, scale 50 μm. 33. Cross section of blade, scale 10 μm.

bean to Uruguay, France to Portugal; elsewhere, widespread in temperate and tropical seas.

Kapraun (1984) reported frond thicknesses of 40–50 μm in North Carolina plants.

Ulva rigida C. Agardh 1822, p. 410.

Figures 31–33

Plants to 1 m or more in length, with median stipe or often unattached; lanceolate blades become lobed with age, often richly perforate; 38–42 μm thick in upper parts, 48–76 μm in central parts, and up to 200 μm thick at the base; margins bear microscopic teeth; cells of blade in ordered rows in upper parts, rectangular to rounded (to polygonal), (11–)14(–17) μm wide, (15–)18(–22) μm long in mature cells; cells larger (29 by 16 μm) in young plants; cells slightly taller than broad, globose in upper margins, to 21 μm tall; to 70 μm tall and tapering toward the top in the lower parts of blades; position of plastids variable; (one–) two(–eight) pyrenoids per cell.

Kapraun 1984; Richardson 1986, 1987. As *Ulva lactuca* var. *rigida*, Hoyt 1920; Humm 1952; Chapman 1971; Brauner 1975.

In lower intertidal and subtidal on jetties and shell material and free floating in sounds, spring–fall.

Distribution: Canadian Arctic to Virginia, North Carolina, Georgia, northeast-

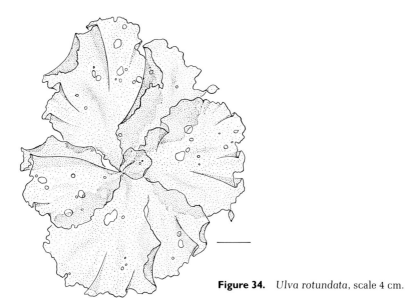

Figure 34. *Ulva rotundata*, scale 4 cm.

ern Florida, USSR to Portugal; elsewhere, widespread in temperate and tropical seas.

Reports of *Ulva lactuca* (Curtis 1867; Johnson 1900; Hoyt 1920; Williams 1948a, 1949; Stephenson and Stephenson 1952; Earle and Humm 1964; Brauner 1975; Schneider 1976) can probably be referred to *U. rigida* in most cases. Specimens from the region are generally less than 10 cm tall, 50–60 μm thick, have marginal, macroscopic teeth, and plastids have two to three (occasionally four) pyrenoids.

Ulva rotundata Bliding 1968, p. 566, figs 19–22.

Figure 34

Plants to 20 cm tall, attached by a distinct median stipe, blade initially rounded, becoming deeply lobed, or split; cells in surface view polygonal, angular, 30–38(–45) μm in longest dimension in young plants, rounder, 20 by 26 μm in older plants; blades to 56 μm thick in upper parts to 75 μm thick in lower parts; cell lumens to 30 μm tall, cells square or wider than tall in cross section; plastids lying beneath outer faces of cells or against side walls, generally two pyrenoids, sometimes one or three.

Rhyne 1973; Kapraun 1984; Richardson 1986, 1987.

Distribution: Virginia, North Carolina, South Carolina, Georgia, Norway, France, Spain.

Kapraun (1984) described plants from North Carolina with cells 10–14 by 16–20 μm and blades 40–50 μm thick. Reports of *Ulva latissima* (Bailey 1851; Harvey 1857; Hoyt 1920; Chapman 1971) probably refer to *U. rotundata*.

CLADOPHORALES

Plants either dividing by cross walls to become filamentous and uniseriate with filaments simple or branched, or cells dividing internally by segregative division with subsequent internal or external growth of cells to form clusters of cells or branched filaments; cells large, multinucleate, often with a large central vacuole; plastids numerous, angular, often united to form a reticulum, or fragmented, with one or more pyrenoids; cell walls containing crystalline cellulose I in lamellae of parallel fibrils which form a crossed pattern with adjacent lamellae; gametes biflagellate, zoospores bi- to quadriflagellate, with flagellar bases lacking terminal caps and the proximal sheaths reduced or absent; motile cells released through exit papillae closed by plugs of mucilage; mitosis and cytokinesis distinctly separated; life history usually diplohaplontic, with isomorphic phases; primarily marine.

Key to the families

1. Filaments united in solid or netlike blades 2
1. Filaments not united to form blades 3
2. Blades stalked . . . . Boodleaceae
2. Blades not stalked . . . . Anadyomenaceae
3. Plants endo- or epiphytic; with aseptate hairs . . . . Chaetosiphonaceae
3. Plants not endo- or epiphytic; no aseptate hairs 4
4. Plants always filamentous; cells formed by cross walls . . . . . Cladophoraceae
4. Plants secondarily filamentous or not filamentous; cells formed by internal, segregative cell division . . . . Valoniaceae

Cladophoraceae

Filaments of plants branched or unbranched; if branched, not forming blades; gametes isogamous.

Key to the genera

1. Filaments unbranched or with few-celled rhizoidal branches 2
1. Filaments branched with regularity . . . . Cladophora
2. Cells large or small; attached by a single holdfast cell, or if unattached the cell diameter is greater than 30 μm . . . . Chaetomorpha
2. Cells small, 30 μm diameter or less; unattached or attached by lateral rhizoidal cells . . . . Rhizoclonium

Chaetomorpha Kützing 1845, nom. cons.
Plants unbranched, uniseriate filaments unattached or attached by single basal holdfast cells; growth diffuse; cells large, in many species in excess of 100 μm

diameter; plastids reticulate, pyrenoids numerous; life history diplohaplontic, isomorphic.

The systematics of the genus in the region is unsatisfactory. Local specimens are often at variance with the species to which they have traditionally been assigned insofar as those species have been delimited in recent studies (Blair 1983; Blair et al. 1982). Plants have here been assigned to taxa primarily on the basis of cell diameters, cell length/width ratios, and on the presence or absence of a basal attachment cell, but with the recognition that some distinctions are arbitrary and traditional and the placement into species is less than satisfactory.

Key to the species

1. Filaments' average diameter greater than 100 µm 2
1. Filaments' average diameter less than 100 µm 6
2. Filament diameter greater than 500 µm 3
2. Filament diameter less than 500 µm 4
3. Larger cells distinctly swollen .... C. aerea (in part)
3. Larger cells cylindrical .... C. crassa
4. Plants unattached, 80–180 µm diameter .... C. brachygona
4. Plants attached or unattached, cells mostly 150–500(–600) µm diameter ... 5
5. Cell length often one-half cell diameter after division, plants usually attached
   .... C. aerea (in part)
5. Cell length greater than one-half cell diameter after division, plants usually unattached .... C. linum
6. Filament diameter 10–27(–40) µm, basal cell less than 100 µm long ....
   .... C. minima
6. Filament diameter (32–)65–86(–120) µm, basal cell, if present, more than 100 µm long .... C. gracilis

Chaetomorpha aerea (Dillwyn) Kützing 1849, p. 379.

    Conferva aerea Dillwyn 1806 [1802–1809], pl. 80.

    Figures 35 and 36

    Plants attached, 10–15(–30) cm tall, bright to dark green, filaments stiff and straight; solitary or growing in clumps; basal cells to 130–150 µm diameter above, tapering below and attached by single, discoid, basally lobed or fringed holdfast cells; cylindrical vegetative cells 125–400 µm diameter, 0.5–2 diameters long, little constricted at septa; fertile cells to 700 µm diameter, cask shaped to subglobose, walls conspicuously thickened.

    Williams 1948a, 1949; Stephenson and Stephenson 1952; Taylor 1960; Aziz and Humm 1962; Earle and Humm 1964; Chapman 1971; Brauner 1975. As *C. melagonium sensu* Hoyt 1920, Searles and Schneider 1978; Kapraun 1984.

    In sounds and on jetties, year-round.

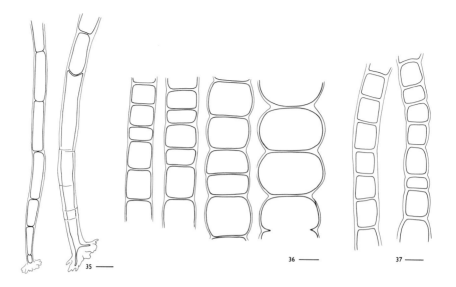

Figures 35, 36. *Chaetomorpha aerea*. 35. Basal portions of filaments, scale 100 μm. 36. Median portions of filaments, scale 200 μm.
Figure 37. *Chaetomorpha brachygona*, scale 50 μm.

Distribution: Newfoundland to Virginia, North Carolina, South Carolina, Georgia, Bermuda, southern Florida, Caribbean, Brazil, Norway to Portugal; elsewhere, widely distributed in temperate and tropical seas.

The algae described here have been reported as *Chaetomorpha melagonium* but are probably not that northern cold-water species. They are similar to plants collected throughout the eastern United States and south into the Caribbean that fit the general characteristics of *C. aerea* as delineated in terms of cell diameter and length/width ratios by Blair (1983).

Chaetomorpha brachygona Harvey 1858, p. 87, pl. 46a.
   Figure 37
   Plants unattached, entangled, soft, flexuous, 80–180 μm diameter; cells 60–420 μm long, length/width is 1(–4), cylindrical or slightly swollen, walls thickened.
   Hoyt 1920; Williams 1948a, 1949, 1951; Taylor 1960.
   Tangled with other algae on coastal jetties and in the drift along the beaches and in the sounds, late spring to early fall.
   Distribution: Newfoundland to New Hampshire, North Carolina, Bermuda, southern Florida, Gulf of Mexico, Caribbean, Brazil, West Africa, Central Pacific islands, British Columbia.

Chaetomorpha crassa (C. Agardh) Kützing 1845, p. 204.
   Conferva crassa C. Agardh 1824, p. 99.
   Figure 38

Plants unattached, dark green; filaments (300–)500–550(–700) μm diameter, cells 1–2 diameters long.

Schneider and Searles 1973; Schneider 1976.

Growing on soft bottom in Onslow Bay from 24 to 50 m, June–September.

Distribution: North Carolina, southern Florida, Caribbean, West Africa, southern Africa, Central Pacific islands, Japan, Pacific Mexico.

Taylor (1960) indicated that this species has thick walls, but walls in our plants are thin. Parke and Dixon (1976) suggest that this species is possibly *Chaetomorpha linum*.

Chaetomorpha gracilis Kützing 1843, p. 259.

Figures 39 and 40

Attached plants to 12 cm long, rigid, straight; cells (32–)56–86(–120) μm diameter, length/width is (0.5–)1–2 above to 2.6 near base; basal cell 175–360 μm or to 4 diameters long; cells of unattached plants 32–70 μm diameter, 2–4 diameters long.

Blomquist and Pyron 1943; Brauner 1975. As *Chaetomorpha linum sensu* Hoyt 1920. As *C. linum* f. *aerea sensu* Hoyt 1920. As *Rhizoclonium tortuosum sensu* Taylor 1957, 1960.

Figure 38. *Chaetomorpha crassa*, apical and median portions, scale 0.5 mm.
Figures 39, 40. *Chaetomorpha gracilis*. 39. Base of filaments, scale 100 μm. 40. Median portions of filaments, scale 100 μm.

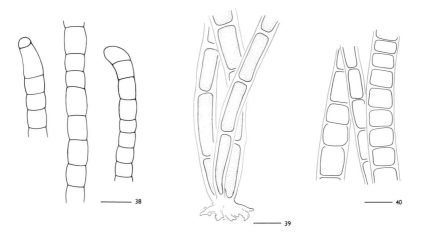

Attached plants on intertidal rocks, on *Zostera marina* blades, buoys, and as unattached plants on mud in Newport River estuary, North Carolina, spring–summer.

Distribution: North Carolina, Bermuda, southern Florida, Caribbean, Brazil.

The average diameter of the filaments falls within or near the limits previously described (Taylor 1960), but the upper limits (120 μm) are well in excess of the upper limit of 70 μm diameter listed by Taylor. Pankow (1971) suggests that *Chaetomorpha gracilis* may be a growth form of *C. linum*.

Chaetomorpha linum (O. F. Müller) Kützing 1845, p. 204.

    Conferva linum O. F. Müller 1778, p. 7, pl. 771, fig. 2.

    Figures 41 and 42

    Plants unattached, to 10 cm long, yellow-green, stiff, curled, entangled, 80–375(–434) μm diameter; cells (0.75–)1–2(–5) diameters long, little constricted.

    Williams 1948a, 1949, 1951; Taylor 1957, 1960; Brauner 1975; Kapraun 1984.

    Drift plants in sounds, summer.

    Distribution: Labrador to Virginia, North Carolina, Bermuda, southern Florida, Caribbean, Brazil, Norway to Portugal; elsewhere, widespread in temperate and tropical seas.

Figures 41, 42. *Chaetomorpha linum.* 41. Unattached tangle of filaments, scale 1 cm. 42. Median portions of filaments, scale 0.3 mm.
Figure 43. *Chaetomorpha minima,* basal and median portions of filaments, scale 20 μm.

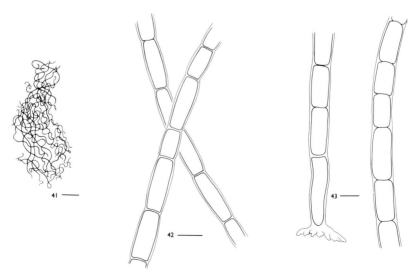

Chaetomorpha minima Collins et Hervey 1917, p. 41, pl. 1, figs 5–7.

Figure 43

Plants to 10 mm tall; cells cylindrical, 10–27(–40) μm diameter, 1.6–5 diameters long; basal cells of uniform diameter or slightly tapered and to 85 μm long; wall laminate, sometimes constricted at the septa; zoospores formed by any cell, escaping by a lateral opening.

On culms of *Juncus* in a salt marsh near Beaufort, N.C., May.

Distribution: Maine, New Hampshire, North Carolina (new record), Bermuda.

The original description of plants from Bermuda indicated cell diameters of 10–20 μm and length/width ratios of 2–4. Blair (1983) included plants from Maine with diameters to 40 um and lengths to 10 mm in the species. North Carolina plants are up to 27 μm diameter, with cells 1.6–5 diameters long.

Species Excluded

Chaetomorpha californica Collins.

There are no voucher specimens for the report of Aziz and Humm (1962). Filaments of *Chaetomorpha californica* resemble those of *C. minima* in being slender (mostly 20–27 μm diameter), but are distinguished by a longer (100–200 μm) and more constricted basal cell.

Cladophora Kützing 1843, nom. cons.

Plants erect, attached or free floating, sparingly to profusely branched; basal attachment by creeping rhizoids or disc-shaped holdfasts; cell division apical, lateral, or intercalary; plastids parietal, numerous, densely packed or united in a reticulate mesh, some with a pyrenoid; asexual reproduction by bi- or quadriflagellate zoospores or fragmentation; sexual reproduction with regular alternation of isomorphic generations, meiosis preceding spore formation; most species also form thick-walled, starch-filled, swollen akinetes.

Species distinctions in the sections of the genus are in large part quantitative, with variable expression within species and overlap among species. The species descriptions are adapted from those of van den Hoek (1982) for *Cladophora* from the American shores of the North Atlantic. In these descriptions dimensions of cells and length/width ratios (l/w) of cells are given according to the following notation: diameter = (a–b)–(c–d) and l/w = (e–f)–(g–h), where (a–b) and (e–f) are the ranges for the thinnest plants and (c–d) and (g–h) are ranges for the thickest plants investigated. The notation (x–)(a–b)–(c–d)(–z) indicates that there were a very few cells with divergently thinner or thicker diameters.

Key to the species

1. Cell division predominantly intercalary 2
1. Cell division predominantly but never exclusively apical 3
2. Main axes long, mostly long celled, unbranched or infrequently pseudo-dichotomously branched; lateral branches initially laterally inserted; loose lying in brackish water or mud flats .... *C. liniformis*
2. Main axes not long, not long celled, with pseudodi- or pseudotrichotomous branching; lateral branches apically inserted; in various habitats 4
3. Intercalary divisions in middle as well as basal parts of plants 8
3. Intercalary divisions of vegetative cells restricted to occasional divisions in basal parts of plants 14
4. Apical cell diameter greater than 80 µm, ultimate branches 100 µm or greater and main branches 240–400 µm .... *C. hutchinsiae*
4. Apical cell diameter 80 µm or less, ultimate branches 100 µm or less, and main branches 200 µm or less 5
5. Apical cell diameter (20–)25–80 µm, ultimate branches (20–)25–100 µm, and main branches (50–)70–200 µm..................................... 6
5. Apical cell diameter 10–25(–40) µm, ultimate branches 10–25(–50) µm, and main branches 15–60(–135) µm 7
6. Plants up to 20–100 cm long, forming hairlike or ropelike bundles, maximum of one branch per node .... *C. ruchingeri*
6. Plants forming dense, penicillate tufts, two (to three) branches per node .... *C. sericea*
7. Distal parts of plants not acropetally organized, unbranched, straight, *Rhizo-clonium*-like, or scarcely branched, or lined with rows of more or less scattered, straight branchlets of varying age intercalated between older branch-lets .... *C. montagneana*
7. Distal parts of plants acropetally organized to irregular, densely branched; laterals bent and sometimes sickle shaped, but also straight, mostly in unilateral rows of straight branchlets of varying age intercalated between older branches .... *C. albida*
8. Maximum diameter of apical cells to 32 µm................. *C. dalmatica*
8. Maximum diameter of apical cells greater than 32 µm 9
9. Diameter of main axes to 100 µm .... *C. laetevirens* (in part)
9. Diameter of main axes greater than 100 µm 10
10. Diameter of main axes to 120 µm .... *C. vadorum* (in part)
10. Diameter of main axes greater than 120 µm 11
11. Ratio of apical cell diameter to that of main branch diameter 4 or greater .... *C. vagabunda* (in part)
11. Ratio of apical cell diameter to that of main branch equal to or less than 4

Cladophora albida (Nees) Kützing 1843, p. 267.

Annulina albida Nees 1820, index [1].

Figures 44 and 45

Plants pale or light green to grass green or dark green, forming spongy globules 1–3 cm high in exposed habitats or more delicate, penicillate plants 7(–15) cm tall in sheltered habitats; attached by branched rhizoids formed by basal and near basal cells; organization acropetal to irregular; bent or sickle-shaped laterals, occurring in vigorously growing acropetally organized upper axes and main laterals; lower main laterals with irregular organization, frequent intercalary divisions, and mostly unilateral branches of varying ages, the younger ones intercalated among older ones; in acropetally organized plants intercalary divisions appearing one to three cells from the apex; in plants from sheltered habitats organization irregular throughout; axes and main laterals produce unilateral and multilateral rows of branches, younger branches intercalated among older ones; old thick-walled filaments proliferate by giving off rows of almost equally long opposite branches; fertile branches end in rows of short, swollen zooidangia; transverse cell divisions predominantly intercalary; basal walls of lateral branches initially oblique, later may become almost horizontal; basal cells of lateral may become partly fused with cells of axis or main laterals; nodes may bear one to two (to three) laterals; angle of branching 45° or somewhat more; diameter of main axes two to three (to four) times that of apical cells; apical cells gradually tapering to obtuse tips; apical cells (12–18)–(24–40) μm diameter, l/w (2.5–7)–(1–3); ultimate branches (12–20)–(25–40) μm diameter, l/w (3–7)–(0.8–2); main axes (20–40)–(30–80) μm diameter, l/w (2–7)–(2–5).

As *C. glaucesens*, Taylor 1957, 1960. As *C. glaucesens* var. *pectinella*, Harvey

Figures 44, 45. *Cladophora albida.* 44. Branching patterns of filaments, scales 0.5 cm. 45. Branch apex, scale 100 μm.
Figure 46. *Cladophora dalmatica,* scale 100 μm.

1858. As *C. refracta,* Harvey 1858; Humm 1952; Stephenson and Stephenson 1952; Chapman 1973; van den Hoek 1982; Kapraun 1984. As *C. delicatula sensu* Stephenson and Stephenson 1952.

Intertidal, winter and spring.

Distribution: New Brunswick to Delaware, North Carolina, South Carolina, Georgia, northeastern Florida to Gulf of Mexico, Norway to Portugal; elsewhere, widespread in temperate and tropical seas.

Cladophora dalmatica Kützing 1843, p. 268.

Figure 46

Plants grass green, light, pale, or sometimes dark green, in exposed sites forming spongy pompoms 0.5–2 cm tall, or intergrading with taller, 10(–20) cm, plants from more sheltered habitats, with terminal penicillate tufts of filaments; organization of terminal branch systems acropetal with apices straight or slightly curved, sometimes bent back, sometimes sickle shaped; main axes pseudodi- or pseudotrichotomously branched; transverse cell divisions apical and intercalary, the first intercalary divisions four to eight cells from the apices and more common toward plant bases; secondary branches forming some distance from apices at intercalary crosswalls; sporulation temporarily reducing plants to main axes; first lateral branches initiated two or three cells from apices, one to four additional branches may be added with increasing distance from apices; branching from distal ends of parent cells, not serial; basal walls of branches initially oblique, later becoming almost horizontal; angle of branching 45° or more, bases of older laterals generally partly fused with adjacent cell of axis; diameter of basal parts of main axes (2.5–)3–5(–7) times that of apical cells; apical cells cylindrical to slightly tapering, or often distinctly club shaped when forming zooidangia; apical cells (14–18)–(21–32) μm diameter, l/w (4.5–13)–(2–6); in fertile, robust plants, swollen apical zooidangia to 50 μm diameter; ultimate branches (14–27)–(21–42) μm diameter, l/w (5–10)–(3–5); main axes (25–65)–(60–300) μm diameter, l/w (8–14)–(2.5–10); thickness of cell walls in ultimate branches 0.5–1 μm, 1–4 μm in main axes.

Kapraun 1984; Searles 1987, 1988.

On jetties and floating docks near Wilmington, late winter–spring; from 23 to 30 m in Onslow Bay; 17 to 21 m on Gray's Reef, June–August.

Distribution: Massachusetts to Virginia, North Carolina, Georgia, southern Florida, Caribbean, Baltic to Spain, West Africa, Central Pacific islands.

Kapraun (1984) reported diameters of the main axes as 250–300 μm, well in excess of the maximum 65 μm reported by van den Hoek (1982) for American material and slightly exceeding the 250 μm van den Hoek reported (1963) for European material.

Cladophora hutchinsiae (Dillwyn) Kützing 1845, p. 210.

Conferva hutchinsiae Dillwyn 1809, pl. 109.

Plants to 35 cm long; growth of young plants slightly acropetal but becoming primarily intercalary; main axes pseudodichotomously branched, bearing branches of different lengths initiated by new intercalary cells; branch insertion initially oblique, later becoming almost horizontal and appearing dichotomous; older axial cells sometimes forming second or third branches; zooidangia formed in long, moniliform series of barrel-shaped cells in ultimate branches; sporulation reducing plants to pseudodichotomously branched main axes; apical cells cylindrical or slightly tapering, (90–140)–(160–195) μm diameter, l/w 1–4; ultimate branches (100–150)–(170–325) μm diameter, l/w 1–4; main axes 240–400 μm diameter, l/w 1–3.5.

Richardson 1986, 1987.

On floating docks, February–May.

Distribution: Georgia, USSR to Portugal, Mediterranean, southern Africa, British Columbia, Washington.

Cladophora laetevirens (Dillwyn) Kützing 1843, p. 267.

Conferva laetevirens Dillwyn 1805 [1802–1809], pl. 48.

Figures 47–49

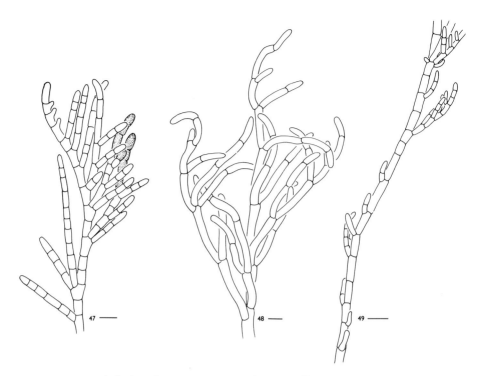

Figures 47–49. *Cladophora laetevirens.* 47. Apical portion of branch with sporangia (stippled) and mostly unilateral branching, scale 100 µm. 48. Apical part of branch with pseudo-, di-, and trichotomous branching, scale 100 µm. 49. Median portion of filament with intercalary divisions and new lateral branches, scale 200 µm.

Plants grass green, light green, to pale green, forming spongy pompoms 0.5–5 cm tall or dense penicillate plants on exposed intertidal sites, or plants up to 20 cm tall with main filaments ending in penicillate tufts in sheltered or submerged sites; plants attached by rhizoids from basal and adjacent cells; organization of pseudodi- or pseudotrichotomously branched main filaments ending in acropetally organized branch systems; branches sickle shaped, bent back, or with straight or nearly straight filaments; growth apical; intercalary divisions increasingly common toward plant bases, the first intercalary divisions five to nine (to fourteen) cells below apices, with secondary branches arising at intercalary cell walls; terminal branch systems form zooidangia and disintegrate, reducing plants to robust main axes that continue growth by intercalary division, giving off robust branches in irregular sequence; new cells cut off acropetally producing single branches at their apical poles, one to four (up to six) cells from apices; second, third, or fourth branches may subsequently form, often on the same side but not serially; branches inserted with oblique basal walls at angles of 45° or more except when more appressed in densely tufted plants; basal walls more horizontal with age, basal cells partly fusing with main axes; in acropetal systems, cells in basipetal direction distinctly longer and broader, most remaining cylindrical or broadening at apical poles; old proliferating axes with many intercalary divisions often distally broadened; basal parts of main axes 1.5–3.5(–5) times diameter of the apical cells in acropetally organized plants; in plants reduced to distally broadened main axes this ratio is reduced to 0.6–2; apical cells cylindrical with rounded tips, rarely somewhat tapered; zooidangia distinctly club shaped; apical cells in acropetally organized plants (37–53)–(51–105) μm diameter, l/w (3.5–11)–(2–4); ultimate branches in acropetally organized plants (48–70)–(51–105) μm diameter, l/w (4–9)–(2–4.5); main axes (53–100)–(150–205) μm diameter, l/w (3–9)–(4–11); in plants with distally broadened main axes with proliferating branches the apical cells' diameter varies from 50 to 150 μm and the l/w from 1 to 4; wall thickness in ultimate branches 0.5–1 μm, 1–4 μm in main branches.

Searles 1987, 1988.

Rare, Gray's Reef from 17 to 21 m, August.

Distribution: Georgia, Bermuda, Caribbean, British Isles to Portugal, West Africa, Central Pacific islands, British Columbia, Washington.

Cladophora liniformis Kützing 1849, p. 405.

Figure 50

Plants pale to dark green, to 10 cm long, forming loose-lying mats; branching mostly pseudodichotomous and infrequent or without pseudodichotomies and with or without irregularly scattered branchlets or rows of branchlets of almost equal length; cell divisions mainly intercalary, less frequently by division of long apical cells; one, rarely two, branches per node; branches initially laterally inserted, cut off from parent cells by steeply inclined walls; in older branches, lateral walls oblique or almost horizontal and branching appearing pseudodichotomous; cells cylindrical unless transformed into club-shaped akinetes; apical cells (15–20)–(33–45) μm diameter, l/w (8–10)–(8–20), l/w in akinete filaments reduced to 4.5; ultimate branches (16–32)–(40–75) μm diameter, l/w (4–8)–(5–15); main axes (32–48)–(60–85) μm diameter, l/w (3.5–20)–(5–15); cell wall thickness in ultimate branches less than 1 μm, in akinetes 2–4 μm; l/w of rapidly growing, almost unbranched filaments ca. 2–4 over long stretches.

Van den Hoek 1982. As *Rhizoclonium hookeri sensu* Blomquist and Pyron 1943; Schneider and Searles 1975.

In brackish water, Newport River and Shackleford Bank, June.

Distribution: Quebec, Massachusetts to Virginia, North Carolina, USSR to Spain.

Cladophora montagneana Kützing 1847, p. 166.

Figure 51

Plants dark green to dull green or grass green; one to several (–30) cm tall; small plants often compact, stiff, thick walled; larger vigorously growing plants

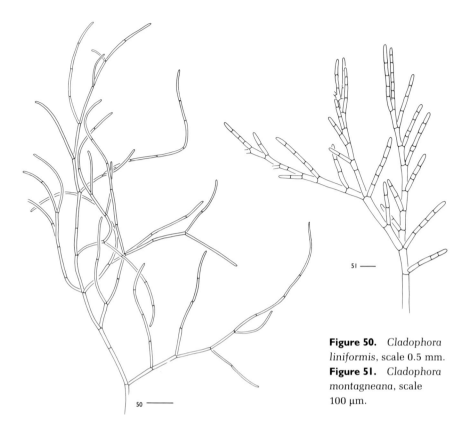

Figure 50. *Cladophora liniformis*, scale 0.5 mm. **Figure 51.** *Cladophora montagneana*, scale 100 μm.

delicate, forming lubricous tufts; organization irregular, nonacropetal; main axes with one to several strongly developed main laterals forming pseudodichotomies; axes and main laterals unbranched over long stretches; lateral branchlets (if present) short, straight, forming unilateral rows or growing in all directions; transverse cell divisions almost exclusively intercalary, young branches intercalated among older ones, one to three branches per node; first branches apical, others opposite, adjacent, or sometimes serial; basal walls initially steeply inclined to oblique, later almost horizontal; bases of laterals often partly fused with adjacent cells of axes; angles of young branches mostly 45° or more; main laterals mostly more or less appressed; diameter of main axes (1.5–)3–4(–7) times diameter of apical cells in small, stiff plants, (1.3–)2–3(–4) in long, vigorously growing plants; apical cells short, distinctly tapering with obtuse tips to gradually tapering, or cylindrical to somewhat swollen preceding sporulation; zooidangia swollen, short, often forming long, moniliform rows; small, stiff plants formed mostly of akinetes, the cells more or less club shaped and thick walled; apical cells in small, stiff plants (13–25)–(18–30)(–42) μm diameter, l/w (1.5–3)–(3–5); ultimate branches (13–28)–(18–30)(–53) μm diameter, l/w (1–2.5)–(2–4); main axes (36–55)–(25–50)(–135) μm diameter, l/w (0.8–3)–(2–4); thickness of walls in ultimate branches 1–3 μm, 4–7 μm in main axes; apical cells in vigorously growing plants (10–14)–(16–36) μm diameter, l/w (4–10)–(2–8); ultimate branches (10–14)–(16–36) μm diameter, l/w (3.5–6)–(1.5–4); main axes

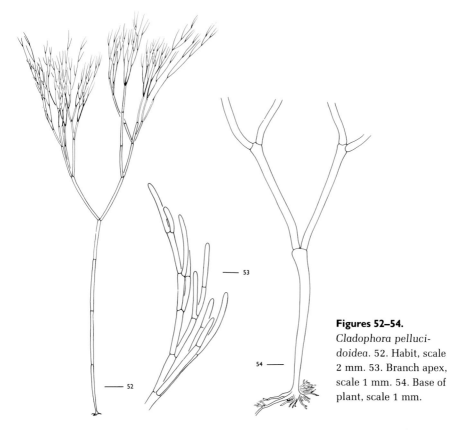

Figures 52–54.
Cladophora pelluci-doidea. 52. Habit, scale 2 mm. 53. Branch apex, scale 1 mm. 54. Base of plant, scale 1 mm.

(20–41)–(20–52)(–90) μm diameter, l/w (2–4)–(1.5–5); thickness of cell walls 0.5–1 μm in ultimate branches, ca. 1–4(–8) μm in main axes.

Van den Hoek 1982; Richardson 1986, 1987. As *C. delicatula*, Taylor 1960.

Intertidal on Cape Lookout and Radio Island, N.C., jetties, August; coastal Georgia, spring.

Distribution: North Carolina, Georgia, southern Florida, Caribbean, West Africa.

Cladophora pellucidoidea van den Hoek 1982, p. 179, figs 358–362, 365.
Figures 52–54

Plants grass green, 1.5–6 cm tall, forming loosely or closely branched, penicillate tufts; attached by branched rhizoids forming at bases of holdfast cells; organization acropetal, without distinct main axes, but with many pseudodichotomies and some tri- or tetrachotomies; cell divisions apical except at bases where basal cells may divide to form two to three cells; first laterals usually formed by subapical cells at distal end, second, or rarely third or fourth, laterals initiated farther from apices; axes and first and second laterals forming pseudodi- or pseudotrichotomies; first and second laterals neither opposite or serial; basal crosswalls of laterals feebly inclined or almost horizontal, branch angles 30° or less in distal branches; in older basal regions insertion of branches lateral to apical and branch angles becoming 40° or more; basal cells not or only slightly

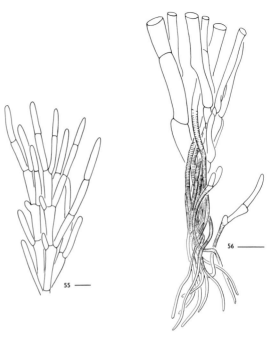

Figures 55, 56.
Cladophora prolifera.
55. Branch apex, scale
0.5 mm. 56. Basal portion,
scale 1 mm.

fused with adjacent axes; basal cells cylindrical, not club shaped; basal cells 1.5–3.5 times diameter of apical cells; apical cells cylindrical or tapering, tips obtuse; apical cells (50–80)–(70–95) μm diameter, l/w (9–22)–(16–32); ultimate branches (50–80)–(70–95) μm diameter, l/w (9–15)–(10–22); main axes (70–130)–(85–260) μm diameter, l/w (7–30)–(8–84); apical cell walls ca. 1.5–4 μm thick, basal cell walls 8–16 μm thick.

Van den Hoek 1982; Searles 1987, 1988. As *C. pseudopellucida sensu* Searles 1981. As *Cladophoropsis membranacea sensu* Schneider and Searles 1975, pro parte; Schneider 1976, pro parte.

Known only from offshore waters in Georgia and the Carolinas from 16 to 20 m, May–September.

Distribution: North Carolina, Georgia, Curaçao.

Cladophora prolifera (Roth) Kützing 1843, p. 271.
    Conferva prolifera Roth 1797, p. 182, pl., fig. 2.
    Figures 55 and 56
Plants dark green, 2–10(–25) cm tall, forming coarse tufts of densely fasciculate plants; organization acropetal, with numerous pseudodi-, pseudotri-, and pseudotetrachotomies; growth apical, intercalary divisions restricted to formation of zooidangia and occasional divisions in basal parts of plants; lateral branches often initiated from subapical cells, second, third, and, rarely, fourth laterals initiated farther from apices; branches forming at distal ends of parent cells, with steeply inclined basal walls, angles of branching 45° or less, bases of older laterals partly fusing with adjacent cells of axes; cells in basipetal direction longer and increasingly club shaped; old cells in basal regions producing rhizoids with annular constrictions; rhizoids growing down parallel to

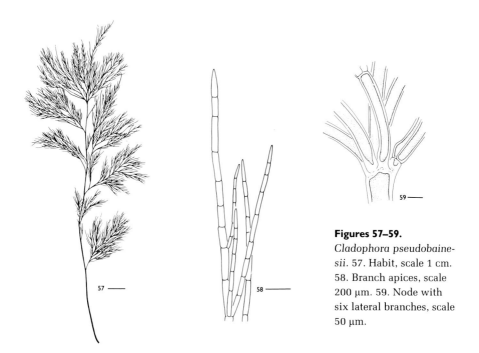

Figures 57–59.
Cladophora pseudobaine-sii. 57. Habit, scale 1 cm. 58. Branch apices, scale 200 µm. 59. Node with six lateral branches, scale 50 µm.

axes and ramifying; thickest axes (1–)2–4 times diameter of apical cells; apical cells cylindrical to gradually tapering, tips rounded; apical cells (95–120)–(170–240) µm diameter, l/w (2–5)–(3–5.5); ultimate branches (95–120)–(170–240) µm diameter, l/w (2.5–4)–(3.5–7); main axes (basal region) (240–345)–(275–800) µm diameter, l/w (7–16)–(4.5–9); thickness of cell wall of ultimate branches ca. 2–10 µm, in main axes (basal region) 15–40 µm.

Hoyt 1920; Williams 1948a; Taylor 1960; Schneider 1976; van den Hoek 1982; Kapraun 1984. As *C. pellucida sensu* Williams 1951.

Rare in shallow water, common offshore on rock outcroppings in Onslow Bay and Long Bay from 16 to 45 m, May–December.

Distribution: North Carolina, South Carolina, southern Florida, Caribbean, Brazil, British Isles, France to Portugal, Azores, West Africa, Central Pacific islands, Pacific Mexico.

Van den Hoek (1963, 1982) reported plants from Europe with main axes to 600 µm and plants from the western Atlantic with main axes to 345 µm; Kapraun (1984) reported plants from North Carolina with main axes 600–800 µm.

Cladophora pseudobainesii van den Hoek et Searles 1988, p. 521, figs 1–3.
Figures 57–59

Plants medium green, 1–3.5 cm tall, erect with single main basal axes, sometimes pseudodichotomously branched in their basal regions, and bearing delicate, fastigiate tufts of densely branching filaments; cell divisions primarily apical; internodes near bases may divide into two to six cells by intercalary cell divisions; secondary branches occasionally produced from intercalary cross walls; two to four cells from apices, new cells forming single branches at apical poles; with increasing distance from apices, cells may form second, third, fourth, and, more rarely, fifth or sixth branches; branches apically inserted at acute angles (30° or less) with feebly inclined to almost horizontal cross walls; basal cells of laterals not, or only slightly, fused at their basal poles to adjacent cells of axes; apical cells 12–20 μm diameter, l/w 7–22, cylindrical below, tapering near mucronate tips; ultimate branch cells 14–24 μm diameter, l/w 5–11; basal stipe cells 100–180 μm diameter, l/w 5–18; ratio of basal stipe cells to apical cell diameter ca. 7–15; walls 0.5–1 μm thick above, to 25 μm below; upper cells transformed into zooidangia, opening by single pores above cell midpoints, empty zooidangia breaking off near bases.

Van den Hoek and Searles 1988. As *Cladophora* sp., Searles 1987, 1988.

Known only on Gray's Reef and the Snapper Banks from 17 to 35 m, July–August.

Distribution: Endemic to Georgia as currently known.

Cladophora ruchingeri (C. Agardh) Kützing 1845, p. 211.
Conferva ruchingeri C. Agardh 1824, p. 112.
Figure 60

Plants grass green, dark green, or glaucous, 10–40(–100) cm long, attached by rhizoids from basal cells and other cells in basal regions; forming long, undulating, coarse, hairlike or ropelike strands; organization irregular, nonacropetal; distinct main axes with strongly developed laterals forming pseudodichotomies; axes and main laterals having unilateral rows of short, straight branchlets, or branchlets growing in all directions; long sections of vigorously growing axes and main laterals with few or no branchlets; filaments ending in long rows of only slightly swollen zooidangia; transverse divisions almost exclusively intercalary; many new cells forming single branches apically, new branches form at younger nodes, intercalated between older nodes and older branches; basal walls of branches initially oblique, becoming almost horizontal; basal cells of old branches partly fused with adjacent cells of main axes; angles of branching small, 45° or less; main laterals more or less appressed; main axes 2.5–4 times the diameter of apical cells; apical cells cylindrical with tapering, obtuse tips or tapering from bases; apical cells (24–32)–(39–65) μm diameter, l/w (6–11)–(4–7); ultimate branches (24–48)–(42–75) μm diameter, l/w (3–8)–(2–5); main axes (50–80)–(70–110) μm diameter, l/w (3.5–8)–(24); thickness of cell walls of ultimate branches ca. 2–4 μm, in main axes 2–6 μm.

Chapman 1973; Kapraun and Zechman 1982; van den Hoek 1982; Kapraun 1984; Richardson 1987.

On ocean jetties, year-round.

Distribution: Maritimes to Connecticut, North Carolina, Georgia, southern Florida, Gulf of Mexico, Netherlands, West Africa, Mediterranean.

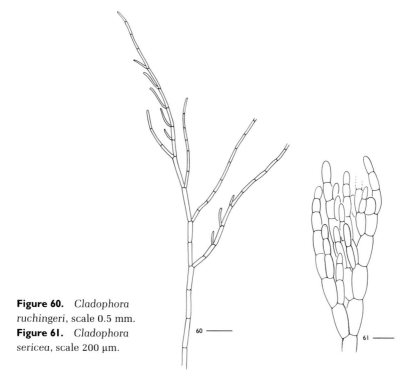

Figure 60. *Cladophora ruchingeri*, scale 0.5 mm.
Figure 61. *Cladophora sericea*, scale 200 μm.

60 ⎯⎯⎯⎯

61 ⎯⎯

Cladophora sericea (Hudson) Kützing 1843, p. 264.

Conferva sericea Hudson 1762, p. 485.

Figure 61

Plants light or pale green to grass green, forming tufts to 20(–30) cm long, attached by basal rhizoids; organization acropetal to irregular, distinct main axes forming pseudodichotomies with several main laterals; upper main laterals sometimes acropetally organized, lower main laterals and some upper main laterals with irregular organization due to frequent intercalary growth, the younger branchlets inserted among older branchlets; irregular organization more pronounced in plants from sheltered habitats; fertile plants with moniliform rows of barrel shaped zooidangia formed by frequent intercalary division of distal parts of filaments; transverse cell divisions predominantly intercalary; basal walls of lateral branches initially oblique, becoming almost horizontal; basal cells of laterals partly fusing with cells of main axes or main laterals; nodes forming two or three (sometimes four) laterals; angles of branching 30°–45°; main axis (2–) 2.5–4(–5.5) diameter of apical cells; apical cells distinctly tapering to almost conical, apical cells (20–30)–(40–70) μm diameter, l/w (3–9)–(3–7); ultimate branches (20–45)–(51–85) μm diameter, l/w (3–10)–(2–5); main axes (50–100)–(120–170) μm diameter, l/w (3.5–8)–(2–5.5); walls of ultimate branch cells ca. 0.5–2 μm thick, 2–6 μm in cells of main axes.

Chapman 1973; van den Hoek 1982; Kapraun 1984; Richardson 1987. As *C. gracilis*, Aziz and Humm 1962; Schneider 1976. As *C. flexuosa*, Taylor 1957; Brauner 1975.

Intertidal, spring–summer; offshore in Raleigh Bay and Onslow Bay from 15–50 m, June–November.

Distribution: Labrador to New Jersey, North Carolina, Georgia, USSR to Portugal, Australia, Central Pacific islands, Alaska to California.

Kapraun (1984) included plants with main axes to 250 μm diameter, too large to be included in this species as delineated by van den Hoek (1982).

Cladophora vadorum (Areschoug) Kützing 1849, p. 402.
  Conferva vadorum Areschoug 1843, p. 269.
  Figure 62

Plants grass green, light, or pale green, forming free-floating, dense masses or attached and forming coarse, flexuous bushes 5–30 cm tall; growth by apical and intercalary cell divisions; initial acropetal organization modified by insertion of secondary laterals at points of intercalary divisions along axes bearing laterals from segments cut off by apical cells; main branching pseudodichotomous; terminal branch systems vaguely acropetal to irregular, terminal branches straight or only weakly curved and unilateral or multilateral; laterals from apical cell segments initiated three to eight (up to eighteen) cells from apices; second or, rarely, third laterals may be formed by each cell; branching from distal ends of parent cells; basal cells of laterals with oblique basal walls, becoming almost horizontal; angles of branching greater than 45°, bases of older laterals partially fused with adjacent cells of axes; cells in broadest parts of main axes (1.5–)2–4(–4.5) times diameter of apical cells; apical cells long, cylindrical, with rounded to tapered tips; apical cells (30–50)–(35–60) μm diameter, l/w (11–21)–(4–20); ultimate branches (30–60)–(42–90) μm diameter, l/w (6–20)–(3–10); main axes (75–120)–(70–330) μm diameter, l/w (2.5–10)–(3–8).

Chapman 1973; van den Hoek 1982; Richardson 1987.

From brackish North Carolina ponds, summer, and from coastal Georgia, winter–spring.

Distribution: North Carolina, Georgia, southern Florida, Caribbean, Norway to France, West Africa, Australia.

Cladophora vagabunda (Linnaeus) van den Hoek 1963, p. 144.
  Conferva vagabunda Linnaeus 1753, p. 1167.
  Figures 63–65

Plants grass green, light, pale, or sometimes dark green, forming spongy pompoms 0.5–4 cm tall in exposed intertidal habitats, or plants 30 cm long with distinct axes ending in tufts or fascicles of filaments in submerged moderate conditions, or floating plants without distinct tufts or fascicles in very sheltered waters; organization with pseudodi-, pseudotri-, or sometimes pseudo-

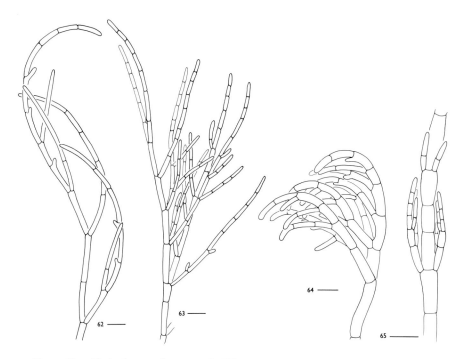

Figure 62. *Cladophora vadorum*, scale 200 μm.
Figures 63–65. *Cladophora vagabunda*. 63. Upper portion of filament, scale 250 μm. 64. Reflexed apex, scale 150 μm. 65. Median portion with lateral branches, scale 0.5 mm.

tetrachotomously branching main axes ending in acropetally organized branch systems where filaments may either be bent backward, sickle shaped, or straight; growth apical, with intercalary divisions toward bases of plants; intercalary divisions appearing nine to thirteen cells from apices of plants from exposed habitats, or three to six cells from apices of plants from sheltered habitats and thus showing less acropetal organization; secondary branches forming at newly intercalated cross walls; short, swollen zooidangia forming by subdivision of terminal branch systems; sporulation followed by proliferation of new branch systems from remaining, robust main axes that continue growth by intercalary division; new cells cut off by apical cells form new branches at their distal ends; in acropetally organized systems, this occurs two to three cells from apices; in irregularly organized systems this may be delayed until four to twenty-one cells from apices; one to four (to five) laterals may be produced by cells in acropetally organized systems, one to two in irregularly organized systems; branching always at distal ends of cells, often on the same sides, not serial; basal walls of laterals change from oblique to almost horizontal; angle of branching 25°–45° where branches are straight, or 45°–60° where branches bend back or are sickle shaped and organization is acropetal; basal cells of older laterals partly fusing with adjacent cells of axes; cells in basipetal direction distinctly longer and broader; akinetes club shaped; main axes (2.5–)4–6(–9.5) times diameter of apical cells in broadest basal parts of acropetally organized plants, 2.5–6 times that of apical cells in absence of acropetal organization; apical cells cylindrical, tips mostly tapered to nearly conical, distinctly swollen when forming zooidangia; apical cells (24–50)(–60)–(48–75)(–135) μm diameter in plants with acropetal organization, l/w (4–6)(–9)–(1.5–4); ultimate branchlets (24–66)–(50–90)–(–160) μm diameter, l/w (2.5–4)(–8)–(1.5–4); main axes (120–210)–(170–250)(–350) μm diameter, l/w (2.5–6)–(2–7); in plants reduced to main axes bearing robust proliferations, diameter of apical cells reaching 70–140 μm, l/w 1.5–3, ultimate branchlets 70–160 μm diameter, l/w 1.5–2; main axes 200–300 μm diameter, l/w 1–3.5; apical cells of plants with less strict acropetal to irregular organization (20–36)–(42–60) μm diameter, l/w (5–11)–(2–10); ultimate branches (20–36)–(48–90) μm diameter, l/w (3–10)–(2–6); main axes (120–180)–(150–190) μm diameter, l/w (4–12)–(3.5–10); cell wall thickness in ultimate branchlets ca. 0.5–1 μm, up to 7 μm in main axes.

Chapman 1973; Hall and Eiseman 1981; van den Hoek 1982; Kapraun 1984; Richardson 1987. As *C. crystallina*, Hoyt 1920; Taylor 1960; Brauner 1975; Schneider 1976. As *C. fascicularis* Hoyt 1920; Williams 1948a, 1949; Taylor 1960. As *C. flexuosa sensu* Hoyt 1920. As *Cladophoropsis membranacea sensu*, Schneider and Searles 1975, *pro parte*; Schneider 1976, *pro parte*.

From inlets and offshore in Onslow Bay and Long Bay from 16 to 60 m, May–September; Indian River estuary, year-round.

Distribution: Newfoundland to Virginia, North Carolina, Georgia, northeastern Florida, Gulf of Mexico, Caribbean to Uruguay, Baltic to Spain, West Africa, Central Pacific islands.

Rhizoclonium Kützing 1843

Plants unbranched, uniseriate filaments, usually unattached or with a few short, rhizoidal branches; growth diffuse; cells typically small, less than 100 μm diameter, plastids reticulate, with pyrenoids; reproduction by fragmentation, zoospores, or gametes; life history diplohaplontic, phases isomorphic; freshwater and marine.

Plants of this genus are easily confounded with free-floating *Chaetomorpha* and infrequently branched *Cladophora* plants.

Rhizoclonium riparium (Roth) Harvey 1849 [1846–1851], pl. 238.
  Conferva riparia Roth 1806, p. 216.
  Figure 66

Plants entangled, yellowish green, branched or unbranched; cells 10–30 μm diameter, 1–6 diameter long.

Hoyt 1920; Chapman 1971; Kapraun 1984; Richardson 1987. As *Rhizoclonium kerneri*, Chapman 1971; Richardson 1987.

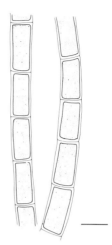

Figure 66. *Rhizoclonium riparium*, scale 25 μm.

In estuaries on *Spartina* culms and on mud, fall–winter.

Distribution: Canadian Arctic to Virginia, North Carolina, Georgia, Bermuda, southern Florida, USSR to Portugal; elsewhere, in temperate and tropical seas.

The description given here includes the range of cells attributed to *Rhizoclonium riparium* and *R. kerneri* Stockmayer. Kapraun's (1984) North Carolina plants fell between the maximum 15 μm diameter of the latter species and the minimum of 20 μm diameter by which *R. riparium* is usually delimited, and he considered the species conspecific; Hoyt (1920) recorded plants of 20–25 μm diameter. North Carolina specimens have the same dimensions as the filaments of *Chaetomorpha minima* and could be unattached plants of that species.

Valoniaceae

Plants lacking central axes, composed of one or (often) many subspherical to clavate vesicles; forming cushions of indefinite size anchored on lower surfaces by hapteroid rhizoids; daughter vesicles produced endogenously or exogenously by parent vesicles; when endogenous, vesicles pressed against one another to form pseudoparenchymatous tissue; minute lenticular cells formed parietally along vesicle walls and often developing hapteroid extensions by means of which vesicles are held securely to one another or to the substratum.

Valonia C. Agardh 1822
Plants composed of one to many subspherical to clavate vesicles occurring singly or in clusters; daughter vesicles produced exogenously by parent vesicles and of like form and size; where known, sexual reproduction by isogametes.

Valonia utricularis (Roth) C. Agardh 1822, p. 431.

Conferva utricularis Roth 1797, p. 160, pl. 1, fig. 1.

Plants attached, spreading or creeping among other algae, in part becoming erect, to 5 cm tall; composed of short, branching filaments; segments cylindrical to clavate, sometimes arcuate, 1.0–2.5 mm diameter, 5–20 mm long.

Williams 1948a.

In lower intertidal on Cape Lookout jetty, spring.

Distribution: North Carolina, Bermuda, southern Florida, Gulf of Mexico, Caribbean, Brazil, Portugal, Mediterranean, Azores, Madiera, Indian Ocean, Central Pacific islands.

Species Excluded

Valonia aegagropila C. Agardh 1822, p. 429.

Known only from plants cast ashore (Blomquist and Pyron 1943).

Dictyosphaeria Decaisne 1842

Species Excluded

Dictyosphaeria cavernosa (Forsskål) Børgesen 1932, p. 2., pl. 1, fig. 1.

Known only from plants cast ashore (Blomquist and Pyron 1943).

Anadyomenaceae

Plants with branched filaments united in a single plane to form foliose blades in an open or closed reticulum.

Key to the genera

1. Filaments forming an open reticulum .... Microdictyon
1. Filaments forming a closed reticulum .... Anadyomene

Anadyomene Lamouroux 1812

Plants foliose, with inconspicuous stalks, attached by rhizoids from lower stalk and larger rib cells; blades derived from radiating system of large cells that form the ribs of the blades, spaces between ribs filled by additional smaller cells; plastids polygonal; gametes anisogamous.

Anadyomene saldanhae Joly et Oliveira Filho 1969, p. 30.

Figures 67 and 68

Plants erect; blades eperforate, foliose, fan-shaped, to 9 cm tall; rhizoids not forming stipes; main ribs uniseriate, visible to the naked eye, 250–350 µm diameter, branching polychotomous, the ultimate spaces filled by smaller ellipsoidal cells in random arrangement, marginal cells spherical.

As *Anadyomene stellata*, Blomquist and Pyron 1943; Schneider and Searles 1973; Schneider 1976.

Rare, collected once in the drift and once offshore in Onslow Bay from 23 m, August.

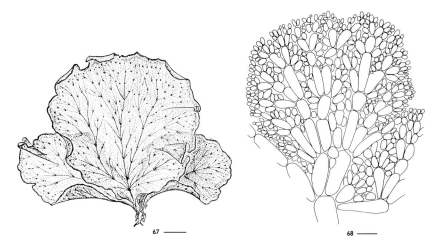

Figures 67, 68. *Anadyomene saldanhae.* 67. Habit, scale 0.5 cm. 68. Cellular detail of blade, scale 250 µm.

Distribution: North Carolina, Bermuda, southern Florida, Bahamas, Caribbean, Brazil.

Microdictyon Decaisne 1841

Plants with one or more nonstipitate, reticulate blades with spaces between the anastomosing filaments; anastomosis accomplished by either unmodified vegetative cells or by cells with annulate or crenulate attachment pads at their free ends; daughter cells cut off in opposite, alternate, subsecund, or fan-shaped arrangement.

Microdictyon boergesenii Setchell 1925, p. 106.

Figure 69

Blades to 6 cm tall and broad, margins lacerate, but if complete, fringed with short projecting branchlets; primary filaments 145–240 µm diameter, slightly swollen, 2.5–3.5 diameters long, walls 2–3 µm thick, ultimate ramifications with cells 45–110 µm diameter, 2–3 diameters long; branching angle wide but projecting upward at less than 90°; meshes more trapezoid than rectangular, 0.3–0.4 mm; anastomosing of filaments accomplished without specialized attachment pads.

Schneider and Searles 1973; Schneider 1976.

Uncommon, on shells in Onslow Bay from 38 to 80 m, August.

Distribution: North Carolina, Bermuda, Caribbean, Brazil, Canary Islands, Central Pacific islands.

North Carolina specimens have primary filaments of 180–240 µm diameter, slightly larger than the maximum originally reported by Setchell (1925); they are small single blades rather than the crowded concrescent specimens reported by Taylor (1960) to the south.

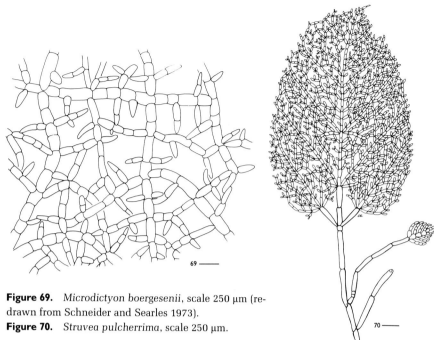

Figure 69. *Microdictyon boergesenii*, scale 250 μm (re-drawn from Schneider and Searles 1973).
Figure 70. *Struvea pulcherrima*, scale 250 μm.

Boodleaceae

Plants reticulate, forming either clusters of stipitate blades with anastomoses restricted to a single plane, or amorphous spongiose mats with anastomoses occurring in more than one plane; axial symmetry of opposite branching established early in development and continuing throughout the life of plants or lost as development continues.

Struvea Sonder 1845, nom. cons.
Plants with reticulate blades, monosiphonous stalks, and branched, prostrate rhizomes; blades single layered, produced by lateral branches cut off in regular, opposite patterns and anastomosing in a single plane; biflagellate and quadriflagellate swarmers forming in unmodified blade cells.

Struvea pulcherrima (J. E. Gray) Murray et Boodle 1888, p. 281.
   Phyllodictyon pulcherrima J. E. Gray 1866, p. 70.
   Figure 70
   Plants to 15(–28) cm tall with several stalks 1–1.25 mm diameter arising from branched, prostrate rhizoidal bases; lower stalks unbranched or occasionally dichotomously branched, 1.5–5 cm tall, upper stalks producing a series of opposite branches; opposite branches forming reticulate blades, the lower blades sometimes remaining separate but more commonly joined with the upper branches to form a single blade; shape of blades variable, 10–12(–25) cm long, to 15(–20) cm broad; lower pair of branches usually pedicellate, pedicels to 1 cm long; remaining branches usually sessile, united in the common blade.

Humm and Cerame-Vivas 1964; Schneider 1976. As *Struvea ramosa*, Schneider and Searles 1973; Schneider 1976; Blair and Hall 1981.

At times abundant in Onslow Bay and Long Bay from 21 to 100 m, April–December.

Distribution: North Carolina, South Carolina, Bermuda, southern Florida, Caribbean, Central Pacific islands.

Struvea ramosa Dickie in Hooker has been distinguished from *S. pulcherrima* by the formation of separate paired blades in addition to the terminal blade. Plants in the Carolinas show both conditions, with intermediates between the extremes occurring in the same collections. We therefore treat both forms as representative of a single species.

Chaetosiphonaceae

Endophytic or epiphytic plants; multinucleate and possessing aseptate hairs.

Blastophysa Reinke 1888

Plants composed of flattened, irregularly lobed or vesicular, ellipsoid or irregularly shaped cells, single or, more commonly, several united by tubular connections; growth by long tubular aseptate filaments that swell at tips to form cells that then separate from empty tubes by walls; colorless hairs formed singly or in clusters on cells; plastids angular, parietal, some with pyrenoids; reproduction by quadriflagellate zoospores.

Blastophysa rhizopus Reinke 1888, p. 241.

Figures 71 and 72

Endophytic plants with ellipsoid, obovoid, or irregularly swollen vesicular cells; epiphytic plants with flattened and irregularly lobed cells; vesicular cells 80 μm diameter, 150 μm long; tubular connections 5–6 μm diameter; hairs single on the cells or in clusters, bases of hairs sometimes enlarged.

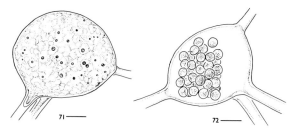

Figures 71, 72.
Blastophysa rhizopus. 71. Immature sporangium, scale 20 μm. 72. Sporangium with spores, scale 20 μm.

Searles and Leister 1980.

Growing in tissue of *Predaea feldmannii*, causing the red gelatinous host to appear greenish; from depths to 26 m in Onslow Bay, summer.

Distribution: Maritimes, Massachusetts, North Carolina, Bermuda, Caribbean, Norway to British Isles, Mediterranean.

CAULERPALES

Plants acellular coenocytes, mostly aseptate except where reproductive structures are delimited; branches commonly tubular, uniaxial, and freely branched, or multiaxial complexes of interwoven and united filaments; if symmetrical, not basically with radial symmetry, but secondarily so in some *Udotea* and *Bryopsis* species; cell walls of mannan or xylan with or without cellulose, some calcified; plastids numerous, discoid, with or without pyrenoids; asexual reproduction by biflagellate zoospores or aplanospores; sexual reproduction mostly anisogamous; basal bodies of flagella with simple, overlapping terminal caps and rudimentary proximal sheaths, life histories diplontic or diplohaplontic and with heteromorphic generations; "diploids" are in some cases heterocaryotic, karyogamy being delayed until long after plasmogamy.

The order is broadly interpreted here; some of the families listed here are considered orders in other treatments.

Key to the families

1. Tubular branches united into compact, integrated plants 2
1. Tubular branches not united into compact, integrated plants 3
2. Gametangia compound, with one discharge tube for several gametangial branches; leucoplasts present . . . . Udoteaceae
2. Gametangia simple; leucoplasts not present . . . . Codiaceae
3. Plants living on sandy or hard substrate 4
3. Plants boring in carbonate shells and coral, or endophytic . . . . Ostreobiaceae
4. Plants with decumbent rhizomes, attaching rhizoids, cytoplasm traversed by trabeculae . . . . Caulerpaceae
4. Plants lacking rhizomes, rhizoids, and trabeculae . . . . Bryopsidaceae

Codiaceae

Plants spongy, erect, or matlike, formed by interwoven tubular filaments; erect plants simple or branched; tubular filaments multinucleate, branched sympodially, apices differentiated to form palisadelike tissue of utricles; plastids discoid, pyrenoids lacking; plant interiors complexes of slender, interwoven,

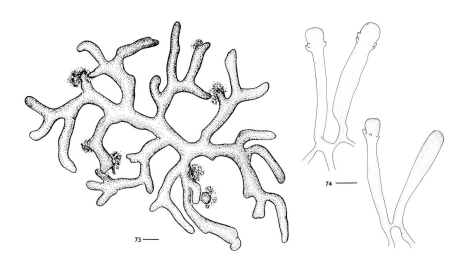

Figures 73, 74. *Codium carolinianum*. 73. Habit, scale 1 cm. 74. Utricles, scale 200 μm.

colorless filaments; walls contain mannan; asexual reproduction by partheno-genesis; sexual reproduction anisogamous, biflagellate gametes formed by meiosis in simple gametangia produced laterally on utricles; dioecious, life history diplontic.

This family contains a single genus.

Codium Stackhouse 1797

Key to the species

1. Plants irregularly branched, partially procumbent, the branches attached to the substratum . . . . *C. carolinianum*
1. Plants dichotomously branched, erect, the branches free of the substratum 2
2. Utricle tips apiculate . . . . *C. fragile* spp. *tomentosoides*
2. Utricle tips not apiculate 3
3. Branches generally 2.5–4 mm diameter . . . . *C. isthmocladum*
3. Branches generally 4 mm diameter or larger 4
4. Utricles mostly 220–500 μm diameter, 1100–1750 μm long . . . . *C. decorticatum*
4. Utricles mostly 110–260 μm diameter, 650–1150 μm long . . . . *C. taylorii*

Codium carolinianum Searles 1972, p. 19, figs 1a, 2.

Figures 73 and 74

Plants procumbent, irregularly branched, anastomosing, and repeatedly attached to substratum; branches terete to slightly flattened, 4–11 mm diameter, utricles clavate or subcylindrical, 75–210 μm diameter, 730–1510 μm long, apices truncate, rounded, or occasionally slightly mammillate; utricle walls to 2 μm thick, 8–10 μm thick at apices; hairs infrequent, to five per utricle, originating 140–175 μm below utricle apex; gametangia 195–260 μm long, one to three

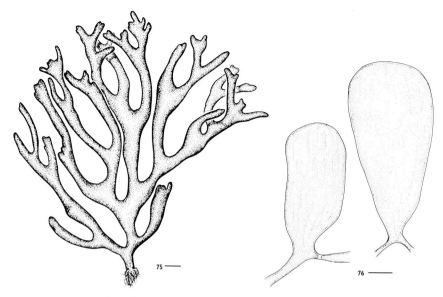

Figures 75, 76. *Codium decorticatum.* 75. Habit, scale 1 cm. 76. Utricles, scale 200 μm.

per utricle, borne 240–350 μm below apices of utricles on pedicels 14 μm long.

Searles 1972; Schneider 1976; Blair and Hall 1981.

Common on rock outcrops in Onslow Bay from 15 to 37 m, year-round, but dying back in winter. Common on some hard-bottom areas with little relief, the intertwined branches forming mounds that are often partly covered by shifting sand.

Distribution: North Carolina (type locality), South Carolina, Florida Gulf coast, Puerto Rico.

This is the dominant green alga in biomass in Onslow Bay (Schneider and Searles 1979).

Codium decorticatum (Woodward) Howe 1911, p. 494.

Ulva decorticata Woodward 1797, p. 55.

Figures 75 and 76

Plants erect, to 0.1–1(–3) m long, sparingly to profusely dichotomously branched; branches terete to flattened, flattening common at dichotomies, but entire branches flattened in sparingly branched individuals, 6–25 mm diameter, flattened branches to 6(–9) cm broad at dichotomies; utricles cylindrical or clavate (115–)220–500(–850) μm diameter, (790–)1100–1750(–2000) μm long; apices rounded, truncate, or depressed; utricle sidewalls 1.5–2.0 μm thick, to 4–8 μm thick at apices; medullary filaments (23–)35–85(–108) μm diameter; gametangia lanceolate-ovoid, (58–)70–125 μm diameter, (144–)185–300(–390) μm long, to seven per utricle, on pedicels 10–15 μm long on protuberances 430–650 μm below utricle apices.

Hoyt 1920; Williams 1948a, 1948b, 1949, 1951; Stephenson and Stephenson 1952; Taylor 1960; Schneider 1976. As *C. tomentosum,* Hoyt 1920.

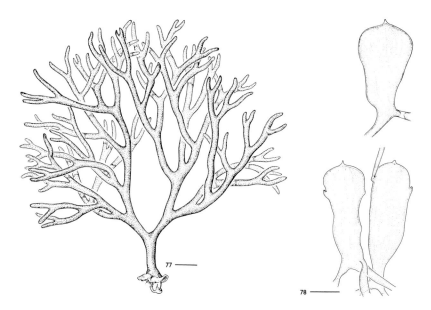

Figure 77, 78. *Codium fragile* subsp. *tomentosoides*. 77. Habit, scale 1 cm. 78. Utricles, scale 200 µm.

Common in sheltered sounds and exposed jetties, year-round, and uncommon offshore in Onslow Bay from 14 to 25 m, May–August.

Distribution: North Carolina, South Carolina, Bermuda, southern Florida, Caribbean to Argentina, Spain, Portugal, Mediterranean, Azores, West Africa, Indian Ocean, ?Pacific Mexico.

Young plants and plants in turbulent oceanic waters tend to be closely branched and terete. Plants from quiet, low-salinity environments are flattened, infrequently branched, and of great size.

Codium fragile subsp. *tomentosoides* (van Goor) Silva 1955, p. 567.

Codium mucronatum J. Agardh var. *tomentosoides* van Goor 1923, p. 134, fig 1c.

Figures 77 and 78

Plants erect, abundantly dichotomo-fastigiately branched, 15–25(–50) cm tall, branches terete, 3–8(–10) mm diameter; utricles narrowly to broadly clavate or cylindrical, (105–)165–325(–400) µm diameter, 550–1050 µm long, apices apiculate; hairs or hair scars one to two per utricle, 130–260 µm below apices; gametangia ovoid, oblong, or fusiform, 72–92 µm diameter, 260–330 µm long, borne on protuberances near middle of utricles; heterothallic.

Searles, Hommersand, and Amsler 1984.

On jetties and seawalls in the sounds in all seasons, growing intermixed with *Codium decorticatum*.

Distribution: Maine to New Jersey, North Carolina, Norway to Spain, Mediterranean, Australia, Central Pacific islands, Japan, Alaska to Pacific Mexico, western South America.

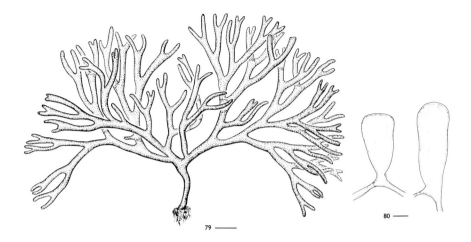

Figures 79, 80. *Codium isthmocladum.* 79. Habit, scale 1 cm. 80. Utricles, scale 100 μm.

This weedy subspecies appeared in Long Island Sound in the late 1950s and by the 1980s had spread to North Carolina, currently the southern limit of distribution on this coast.

Codium isthmocladum Vickers 1905, p. 57.

Figures 79 and 80

Erect branches from crustose base, to 20 cm high, dichotomously branched to twelve orders, at times proliferous; branches terete or occasionally flattened in middle and lower parts of plant, at times markedly constricted above dichotomies, (2–)2.5–4(–5) mm diameter; utricles subcylindrical to clavate or pyriform, 120–350(–475) μm diameter, 460–850 μm long, often constricted 130–260 μm below apices; apices rounded or truncate; utricle sidewalls 1.5 μm thick, apices slightly (4–8 μm) to markedly (–110 μm) thick, incrassate apices lamellate; hairs or hair scars common, few to several per utricle, 40–120 μm below utricle apices; medullary filaments 26–42 μm diameter; gametangia lanceolate-ovoid to oblanceolate-ovoid, frequently longitudinally asymmetrical, (48–)65–130 μm diameter, 180–280 μm long, two to several per utricle on short (ca. 5–7 μm) pedicels near middle of utricles.

Blomquist and Pyron 1943; Williams 1948a, 1948b, 1949, 1951; Taylor 1960; Schneider 1976; Searles 1987, 1988.

On coastal jetties and in sounds in summer and fall, common offshore in Onslow Bay and Long Bay from 14 to 45 m year-round, and from 17 to 35 m on Gray's Reef and Snapper Banks, March–September.

Distribution: North Carolina, South Carolina, Georgia, Bermuda, southern Florida, Gulf of Mexico, Caribbean, Brazil.

This species is the most frequently collected *Codium* in offshore waters (Schneider 1976), but it is second to *C. carolinianum* in total biomass in Onslow Bay (Schneider and Searles 1979).

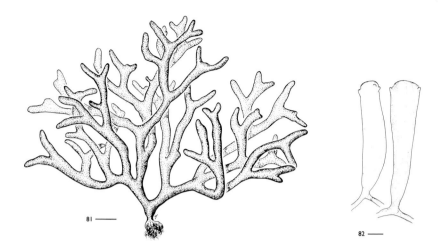

Figures 81, 82. *Codium taylorii.* 81. Habit, scale 1 cm. 82. Utricles, scale 100 μm.

Codium taylorii Silva 1960, p. 510.

Figures 81 and 82

Plants erect, to 15 cm tall, divaricately dichotomously branched to seven to nine orders, sometimes appearing cervicorn due to uneven development of the dichotomies; branches (3–)4–8(–25) mm broad, 3–4 mm thick, tips obtuse; utricles cylindrical or clavate (55–)110–260(–380) μm diameter, (550–)650–1150(–1450) μm long, apices slightly rounded to subtruncate; end walls to 23 μm thick; hairs or hair scars 50–105 μm below ends of utricles, two to several per utricle; medullary filaments 15–35 μm diameter; gametangia ellipsoid or cylindrical, 45–85 μm diameter, 200–300(–350) μm long, one or two per utricle, on stalks to 8 μm long, 275–430 μm below ends of utricles.

Searles, Hommersand, and Amsler 1984.

Rare, on rock ledges off Cape Fear from 27 m, August–September.

Distribution: North Carolina, Bermuda, southern Florida, Gulf of Mexico, Caribbean, Brazil, Israel, Canary Islands, West Africa.

Udoteaceae

Tubular, multinucleate, branched filaments, unorganized or more typically interwoven or laterally united to form large macroscopic plants; with or without specialized photosynthetic utricles; walls composed of xylan, many also producing deposits of calcium carbonate; two types of plastids formed: chloroplasts and colorless, starch-forming leucoplasts; life history diplohaplontic, anisogamous gametes formed in clusters of gametangia that have common discharge tubes; no asexual reproductive cells formed, but new plants vegetatively arising from rhizoids.

Key to the genera

1. Filaments forming loose, undifferentiated mats of filaments ... Boodleopsis
1. Filaments united into macroscopic plants with stipes and expanded blades
.... 2
2. Plants calcified .... Udotea
2. Plants uncalcified .... Avrainvillea

Avrainvillea Decaisne 1842

Plants with masses of rhizoidal filaments anchoring stalked or sessile masses of soft, spongy filaments united to form the erect, photosynthetic, often bladelike parts of plants; not calcified; dark green, yellowish green, or brownish green; filaments dichotomously branched, sometimes constricted above dichotomies, often torulose or moniliform.

Avrainvillea longicaulis (Kützing) Murray et Boodle 1889, p. 70.
   Rhipilia longicaulis Kützing 1858, p. 13, pl. 28, fig. 2.
   Figure 83
   Plants 6–15 cm tall, sometimes gregarious; stipes compressed or terete, 3–10 cm long, 4–10 mm diameter; blades cuneate-rounded at bases, above round, obovate, or spatulate, 6–11 cm long, 5–14 cm broad, not zonate, either smooth or hairy; medullary filaments 28–60 μm diameter in stipes, 28–35 μm in blades, constricted above the forkings, otherwise of uniform diameter; outer filaments thinner and sometimes moniliform.
   Taylor 1960; Schneider 1976. As *A. mazei sensu* Williams 1951.
   Common in Onslow Bay, growing on sand-covered rock outcroppings from 15 to 32 m, June–November.
   Distribution: North Carolina, Bermuda, southern Florida, Caribbean, Central Pacific islands.
   Inclusion of North Carolina plants in this species requires emendation of the description of the species presented by Olsen-Stojkovich (1985) in her monograph of *Avrainvillea*. Local plants are larger than the 12-cm-tall, 3–8-cm-wide plants she reported. North Carolina plants are up to 15 cm tall and blades are up to 14 cm broad. Stipes are at most 4 cm long with medullary filaments of greater diameter—50–60 μm—than previously noted. Stipes are covered by an inter-digitating cortex of moniliform, dichotomously branched filaments. Blades up to 11 cm long and 14 cm broad are composed of loosely associated filaments, 28–35 μm diameter, constricted above dichotomies and elsewhere torulose but not moniliform.

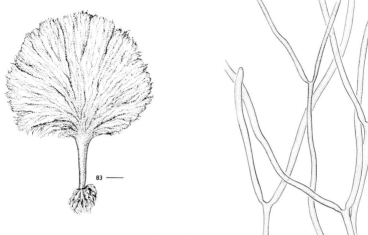

Figure 83. *Avrainvillea longicaulis*, scale 1 cm.
Figure 84. *Boodleopsis pusilla*, scale 100 μm.

Boodleopsis A. et E. S. Gepp 1911
Plants filamentous, uncalcified; the filaments loosely interwoven and forming mats; filaments di- to trichotomously or irregularly branched, constricted or not constricted above forks; reproducing by biflagellate and stephanokont cells.

Boodleopsis pusilla (Collins) Taylor, Joly, et Bernatowicz 1953, p. 105.
    Dichotomosiphon pusillus Collins 1909, p. 431.
    Figure 84
    Lower filaments to 90 μm diameter, often colorless, of irregular contour, not sharply constricted, though frequently showing cross walls, often attenuated into rhizoids; upper filaments more frequently branched, 23–48 μm diameter, strongly constricted at bases of branches and elsewhere; coral-like masses of short, unconstricted filaments often produced on upper branches; pyriform to subspherical sporangiumlike structures sometimes present, 60–170 μm diameter, 82–260 μm long; akinetes of very irregular form produced in culture.
    Searles and Leister 1980; Searles 1987, 1988.
    On rocks in depths to 35 m on the Carolina and Georgia continental shelf, June–August.
    Distribution: North Carolina, Georgia, Bermuda, Caribbean, Brazil, Central Pacific islands.
    Collections in the region have not demonstrated the cross walls, the coral-like masses of filaments, or the sporangiumlike structures reported for the species. Small, lateral, pricklelike structures are present in some specimens and trichotomies are frequent.
    Calderón-Sáenz and Schnetter (1989) studied plants identified as this species from the supralittoral in Colombia—plants with slender (to 20 μm diameter), erect filaments. They reported formation of motile, biflagellate, and stephanokont cells formed in separate gametangia and sporangia on the same plants.

Udotea Lamouroux 1812

Erect, calcified plants anchored by rhizoidal bases, with fan- or funnel-shaped blades supported by stalks; stalks and blades composed of laterally joined, dichotomously branched filaments constricted at bases of branches as well as along internodes; distinctive lateral branch filaments forming cortical layers on stalks and in some cases over blades; reproduction anisogamous, plants dioecious.

Key to the species

1. Blade without corticating filaments .... *U. cyathiformis*
1. Blade corticated by branched filaments .... *U. flabellum*

Udotea cyathiformis Decaisne 1842, p. 106.

Figures 85 and 86

Plants 6–13 cm tall; blades whitish green, forming shallow cups or funnels; cups symmetrical or asymmetrical and sometimes incomplete, 3–8 cm diameter, texture thin, readily split, the surface smooth except at outer edges, zonate growth bands sometimes evident, composed of straight, nearly parallel filaments running radially from the top of stalks; filaments 30–50(–135) μm diameter, without corticating filaments or projections; stalks terete, 2–4 mm diameter, to 1.5 cm long, surfaces smooth, corticating filaments closely packed, ends truncate, not greatly thickened.

Hoyt 1920; Taylor 1960; Schneider 1976. As *Udotea conglutinata sensu* Williams 1951.

Common in Onslow Bay from 14 to 48 m, May–November.

Distribution: North Carolina, Bermuda, southern Florida, Caribbean, Brazil.

A frequently encountered alga on flat, sand-covered rocks in offshore North Carolina waters. Often occurring in small stands, probably resulting from vegetative propagation by rhizoidal growth through the sand.

North Carolina plants assigned to *Udotea conglutinata* (Ellis et Solander) Lamouroux (1816) appear to be specimens of *U. cyathiformis* in which the peltate blade is incomplete or split.

Udotea flabellum (Ellis et Solander) Howe 1904, p. 94.

Corallina flabellum Ellis et Solander 1786, p. 124, pl. 24.

Figure 87

Plants to over 20 cm tall, bright or dark green; blades broadly fan shaped, green to dark green, strongly zonate, flexible, although strongly calcified, 5–21 cm long and 5–30 cm wide, cordate below, rounded above, often reniform and proliferous, free margins irregularly crenate, surfaces often becoming folded and

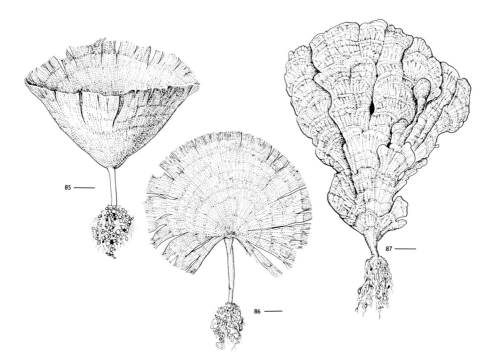

Figures 85, 86. *Udotea cyathiformis*. 85. Funnel-shaped blade, scale 0.5 cm. 86. Split, fan-shaped blade, scale 1 cm.
Figure 87. *Udotea flabellum*, scale 1 cm.

ribbed, sometimes cleft into several wedge-shaped segments; surfaces continuous, smooth, corticated with densely branched, truncate filaments with thickened ends; stalks 1–2 cm long, arising from masses of rhizoidal filaments, to 1 cm wide and flattened above, smaller and terete below, corticating filaments similar to but longer than those of blades.

Williams 1951; Schneider 1976.

Infrequent in Onslow Bay from 35 to 40 m, May–September.

Distribution: North Carolina, Bermuda, southern Florida, Caribbean, Brazil, Indian Ocean, Central Pacific islands.

Caulerpaceae

Plants mostly with stolons or rhizomes, anchoring rhizoids and erect photosynthetic branches; rhizome, stolon, and branch cytoplasm traversed by trabeculae; walls composed of xylan; erect branches of diverse morphology, simple or branched with radial or bilateral symmetry; chloroplasts and leucoplasts both present; life history diplontic; sexual reproduction anisogamous; gametes forming by meiosis in undifferentiated parts of branches and discharged through papillae.

This family contains a single genus.

Caulerpa Lamouroux 1809c

Key to the species

1. Erect photosynthetic branches having the form of flat entire blades, occasionally proliferous from the stalk or blade .... *C. prolifera*
1. Erect photosynthetic branches pinnately or radially branched 2
2. Branchlets pinnate .... *C. mexicana*
2. Branchlets radial .... *C. racemosa* var. *laetevirens*

Caulerpa mexicana Sonder ex Kützing 1849, p. 496.
　Figure 88
　Erect branches to 25 cm tall, 5–16 mm wide, simple or occasionally forked, pinnately divided, with two rows of pinnules; pinnules acuminate, flat, linear or tapering, oval to oblong, or arcuate, basally narrowed; rhizomes 0.5–1.25 mm diameter.
　Searles 1987, 1988.
　On rock outcroppings of Gray's Reef from 17 to 21 m, July–August.
　Distribution: Georgia, Bermuda, southern Florida, Gulf of Mexico, Caribbean, Brazil, Indian Ocean, Central Pacific islands.

Caulerpa prolifera (Forsskål) Lamouroux 1809d, p. 332.
　Fucus prolifer Forsskål 1775, p. 193.
　Figure 89
　Erect branches bearing single, flat, oval to linear-oblong entire blades, 3–15 cm long, 3–15 mm wide, on stalks 0.5–1 mm long; occasionally with secondary blades from stalks or blade faces or margins.
　Hoyt 1920; Williams 1951; Taylor 1960; Schneider 1976; Kapraun 1984.
　Common in Onslow Bay from 15 to 48 m, May–December; infrequent in sounds and on jetties year-round.

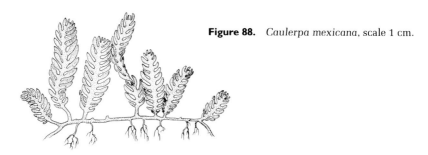

Figure 88. *Caulerpa mexicana*, scale 1 cm.

Figure 89. *Caulerpa prolifera*, scale 1 cm.
Figure 90. *Caulerpa racemosa* var. *laetevirens*, scale 0.5 cm.

Distribution: North Carolina, Bermuda, southern Florida, Gulf of Mexico, Caribbean, Brazil, Mediterranean, Central Pacific islands.

Caulerpa racemosa var. *laetevirens* (Montagne) Weber-van Bosse 1898, p. 366.
   Caulerpa laetevirens Montagne 1845, p. 16.
   Figure 90
   Erect branches 0.5–5 cm tall; branchlets peltate, with slender pedicels 1–2 mm long supporting cup-shaped or flattened discs 1–2 mm thick, 1.5–8 mm diameter, with entire margins, or branchlets subcylindrical to clavate, the apices rounded.
   As *C. racemosa* var. *peltata*, Schneider 1975b, 1976.
   Rare, collected attached once, in Onslow Bay from 37 m, July.
   Distribution: North Carolina, Bermuda, southern Florida, Gulf of Mexico, Caribbean, Brazil; elsewhere, widespread in tropical seas.
   Plants with peltate branchlets have traditionally been placed in *Caulerpa peltata* (Lamouroux) Weber-van Bosse (1898) or in *C. ramosa* var. *peltata* (Lamouroux) Eubank (1946). Ohba and Enomoto (1987) presented experimental observations indicating that the peltate form is a response of var. *laetevirens* to low light flux.

Bropsidaceae

Plants with branched, tubular, multinucleate filaments or simple vesicles; filaments not united into complex macroscopic plants; chloroplasts numerous, discoid or elliptical, with or without pyrenoids; cell walls of gametophytes contain cellulose and xylan; the walls of sporophytes contain mannan; life histories diplontic or diplohaplontic and the generations heteromorphic.
   The family is represented in the flora by plants assigned to *Bryopsis* and *Derbesia*, which, on the basis of their respective type species, may be retained as distinct genera. The type species of *Bryopsis* has only one macroscopic phase,

a siphonous gametophyte, the sporophyte being represented by a protonemal stage. The type species of *Derbesia*, on the other hand, has a life history in which a siphonous sporophyte alternates with a vesicular gametophyte originally described as a species in an independent genus, *Halicystis*. One species of *Derbesia*, however, has been found to be the sporophyte in a life history in which the gametophyte is recognizable as a previously described species of *Bryopsis*. This species has been segregated into the genus *Bryopsidella*. The life histories of *Bryopsis* and *Derbesia* species in this flora have not been determined, nor have *Halicystis* stages been reported.

Key to the genera

1. Plants with erect axes bearing short, determinate branchlets (pinnules) . . . . *Bryopsis*
1. Plants irregularly branched, with no differentiation of main axes and branchlets . . . . *Derbesia*

Bryopsis Lamouroux 1809b

Plants erect, clumps of percurrent axes with one to several orders of branching; axes with pinnate, radial, or unilateral branchlets (pinnules); plastids numerous, discoid, with one to several pyrenoids; pinnules cut off by basal cross walls functioning as gametangia or as vegetative propagules.

Key to the species

1. Plants light green to olive green, erect axes with pinnules in two rows, forming triangular to lanceolate fronds, or pinnules irregularly disposed along and around the axes . . . . *B. plumosa*
1. Plants dark green, erect axes with pinnules in two rows, all pinnules except those at tips of axes approximately the same length to form linear-lanceolate fronds . . . . *B. pennata*

Bryopsis pennata Lamouroux 1809d, p. 333.

Figures 91 and 92

Plants growing in tufts, dark green, attached by rhizoids, to 7 cm tall, the tips sometimes arcuate; pinnules distichous and, except for those at tips of fronds, mostly of equal length; fronds linear-lanceolate or oblong; main axes 240–360 μm diameter, fronds 5–8 mm broad, branchlets 75–150 μm diameter.

Blomquist and Humm 1946; Williams 1948a; Taylor 1960; Schneider 1976; Searles 1987, 1988; Kapraun 1984.

On ocean jetties, summer; rare offshore in Onslow Bay from 34 m, June; offshore in Georgia from 17 to 35 m, July–August.

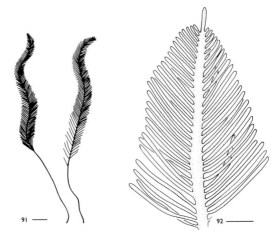

Figures 91, 92. *Bryopsis pennata.* Habit of two blades, scale 250 μm. 92. Blade apex, scale 100 μm.

Distribution: North Carolina, Georgia, Bermuda, southern Florida, Gulf of Mexico, Brazil, France to Portugal; elsewhere, widespread in temperate and tropical seas.

Bryopsis plumosa (Hudson) C. Agardh 1822, p. 448.

Ulva plumosa Hudson 1778, p. 571.

Figure 93

Plants light to olive green, in erect clumps, to 12 cm tall; primary axes frequently forming secondary axes from older laterals; pinnules forming triangular fronds above, with axes naked below, or unilateral pinnules, or irregularly and radially distributed along and around axes.

Bailey 1848; Harvey 1958; Melvill 1875; Hoyt 1920; Taylor 1960; Kapraun 1984; Richardson 1986, 1987. As *B. hypnoides*, Blomquist and Humm 1946; Williams 1948a, 1949, Stephenson and Stephenson 1952; Taylor 1960; Kapraun 1984. As *B. hypnoides* f. *prolongata*, Chapman 1971; Richardson 1987.

On jetties and seawalls year-round in the Carolinas and Georgia, but most abundant fall to spring.

Distribution: Newfoundland to Virginia, North Carolina, South Carolina, Georgia, USSR to Portugal, Mediterranean, Azores, southern Africa, Indian Ocean, Australia, Japan, British Columbia to Pacific Mexico.

Shevlin (personal communication) thinks that *Bryopsis hypnoides* Lamouroux is probably a growth form of *B. plumosa*. It has been described as dull or dark green with pinnules as slender as 40–80 μm surrounding the branch axes in an irregular radial pattern.

Derbesia Solier 1846

Sporophytes with tufted, tubular filaments, dichotomously or laterally branched; multinucleate, with many discoid to ellipsoid or spindle-shaped plastids with or without pyrenoids; walls containing xylan; sporangia lateral on

Figure 93. *Bryopsis plumosa*, habit of two plants, scale 0.5 cm.

erect filaments; zoospores stephanokont; gametophytes subspherical, with many plastids like those of sporophytes; gametes anisogamous, biflagellate.

Key to the species

1. Filament diameter less than 40 μm, plastids lacking pyrenoids, sporangia cylindrical to clavate .... *D. marina*
1. Filament diameter greater than 40 μm, plastids with pyrenoids, sporangia not cylindrical or clavate .... *D. turbinata*

Derbesia marina (Lyngbye) Solier 1846, p. 453.
    Vaucheria marina Lyngbye 1819, p. 79, pl. 22:A.
    Figure 94
    Plants forming tufted clumps 1–3 cm high attached by contorted creeping basal filaments; branching lateral and dichotomous; filaments (11–)25–40(–70) μm diameter; plastids ovoid discs, (1–)2.5(–4) μm wide, (3–)5.5(–8) μm long, without pyrenoids; sporangia clavate, subcylindrical, or ovoid, with rounded apices, (38–)53–110(–150) μm diameter, (85–)120–255(–333) μm long, pedicels at septa (10–)12–19(–35) μm diameter, sporangia at maturity containing sixteen to thirty-two zoospores.
    Sears and Wilce 1970; Searles 1987, 1988. As *D. vaucheriaeformis sensu* Williams 1948a; Schneider 1976. As *D. lamourouxii sensu* Williams 1948a.
    Growing from the shallow subtidal on exposed coastal jetties year-round, and

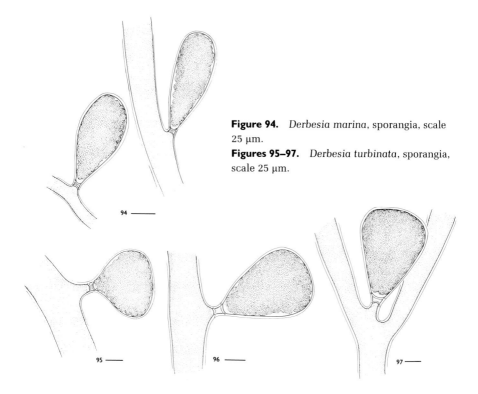

Figure 94. *Derbesia marina*, sporangia, scale 25 μm.

Figures 95–97. *Derbesia turbinata*, sporangia, scale 25 μm.

offshore in the Carolinas and Georgia from 17 to 60 m, June–December.

Distribution: Newfoundland to Connecticut, North Carolina, Georgia, Bermuda, USSR to France, Australia, Japan, Alaska to California.

Identified by Sears and Wilce as *Derbesia marina*, the local plants are in most respects smaller than those recorded elsewhere. Filaments range from 11 to 31 μm diameter and sporangia from 38 to 53 μm diameter and 85 to 120 μm long. This species appears to be the commonest *Derbesia* in the flora. Elsewhere it is associated with a *Halicystis* stage. Previous records of *D. vaucheriaeformis* and *D. lamourouxii* appear to have been based on this species, except for Brauner (1975). The filaments of Brauner's alga are broader than 40 μm but are sterile and not determinable to species or genus and may have been a species of *Vaucheria*.

Derbesia turbinata Howe et Hoyt 1916, p. 106, pl. 11, figs 10–16.

Figures 95–97

Filaments dark green or olive green, more or less creeping, forming tangled tufts to 9 cm long, basal parts sometimes here and there forming cysts; filaments (16–)40–75(–100) μm diameter, sparingly subdichotomous but more often laterally branched, lateral branches usually without basal cross walls, dichotomous branches with or without one or two cross walls above dichotomies; plastids with pyrenoids, orbicular, elliptic, or ovate, 4–5 μm wide and 6–8 μm long, or tapering sharply at each end; sporangia usually obconic or pyriform, less commonly turbinate, obovoid, pestle shaped, or spherical, (55–)115–190 μm long, excluding pedicels, 85–165 μm diameter, apices rounded or flat; pedicels mostly 15–33(–75) μm long, 12–22 μm diameter; zoospores unknown.

Howe and Hoyt 1916; Collins 1918; Hoyt 1920; Searles 1987, 1988. As *D.* sp. Sears and Wilce 1970.

From offshore rocks in Onslow Bay and Georgia Snapper Banks, from 17 to 35 m, July–August.

Distribution: North Carolina (type locality), Georgia.

In our collections, any single plant had sporangia that were either clearly longer than broad or else were broader than long. The latter plants were considered distinct in earlier publications (Searles 1987, 1988) but are here included in Howe and Hoyt's species.

Ostreobiaceae

Plants endophytic or endozoic in calcareous substrates; filaments with irregular diameters and branching, generally cylindrical or inflated, at times constricted, septate, and anastomosing; plastids discoid and lack pyrenoids.

This family contains a single genus.

Ostreobium Bornet et Flahault 1889

Ostreobium quekettii Bornet and Flahault 1889, p. clxi, pl. 9, figs 5–8.

Figure 98

Filaments slender, mostly 4–5 μm diameter, twisted in the substrate, inflated to 40 μm diameter locally, reduced to 2 μm diameter at branch tips.

Aziz and Humm 1962.

In shells in tide flats and in coral on jetties in sounds; July.

Distribution: Canadian Arctic to Connecticut, North Carolina, Colombia, Norway to Spain, British Columbia to Pacific Mexico.

Figure 98. *Ostreobium quekettii*, scale 10 μm.

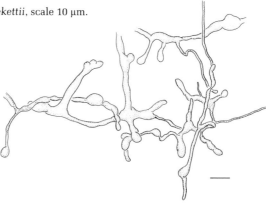

DASYCLADALES

Species Excluded

Batophora oerstedii var. *occidentalis* (Harvey) Howe

Cast ashore as drift after storm (Blomquist and Pyron 1943). There are no records of attached plants.

CHLOROPHYCEAE

Species Excluded

Uronema curvata Printz

There are no voucher specimens from Williams's (1948b) record of this alga and it has not been re-collected in the region.

CHRYSOPHYTA

Plants marine or freshwater; unicellular, colonial, multicellular, or acellular; chlorophyll *a* often associated with chlorophyll *c*, often with abundant xanthophylls; endoplasmic reticulum often fused with plastid membranes; stored carbohydrates beta-linked glucose (chrysolaminarin).

XANTHOPHYCEAE

Chlorophyll pigments usually dominant; plants typically green or yellow-green; chlorophyll *c* present, but not in large quantities; cell walls cellulosic; motile cells with pairs of flagella of different length, one typically hairy, and one smooth.

VAUCHERIALES

Vegetative plants acellular; asexual reproduction by zoospores, autospores, aplanospores; isogamous, anisogamous, or oogamous.

Vaucheriaceae

Plants with branching coenocytic filaments.

Vaucheria De Candolle 1801

Branched filaments making cross walls only with the formation of reproductive structures; peripheral cytoplasm containing many discoid plastids and nuclei; asexual reproduction by multinucleate compound zoospores bearing many pairs of smooth but unequal-length flagella, one compound zoospore formed per sporangium; sexual reproduction oogamous, one egg and zygospore formed per oogonium; sperm biflagellate, one flagellum bearing two rows of stiff hairs, the other smooth, both laterally inserted; meiosis occurring at germination of the zygote.

The following account is based on the work on the marine and brackish species of North Carolina by Ott and Hommersand (1974).

Key to the species

1. Plants with empty spaces cut off below the antheridia 2
1. Plants without empty spaces below the antheridia 8
2. Androphores separated from the main filaments by empty spaces 3
2. Androphores not separated from the main filaments 4
3. Oogonia sessile, bases of oogonia equal to or slightly smaller than diameter of
   oogonia .... V. acrandra

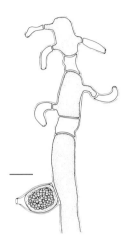

Figure 99. *Vaucheria acrandra*, filament bearing oogonium and antheridia, scale 50 μm (redrawn from Ott and Hommersand 1974).

3. Oogonia pedicellate, pedicel narrower than oogonium .... *V. adela*
4. Filament diameter 32 μm or less .... *V. minuta*
4. Filament diameter greater than 32 μm 5
5. Filament diameter 72 μm or greater; oogonia with separate protoplasmic mass in oogonia below zygospore .... *V. litorea*
5. Filament diameter less than 72 μm; oogonia without separate protoplasmic mass 6
6. Oogonia with three to six tubular papillae forming crowns ..... *V. coronata*
6. Oogonia without crowns 7
7. Oogonia opening by prominent hooked beaks .... *V. nasuta*
7. Oogonia without hooked beaks .... *V. longicaulis*
8. Antheridia sessile, ovoid .... *V. velutina*
8. Antheridia pedicellate or terminating a straight branch, cylindrical or ovoid 9
9. Mature zygospores dark green .... *V. arcassonensis*
9. Mature zygospores reddish .... *V. erythrospora*

Vaucheria acrandra Ott et Hommersand 1974, p. 373, figs 1–4.
    Figure 99
    Filaments 45–78 μm diameter; monoecious, with androphores borne terminally; one to five antheridia borne laterally on an androphore, sessile or pedicellate, hooked or straight, 65–100 μm long, opening by terminal pores; androphores produced from the main filament, solitary or occasionally two or more in series, terminal, cylindrical, 123–225 μm long, 45–70 μm diameter, separated from the main filament or adjacent androphores by an empty space; oogonia solitary on the filament, sessile, attached by a broad base, ovoid to concavo-convex, erect or slightly oblique, 90–150 μm long, 65–105 μm diameter, opening by a vertical to horizontal terminal pore, with or without short beaks; zygospores ovoid to subspherical, almost filling the oogonium, 70–142 μm long, 65–97 μm diameter; mature zygospores dark yellowish brown.
    Ott and Hommersand 1974.

Figure 100. *Vaucheria adela*, filament bearing oogonium and antheridia, scale 50 μm (re-drawn from Ott and Hommersand 1974).

Figure 101. *Vaucheria arcassonensis*, filament bearing oogonium and antheridium, scale 50 μm (redrawn from Ott and Hommersand 1974).

Reported only from Fort Fisher, N.C., where it formed small mats on salt marsh mud, January.

Distribution: Endemic to North Carolina as currently known.

Vaucheria adela Ott et Hommersand 1974, p. 374, figs 5–8.

Figure 100

Filaments short and frequently branched, 20–45 μm diameter; monoecious, with androphores borne terminally; antheridia solitary or sometimes two or three on an androphore, borne on short to exceedingly long pedicels (10–130 μm), straight, slightly hooked, or coiled, lanceolate to oblanceolate, 58–85 μm long, 15–30 μm diameter, opening by single terminal pores; androphores produced from the main filament, terminal or rarely intercalary, cylindrical, 50–450 μm long and 20–45 μm diameter, separated from main filaments by empty spaces; oogonia solitary on filaments, subsessile to pedicellate, ovoid to obliquely ovoid, 97–130 μm long, 45–72 μm diameter, opening by narrow vertical or slightly oblique beaks; zygospores ovoid, almost filling the oogonium, 58–90 μm long, 45–72 μm diameter, mature zygospores pale brown with single reddish-brown spots.

Ott and Hommersand 1974.

Forming small mats on intertidal mud among other species of *Vaucheria* in shade of *Juncus* and *Spartina*, Fort Macon (type locality) and Fort Fisher, January–February.

Distribution: Endemic to North Carolina as currently known.

Vaucheria arcassonensis Dangeard 1939, p. 216, figs 10m–10r, 11a–11e.

Figure 101

Filaments 20–65 μm diameter; dioecious or occasionally monoecious, one or two antheridia, adjacent to oogonia, rarely bifurcate, on pedicels of vary-ing length and highly contorted or coiled, tubular and sometimes approaching 200 μm long, opening by single small terminal pores; oogonia solitary or clus-tered in groups of two to four, borne on curved pedicels of varying length, rarely

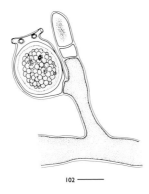
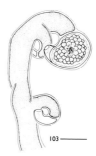

Figure 102. *Vaucheria coronata*, filament bearing oogonium and antheridium, scale 50 μm (redrawn from Ott and Hommersand 1974).

Figure 103. *Vaucheria erythrospora*, filament bearing oogonium and antheridia, scale 50 μm (redrawn from Ott and Hommersand 1974).

sessile, ovoid to cylindrical, usually slightly curved and opening by single terminal pores which can be horizontal, vertical, or incurved; zygospores ellipsoid to ovate, filling all but the base and apex of oogonia, 65–149 μm long, 58–84 μm broad; mature zygospores dark green with conspicuous oil droplets, not persisting, leaving behind old oogonium walls.

Whitford and Schumacher 1969; Blum 1972; Ott and Hommersand 1974. Found growing under *Juncus* and *Spartina* throughout North Carolina on intertidal marsh mud, December–March.

Distribution: New England to North Carolina, Atlantic Europe.

The most abundant of the *Vaucheria* species.

Vaucheria coronata Nordstedt 1879, p. 177, pl. 1, figs 1–9.

Figure 102

Filaments 40–65 μm diameter; monoecious or dioecious, antheridia solitary and terminal on branches with or without associated oogonia, separated from filaments by empty spaces 12–15 μm long, or antheridia intercalary, two to six in series on main filaments, each separated by an empty space; antheridia 40–105 μm long and 30–52 μm diameter, with one (rarely two) lateral discharge papillae; oogonia borne on the antheridial branches or on unisexual filaments, solitary, sessile or subsessile, obovoid to obliquely obovoid, vertical or bent horizontally toward the distal end of main filaments, 94–165 μm long, 76–120 μm diameter, opening by three to six tubular papillae forming a corona; zygospores spherical to subspherical, not filling oogonia, 55–130 μm diameter, pale brown at maturity with two to three reddish-brown spots.

Whitford and Schumacher 1969; Blum 1972; Ott and Hommersand 1974.

Throughout North Carolina salt marshes on intertidal mud under *Juncus* and *Spartina*, January–April.

Distribution: New England to North Carolina, Atlantic Europe.

Vaucheria erythrospora Christensen 1956, p. 275, figs 1, 2.

Figure 103

Filaments 25–40 µm diameter; monoecious, antheridia single, pedicellate, and circinate, 16–26 µm diameter, opening by single terminal pores; oogonia borne on the end of short or long lateral branches, solitary or occasionally in pairs, ovoid to concavo-convex, 80–145 µm long, 58–80 µm diameter, opening by single terminal pores on papillae or beaks directed toward the associated antheridia; zygospores ovoid to globose, not filling the oogonial beaks, zygospore walls with a reddish layer at maturity.

Blum 1972; Ott and Hommersand 1974.

Near the high-tide line of marshes on Topsail Island exposed to saltwater flooding at extremely high tides, November–March.

Distribution: North Carolina, Atlantic Europe, Japan.

Vaucheria litorea Hofman ex C. Agardh 1822, p. 463.

Figures 104 and 105

Filaments 72–97 µm diameter; dioecious, antheridia 525–750(–1450) µm long, 65–74 µm diameter, at ends of short or long branches and separated from main filaments by empty spaces, cylindrical and elongate, with one terminal and two to six conical lateral discharge papillae; oogonia at ends of reflexed branches, occasionally having a sympodial-secund arrangement and usually separated

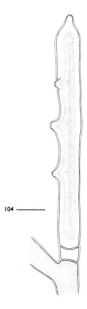
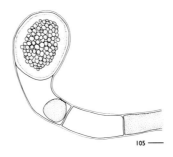

Figures 104, 105. *Vaucheria litorea.* 104. Filament bearing terminal antheridium, scale 100 µm (redrawn from Ott and Hommersand 1974). 105. Filament bearing terminal oogonium, scale 50 µm (redrawn from Ott and Hommersand 1974).

from the main filament by empty spaces 50–195 µm long or, rarely, with empty spaces lacking; the oogonia clavate to obovoid, 262–510 µm long, 170–220 µm diameter, opening by single terminal pores and with separate dense protoplasmic masses below the zygospores; zygospores globose to subglobose, occupying upper ends of, and not filling, the oogonia, 175–220 µm long, 162–214 µm diameter, extremely dense at maturity and lacking conspicuous oil droplets.

Whitford and Schumacher 1969; Blum 1972; Ott and Hommersand 1974.

Found floating in a brackish pond and on mud in flooded salt marsh depression on Topsail Island, November–March.

Distribution: New England, North Carolina, Atlantic Europe.

Vaucheria longicaulis Hoppaugh 1930, p. 332, figs 1–4.

Figures 106 and 107

Filaments 33–45 µm diameter; dioecious, antheridia 275–420 µm long, 35–60 µm diameter, terminal at the end of straight branches, developing sympodially and usually separated from main filaments by empty spaces 6–35 µm long, cylindrical and elongate, with one terminal and one to four lateral discharge papillae; oogonia terminal on straight branches, becoming lateral by sympodial growth, pyriform, 295–420 µm long, 95–135 µm diameter, opening by single terminal pores; zygospores globose to subglobose, occupying upper ends and not filling the oogonia, 97–130 µm diameter, dark green at maturity and very dense.

Blum 1972; Ott and Hommersand 1974.

Forming abundant mud-trapping mounds on intertidal rocks of Radio Island jetty within the Beaufort, N.C., estuary, January–April.

Distribution: North Carolina, Brazil, Pakistan, Japan, Washington, California.

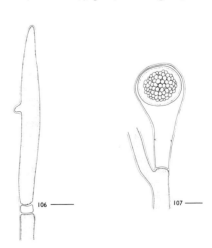

Figures 106, 107.
Vaucheria longicaulis.
106. Filament bearing terminal antheridium, scale 50 µm (redrawn from Ott and Hommersand 1974).
107. Filament bearing terminal oogonium, scale 50 µm (redrawn from Ott and Hommersand 1974).

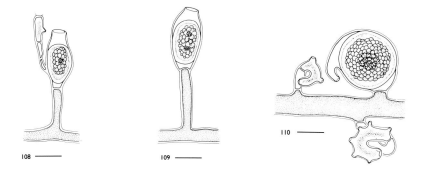

Figures 108, 109. *Vaucheria minuta.* 108. Filament bearing oogonium and antheridium, scale 50 μm (redrawn from Ott and Hommersand 1974). 109. Filament bearing oogonium, scale 50 μm (redrawn from Ott and Hommersand 1974).

Figure 110. *Vaucheria nasuta,* filament bearing antheridia and oogonium, scale 50 μm (redrawn from Ott and Hommersand 1974).

Vaucheria minuta Blum et Conover 1953, p. 399, figs 12–29.

Figures 108 and 109

Filaments 13–32 μm diameter; monoecious, occasionally dioecious, antheridia terminal or attached to sides or bases of oogonia, cylindrical, 33–62 μm long, 13–22 μm diameter, separated by empty spaces, opening by single lateral pores; oogonia solitary on lateral branches, either subterminal to antheridia or terminal if dioecious, ovoid to cylindrical, 78–105 μm long, 42–52 μm diameter, opening by single terminal pores; zygospores ovoid to cylindrical, filling the sides but not the top or bottom parts of oogonia, 39–90 μm long, 38–52 μm diameter, dark green at maturity with one to two reddish-brown spots.

Whitford and Schumacher 1969; Blum 1972; Ott and Hommersand 1974.

On intertidal mud throughout North Carolina mixed with other *Vaucheria* species under *Juncus* and *Spartina*, January–March.

Distribution: New England, North Carolina, Atlantic Europe.

Vaucheria nasuta Taylor et Bernatowicz 1952, p. 408, pl. 1, figs 13, 14; pl. 2, figs 1–14.

Figure 110

Filaments 32–52 μm diameter; dioecious, occasionally monoecious, antheridia in clusters of two to six adjacent to oogonia or sometimes occurring in unisexual clusters of two to eight, sessile to subsessile, 45–95 μm long and 15–25 μm diameter, antheridia borne on stalks having one to two empty spaces, with one to six lateral papillae and with or without terminal papillae, antheridia usually curved, cylindrical, or slightly inflated, commonly tapering toward tips; oogonia solitary or clustered in groups of two to three, sessile to subsessile or occasionally stipitate, subspherical to ovoid, erect or inclined obliquely, sometimes nearly parallel to the filament, 75–162 µm long, 65–130 µm diameter, opening by prominent reflexed beaks 21–28 µm long and 18–28 µm diameter; zygospores subspherical to ovoid, not completely filling the oogonia, 65–124 µm long, 66–117 µm diameter, mature zygospores pale brown.

Blum 1972; Ott and Hommersand 1974.

Mixed with other *Vaucheria* species on intertidal mud under *Juncus* and *Spartina* throughout the Carolinas, January–April.

Distribution: Massachusetts, North Carolina, South Carolina, Bermuda, California.

Vaucheria velutina C. Agardh 1824, p. 312.

Figure 111

Filaments 38–72 µm diameter; monoecious, antheridia seriate on one or both sides of main filaments, two to seven together, sessile or shortly pedicellate, ovoid, 40–91 µm long, 40–58 µm diameter, with single terminal discharge papillae, antheridia collapsing after discharge; oogonia occurring singly or in pairs, sessile or shortly pedicellate, obovoid to pyriform, 208–260 µm long, 143–182 µm diameter, inclined obliquely toward the associated antheridia; mature oogonia deciduous, each opening by terminal pores and occasionally forming tubular papillae to 26 µm long, 22 µm diameter; zygospores spherical to subspherical, not filling oogonia, 123–182 µm long, 125–162 µm diameter, pale brown at maturity.

As *V. thuretii*, Blum 1972; Ott and Hommersand 1974.

Forming unialgal mats in the littoral or supralittoral fringe along the northern coast of North Carolina, December–May.

Distribution: New England, North Carolina, southern Florida, Gulf of Mexico, Europe, West Africa, Asia, Pacific United States.

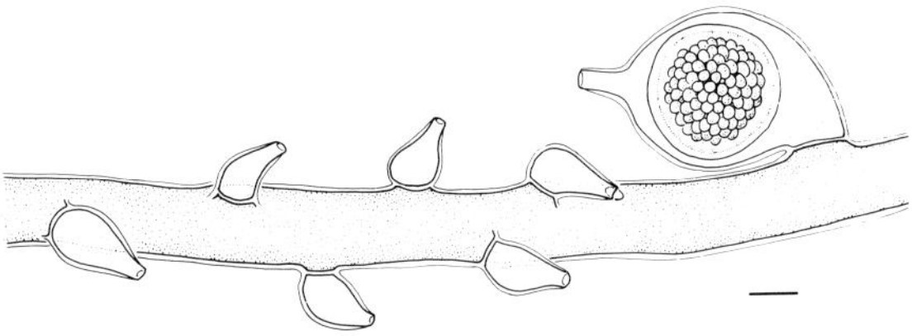

Figure III. *Vaucheria velutina*, filament bearing antheridia and oogonium, scale 50 µm (redrawn from Ott and Hommersand 1974).

PHAEOPHYTA

Plants mostly marine, composed of branched filaments or parenchymatous; filamentous plants often heterotrichous, but in some taxa erect or basal systems reduced or absent; filaments may be aggregated to form tagmatic structures; distribution of cell division may be diffuse or concentrated in intercalary, apical, or marginal meristems; cells uninucleate; plastids single to numerous, discoid, band shaped, rod shaped, or platelike, with or without pyrenoids; chlorophylls *a* and *c* present with fucoxanthin; carbohydrates stored as beta-linked glucose (laminarin) forming droplets in the cytoplasm; fucosin vesicles often clustered in the cytoplasm around nuclei; flagella usually paired, with a forward-directed, hairy flagellum and a trailing, smooth flagellum; life histories either diplohaplontic with gametophytes and sporophytes isomorphic or heteromorphic, or diplontic; many taxa reproduce only asexually.

Along this coast brown algae occur from the intertidal to depths of over 90 m. Some pelagic species of *Sargassum* constitute the primary vegetation of the Sargasso Sea, drifting offshore beyond the Florida Current–Gulf Stream current system. Shifts in winds and currents bring large numbers of these pelagic plants into near-shore waters where they drift onto the beaches.

A benthic species of *Sargassum*, *S. filipendula*, is the most important contributor to the standing crop of seaweeds on the local continental shelf. Brown algae are important economic resources in other parts of the world, but they are not used commercially on this coast, although an attempt was made in the 1970s to harvest the pelagic *Sargassum* species for commercial use. Brown algae are, however, important in providing habitats for a variety of fishes, invertebrate animals, and epiphytic algae by forming much of the three-dimensional miniature forests on rocky areas in the intertidal and subtidal and by providing substantial quantities of organic matter for the food chains.

Although some authorities have recognized several classes within the Phaeophyta, only a single class is acknowledged here.

PHAEOPHYCEAE

Key to the orders

1. Plants composed of uniseriate filaments, sometimes united as pseudoparenchyma 2
1. Plant parenchymatous or with at least some longitudinal cell divisions ... 5
2. Plants erect, freely branched; generations isomorphic or nearly so .... Ectocarpales (in part)
2. Plants creeping, tufted, or if erect, filaments aggregated; generations isomorphic or heteromorphic 3

3. Plants discoid, erect filaments of approximately equal length, united to form a pseudoparenchyma; generations isomorphic or nearly so .... Ectocarpales (in part)
3. Plants either not discoid, or if discoid, the filaments not closely united in a pseudoparenchyma; generations heteromorphic 4
4. Plants discoid, tufted, or erect, and if erect, either with multiaxial trichothallic or apical growth or uniaxial growth from an apical cell; sexual reproduction isogamous or anisogamous .... Chordariales
4. Plants erect, uniaxial, growth trichothallic from a terminal hair; sexual reproduction oogamous .... Desmarestiales
5. Growth apical or marginal 6
5. Growth diffuse or intercalary 8
6. Reproductive cells in conceptacles .... Fucales
6. Reproductive cells exposed 7
7. Plants becoming flattened, bladelike .... Dictyotales
7. Plants forming cylindrical branches .... Sphacelariales
8. Branch tips swollen, terminating in tufts of photosynthetic hairs .... Sporochnales
8. Branch tips not swollen, lacking a terminal tuft of hairs 9
9. Plastids single, parietal, platelike; plurilocular structures uniseriate .... Scytosiphonales
9. Plastids numerous, discoid; plurilocular structures parenchymatous .... Dictyosiphonales

ECTOCARPALES

Plants filamentous; filaments branched, uniseriate, in tufts, endophytic, or united to form crusts; growth diffuse or trichothallic; generations isomorphic or nearly so, reproduction by unilocular sporangia or plurilocular sporangia or gametangia, rarely monosporangia.

The interpretation of the order here is narrow. For a broader definition see Russell and Fletcher (1975).

Key to the families

1. Plants crustose, with prostrate filaments bearing closely packed erect filaments .... Ralfsiaceae
1. Plants not crustose, the filaments forming tufts or endophytic and ramifying in host tissue, or prostrate but without closely packed erect filaments .... Ectocarpaceae

Ectocarpaceae

Filaments endophytic, epiphytic, epizoic, or saxicolous, with erect and prostrate filaments, but not forming pseudoparenchymatous crusts; often forming tufts or partially endophytic and ramifying through host; plastids rod shaped, band shaped, or discoid; life histories diplohaplontic and generations isomorphic or slightly heteromorphic, all phases conforming to the family in terms of morphology. (This family treatment was contributed by C. D. Amsler.)

Key to the genera

1. Plants minute epiphytic discs or endophytic filaments 2
1. Plants larger, not discoid, and not primarily endophytic 4
2. Plants forming irregular discs . . . . Phaeostroma
2. Plants not forming discs 3
3. Erect hairlike filaments subtended by basal meristems . . . . Streblonema
3. Erect hairlike filaments without basal meristem . . . . Herponema
4. Plastids rod shaped or band shaped, few to numerous 5
4. Plastids discoid, numerous 6
5. Plastids band shaped, parietal . . . . Ectocarpus
5. Plastids rod shaped, forming one to two stellate clusters . . . . Bachelotia
6. Plurilocular sporangia grouped in irregularly shaped sori . . . . Botrytella
6. Plurilocular sporangia not in sori 7
7. Plants sparsely branched, branches often short, inserted at right angles to axial filament . . . . Acinetospora
7. Plants without numerous short branches inserted at right angles, or if so, the branches restricted to the lower parts of the plant . . . . Hincksia

Acinetospora Bornet 1891

Plants tufted, epiphytic, saxicolous, or on a wide variety of substrates; sparsely and irregularly branched, the branches often perpendicular to the axial filament; characterized by short branches that may be simple cellular extensions of axial cells or comprise a few cells; plastids numerous, discoid; meristems intercalary, scattered throughout the main axes; unilocular sporangia pedicellate or sessile, spherical to subspherical; plurilocular sporangia conical with large locules, sessile or on short pedicels; monosporangia pedicellate, spherical, producing a single nonmotile spore.

Acinetospora crinita (Carmichael ex Harvey) Kornmann 1953, p. 205, figs 1–14.
    Ectocarpus crinitus Carmichael ex Harvey in Hooker 1833, p. 326, pl. 330.
    Figures 112 and 113
    Plants forming tangled filamentous tufts to 8 cm and floating mats of en-

tangled, loose filaments to 1 m; main filaments 15–32(–56?) µm diameter, cells 1–6 diameters long; branching irregular, with a few long branches forming wide angles with the main axis; meristems scattered throughout plant; cells with numerous discoid plastids, occasionally concentrated in the cell center; unilocular sporangia pedicellate, rarely sessile, spherical to subspherical, 25–50 µm diameter; plurilocular sporangia sessile or on short pedicels, conical, 75–128 µm long, 24–42 µm wide; monosporangia sessile or on short pedicels, spherical to subspherical, 31–53 µm diameter.

Brauner 1975; Kapraun and Zechman 1982; Kapraun 1984; Amsler 1984a. As *A. pusilla*, Blomquist 1955; Taylor 1960. As *Ectocarpus breviarticulatus sensu* Williams 1948a.

On jetties and in sounds, all seasons, but most abundant in protected areas with weak currents, December–March.

Distribution: North Carolina, Caribbean, Gulf of Mexico, Greenland to USSR to Portugal, Canary Islands, southern Africa, Australia.

The Wrightsville Beach, N.C., population reproduces solely by fragmentation (Amsler 1984a), although the Beaufort, N.C., population produces unilocular sporangia (Brauner 1975, cf. Amsler 1984a). Farther south only monosporangia are reported.

Bachelotia (Bornet) Kuckuck ex Hamel, 1939
Plants filamentous, uniseriate, epiphytic or saxicolous, attached by prostrate filaments; branching sparse to moderate; plastids rod shaped, in one to two stellate clusters per cell; unilocular and plurilocular sporangia intercalary.

Bachelotia antillarum (Grunow) Gerloff 1959, p. 38.
Ectocarpus antillarum Grunow 1867, p. 46.
Figures 114 and 115
Plants filamentous, epiphytic or saxicolous, forming entangled mats attached by prostrate filaments; main axes 11–52 µm diameter, cells 0.5–2(–3) diameters

Figures 112, 113.
Acinetospora crinita. 112. Prostrate branches, scale 20 µm. 113. Unilocular sporangium, scale 20 µm.

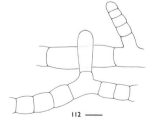
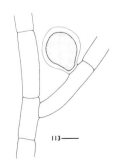

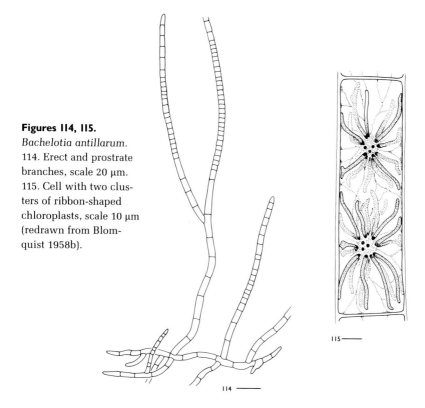

Figures 114, 115.
Bachelotia antillarum.
114. Erect and prostrate branches, scale 20 μm.
115. Cell with two clusters of ribbon-shaped chloroplasts, scale 10 μm (redrawn from Blomquist 1958b).

long; branching sparse to moderate with numerous spurlike branchlets; meristematic regions well defined in lower portions.

Kapraun 1984. As *B. fulvescens*, Blomquist and Humm 1946; Williams 1948a; Taylor 1960. As *Pilayella antillarum*, Blomquist 1958b.

Occasional in inshore waters, year-round.

Distribution: North Carolina, Bermuda, Gulf of Mexico, Caribbean, Brazil, France to Portugal, West Africa, southern Africa, Indian Ocean, Australia.

Botrytella Bory 1822

Plants forming abundantly branched filamentous tufts; filaments uniseriate, with lateral and terminal colorless hairs from basal meristems; plastids numerous, discoid; plurilocular sporangia clustered in irregular sori.

Botrytella micromora Bory 1822, p. 426.

Figure 116

Plants epiphytic or saxicolous; forming filamentous tufts to 4(–20) cm high; main axes 40–68 μm diameter; cells 1–3 diameters long; growth generally diffuse, but colorless hairs arising from distinct meristems; branching spiral to alternate or irregular; plurilocular sporangia conical, 20–40 μm long, 12–20 μm wide, clustered, typically fused into sori; unilocular sporangia at times occurring with plurilocular sporangia in sori.

As *Sorocarpus micromorus*, Kapraun and Zechman 1982; Kapraun 1984. As

S. *uvaeformis*, Williams 1948a; Taylor 1960.

On Cape Lookout and Wrightsville Beach jetties, March–April.

Distribution: Newfoundland to Virginia, North Carolina, Norway, Greenland, British Isles, Australia, Japan.

Ectocarpus Lyngbye 1819, nom. cons.

Erect, moderately to densely branched uniseriate plants; branching variable; epiphytic, epizoic, or saxicolous, attached by rhizoids, prostrate filaments, or both; plastids parietal, bandlike or ribbonlike, one to several per cell, usually with several pyrenoids per plastid; growth diffuse; plurilocular sporangia sessile or pedicellate; unilocular sporangia uncommon in most species, typically sessile, not clustered.

Key to the species

1. Plants minute, erect filaments simple or branching only near the base; less than 5 mm high; plurilocular sporangia borne in lower portions only . . . . . .
. . . . *E. elachistaeformis*
1. Plants macroscopic; highly branched erect filaments to 10 mm or more long; plurilocular sporangia scattered 2
2. Branchlets in at least distal portions clustered; plurilocular sporangia usually pedicellate, to 150 µm long . . . . *E. fasciculatus*
2. Branchlets not clustered; plurilocular sporangia sessile or pedicellate, hair-tipped or not, but usually at least 300 µm long . . . . *E. siliculosus*

Figure 116. *Botrytella micromora*, filament with plurilocular structures, scale 100 µm (redrawn from Williams 1948a).
Figure 117. *Ectocarpus elachistaeformis.* Erect branches, scale 100 µm.
Figure 118. *Ectocarpus elachistaeformis.* Plurilocular reproductive structures, scale 25 µm.

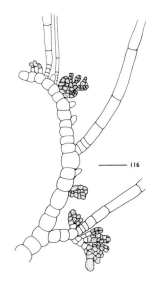

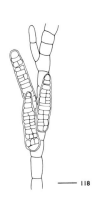

117 ——

Ectocarpus elachistaeformis Heydrich 1892, p. 479, pl. 25.

Figures 117 and 118

Plants minute epiphytes to 1.5 mm high; prostrate filaments attached by haptera; growth diffuse, branching sparse, primarily from the base; main axes 10–18 μm diameter, cells 1–3 diameters long below, longer and narrower above; plastids parietal, band shaped; plurilocular sporangia fusiform to elongate and conical, 63–80(–200) μm long, 15–25 μm wide, sessile or on one-celled pedicels, borne on or near prostrate filaments.

Williams 1948a; Aziz and Humm 1962; Brauner 1975; Kapraun 1984.

Epiphytic on larger algae and sea grasses in shallow water, July–November.

Distribution: Virginia, North Carolina, Bermuda, southern Florida, Gulf of Mexico, Caribbean, Brazil, West Africa, Indian Ocean.

This species is similar in habit to forms of *Hincksia irregularis*; it can be distinguished by examination of plastids. Although it has also been placed in the genus *Feldmannia* as *F. elachistaeformis* (Heydrich) Pham-Hoang, it does not have the discoid plastids of that genus.

Ectocarpus fasciculatus Harvey 1841, p. 40.

Figures 119 and 120

Plants forming filamentous tufts to 8 cm high; epiphytic or saxicolous, attached by rhizoids and prostrate filaments; main axes 30–70 μm diameter, cells 1–2 diameters long; plastids parietal, band shaped; fasciculate above, often irregular below; plurilocular sporangia pedicellate, fusiform, 70–150 μm long, 18–40 μm wide.

Williams 1948a; Brauner 1975; Kapraun and Zechman 1982; Kapraun 1984.

On jetties and in sounds, winter–spring.

Distribution: Canadian Arctic to Delaware, North Carolina, Norway to Portugal, Australia.

Ectocarpus siliculosus (Dillwyn) Lyngbye 1819, p. 131. pl. 43.

Conferva siliculosa Dillwyn 1809, p. 69, suppl. pl. E.

Figures 121 and 122

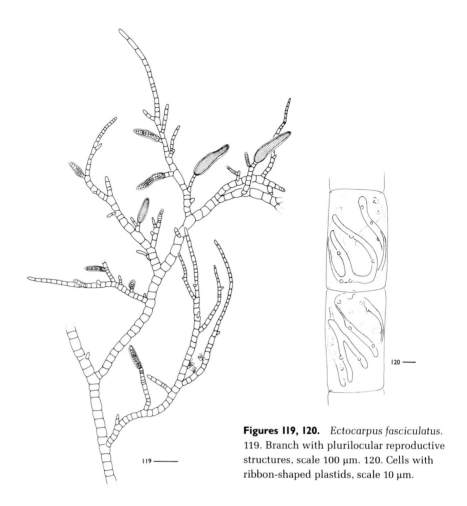

Figures 119, 120. *Ectocarpus fasciculatus.* 119. Branch with plurilocular reproductive structures, scale 100 μm. 120. Cells with ribbon-shaped plastids, scale 10 μm.

Plants forming filamentous tufts to 30 cm high and, in sheltered waters, entangled mats over 1 m long; epiphytic, epizoic, or saxicolous, attached by prostrate filaments; main axes 40–60 μm diameter; cells 1–5 diameters long, with parietal, band-shaped plastids; branching alternate, secund, or nearly pseudoparenchymatous; plurilocular sporangia usually pedicellate, elongate, conical, 50–400(–600) μm long, 12–50 μm wide, sometimes with a conspicuous terminal hair; unilocular sporangia oval, sessile or pedicellate, 30–140 μm long, 20–50 μm wide.

Bailey 1848; Harvey 1853; Johnson 1900; Hoyt 1920; Williams 1948a; Taylor 1960; Chapman 1971; Brauner 1975; Kapraun 1984; Richardson 1987. As *Ectocarpus confervoides*, Hoyt 1920; Taylor 1960; Brauner 1975; Schneider 1976. As *E. viridis*, Bailey 1848; Harvey 1853.

Common on jetties and in sounds, winter–spring; rare offshore in Onslow Bay, from 23 m, August.

Distribution: Canadian Arctic to Virginia, North Carolina, South Carolina, Georgia, Bermuda, southern Florida, Gulf of Mexico, Brazil, USSR to Portugal, southern Africa, Australia, Japan, Alaska to California, western South America.

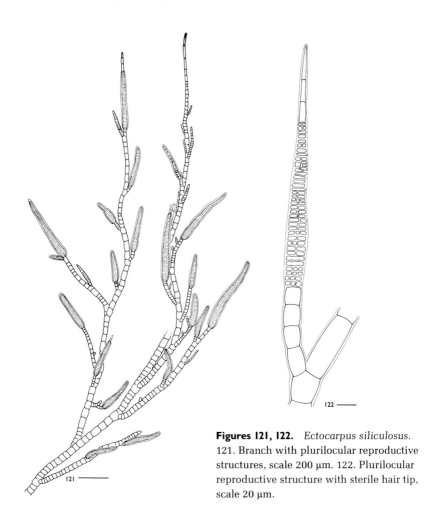

Figures 121, 122. *Ectocarpus siliculosus.*
121. Branch with plurilocular reproductive structures, scale 200 μm. 122. Plurilocular reproductive structure with sterile hair tip, scale 20 μm.

Russell (1966, 1967) demonstrated that all previously described British *Ectocarpus* species are forms of two large polymorphic species, *E. siliculosus* and *E. fasciculatus.*

Herponema C. Agardh 1822
Plants completely or partially endophytic, with irregularly branched uniseriate filaments; not forming discs; growth diffuse; plastids band shaped; sporangia terminal on short erect filaments or from the base of erect filaments.

Herponema solitarium (Sauvageau) Hamel 1939, p. 68.
  Ectocarpus solitarius Sauvageau 1892, p. 97. pl. 3, figs 24–27.
  Figures 123 and 124
  Plants mostly endophytic, filaments intercellular, 10–15 μm diameter, cells 1–3 diameters long; erect hairlike filaments without basal meristems, 9–15 μm diameter, cells 2–4 diameters long; plurilocular sporangia ovoid to cylindrical, 25–105 μm long, 14–45 μm wide.

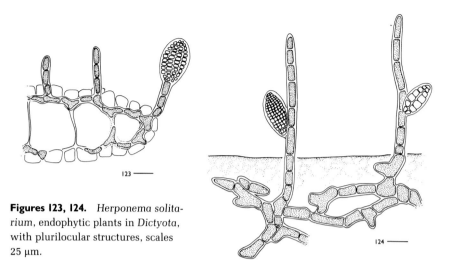

Figures 123, 124. *Herponema solita-rium*, endophytic plants in *Dictyota*, with plurilocular structures, scales 25 μm.

As *Streblonema solitarium*, Hoyt 1920. As *Ectocarpus solitarius*, Aziz and Humm 1962.

Epiphytic on *Dictyota menstrualis*, in sounds and offshore in Onslow Bay, from 25 m, July–August.

Distribution: North Carolina, British Isles, France, Spain.

Hincksia J. E. Gray 1864

Erect, moderately to densely branched, uniseriate plants; branching usually ir-regular; epiphytic, epizoic, or saxicolous; attached by rhizoids, prostrate fila-ments, or both; plastids numerous, discoid, with none to two pyrenoids; meri-stems poorly defined, scattered throughout the plant or well defined, subtending hairlike filaments; plurilocular sporangia usually sessile; unilocular sporangia uncommon in most species, typically sessile.

The species included here have previously been placed in the genus *Giffordia*, but Silva et al. (1987) pointed out the priority of Gray's genus *Hincksia*.

Key to the species

1. Plants sparsely branched below with short branches inserted at right angles to the axial filament; highly branched above, branches inserted in the middle of axial cells; plants from offshore waters .... *H. onslowensis*
1. Plants not as above 2
2. Plurilocular sporangia cylindrical .... *H. mitchelliae*
2. Plurilocular sporangia conical or ovoid 3
3. Branching often opposite or nearly so .... *H. granulosa*
3. Branching not opposite 4
4. Plurilocular sporangia usually ovoid, sometimes borne in pairs or whorls of three from a cell; sparsely branched, with sporangia throughout at least lower 60 percent of plant .... *H. ovata*
4. Plurilocular sporangia conical to fusiform, restricted to base in sparsely

branched plants or throughout, often in short secund series near branch bases in more highly branched plants, never in pairs or whorls from single cells
.... *H. irregularis*

Hincksia granulosa (J. E. Smith) Silva in Silva, Meñez et Moe 1987, p. 130.
   Conferva granulosa J. E. Smith 1811 [1811–1812], pl. 2351.
   Figures 125–128
   Plants epiphytic, epizoic, or saxicolous, to 10(–25) cm tall, attached by rhizoids that sometimes completely corticate lower filaments; main axes 21–100 μm diameter; cells 1–2 diameters long; meristems indistinct; branching opposite to spiral or secund; plurilocular sporangia typically broadly ovoid, often asymmetrical, sessile, 60–100(–165) μm long, 30–60 μm wide; local material with conical to fusiform plurilocular sporangia, 33–100 μm long, 16–30 μm wide; unilocular sporangia reported to be sessile, asymmetrical, globose to subglobose.
   In Beaufort and Wrightsville Beach, N.C., area sounds, October, ?December.
   Distribution: Newfoundland to New Jersey, North Carolina (new record), Norway to Portugal, Australia, British Columbia to Pacific Mexico.
   The local North Carolina material is placed in *Hincksia granulosa* because of

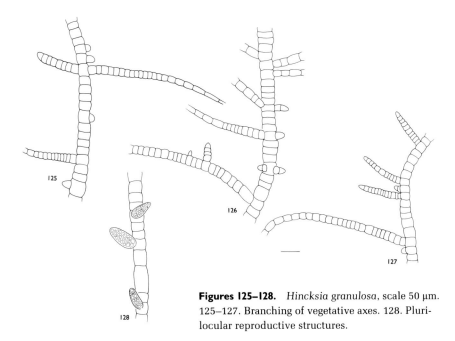

Figures 125–128. *Hincksia granulosa*, scale 50 μm. 125–127. Branching of vegetative axes. 128. Plurilocular reproductive structures.

the abundance of short, spinelike branches and characteristic opposite branching. The plurilocular sporangia are much narrower than typical forms, but most are within the range of variability illustrated by Cardinal (1964, figs 20D–20F). The main axes are also narrower than in other reports. Kuckuck (1961) described three varieties of the species (as *Giffordia granulosa*) based on branching habit, which Cardinal (1964) and Clayton (1974) observed to intergrade in nature. The North Carolina plants would be placed in var. *laeta* (C. Agardh) Kuckuck, while the nearest populations in New England appear from published illustrations to be similar to var. *eugranulosa* Kuckuck. These differences might be phenotypic at the edge of the geographic range of the species.

Fertile specimens of this species have only been collected once, in a Beaufort, N.C., mud flat (Duerr and Schneider, unpublished). Similar, sterile material has also been collected at Wrightsville Beach, N.C. (Amsler, unpublished). Its identity remains tentative pending detailed investigation.

Hincksia irregularis (Kützing) Amsler comb. nov.

Ectocarpus irregularis Kützing 1845, p. 234.

Figures 129–131

Plants forming filamentous tufts to 5 cm; epiphytic, epizoic, or saxicolous; attached by prostrate filaments or rhizoids; main axes 25–50 μm diameter; cells 0.5–4 diameters long below, to 10 diameters long above; plastids numerous, discoid; branching irregular, sometimes secund, alternate, or opposite in parts of larger plants; branches in larger plants sometimes forming wide angles with small axial cells; meristems distinct or not, scattered, sometimes subtending hairlike filaments; plurilocular sporangia conical to fusiform, pedicellate or more often sessile, 40–200 μm long, 24–40 μm diameter; unilocular sporangia rare, sessile, ovoid, 50–70 μm long, 25–45 μm wide.

As *Giffordia irregularis*, Amsler 1983, 1984b. As *Ectocarpus coniferus*, Blomquist and Humm 1946. As *G. conifera*, Taylor 1960; Brauner 1975; Kapraun 1984; Richardson 1986, 1987. As *E. rallsiae sensu* Williams 1948. As *G. rallsiae sensu* Taylor 1960; Brauner 1975; Kapraun and Zechman 1982; Kapraun 1984. As *Acinetospora crinita sensu* Schneider 1976, pro parte.

In inshore waters all seasons, rare in winter, common in summer; offshore in Onslow Bay to 28 m, August.

Distribution: North Carolina, Georgia, Bermuda, southern Florida, Gulf of Mexico, Caribbean, Brazil, West Africa, southern Africa, Indian Ocean, Australia, Central Pacific islands, Viet Nam, California, Pacific Mexico.

This species includes morphological forms that have been called *Giffordia rallsiae* (Vickers) Taylor, *G. conifera* (Børgesen) Taylor, *G. irregularis* (Kützing) Joly, and *Feldmannia irregularis* (Kützing) Hamel in the western North Atlantic.

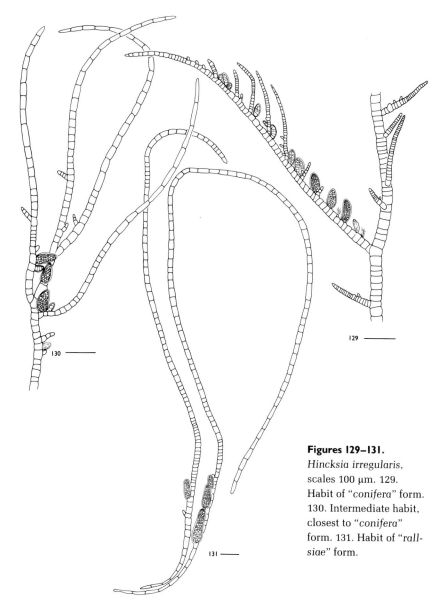

Figures 129–131.
Hincksia irregularis,
scales 100 μm. 129.
Habit of "*conifera*" form.
130. Intermediate habit,
closest to "*conifera*"
form. 131. Habit of "*rall-siae*" form.

Based on extensive observation of plants collected in North Carolina and on culture studies, Amsler (1985) concluded that *G. conifera* and *G. rallsiae* were conspecific because they clearly intergraded in nature and had the same phenology, growth, and reproductive responses to light and temperature in culture, and the same morphology and development in culture. Sauvageau (1933) and Børgesen (1941) considered *Ectocarpus coniferus* and *E. rallsiae* to be conspecific with *E. irregularis*. More recent workers have placed *E. irregularis* in either *Hincksia* (as *Giffordia*, Joly 1965) or *Feldmannia* (Hamel 1939). Unfortunately, type material of *E. irregularis* cannot be located and is poorly illustrated (Børgesen 1914; Clayton 1974). Silva et al. (1987) replaced the genus name *Giffordia* with the earlier name *Hincksia* established by J. E. Gray (1864) but chose to retain Kützing's species *irregularis* in the genus *Feldmannia*. *Feldmannia* was established by Hamel (1939) to include plants with meristematic zones close to the base which subtend unbranched, sterile filaments. Plants previously referred to *G. conifera* are densely branched and bear sporangia throughout. Only the more extreme morphological forms previously included in *G. rallsiae* have meristems restricted to the base with no distal branches or sporangia (Amsler 1983). *Hincksia ovata* (Kjellman) Silva can also have unbranched sterile filaments subtended by meristems. Consequently, *irregularis* is best placed in *Hincksia*. This necessitates the new combination *Hincksia irregularis*. Amsler (1983) demonstrated that this species (as *Giffordia*) might exist as a warm-adapted phase with tropical affinities throughout much of the year with a cold-adapted phase in winter. Unilocular sporangia have been observed only in offshore material and in winter-phase plants from culture.

Hincksia mitchelliae (Harvey) Silva in Silva, Meñez, et Moe 1987, p. 73.
    Ectocarpus mitchelliae Harvey 1852, p. 142, pl. 12G, figs 1–3.
    Figures 132–134
    Plants epiphytic or saxicolous, forming discrete plants to 12 cm long or entangled mats to several dm long; attached by few prostrate filaments or rhizoids; branching dense, irregular to spiral; upper branches sometimes ending in hairlike filaments; main axes (17–)31–50 μm diameter, cells 1–4 diameters long; meristematic regions short, intercalary, scattered throughout the plant, or indistinct; plurilocular sporangia sessile or rarely pedicellate, cylindrical or occasionally tapering slightly at apices, 42–220 μm long, 15–36 μm wide, locule dimensions usually 4–6 μm; plurilocular structures of similar overall dimensions but with larger (megasporangia) and smaller (microsporangia) locules reported; unilocular sporangia rare, ovoid, sessile, or on one-celled pedicels, 50–100 μm long, 25–50 μm wide.

    As *Giffordia mitchelliae*, Taylor 1960; Brauner 1975; Kapraun and Zechman 1982; Kapraun 1984; Amsler 1985; Searles 1987, 1988. As *Ectocarpus mitchelliae*, Hoyt 1920; Williams 1948a. As *E. duchassaingianus*, Hoyt 1920. As *G. duchassaigniana*, Taylor 1960. As *G. indica sensu* Brauner 1975; Schneider 1976.

    On jetties and in sounds year-round, reaching maximum size and abundance in spring and fall, rare in winter; rare offshore in Onslow Bay from 29 m, May; on Gray's Reef from 17 to 21 m, July.

    Distribution: Maritimes to Virginia, North Carolina, Georgia, Bermuda, southern Florida, Gulf of Mexico, Caribbean, Brazil, British Isles to Portugal; elsewhere, widespread in temperate and tropical seas.

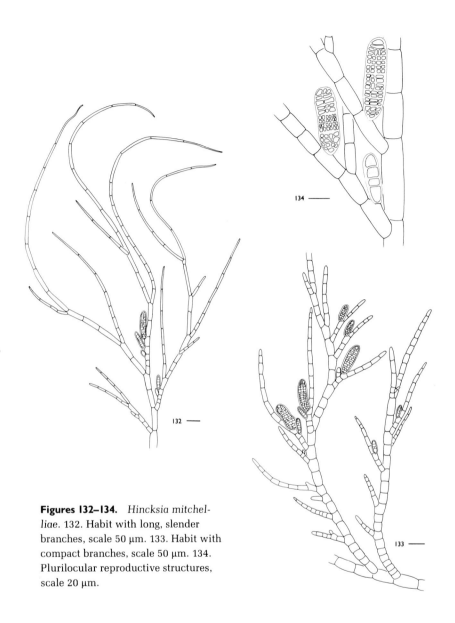

Figures 132–134. *Hincksia mitchelliae.* 132. Habit with long, slender branches, scale 50 μm. 133. Habit with compact branches, scale 50 μm. 134. Plurilocular reproductive structures, scale 20 μm.

Hincksia onslowensis (Amsler and Kapraun) Silva in Silva, Meñez, et Moe 1987, p. 130.

Giffordia onslowensis Amsler and Kapraun 1985, p. 94, figs 1, 2.

Figures 135–139

Plants epiphytic, forming entangled tufts to 5 cm; lower filaments sparsely and irregularly branched, 20–50 μm diameter, cells to 4 diameters long, branches often short, arising from the center of axial cells; upper filaments often profusely

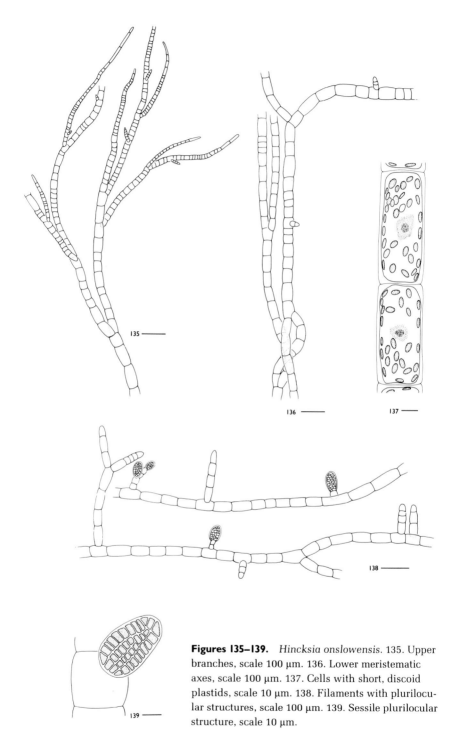

Figures 135–139. *Hincksia onslowensis.* 135. Upper branches, scale 100 μm. 136. Lower meristematic axes, scale 100 μm. 137. Cells with short, discoid plastids, scale 10 μm. 138. Filaments with plurilocular structures, scale 100 μm. 139. Sessile plurilocular structure, scale 10 μm.

branched, 4–22 µm diameter, arising from the upper parts of axial cells; meristematic regions short, intercalary, scattered, frequently indistinct; plurilocular sporangia ovoid to short conical, 22–52 µm long, 10–20 µm wide, single or paired, sessile or on one- to several-celled pedicels.

As *Giffordia onslowensis*, Amsler and Kapraun 1985; Searles 1987, 1988. As *Acinetospora crinita sensu* Schneider 1976, pro parte.

On offshore rock outcroppings, 17–50 m, June–September.

Distribution: North Carolina (type locality Onslow Bay), Georgia.

Hincksia ovata (Kjellman) P. C. Silva in Silva, Meñez, et Moe 1987, p. 130.
Ectocarpus ovatus Kjellman 1877, p. 35.
Figure 140

Plants usually epiphytic, occasionally saxicolous, forming filamentous tufts to 2(–6) cm tall; attached by prostrate filaments or rhizoids; main axes 19–56 µm diameter; cells 1–3 diameters long; meristems poorly defined and scattered, or long and clearly defined, subtending hairlike filaments; branching sparse, primarily from bases; plurilocular sporangia conical to ovoid, occurring singly, oppositely, or in whorls of three, 28–110 µm long, 13–50 µm wide, sessile or on one-celled pedicels, often in long series; unilocular sporangia rare, sessile, globose to ovoid, 20–46 µm long, 19–36 µm wide.

As *Giffordia ovata*, Amsler 1984b; Kapraun 1984. As *G. intermedia*, Kapraun and Zechman 1982.

On coarser algae, inshore in the Wilmington area, October–June.

Distribution: Canadian Arctic to Maritimes, North Carolina, USSR to Portugal, Australia, Alaska to Washington.

Phaeostroma Kuckuck in Reinbold 1893a
Plants filamentous, epiphytic, forming irregular, discoid, usually single-layered colonies; central filaments usually prostrate, sometimes crowded and becoming one to three layered; basal meristematic regions giving rise to colorless hairs; cells with several platelike plastids; sporangia arising from transformation of vegetative cells.

Phaeostroma pusillum Howe et Hoyt 1916, p. 109, pl. II, figs 1–9.
Figure 141

Plants epiphytic, forming irregular discoid colonies, to 1 mm diameter; vegetative cells rectangular to irregular, 5–10 µm diameter, 10–16 µm long; short hairs occasional from basal meristems; plurilocular sporangia conical to ovoid, sessile, 22–40 µm long, 15–20 µm wide; unilocular sporangia globose to ovoid, 8–

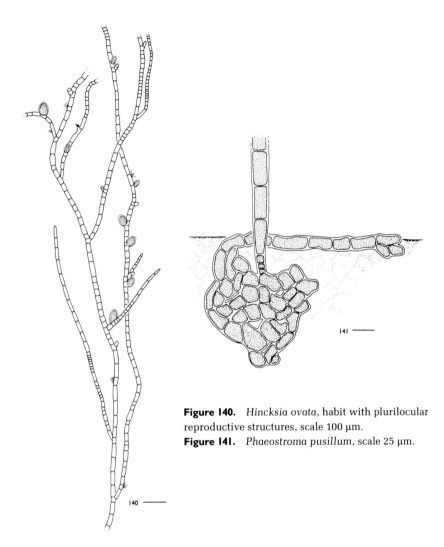

Figure 140. *Hincksia ovata*, habit with plurilocular reproductive structures, scale 100 μm.
Figure 141. *Phaeostroma pusillum*, scale 25 μm.

16(–25) μm diameter or, by previous division of a sporangial initial, forming almost mulberry-shaped sori, 16–48 μm diameter.

Howe and Hoyt 1916; Hoyt 1920; Taylor 1960; Brauner 1975; Kapraun 1984.

Rare, in inshore waters, January–August.

Distribution: North Carolina (type locality), Bermuda, southern Florida, Bahamas, Gulf of Mexico.

Streblonema Derbès et Solier 1851

Plants completely or partially endophytic, with irregularly branched filaments; not forming a disc; growth mostly diffuse, but colorless, erect hairs formed by basal meristems; plastids band shaped, one to several per cell, pyrenoids absent; plurilocular sporangia multiseriate or uniseriate, cylindrical or ovoid; unilocular sporangia ovoid to globose, rare.

Key to the species

1. Plurilocular sporangia uniseriate or partially biseriate at maturity..........
....S. oligosporum
1. Plurilocular sporangia at least partially multiseriate at maturity............
.... S. invisibile

Steblonema invisibile Hoyt 1920, p. 441, figs 16–19.
   Figures 142–145
   Plants endophytic or epiphytic; filaments intercellular, 5–8 μm diameter, irregularly swollen, sparsely branched; plurilocular sporangia lanceolate, 25–55 μm long, 11–17 μm wide; formed in patches on upright branches; unilocular sporangia unknown.
   Hoyt 1920. As *Entonema oligosporum sensu* Brauner 1975.
   Endophytic in *Meristotheca floridana* and epiphytic on *Zostera marina*, August–October.
   Distribution: Endemic to North Carolina as currently known.

Steblonema oligosporum Strömfelt 1884, p. 133, pl. 1.
   Plants minute, primarily endophytic; basal filaments 4–6 μm diameter,

Figures 142–144. *Streblonema invisibile*, scale 20 μm. 142, 143. Endophytic plants (redrawn from Hoyt 1920). 144. Epiphytic plant.
Figure 145. *Streblonema oligosporum*, endophytic in *Gracilaria*, scale 20 μm (redrawn from Aziz and Humm 1962).

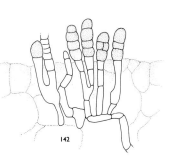

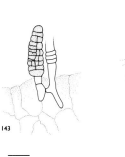

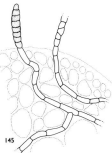

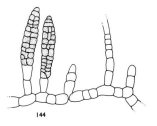

branched, spreading over host cuticle or among cortical cells; erect filaments short, terminating in colorless hairs; plurilocular sporangia uniseriate or partially biseriate, cylindrical to ovoid, obtuse, 25–70 μm long, 8–20 μm wide, sessile or on one- to three-celled pedicels.

Kapraun 1984. As *Entonema oligosporum*, Aziz and Humm 1962.

On jetties and in sounds, fall–spring.

Distribution: Virginia, North Carolina, Texas, Atlantic Europe.

Ralfsiaceae

Plants crustose, saxicolous, pseudoparenchymatous; horizontal basal layer often several cells thick and bearing erect, closely packed, usually firmly adherent filaments of equal length; plants generally lacking rhizoids; growth marginal or intercalary; plastids platelike and one per cell, or discoid and numerous; with or without pyrenoids; life history generally direct, without evidence of sexuality.

Taxonomic treatment of this group is not generally agreed upon. Nelson (1982) recommended recognition of a single family in the order Ectocarpales—the policy followed here—whereas Womersley (1987) considered the family a member of the Chordariales. On the other hand, it was given ordinal status by Nakamura (1972) and Tanaka and Chihara (1982) and separated into two to four families.

Pseudolithoderma Svedelius in Kjellman and Svedelius 1911
Plants crustose, completely adherent; erect filaments closely packed, with or without terminal ascocystlike cells, each cell with several plastids; plurilocular and unilocular sporangia produced by transformation of terminal surface cells.

Pseudolithoderma extensum (P. Crouan et H. Crouan) Lund 1959, p. 84.

Ralfsia extensa P. Crouan et H. Crouan 1852.

Figure 146

Plants forming thin, tightly packed, adherent, dark brown, discoid crusts to 8 cm diameter; erect filaments 8–12 μm diameter, to 150 μm long, from a poorly defined basal system; unilocular and plurilocular sporangia on the same or separate crusts, unilocular sporangia globose, plurilocular sporangia uniseriate or biseriate.

As *Lithoderma fatiscens* sensu Williams 1948a, 1949, 1951.

Subtidal, saxicolous and on oysters of Cape Lookout jetty and offshore rock outcroppings, year-round.

Distribution: Canadian Arctic to Massachusetts, North Carolina, USSR to France.

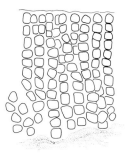

Figure 146. *Pseudolithoderma extensum*, section of sterile crust, scale 10 μm.

This species has not been re-collected in the region, and only sterile material remains from Williams's (1948a, 1948b, 1951) collections. His observation of terminal unilocular and plurilocular sporangia suggests that the plants may be representative of *Pseudolithoderma extensum* as defined by Lund (1959), although the presence of several plastids in the cells is difficult to verify in Williams's voucher specimen.

Questionable Record

Ralfsia expansa J. Agardh 1848, p. 63.

Williams (1948a, 1949) reported this species from the upper intertidal at Cape Lookout. It has not been re-collected and no voucher specimens remain. It may be that the plants seen by Williams were the crustose phase of a member of the Scytosiphonales.

CHORDARIALES

Life history usually heteromorphic; sporophyte generation generally macroscopic and pseudoparenchymatous with multiaxial or uniaxial organization; gametophyte generation smaller, filamentous, creeping; sporophytes with scattered unilocular sporangia; presence of plurilocular sporangia variable among the taxa; gametes unequal; fusion of zooids from unilocular sporangia or direct development of zooids from unilocular or plurilocular sporangia of macrothallus into new macrothalli is reported; plastids one to many, platelike, elongate, or discoid, often with pyrenoids.

Key to the families

1. Plants of both generations discoid, creeping . . . . Myrionemataceae
1. Sporophytes tufted or erect 2
2. Sporophytes small, pulvinate tufts or subspherical, globose, or amorphous, and spongy . . . . Leathesiaceae
2. Sporophytes erect, branched 3
3. Growth of axial filament trichothallic . . . . Chordariaceae
3. Growth of axial filaments apical . . . . Spermatochnaceae

Myrionemataceae

Plants small, discoid, usually epiphytic, basal filaments radiating, closely packed, with each cell able to bear either a short unbranched or weakly branched vertical photosynthetic filament, endogenous hair, or a reproductive structure; photosynthetic filaments of equal length; one to eight plastids per cell, parietal, discoid, or irregular, with single pyrenoids.

Key to the genera

1. Basal filaments uniseriate .... Myrionema
1. Basal filaments partly biseriate .... Hecatonema

Hecatonema Sauvageau 1897

Plants discoid, crustose, or pulvinate; basal filaments partly biseriate by occasional or frequent longitudinal divisions parallel to the substratum, sometimes producing rhizoids from basal layers; some cells of the basal system bear simple or branched erect filaments of different lengths, colorless hairs, ascocysts, or reproductive structures; plastids discoid, one to several per cell; plurilocular structures sessile or stalked, multiseriate; unilocular sporangia unknown in most species.

Key to the species

1. Ascocysts present .... *H. foecundum*
1. Ascocysts absent 2
2. Basal cells 12 μm diameter or larger .... *H. maculans*
2. Basal cells less than 12 μm diameter .... *H. floridanum*

Hecatonema floridanum (W. R. Taylor) W. R. Taylor 1960, p. 241.
  Phycocœlis floridana W. R. Taylor 1928, p. 109, pl. 14, fig. 19.
  Figures 147 and 148
  Plant a minute epiphyte with decumbent branches ramifying over the host or closely associated to form discs; discs becoming two cells thick, giving rise to assimilative filaments, gametangia, and (rarely) hairs; basal cells 7–11 μm diameter and 11–20 μm long; erect filaments 8–12 μm diameter, averaging 12 cells long with cells 17–25 μm long; gametangia 12–30 μm diameter and 50–80 μm long, spindle shaped, obtuse, or acute at apices.
  Growing on plastic ribbon attached to rock, offshore in Onslow Bay, 17–18 m, July.
  Distribution: North Carolina (new record), southern Florida, Brazil, ?West Africa.

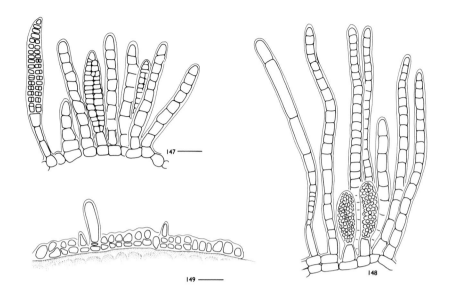

Figures 147, 148. *Hecatonema floridanum*, scale 20 μm. 147. Crust with plurilocular reproductive structures. 148. Crust with unilocular sporangia.
Figure 149. *Hecatonema foecundum*, crust with ascocyst, scale 20 μm.

Hecatonema foecundum (Strömfelt) Loiseaux 1967a, p. 338.

  Phycocelis foecunda Strömfelt 1888, p. 383.

  Figure 149

  Plants epiphytic, discoid, to 0.5–9 mm diameter, cells of basal filaments 9–18 by 5–9 μm diameter, 1–3 diameters long, with single multilobed plastids, basal filaments becoming biseriate, erect filaments of 3–9(–13) cells, to 134 μm tall, cells 6–10(–13) μm diameter, 1–2 diameters long, each with single, large, multilobed plastid with three pyrenoids; hairs frequent, with basal meristems and sheaths; ascocysts stalked or sessile, cylindrical to clavate, 8–12 μm diameter, 36–100 μm long; plurilocular sporangia sessile or on one- to three-celled stalks, subcylindrical, attenuate toward apices, uni- or biseriate, 7–12(–18) μm diameter, 17–50(–70) μm long.

  Brauner 1975; Kapraun 1984. As *Phycocelis foecunda*, Williams 1948a.

  On *Zostera marina* and coarse algae in sounds, December–July.

  Distribution: Canadian Arctic, Rhode Island, Connecticut, North Carolina, Norway to France.

  Loiseaux reported a heteromorphic sexual life history for this species (1967b). Fletcher (1987) recognized the genus *Chilonema* Sauvageau (1897) for *Hecatonema*-like plants in which the erect filaments are of similar length and are simple or rarely branched, and he transferred *Hecatonema foecundum* to that genus. We chose not to segregate such plants from *Hecatonema*, but were *Chilonema* recognized here, it would also include *H. floridanum*.

Hecatonema maculans (Collins) Sauvageau 1897, p. 256.

   Phycocelis maculans Collins 1896, p. 459, pl. 278.

   Plants epiphytic, basal discs less than 2 mm to several cm diameter and bearing tufts of filaments to 1 mm long; radiating filaments compact, cells 12–14(–17) μm diameter, to 25 μm long; erect filaments simple or branched below, 30(–60) cells long, 10–17 μm diameter, cells 1–2 diameters long, bearing colorless hairs; hairs with basal meristems and basal sheaths, 9–12(–22) μm diameter; ascocysts lacking; plurilocular structures either stalked on basal filaments or, more commonly, terminal or lateral on erect filaments, bi- or multiseriate, conical lanceolate or ovate lanceolate, 50–115(–135) μm long, 18–25 μm diameter; unilocular sporangia unknown.

   Brauner 1975; Kapraun and Zechman 1982; Kapraun 1984. As *Ectocarpus terminalis*, Williams 1948a.

   Epiphytic on sea grasses and on other algae, winter–summer.

   Distribution: Canadian Arctic to New Jersey, North Carolina, Norway to France, Indian Ocean, Australia.

   Loiseaux (1969) considered *H. maculans* a phase in the life history of plants in the genus *Myriotrichia*. A direct type of life history with the macrothallus recycled by spores from plurilocular sporangia was reported by Edelstein et al. (1971). Clayton (1974) and Pedersen (1984) indicated that this alga is indistinguishable from juvenile phases of plants traditionally assigned to *Desmotricum*, *Punctaria*, and *Asperococcus*. It is retained here because of the uncertain status of the life history of local plants.

Myrionema Greville 1827

Plants discoid, small, each cell of the uniseriate basal filaments bearing either a short assimilatory filament, ascocyst, hair, or reproductive structure; growth marginal; plurilocular structures uniseriate to biseriate.

Key to the species

1. Ascocysts present . . . . *M. magnusii*
1. Ascocysts absent . . . . *M. strangulans*

Myrionema magnusii (Sauvageau) Loiseaux 1967b, p. 338.

   Ascocyclus magnusii Sauvageau 1927, p. 13.

   Plants forming small discs 0.5(–1.5) mm diameter with peripheral cells rectangular in surface view, 3–7 μm by 6–13(–18) μm, and central, rounded cells 4–9 μm diameter, one to three plastids per cell; discs produce endogenous hairs and ascocysts; hairs to 6–10 μm diameter with a basal meristem and sheath; ascocysts slender, 8–11 μm diameter and 45–150 μm long; plurilocular sporangia uniseriate or (rarely) biseriate and 7–10 μm diameter, 30–50(–70) μm long; unilocular sporangia formed in clusters of two to three, 14–16 μm diameter, to 35 μm long; ascocysts, hairs, and plurilocular and unilocular sporangia sessile or pedicellate on one- to three-celled stalks.

Kapraun 1984. As *M. orbiculare sensu* Brauner 1975.

On *Zostera marina* in sounds, year-round.

Distribution: Newfoundland to Virginia, North Carolina, Bermuda, Norway to Spain, Mediterranean.

Loiseaux (1967a) recognized this as a species distinct from *M. orbiculare*. Parke and Dixon (1968) united the two, but Fletcher (1987) accepted Loiseaux's interpretation (see also Dixon and Russell 1964). Price et al. (1978) accepted the two-species concept and attributed the northeastern United States records of *M. orbiculare* to *M. magnusii* along with those of Bermuda, the United Kingdom, Europe, and the Mediterranean.

Loiseaux (1967b) observed two types of nondiscoid, filamentous plants in culture. Slender plants had filaments 8–10 μm diameter, one to three plastids per cell, hairs, plurilocular sporangia like those of the discs, and, rarely, ascocysts to 10 μm diameter and 30 μm long. Coarser plants had filaments 12–15 μm diameter, five to six plastids per cell, no hairs or ascocysts, and formed plurilocular sporangia 15–20 μm diameter and 100–150 μm long.

Myrionema strangulans Greville 1827, pl. 300.

Plants discoid, to 2 mm diameter, basal cells 5–8.5 μm diameter, 1–3 diameters long; erect filaments closely packed, 5–7(–9) cells (30–100 μm) tall, somewhat clavate, cells cylindrical below and subspherical above, to 6–8(–11) μm diameter; one to three discoid plastids per cell; hairs 6–8 μm diameter, with basal meristem and sheath; reproductive structures sessile on basal filaments, on one- or two-celled pedicels, or lateral on basal cells of erect filaments; plurilocular reproductive structures cylindrical, uniseriate or sometimes biseriate, 7–11 μm diameter, 15–50 μm long, sessile or on one- to two-celled stalk; unilocular sporangia ellipsoid obovoid, 30–40(–65) μm long, 18–27(–35) μm diameter.

Hoyt 1920; Taylor 1960. As *Myrionema vulgare*, Williams 1948a, 1951.

Epiphytic on *Petalonia fascia*, *Calonitophyllum medium*, and oyster shells in shallow water, spring–fall.

Distribution: Canadian Arctic to Virginia, North Carolina, southern Florida, Caribbean, Norway, British Isles to Portugal, Azores, Indian Ocean, Australia, ?Alaska to ?California, western South America.

Loiseaux (1967a) described three types of filamentous plants in the life history in addition to the typical discoid type on which the species is based. These filamentous plants included a diploid, pseudodiscoid plant with slender filaments and uniseriate plurilocular and unilocular sporangia; a coarsely filamentous diploid plant with multiseriate plurilocular sporangia; and a bushy haploid gametophyte with uniseriate plurilocular gametangia. Products of unilocular sporangia in Loiseaux's cultures functioned directly as gametes or germinated to form gametophytes.

Elachistaceae

Questionable Record

Elachista fucicola (Velley) Areschoug 1842, p. 235.

Williams's (1948a) record as *Elachista grevillei* Arnott in Harvey (1857) was tentative, and no specimens are available from his study.

Leathesiaceae

Plants with a compacted, pseudoparenchymatous, colorless base bearing a tuft of assimilatory paraphyses, or plants with spongy, globular, or amorphous shapes, with a medulla of colorless, pseudoparenchymatous medullary tissue and a cortex of smaller assimilatory paraphyses; paraphyses without a trichothallic basal meristem; colorless hairs and mucilage abundant; plastids numerous, irregular plates; unilocular sporangia borne at base of paraphyses.

Key to the genera

1. Plants subspherical, spongy, hollow with age .... Leathesia
1. Plants a tuft of photosynthetic hairs produced by a small, compact mass of endophytic tissue .... Myriactula

Leathesia S. F. Gray 1821

Sporophytes globular, subspherical, or irregular, mucilaginous; medulla composed of large colorless cells that form filamentous to pseudoparenchymatous tissue; cortex composed of smaller, pigmented, moniliform paraphyses of three to six clavate cells, the terminal cell usually enlarged; with or without clusters of colorless hairs on the cortex surface; unilocular sporangia and uniseriate plurilocular sporangia borne at base of cortical filaments; gametophytes filamentous, microscopic.

Leathesia difformis (Linnaeus) Areschoug 1847, p. 376.

Tremella difformis Linnaeus 1755, p. 429.

Figure 150

Sporophytes tan, olive, or yellowish, subspherical, convolute, hollow with

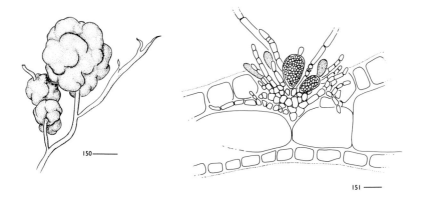

Figure 150. *Leathesia difformis*, epiphytic on *Gracilaria*, scale 1 cm.
Figure 151. *Myriactula stellulata*, endophytic in *Dictyota*, with unilocular sporangia, scale 25 µm.

age, to 12 cm diameter; medulla spongy, cortex crisp, assimilatory filaments 6–13 µm diameter; unilocular sporangia ovoid, 15–25 µm diameter, 35–50 µm long, plurilocular sporangia obtuse cylindrical, 3–16 µm diameter, five to ten cells long.

Hoyt 1920; Taylor 1960; Brauner 1975; Kapraun and Zechman 1982; Kapraun 1984.

Uncommon, on rocks, sea grasses, and coarse algae in sheltered sounds, December–April.

Distribution: Newfoundland to Virginia, North Carolina, USSR to Portugal, Azores, southern Africa, Australia, Alaska to Pacific Mexico, western South America.

Carolina plants, at the southern limit of distribution for the species in the western Atlantic, are small, to 5 cm diameter.

Myriactula Kuntze 1898
Base of plant small, colorless, pseudoparenchymatous, epiphytic or endophytic; emergent tissue bearing paraphyses, hairs, and sporangia; paraphyses gelatinous; sporangia borne at bases of paraphyses, plurilocular structures simple, uniseriate.

Myriactula stellulata (Harvey) Levring 1937, p. 57.
Conferva stellulata Harvey 1841, p. 132.
Figure 151

Plants epi-endophytic, forming erect tufts to 1 mm diameter; endophytic filaments branching between and beneath host cortical cells; small clusters of cells erupting through cortex to produce paraphyses, colorless hairs, and plurilocular and/or unilocular sporangia; paraphyses 4–12(–24) µm diameter, to 285 µm, 12(–17) cells, long; hairs to 9–11 µm diameter; plurilocular sporangia on same or different plants, uniseriate, in clusters, sessile or on short pedicels, clavate,

5–10 μm diameter, 30–50(–80) μm long; unilocular sporangia obovate, 10–18(–33) μm diameter, 25–40(–72) μm long.

As *Elachista stellulata*, Hoyt 1920; Williams 1948a, 1949.

On *Dictyota menstrualis* on exposed jetties and offshore in Onslow Bay from 20 m, summer.

Distribution: North Carolina, Norway to France.

Unilocular sporangia in North Carolina are only 10–18 μm diameter by 25–40 μm long, in contrast with dimensions of 25 μm diameter and 60 μm long reported for plants from Europe (Kuckuck 1929, as *Myriactis stellulata*; Hamel 1935, as *Gonidia stellulata*).

Chordariaceae

Sporangial plants erect, cylindrical, branched, and uniaxial or multiaxial; growth mostly trichothallic with subapical meristems near branch apices; central medullary tissue of colorless, elongate filaments with sharp or gradual transitions to anticlinal, assimilatory filaments; colorless hairs with basal meristems often present; plastids numerous, discoid; unilocular sporangia borne near bases of cortical filaments; plurilocular sporangia present in some genera, borne on cortical filaments; gametophytes (where known) small, either microscopic or macroscopic.

Cladosiphon Kützing 1843

Sporophyte axes cylindrical, branched, multiaxial, becoming pseudoparenchymatous, sometimes becoming hollow; medullary filament growth sympodial, cells cylindrical, four to eight times longer than the diameter, more or less united in a pseudoparenchyma; subcortex of smaller cells, one to three cells thick; assimilatory filaments simple or branched only near their bases; colorless hairs present; unilocular sporangia borne at the base of assimilatory filaments; plurilocular sporangia developed by division and differentiation of terminal parts of assimilatory filaments.

Cladosiphon occidentalis Kylin 1940, p. 27, pl. 3, fig. 8.

Figures 152–154

Sporophytes lax, gelatinous, 10–30 cm long, nearly simple to much branched, attached by small basal discs; branches 1–3 mm diameter, medullary cells 50–130 μm diameter, 200–360 μm long, firmly united, often forming single central hollow cavities; assimilatory filaments 100–300 μm long, basal cells cylindrical, 6–9 μm diameter, 19–28 μm long, distal parts of filaments moniliform, cells 11–15 μm diameter; colorless hairs 9–11 μm diameter, to 150 μm long; unilocular

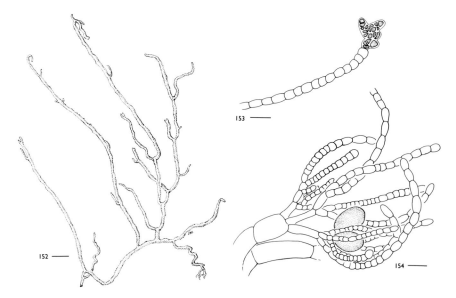

Figures 152–154. *Cladosiphon occidentalis.* 152. Habit, scale 0.5 cm. 153. Assimilatory filament with plurilocular sporangium, scale 25 μm. 154. Assimilatory filaments with unilocular sporangia, scale 25 μm.

sporangia single or in clusters of up to three, oblong ovoid, 25–40 μm diameter, 40–55 μm long; plurilocular sporangia formed by distal cells of assimilatory filaments functioning directly as sporangia or with longitudinal and transverse divisions converting each cell into two to nine locules, distal ends of filaments then becoming oppositely, alternately, or unilaterally branched; sporangial structures to 19 μm wide, 48 μm long.

Earle 1969; Brauner 1975; Kapraun 1984. As *Castagnea zosterae sensu* Hoyt 1920. As *Eudesme zosterae sensu* Taylor 1957, 1960.

Epiphytic on sea grasses and *Sargassum filipendula* in sounds, October–June.
Distribution: North Carolina, Bermuda, Gulf of Mexico, Caribbean.

See Earle (1969) for a discussion of the confusion over this species, *Cladosiphon zosterae* (J. Agardh) Kylin, and *Eudesme virescens* (Carmichael) J. Agardh.

Spermatochnaceae

Plants cylindrical, branched; medullary filaments developing from one to a few apical cells to form a medullary core with one or more central filaments surrounded by a layer of anticlinal cortical assimilatory filaments, or medullary filaments developing secondarily by downgrowth of filaments; unilocular sporangia at the base of assimilatory paraphyses; plurilocular sporangia filamentous, uniseriate, or partially biseriate; gametophytes microscopic.

Key to the genera

1. Branches uniaxial; plants soft, mucilaginous . . . . Nemacystus
1. Branches multiaxial; plant firm in texture . . . . Stilophora

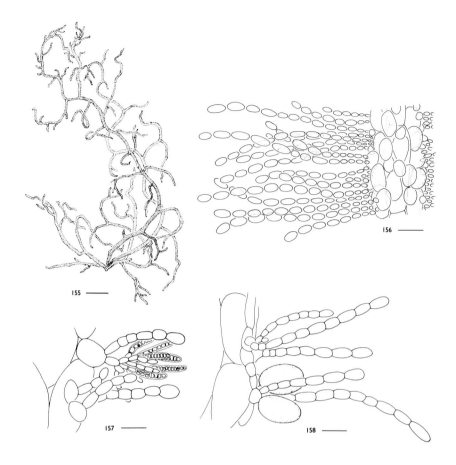

Figures 155–158. *Nemacystus howei.* 155. Habit, scale 0.5 cm. 156. Optical section showing axial cells surrounded by medullary cells and radiating assimilatory filaments, scale 20 μm. 157. Assimilatory filaments bearing plurilocular reproductive structures, scale 20 μm. 158. Assimilatory filaments bearing unilocular sporangia, scale 20 μm.

Nemacystus Derbès et Solier 1850
Sporophytes irregularly or occasionally dichotomously branched, either solid or hollow; a single axial filament with apical growth evident in younger parts; cortical filaments simple and unbranched; medulla developing from filaments growing down from the base of cortical filaments, becoming pseudoparenchymatous; unilocular and plurilocular sporangia produced among bases of assimilatory filaments.

Nemacystus howei (Taylor) Kylin 1940, p. 49.
   Castagnea howei Taylor 1928, p. 111, pl. 15, figs 1–8.
   Figures 155–158
   Plants soft, mucilaginous, in tangled masses to 40 cm or more long; main axes

to 1 mm diameter, branching alternate, abundant to at least three orders; medullary cells 30–166 μm diameter, 132–1410 μm long; uniaxial structure evident behind apices, apical cell buried among cortical filaments; cortical filaments 10–17 μm diameter near apices, 3–8 μm diameter below, eight to thirteen cells long, cylindrical below and moniliform or reniform above, containing several discoid plastids per cell; colorless hairs 8–10 μm diameter at base; unilocular sporangia spherical to ovoid, 18–37 μm diameter, 20–45 μm long, one to two (to six) at base of cortical filament fascicles or borne directly on outer medullary cells; plurilocular sporangia 6–8 μm diameter, 37–78 μm long, uniseriate or biseriate, replacing cortical filaments in fascicles, up to ten per fascicle.

Schneider and Searles 1975; Schneider 1976.

Epiphytic on *Sargassum filipendula* from rock outcroppings in Onslow Bay from 15 to 25 m, May–August, November, and from estuarine jetties adjacent to Onslow Bay, July.

Distribution: North Carolina, Bermuda, southern Florida.

North Carolina specimens are slender with main axes not exceeding 0.5 mm diameter, and assimilatory hairs contain more cells than the eight reported by Taylor (1928, 1960) and Earle (1969).

Stilophora J. G. Agardh 1841, nom. cons.
Sporophytes dichotomously to alternately branched, axes firm, cylindrical, produced by the apical growth of a small group of cells at the apex that give rise to a pseudoparenchymatous medullary core of at most four or five filaments; assimilatory filaments unbranched or, when fertile, branched and producing unilocular and plurilocular sporangia in raised sori scattered on surfaces of branches.

Stilophora rhizodes (Turner) J. Agardh 1848, p. 85.
Fucus rhizodes Turner 1815–1819, pl. 235.
Figure 159

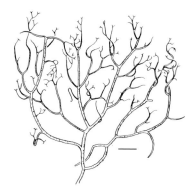

Figure 159. *Stilophora rhizodes*, scale 1 cm.

Plants erect from discoid holdfasts, often entangled, to 30 cm tall, axes to 3 mm diameter; medulla of four or five closely packed filaments; cortex firm, pseudoparenchymatous, assimilatory filaments 75–85 μm long, 3.5 μm diameter below to 9–13 μm above; unilocular sporangia formed at bases of assimilatory filaments, obovate to clavate, 36–56 μm long, 22–32 μm diameter; plurilocular sporangia uniseriate, four to ten cells, 30–50 μm long, 9–12 μm diameter.

Hoyt 1920; Williams 1948a, 1949; Taylor 1957, 1960; Brauner 1975.

In sheltered sounds, rarely on coastal jetties near Beaufort, N.C., April–May.

Distribution: Maritimes to Virginia, North Carolina, Gulf of Mexico, Norway to France, Australia.

DICTYOSIPHONALES

Sporophytes parenchymatous, macroscopic; growth diffuse, intercalary or apical; colorless hairs common; plastids numerous, discoid; unilocular and plurilocular sporangia occur together or separately; gametophytes (where known) small, filamentous.

Key to the families

1. Branches terminating in a uniseriate filament . . . . Striariaceae
1. Branches not terminating in a uniseriate filament . . . . Punctariaceae

Striariaceae

Sporophytes erect, axes cylindrical, either branched or unbranched, the axes terminating in a long, uniseriate, pigmented hair; parenchymatous below with colorless medulla and pigmented outer cortex; plastids without pyrenoids; plurilocular and unilocular sporangia formed from superficial cells.

Key to the genera

1. Sporophytes solid, simple or sparingly branched, to 5(–12) cm tall; medulla in cross section composed of four large cells; unilocular sporangia independent of paraphyses . . . . Hummia
1. Sporophytes hollow, profusely and oppositely branched, to 70 cm tall; medulla in cross section composed of many large, rounded cells; unilocular sporangia aggregated in sori with unicellular paraphyses . . . . Striaria

Hummia Fiore 1975

Sporophytes cylindrical, parenchymatous, unbranched or with a single order of branches; growth trichothallic; medulla formed by four vertical files of large, col-

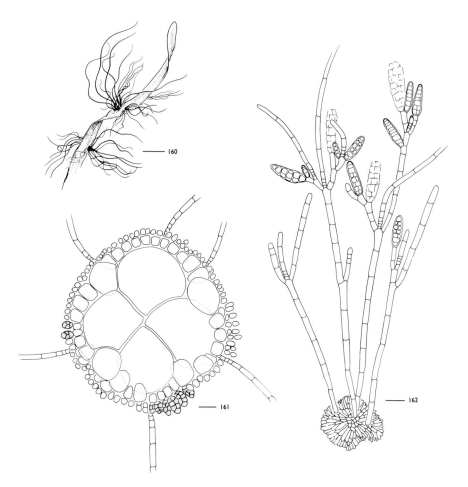

Figures 160–162. *Hummia onusta*. 160. Sporophytes on blade of *Zostera*, scale 1 cm. 161. Cross section of sporophyte, scale 10 μm. 162. Gametophyte, scale 50 μm (redrawn from Blomquist 1954).

orless cells; cortex of one to three layers of small pigmented cells, some of which bear deciduous, colorless hairs; gametophytes ectocarpoid, axes erect, uniseriate, with alternate or opposite branching; arising from discoid or scutate bases; colorless hairs terminal, growth of hairs and filaments trichothallic; plurilocular gametangia stalked; anisogamous and dioecious.

Hummia onusta (Kützing) Fiore 1975, p. 498.
    Ectocarpus onustus Kützing 1849, p. 457.
    Figures 160–162
    Sporophytes filiform, to 5(–12) cm tall, unbranched or with a few short branches at ends of the axes, 0.5–1 mm diameter; unilocular sporangia globose to ovoid, 30–56 μm diameter, 56–84 μm long; plurilocular sporangia stalked, ovate

conical, 15–25 μm diameter, 25–45 μm long; gametophytes 1–5 mm tall, erect filaments 12–15 μm diameter, cells 2–4 diameters long, gametangia cylindrical, 20–32 μm diameter, and 100–135 μm long; microgametangia nearly acute, three to five locules wide; macrogametangia obtuse, two to three locules wide.

Fiore 1975, 1977; Hall and Eiseman 1981; Kapraun 1984; Richardson 1987. As *Ectocarpus subcorymbosus*, Taylor 1957; Chapman 1971. As *Myriotrichia scutata*, Blomquist 1954. As *M. subcorymbosa*, Blomquist 1958a; Brauner 1975. As *Stictyosiphon subsimplex*, Brauner 1975.

Both sporophytes and gametophytes epiphytic on sea grasses in sounds; sporophytes year-round but rare in summer; gametophytes found April–December.

Distribution: Maritimes to Virginia, North Carolina, Georgia, northeastern Florida to Gulf of Mexico.

Striaria Greville 1828
Sporophytes attached by basal discs, axes tubular, branching alternate, opposite, or verticillate; growth trichothallic; medulla one to two layers of large, colorless cells; cortex a layer of small, pigmented cells; unilocular sporangia, hairs, and clavate unicellular paraphyses borne in annular sori giving plants a banded appearance; gametophytes filamentous, creeping; life histories diplohaplontic with fusion of cells from unilocular sporangia also reported as well as apomeiotic development of unilocular sporangia.

Striaria attenuata (C. Agardh) Greville 1828, p. 44, pl. 288.
Solenia attenuata C. Agardh 1824, p. 187.
Figure 163
Sporophytes 10–70 cm tall, branches numerous and tapered at both ends; medullary cells rounded, 50–80 μm diameter; cortical cells quadrangular, 15–32 μm diameter; unilocular sporangia ovoid, 30–60 μm diameter.

Fiore 1969; Brauner 1975; Kapraun 1984.

In lower intertidal and shallow subtidal in sounds and tidal creeks, January–April.

Distribution: Newfoundland to Virginia, North Carolina, USSR to Spain, Mediterranean, Australia, New Zealand, Japan.

Punctariaceae

Sporophytes unbranched and bladelike, saccate, or cylindrical; meristems diffuse or intercalary; plastids either discoid and numerous or single and lobed, most with pyrenoids; colorless hairs usually abundant; unilocular and plurilocular sporangia formed by transformation of vegetative cells; plurilocular

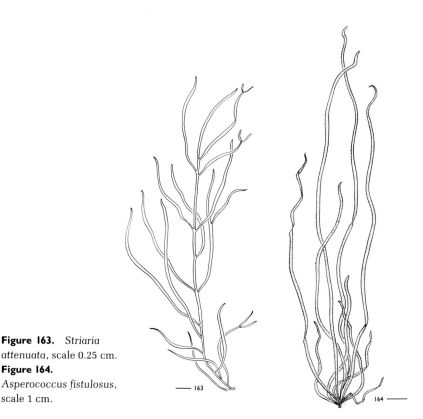

Figure 163. *Striaria attenuata*, scale 0.25 cm.
Figure 164. *Asperococcus fistulosus*, scale 1 cm.

— 163

164 —

sporangia, if present, occurring on same plants as unilocular sporangia; where known, gametophytes filamentous or discoid; gametangia uniseriate.

Key to the genera

1. Plants tubular .... Asperococcus
1. Plants not tubular .... Punctaria

Asperococcus Lamouroux 1813
Sporophytes tubular, attached by basal discs; inner cortex large celled, cells smaller toward the exterior; pigmented cells with many plastids; colorless hairs in tufts; unilocular and plurilocular sporangia interspersed with multicellular paraphyses in sori; gametophytes minute and filamentous, discoid plants bearing erect filaments with plurilocular gametangia.

Asperococcus fistulosus (Hudson) Hooker 1833, p. 277.
 Ulva fistulosa Hudson 1778, p. 569.
 Figure 164
 Sporophytes to 50 cm tall, 2–5(–20) mm diameter, tapering at the base to a narrow stipe; surface cells elongate, 8–26 μm wide, 10–36(–44) μm long, 9–22 μm in cross section; sori elongate, scattered; paraphyses three to several cells long; unilocular sporangia ovoid to globose, sessile or on one-celled stalks, 38–55(–70) μm diameter, 30–70(–85) μm long.
 As *Asperococcus echinatus*, Williams 1948a, 1949.

Rare on rocks and shells in sounds and on ocean jetties in Beaufort, N.C., area; December–April.

Distribution: Canadian Arctic to Virginia, North Carolina, Norway to Spain, Australia.

Punctaria Greville 1830

Sporophytes with one or more stalked blades arising from discoid bases, blades linear to elliptical, two to seven cells thick, inner cells slightly larger than surface cells, marginal cells and/or surface cells bearing tufts of hairs; plastids discoid, a few to several per cell, with pyrenoids; unilocular sporangia scattered, single; plurilocular sporangia multiseriate, clustered or single, angular, apex projecting; gametophytes small discs or branched filaments; gametes isogamous.

Key to the species

1. Hairs restricted to surface of blades .... *P. latifolia*
1. Hairs on margins of blades as well as surfaces.............. *P. tenuissima*

Punctaria latifolia Greville 1830, p. 52.

Figure 165

Sporophytes with blades 10–45 cm long, 2–8(–15) cm wide, lanceolate to oblong, stalks 2–5 mm long, bases of blades cuneate, margins flat or crispate, apices obtuse, color yellowish to olive brown, two to four (up to six) cells thick, totaling 50–120(–160) μm thick; cells in surface view quadrate to rectangular, 15–40 μm across; hairs scattered in patches on blade surface, lacking on margins and apices; plurilocular and unilocular sporangia on the same or different plants;

Figure 165. *Punctaria latifolia*, scale 0.5 cm.

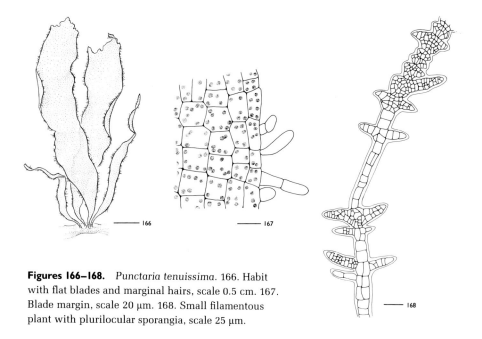

Figures 166–168. *Punctaria tenuissima*. 166. Habit
with flat blades and marginal hairs, scale 0.5 cm. 167.
Blade margin, scale 20 μm. 168. Small filamentous
plant with plurilocular sporangia, scale 25 μm.

plurilocular sporangia square to rectangular in surface view, 14–42(–59) μm by
12–26(–42) μm, in cross section square or, more commonly, conical; unilocular
sporangia rare, orbicular to ovoid, 40–50 by 36–43 μm in surface view; gametophytes comparable to those described as *Hecatonema maculans* (p. 132).

Williams 1948a, 1949; Taylor 1960; Brauner 1975; Kapraun 1984. As *Punctaria
plantaginea*, Humm 1979.

Epiphytic and on rocks, shells, and seawalls in the sounds, January–April.

Distribution: Canadian Arctic to Virginia, North Carolina, Portugal, Australia.

Punctaria tenuissima (C. Agardh) Greville 1830, p. 54.

Zonaria tenuissima C. Agardh 1824, p. 268.

Figures 166–168

Sporophytes with light brown blades, base and apices tapered, linear lanceolate, 2–8(–20) cm long, 2–10(–12) mm wide, membranous, thin, entire or with
small serrations, 22–42(–66) μm thick, one to two (up to four) cells thick; in
surface view, cells 10–27 μm wide, 7–18(–48) μm long, with several discoid
plastids; in plants four cells thick, inner cells colorless, longitudinally elongate, transversely rounded; marginal and surface cells bearing long hairs, (5–)
10–12 μm diameter, with basal meristems and sheaths, hairs later deciduous;
unilocular sporangia to 40 μm diameter, solitary or in clusters of two to four;
plurilocular sporangia single or scattered, often marginal, bases immersed, conical tips projecting, 20–27 μm diameter, 24–36 μm long; gametophytes comparable to *Hecatonema maculans* (p. 132).

As *Desmotrichum undulatum*, Brauner 1975; Kapraun and Zechman 1982;
Kapraun 1984. As *D. balticum*, Brauner 1975.

Epiphytic on sea grasses and on rocks and shells in sounds, November–April. Distribution: Canadian Arctic to Virginia, North Carolina, USSR to Portugal.

When Fletcher (1987) merged *Desmotrichum balticum* (Kützing) Batters and *D. undulatum* (J. Agardh) Reinke into *Punctaria tenuissima*, he noted that these species may in turn be developmental stages of *P. latifolia*.

SCYTOSIPHONALES

Plants bladelike, ribbonlike, cylindrical and hollow or solid, or saccate; growth diffuse or intercalary; plastids single, large, parietal with single pyrenoids; hairs frequent; plurilocular reproductive structures uniseriate or biseriate.

Scytosiphonaceae

Plants bladelike, ribbonlike, tubular, or saccate; plurilocular reproductive structures numerous, borne on erect, macroscopic plants; upright plants alternate with crustose, *Ralfsia*-like plants or with prostrate filaments; unilocular sporangia rare, borne on crusts.

Key to the genera

1. Plants tubular or hollow 2
1. Plants solid . . . . Petalonia
2. Plants branched . . . . Rosenvingea
2. Plants simple, sometimes several from a common base 3
3. Plants saccate . . . . Colpomenia
3. Plants cylindrical . . . . Scytosiphon

Colpomenia (Endichler) Derbès et Solier in Castagne 1851
Plants saccate, globose, or irregularly lobed; initially solid, becoming hollow; walls with large, colorless cells inside grading into small, pigmented surface cells; colorless hairs in clusters, associated with sori of plurilocular reproductive structures and unicellular, colorless paraphyses.

Colpomenia sinuosa (Mertens ex Roth) Derbès et Solier in Castagne 1851, p. 95.
　Ulva sinuosa Roth 1806, p. 327, pl. 12.
　Figures 169–171
　Plants globose, becoming papillate or lobed, to 12(–15) cm diameter; wall 0.3–0.4 mm thick, inner cells to 180 μm diameter, surface cells angular, 4–8(–16) μm across in surface view; plurilocular structures 4–8 μm diameter, 18–30 μm long, biseriate; paraphyses obovoid, to 11 μm diameter, 47 μm tall.

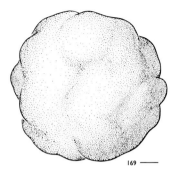

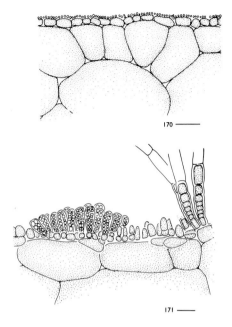

Figures 169–171. *Colpomenia sinuosa.* 169. Habit, scale 1 cm. 170. Partial cross section of vegetative plant, scale 50 μm. 171. Partial cross section with plurilocular structures and trichothallic hairs, scale 10 μm.

Blomquist and Humm 1946; Williams 1948a, 1951; Taylor 1960; Schneider 1976; Searles 1987, 1988.

On rocks and coarse algae; rare on coastal jetties, frequent in Onslow Bay from 15 to 45 m, May–September.

Distribution: North Carolina, Georgia, Bermuda, southern Florida, Gulf of Mexico, Caribbean, Brazil, Norway to Portugal; elsewhere, widespread in tropical and temperate seas.

Petalonia Derbès et Solier 1850, nom. cons.

Erect plants with one or more stalked blades arising from discoid holdfasts; medulla composed of large, colorless cells; cortex cells smaller, pigmented, bearing tufts of hairs; plurilocular structures in fused patches ultimately covering both blade surfaces; crustose plants with consolidated hypothallium and epithallium; unilocular sporangia rare, formed at bases of epithallial filaments.

Petalonia fascia (O. F. Müller) Kuntze 1898, p. 419.

Fucus fascia O. F. Müller 1778, pl. 768.

Figure 172

Blades to 45 cm long, 30 cm wide, linear lanceolate with tapered apices and cuneate, asymmetrical bases, 145–200 μm thick, to 270 μm thick in fertile areas; plurilocular structures uniseriate, six to eight cells long.

Hoyt 1920; Kapraun and Zechman 1982; Taylor 1960; Kapraun 1984; Richardson 1986, 1987. As *Ilea fascia*, Williams 1948a, 1949; Humm 1952.

Blade phase common on rocks and shells in shallow water along the coast and in the sounds, December–May.

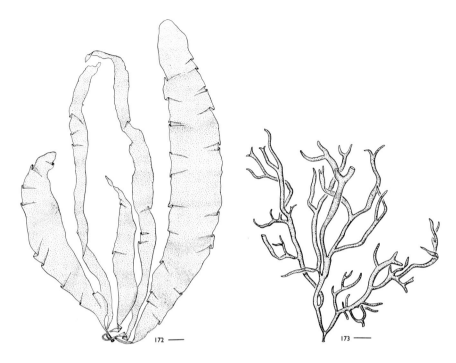

Figure 172. *Petalonia fascia*, scale 1 cm.
Figure 173. *Rosenvingea orientalis*, scale 1 cm.

Distribution: Canadian Arctic to Virginia, North Carolina, South Carolina, Georgia, northeastern Florida, Gulf of Mexico, Norway to Portugal, Azores, southern Africa, Australia, Pacific Mexico.

Rosenvingea Børgesen 1914
Plants tubular and cylindrical or compressed, subsimple to bushy, branching alternate to subdichotomous, attached by basal discs; walls of branches formed of three to four layers of cells, outer cells small and pigmented, inner cells larger and colorless, rhizoidal cells sometimes in central cavities near bases; surfaces bearing hairs singly or in clusters; plurilocular reproductive structures formed in sori by surface cells.

Rosenvingea orientalis (J. Agardh) Børgesen 1914, p. 26.
    Asperococcus orientalis J. Agardh 1848, p. 78.
    Figure 173
    Plants in tufts, to 40 cm tall, cylindrical or sometimes compressed, stipitate; branching dichotomous, subdichotomous, or irregularly alternate; branches slender, to 2 mm diameter, not changing diameter greatly from lower to upper axes, surfaces uneven, tips obtuse; walls of branches three to four cells thick; hairs scattered, 8–9 μm diameter; gametangia in rounded or oval sori, multiseriate, cylindrical to clavate, 5–12 μm diameter, 20–40 μm long or more.
    Hoyt 1920. As *Rosenvingea sanctae-crucis sensu* Taylor 1960; Brauner 1975; Kapraun 1984.

In sounds on jetties and on rocks and shells in quiet bays, August–November.
Distribution: North Carolina, southern Florida, Caribbean, India.

Earle (1969) suggested that *Rosenvingea sanctae-crucis* Børgesen may be synonymous with *R. orientalis* but retained them as separate species and indicated that Hoyt's plants were *orientalis*.

Questionable Record

Rosenvingea intricata (J. Agardh) Børgesen 1914, p. 26.

The identity of plants from the region referred to this species remains in question. Williams's (1951) specimens cannot be located, and Searles's (1987) single collection is an immature plant.

Scytosiphon C. Agardh 1820, nom. cons.
Erect plants forming unbranched, tubular axes from basal discs; growth intercalary near the base; walls of cylinder grading from large, colorless inner cells to small, pigmented outer cells; plurilocular reproductive structures interspersed with single-celled paraphyses.

Scytosiphon lomentaria (Lyngbye) Link 1833, p. 233.
  Chorda lomentaria Lyngbye 1819, p. 74. pl. 18.
  Figures 174 and 175
  Erect axes 15–70 cm long, cylindrical, 5–10 (–30) mm diameter, diameter tapering and sometimes periodically constricted; plurilocular reproductive structures 3–8(–10) μm diameter, to 65 μm long, mostly uniseriate, occasionally biseriate or forked, mixed with paraphyses in sori covering the axis.

S. lomentaria var. *complanatus* Rosenvinge 1893, p. 863.
  Plants slender, axes 1.5–2(–5) mm diameter, 0.5–1 m long, hollow but not constricted, and lacking paraphyses in the sori.

Figure 174. *Scytosiphon lomentaria.* Habit, scale 1 cm.

Taylor 1957, 1960; Kapraun 1984; Richardson 1986, 1987. As var. *complanatus*, Blomquist and Humm 1946; Williams 1948a.

The variety is found intermixed with typical plants, and they are common on coastal jetties and in sounds near inlets in the intertidal and shallow subtidal, December–April.

Distribution: Canadian Arctic to Virginia, North Carolina, South Carolina, Georgia, Bermuda, southern Florida, Brazil, USSR to Portugal, Azores, southern Africa, Australia, Japan, Alaska to Mexico, western South America.

SPHACELARIALES

Plants filamentous, branched, heterotrichous; growth apical, subapical cells not dividing, dividing only longitudinally, or dividing once transversely and then longitudinally to produce parenchymatous tissue; the axes in some genera corticated by downgrowth of rhizoidal filaments to produce a pseudoparenchymatous tissue outside the parenchyma; plastids numerous and discoid, pyrenoids reported for zoospores but not for vegetative cells; sporophytes and gametophytes isomorphic or slightly heteromorphic; unilocular and plurilocular structures on separate plants where known; vegetative reproduction by unicellular or multicellular, deciduous propagules.

Key to the families

1. Subapical cells not dividing transversely .... Choristocarpaceae
1. Subapical cells dividing transversely .... Sphacelariaceae

Figure 175. *Scytosiphon lomentaria.* Cross section with plurilocular structures and hairs, scale 20 μm (redrawn from Blomquist and Humm 1946).

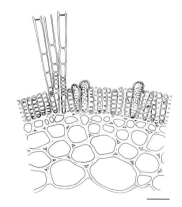

Choristocarpaceae

Filaments uniseriate or subapical, dividing longitudinally but not transversely; walls do not blacken when treated with bleaching solution.

Prud'homme van Reine (1982) did not consider the Choristocarpaceae a member of the Sphacelariales, whereas Henry (1987b) argued for its inclusion in the Sphacelariales when he transferred *Onslowia* to the Choristocarpaceae.

Onslowia Searles in Searles et Leister 1980
Plants endophytic; asexual reproduction by propagules with several, usually four, apical cells.

Onslowia endophytica Searles in Searles et Leister 1980, p. 38, figs 6–15.
Figures 176 and 177
Plants filamentous; filaments uniseriate, occasionally biseriate, 8–11 µm diameter; branching irregularly alternate; emergent filaments forming either multicellular, colorless hairs with basal trichothallic meristems, or stalked, globose cells which divide longitudinally to form four cells, each of which enlarges and then cuts off a single lens-shaped apical cell from its outer face to form rectangular cushion-shaped propagules; unilocular sporangia ovoid, 30–45 µm diameter, 45–60 µm long, sessile or stalked.
Searles and Leister 1980; Searles 1987, 1988.
Growing in *Halymenia floridana*, *Halymenia trigona*, and *Soliera filiformis* in

Figures 176, 177.
Onslowia endophytica, scale 10 µm. 176. Immature propagule on endophytic filament. 177. Lateral view of mature propagule.

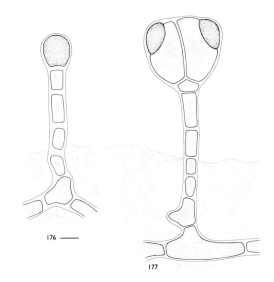

176 ——

177

Onslow Bay and at Gray's Reef from 17 to 32 m, July–September.

Distribution: North Carolina (type locality), Georgia, southern Florida.

Henry (1987b) observed propagules with only two or three apical cells in culture and zoospores from unilocular sporangia developing into additional generations of unilocular-bearing plants.

Sphacelariaceae

Subapical cells dividing once transversely before forming parenchyma by longitudinal divisions; all lateral branches formed from cells below the apical cell; hairs initiated from lenticular cells cut off by apical cells, or less frequently from cells below apical cells; walls blacken when treated with bleaching solution.

Sphacelaria Lyngbye 1819

Plants erect, forming tufts or mats; axes cylindrical, parenchymatous, a few corticated by descending rhizoids; vegetative reproduction by biradiate or triradiate propagules in some species; sexual reproduction isogamous or anisogamous.

Key to the species

1. Propagule arms slender, several times longer than broad .... S. *rigidula*
1. Propagule arms short, propagule broadly triangular in outline ....
.... S. *tribuloides*

Sphacelaria rigidula Kützing 1843, p. 292.

Figures 178–180

Axes erect, 0.5–3 cm tall, basal systems creeping and stoloniferous or penetrating, endophytic; segments as long as or somewhat longer than broad; hairs present, 12–16 μm diameter; propagules slender, biradiate or triradiate, stalks to 24 μm diameter, stalk and arms of similar length, span of arms to 450 μm; unilocular sporangia spherical, 50–70 μm diameter; gametangia dimorphic, one cylindrical, 45–65 μm long, 24–28 μm diameter, with cells about 3 μm diameter, the other more irregularly shaped, 30–60 μm long, 28–40 μm diameter, with cells about 6 μm diameter.

As S. *furcigera*, Blomquist and Humm 1946; Williams 1948a, 1949, 1951; Taylor 1960; Earle and Humm 1964; Brauner 1975; Kapraun and Zechman 1982; Kapraun 1984.

Epi-endophytic on sea grasses and on rock in the lower intertidal and shallow subtidal in sounds and on coastal jetties, year-round.

Distribution: Newfoundland to Virginia, North Carolina, Bermuda, southern

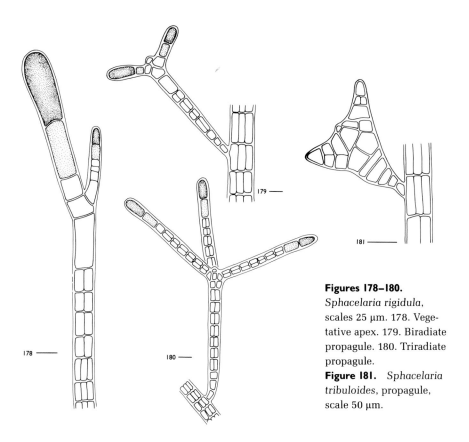

Figures 178–180.
Sphacelaria rigidula,
scales 25 µm. 178. Vege-
tative apex. 179. Biradiate
propagule. 180. Triradiate
propagule.
Figure 181. *Sphacelaria
tribuloides,* propagule,
scale 50 µm.

Florida, Gulf of Mexico, Caribbean, Brazil, Norway to Portugal; elsewhere, widespread in temperate and tropical seas.

Propagules are usually described as biradiate in this species, but in North
Carolina they are commonly triradiate. The generations have been shown to be
slightly heteromorphic with reproduction regulated by temperature and photoperiod changes (van den Hoek and Flinterman 1968; Colijn and van den Hoek
1971).

Sphacelaria tribuloides Meneghini 1840a, p. [2].

Figure 181

Plants tufted, attached by short stolons, erect axes to 1 cm tall, bearing radially disposed branches; segments 25–60 µm diameter, length equal to or slightly
greater than diameter; hairs often present, 10–15 µm diameter; propagules pedicellate, biradiate, broadly triangular in lateral view, 200 µm long, arms barely
extended, span 140–165 µm.

Blomquist and Humm 1946; Williams 1948a, 1951; Taylor 1960; Schneider
1976; Kapraun 1984.

Infrequently collected, usually intermixed with other algae in lower intertidal
and subtidal in sounds and rarely offshore in Onslow Bay from 18 to 19 m, June–
August.

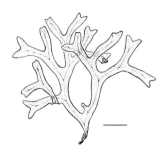

Figure 182. *Dictyopteris delicatula*, scale 1 cm.

Distribution: North Carolina, Bermuda, southern Florida, Gulf of Mexico, Caribbean, Brazil, Baltic to Portugal; elsewhere, widespread in temperate and tropical seas.

DICTYOTALES

Plants parenchymatous, bladelike; attached by rhizoidal filaments; growth from single apical cells, apical groups of cells, or marginal rows of cells; structure with inner layer(s) of large colorless cells covered by one to a few layers of smaller pigmented cells; plastids discoid and numerous; colorless hairs in bands or tufts; sporophytes and gametophytes isomorphic; unilocular sporangia scattered or in sori; meiospores nonmotile, four or eight per sporangium; oogamous, dioecious, oogonia and spermatangia in sori, one egg per oogonium, spermatangia plurilocular; mitospores unknown.

There is a single family in the order.

Dictyotaceae

Key to the genera

1. Growth by a marginal row or cluster of meristematic cells 2
1. Growth by a single apical cell at the branch tip . . . . Dictyota
2. Growth by marginal row of meristematic cells; blades fanlike 3
2. Growth by cluster of meristematic cells; blades strap shaped 5
3. Margin of blade flat 4
3. Margin of blade curled . . . . Padina
4. Plants erect, costate, becoming stalked by loss of blade tissue around costae of lower parts . . . . Zonaria
4. Plants creeping or erect, costae lacking . . . . Lobophora
5. Plants with midrib . . . . Dictyopteris
5. Plants without midrib . . . . Spatoglossum

Dictyopteris Lamouroux 1809b, nom. cons.
Plants dichotomously branched, growth from single clusters of meristematic cells at branch tips behind which midribs are prominent in straplike branches; colorless hairs in clusters on blade surfaces; reproductive cells in sori.

Key to the species

1. Blade width exceeding 5 mm 2
1. Blade to 5 mm wide .... D. delicatula
2. Blades with fine veins, margins dentate .... D. hoytii
2. Blades lacking veins, margins entire .... D. membranacea

Dictyopteris delicatula Lamouroux 1809d, p. 332, pl. 6, fig. 2B.

Figure 182

Plants light brown, erect, to 8 cm tall, or decumbent to some degree and spreading indefinitely; tangled branches attached to substratum and each other; branching irregular or dichotomous, segments of blades to 5 mm broad; blades two cells thick, thicker along margins and midrib, cells of blades in parallel rows at acute angles to midribs; hairs in clusters on only one surface of blades; reproductive cells in sori in single rows along the midrib.

Schneider 1975a, 1976.

Rare, on rock outcrops in Onslow Bay and Long Bay from 35 to 40 m, June–July.

Distribution: North Carolina, South Carolina, southern Florida, Gulf of Mexico, Caribbean, Brazil, West Africa, southern Africa, Indian Ocean, Central Pacific islands.

Dictyopteris hoytii Taylor 1960, pp. 229, 631, pl. 32, fig. 1.

Figure 183

Plants bushy, yellowish brown, to 45 cm tall, anchored by a feltlike cushion of rhizoidal filaments; simple to frequently branched lower axes denuded for 1–6 cm; blades above usually dichotomous to two to six orders, internodes to 8 cm long, sinuses narrow, apices obtuse to retuse; blades strap shaped, 1.4–4 cm wide, 90–150 μm thick, undulate, the margins straight or obscurely crenate, conspicuously aculeate dentate, teeth to 1 mm long, at intervals of 1–2 mm; structurally with two to four layers of large medullary cells and a cortical layer of smaller cells on each face; midribs conspicuous, extending to the branch apices; lateral veins delicate but evident, alternate or opposite at intervals of 2–3 mm, extending obliquely to the margins, composed of small, thick-walled cells in irregular tiers of five to six, midribs to 185 μm thick; hairs in small groups scattered over the blade; tetrasporangia in small sori in lines parallel to the lateral veins; oogonia 65–130 μm diameter, scattered over the blade singly or in clusters of two or three, except for a narrow marginal zone.

Taylor 1960; Schneider 1976; Kapraun 1984; Searles 1987, 1988. As *D. serrata* sensu Hoyt 1920; Blomquist and Pyron 1943; Williams 1951.

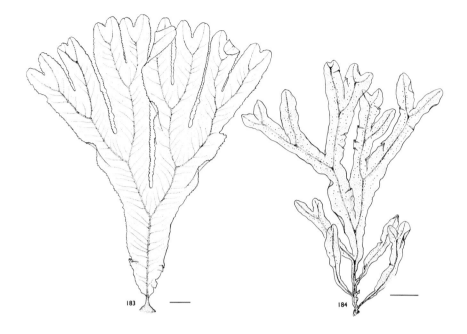

Figure 183. *Dictyopteris hoytii*, scale 2 cm.
Figure 184. *Dictyopteris membranacea*, scale 2 cm.

Common offshore from 14 to 47 m, March–December.

Distribution: North Carolina (type locality), South Carolina, Georgia, southern Florida, Venezuela.

Dictyopteris membranacea (Stackhouse) Batters 1902, p. 54.

Fucus membranaceus Stackhouse 1816, p. 13.

Figure 184

Plants erect, bushy, 10–30 cm tall, irregularly dichotomously branched at intervals of 1–5(–10) cm, sometimes proliferous from the stipe or midribs; segments entire, 5–15 mm broad; midribs precurrent, denuded below to form stipes; lateral veins absent; blades two cells thick except at margins and midribs; sporangia in irregular bandlike sori on each side of midribs; gametangia in larger but less compact groups occupying most of the blade.

Taylor 1960; Schneider 1976; Kapraun 1984. As *D. polypodioides*, Hoyt 1920; Blomquist and Pyron 1943; Williams 1948a, 1949, 1951.

Subtidal in sounds (rarely) and on coastal jetties, and common offshore in Onslow Bay from 14 to 40 m, May–November.

Distribution: North Carolina, South Carolina, southern Florida, Caribbean, Brazil, Pacific Mexico.

Dictyopteris justii Lamouroux 1809d, p. 332, pl. 6, fig. A.

Blair and Hall (1981).

Specimens of this collection have been lost.

Dictyota Lamouroux 1809a, nom. cons.

Plants usually dichotomously branched, with or without marginal proliferations, blades ribbonlike, growth from single lens-shaped apical cells; medulla
one cell thick, cortex a single layer of much smaller cells; colorless hairs in clusters on blades' surfaces; dioecious, antheridia in patches with sterile margins,
oogonia in patches without sterile margins, both exserted; unilocular sporangia
in sori or scattered, four aplanospores per sporangium.

Key to the species

1. Margins of blades aculeate dentate . . . . *D. ciliolata*
1. Margins of blades entire 2
2. Ultimate branches cervicorn to irregular . . . . *D. cervicornis*
2. Branching dichotomous throughout 3
3. Branches to 2 mm wide . . . . *D. pulchella*
3. Branches greater than 2 mm wide . . . . *D. menstrualis*

Dictyota cervicornis Kützing 1859, p. 11, pl. 24, fig. 2.

Figure 185

Plants to 20 cm tall, characterized by asymmetrical, subdichotomous branching, one dichotomy short and spurlike; segments slender, 1–2.5 mm broad,
internodes 1–3.5 cm long; often with proliferations from blade surfaces.

Williams 1951; Taylor 1960; Schneider 1976.

On coastal jetties and uncommon offshore in Onslow Bay from 21 to 40 m,
June–September.

Distribution: North Carolina, Bermuda, southern Florida, Caribbean, Azores,
West Africa, Central Pacific islands.

Dictyota ciliolata Kützing 1859, p. 12, pl. 27, fig. 1.

Figure 186

Plants to 15 cm tall, regularly dichotomously branched, sometimes spirally
twisted, angle of branching narrow, acute, or rounded; internodes to 12 mm
broad below dichotomies, but narrowing to 7 mm just above each fork; internodes 1–5 cm long, terminal segments generally tapering and subacute; margins
sparingly to regularly aculeate dentate, projections typically ascending, broad

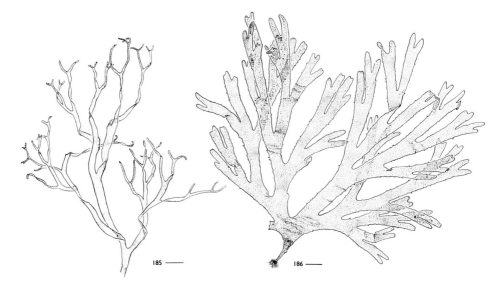

Figure 185. *Dictyota cervicornis*, scale 0.5 cm.
Figure 186. *Dictyota ciliolata*, scale 1 cm.

at bases and cylindrical toward apices, margins also bearing proliferous blades; sporangia in sori.

Schneider and Searles 1973; Schneider 1976.

On jetties in the Beaufort, N.C., area, April–December; common offshore in Onslow Bay from 15 to 50 m, June–August.

Distribution: North Carolina, Bermuda, southern Florida, Gulf of Mexico, Caribbean, Brazil, West Africa, Central Pacific islands.

Earle (1969) questioned whether this is a species distinct from *D. menstrualis* (as *D. dichotoma*) because the characteristic of marginal teeth is highly variable. *Dictyota menstrualis* does not, however, have sporangia in sori.

Dictyota menstrualis (Hoyt) Schnetter, Hörnig, et Weber-Peukert 1987, p. 195, figs 5–6.

Dictyota dichotoma var. *menstrualis* Hoyt 1927, p. 616.

Figures 187–191

Plants to 29 cm tall, branching regularly dichotomous at angles of 15°–45°, blades entire, but margins often proliferous, occasionally twisted; internodes 0.5–2 cm broad, progressively narrower in younger internodes, 1–2.5 cm long; medulla single layered above, becoming multilayered in basal and prostrate regions; sporangia scattered, one (to three), not in sori; producing gametes once per month, at spring tides.

Schnetter et al. 1987. As *D. dichotoma* var. *menstrualis*, Hoyt 1927; Taylor 1960; Brauner 1975; Kapraun and Zechman 1982; Kapraun 1984; Searles 1987, 1988. As *D. dichotoma* sensu Harvey 1852; Johnson 1900; Hoyt 1920; Williams 1948a, 1951; Blomquist and Pyron 1943; Schneider 1976; Richardson 1979; Kapraun and Zechman 1982; Kapraun 1984.

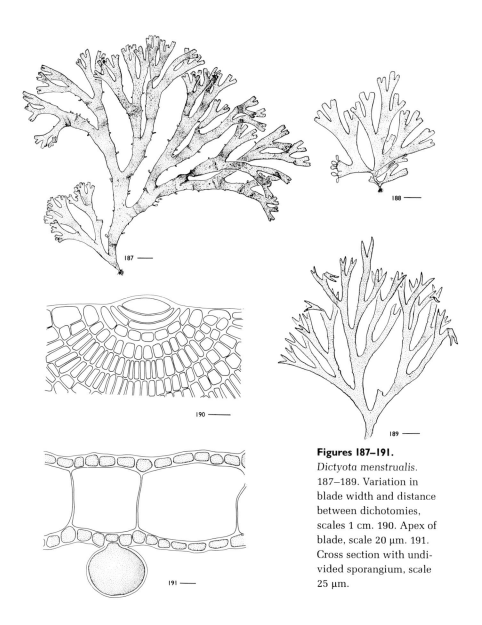

Figures 187–191.
Dictyota menstrualis.
187–189. Variation in
blade width and distance
between dichotomies,
scales 1 cm. 190. Apex of
blade, scale 20 μm. 191.
Cross section with undi-
vided sporangium, scale
25 μm.

In shallow subtidal on coastal jetties and in the sounds, April–December, common offshore from 15 to 63 m, May–October.

Distribution: Virginia, North Carolina (type locality), South Carolina, Georgia,
southern Florida, Gulf of Mexico, Caribbean, Brazil.

Schnetter et al. (1987) recently demonstrated that the western Atlantic plants
previously referred to *Dictyota dichotoma* (Hudson) Lamouroux are distinct
from those in its type locality in the eastern Atlantic in anatomy and chromosome number as well as in the reproductive periodicity (as demonstrated by

Hoyt in 1927). Richardson (1979) showed that plants in North Carolina persist through the cold months as simple germlings without a clearly organized structure.

Dictyota pulchella Hörnig et Schnetter 1988, p. 285, fig. 7.
 Figure 192
 Plants to 7 cm long, spreading or erect, iridescent or brown, branching regularly dichotomous, angle of branches 90°–120°, width of branches to 3 mm below, often abruptly narrowing to filamentous, 0.1–0.2 mm wide in the terminal branches.
 Uncommon offshore in Onslow Bay from 30 to 32 m, September.
 Distribution: North Carolina (new record), Bermuda, southern Florida, Gulf of Mexico, Caribbean, Brazil, West Africa, Indian Ocean, Central Pacific islands, Pacific Mexico.
 This is a northern extension of the range of the alga that was known as *Dictyota divaricata* Lamouroux (1809a).

Lobophora J. Agardh 1894
Plants with fan-shaped blades growing in clusters from matted, rhizoidal bases; growth from marginal rows of initials, margins of blades flat; blades with large-celled medulla grading into smaller-celled subcortex and cortex; sporangia or gametangia in scattered sori on both surfaces of blades; sporangial sori lacking paraphyses; sporangia forming four or eight aplanospores.

Lobophora variegata (Lamouroux) Womersley 1967, p. 221.
 Dictyota variegata Lamouroux 1809a, p. 40.
 Figure 193
 Blades spreading horizontally and closely appressed to substratum, or erect and forming crisp clumps, to 8 cm long, light to dark brown, fan shaped, margins curved to rounded, eight to nine cells thick (300 μm); cells large and colorless internally, and smaller with increasing concentrations of plastids toward the surface, eight surface cells per central medullary cell on one face, four on the other.
 Schneider 1976; Kapraun 1984. As *Zonaria variegata*, Hoyt 1920; Blomquist and Pyron 1943. As *Pocockiella variegata*, Williams 1951; Taylor 1960.
 Common offshore in Onslow Bay and Long Bay from 14 to 48 m, year-round.
 Distribution: North Carolina, South Carolina, southern Florida, Caribbean, Brazil; elsewhere, widespread in temperate and tropical seas.
 In North Carolina only horizontally oriented blades are formed. The erect,

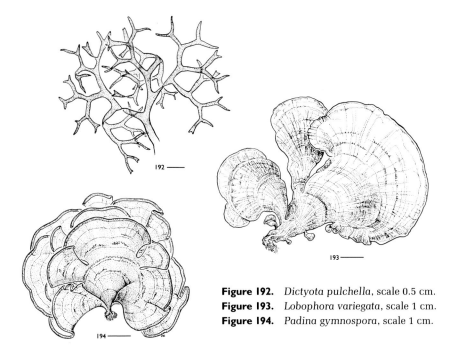

Figure 192. *Dictyota pulchella*, scale 0.5 cm.
Figure 193. *Lobophora variegata*, scale 1 cm.
Figure 194. *Padina gymnospora*, scale 1 cm.

tightly bunched, clump-forming morphology observed in the Caribbean does not appear to occur in the region.

Padina Adanson 1763, nom. cons.
Plants with fan-shaped blades growing in clusters from matted, rhizoidal bases; growth from marginal rows of initials protected by involuted curling of the blade margin; blades split perpendicular to the blade margins, surface of blades calcified in some species; hairs formed in distinct bands parallel to the margins; reproductive cells in linear sori between bands of hairs; generally dioecious, oogonial sori surrounded by sterile cells, as are antheridial sori in some species.

Key to the species

1. Blades to 15 cm tall, bands of hairs conspicuous, centrally four to six cells thick, base six to eight cells thick .... *P. gymnospora*
1. Blades to 25 cm tall, bands of hairs faint, centrally three to four cells thick, base of blade three to six cells thick .... *P. profunda*

Padina gymnospora (Kützing) Sonder 1871, p. 47.
   Zonaria gymnospora Kützing 1859, p. 29, pl. 71, fig. 2.
   Figure 194
   Plants stipitate, 4–22 cm tall, 5–37 cm broad, entire when young, becoming more or less deeply laciniate, the segments cuneate spatulate to fan shaped, sometimes lightly encrusted with lime; zonate, hairlines 1.5–8 mm apart, often inconspicuous in older parts; blades two to three cells thick (50 μm) near margins, four cells thick elsewhere, or six to eight cells thick (150–220 μm) in the

bases; reproductive structures in one or two bands midway between hairlines on both surfaces, but primarily on ventral surface; antheridia naked, in bands 200 μm wide; oogonia 30–65 μm diameter, in bands with a thin, evanescent covering of sterile cells; tetrasporangia in bands 0.5 mm wide, also with sterile coverings.

As *P. vickersiae*, Hoyt 1920; Williams 1948a, 1949, 1951; Taylor 1960; Chapman 1971; Schneider 1976; Kapraun and Zechman 1982; Kapraun 1984; Richardson 1987. As *P. pavonia sensu* Curtis 1867; Johnson 1900.

In shallow subtidal on jetties and seawalls and in Onslow Bay from 14 to 25 m, May–December.

Distribution: North Carolina, Georgia, Bermuda, southern Florida, Gulf of Mexico, Caribbean, Brazil; elsewhere, widespread in tropical seas.

Padina profunda Earle 1969, p. 167, figs 62–68.

Figure 195

Plants 10–25 cm tall, blades entire or split, not calcified, two cells thick near margins, three to four cells thick (125–500 μm) in midblade, three to six cells thick (500–700 μm) at base; cortical cells from midblade 55–74 μm wide and 55–130 μm long in surface view, in cross section 27–37 μm tall; hair zones faint, hairs few, distributed irregularly on both surfaces of blades, 23–28 μm diameter; linear bands, 0.25–20 mm broad, of darkly pigmented, larger cells produced from marginal meristems.

Schneider and Searles 1973; Schneider 1976.

Uncommon on offshore rock outcroppings near edge of continental shelf in Onslow Bay from 45 m, June–August.

Distribution: North Carolina and Dry Tortugas, Florida.

Figure 195. *Padina profunda*, scale 0.5 cm.

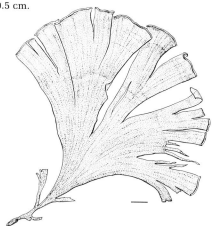

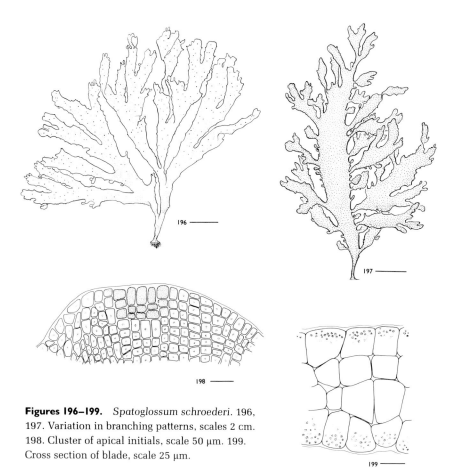

Figures 196–199. *Spatoglossum schroederi.* 196, 197. Variation in branching patterns, scales 2 cm. 198. Cluster of apical initials, scale 50 μm. 199. Cross section of blade, scale 25 μm.

Spatoglossum Kützing 1843

Plants erect, irregularly subpalmately and marginally branched; growth from a cluster of marginal cells at branch tip; blade several cells thick, cells in median layers larger than surface layer; midribs lacking; hairs in tufts; tetrasporangia scattered on both blade surfaces; dioecious, oogonia mostly scattered, antheridia in small scattered sori slightly raised above the blade surface.

Spatoglossum schroederi (C. Agardh) Kützing 1859, p. 21, pl. 51, fig. 1.
    Zonaria schroederi C. Agardh 1824, p. 256.
    Figures 196–199
    Plants 10–30 cm tall with alternate subdichotomous, palmately lobed, or branched segments, 0.5–2.5(–4) cm broad; to 7 cm between branch points; apices rounded; marginally proliferous, the margins undulate, irregularly dentate, the teeth in part acute; older plants stalked.
    Hoyt 1920; Williams 1951; Taylor 1960; Schneider 1976; Searles 1987, 1988.

Common in Onslow Bay from 17 to 46 m, June–November; on Gray's Reef from 17 to 21 m, June–August.

Distribution: North Carolina, Georgia, Bermuda, southern Florida, Gulf of Mexico, Caribbean, Brazil.

Reported to be iridescent elsewhere, but no iridescence has been noted in plants in the region.

Zonaria C. Agardh 1817, nom. cons.
Plants erect, bushy, growing from felted, rhizoidal bases; blades flat, splitting into segments which are stalked below, stalks sometimes continuing into the blades as ribs, accessory blades sometimes arising from stalks; blades transversely banded by rows of hairs; growth from marginal rows of meristematic cells that produce cells in rows parallel to the margin; periclinal divisions producing tiers of cells, the inner cells forming a medulla, the outer cells dividing anticlinally so that each medullary cell is covered by at least two cortical cells; sporangia in sori on both blade surfaces, devoid of paraphyses; each sporangium producing eight aplanospores.

Zonaria tournefortii (Lamouroux) Montagne 1846, p. 42.
  Fucus tournefortii Lamouroux 1805, p. 44.
  Figure 200
  Plants erect, 3–17 cm tall, yellowish brown when young, dark brown when old, reddish when dry; blades cuneate, incised, and becoming fan shaped with

Figure 200. *Zonaria tournefortii*, scale 1 cm.

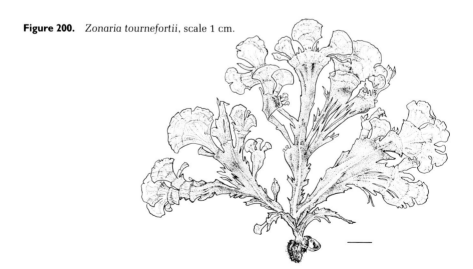

vague concentric zonation parallel to the growing margins; stalk subterete, continuing into blades and diverging into tapering ribs; sori forming irregular scattered patches over surfaces of blades.

Williams 1951; Taylor 1960; Schneider 1976; Kapraun 1984. As *Zonaria flava*, Hoyt 1920; Williams 1948a.

Frequent offshore in Onslow Bay and Long Bay from 14 to 93 m, May–December.

Distribution: North Carolina, South Carolina, Brazil, Mediterranean, South Africa.

This offshore species is common in the drift on coastal beaches.

SPOROCHNALES

Sporophytes with pseudoparenchymatous axes originating from an intercalary meristem formed at the base of a terminal tuft of photosynthetic hairs; reproduction by unilocular meiosporangia; gametophytes microscopic, filamentous, oogamous.

There is a single family in the order.

Sporochnaceae

Sporochnus C. Agardh 1817
Sporophytes erect; axes cylindrical, with many lateral secondary and tertiary branches bearing short, pedicellate branches ending in tufts of hairs; unilocular sporangia borne laterally on short filaments covering part of the surface of the ultimate branches.

Sporochnus pedunculatus (Hudson) C. Agardh 1820, p. 149.
Fucus pedunculatus Hudson 1778, p. 587.
Figure 201
Sporophytes to 30 cm tall; secondary branches to 20 cm long; pedicels of ultimate branchlets 1(–2) mm long, 0.5 mm diameter, with terminal tufts of golden brown hairs to 4 mm long; sori of unilocular sporangial filaments forming ovoid or spindle-shaped swellings to 2 mm long on branchlets next to terminal tufts of hairs.

Hoyt 1920; Taylor 1960; Schneider 1976; Kapraun 1984.

Offshore in Onslow Bay from 14 to 45 m, June–November.

Distribution: North Carolina, Bermuda, southern Florida, Gulf of Mexico, Caribbean, Brazil, Norway to Portugal, Pacific Mexico.

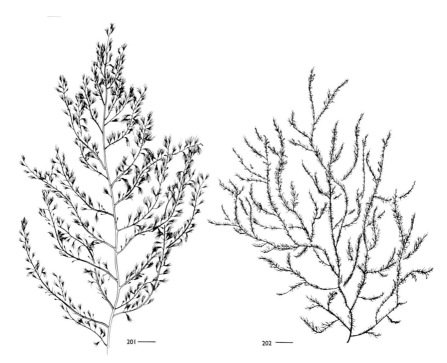

Figure 201. *Sporochnus pedunculatus*, scale 0.5 cm.
Figure 202. *Arthrocladia villosa*, scale 0.5 cm.

DESMARESTIALES

Sporophytes large, uniaxial, pseudoparenchymatous; axes cylindrical, compressed or bladelike; growth from an apical hair with trichothallic growth, cortex formed by descending rhizoidal filaments produced from bases of lateral photosynthetic hairs; sporangia unilocular only; gametophytes microscopic, filamentous, dioecious or monoecious, oogamous.

Arthrocladiaceae

Plant axes cylindrical; unilocular sporangia in chains formed on photosynthetic filaments; gameophytes monoecious.

Arthrocladia Duby 1830

Sporophytes with cylindrical axes, bushy, clothed by whorls of branched or simple villose, photosynthetic filaments; axial filament large; cortex with large, colorless inner cells and smaller, pigmented outer cells; unilocular sporangia in moniliform chains replacing lower branches of photosynthetic filaments.

Arthrocladia villosa (Hudson) Duby 1830, p. 971.
  Conferva villosa Hudson 1778, p. 603.
  Figure 202
  Plants to 40 cm tall, with open, infrequent alternate or opposite branching; axes to 1 mm diameter and clothed in golden brown photosynthetic filaments

to 4 mm long; filaments branched oppositely below, alternately above, to 21 μm diameter below, 7 μm above; unilocular sporangia alternately or unilaterally borne on the lower filament cells in chains of fifteen to twenty, sporangia 11–15 μm diameter, 5–9 μm long.

Harvey 1853; Williams 1948a; Taylor 1960; Schneider 1976; Kapraun 1984.

In the shallow subtidal on coastal jetties (rare) and offshore in Long Bay and Onslow Bay from 15 to 32 m, March–August.

Distribution: Massachusetts to Virginia, North Carolina, South Carolina, Baltic to Portugal.

FUCALES

Plants with parenchymatous growth, medulla secondarily becoming filamentous; growth in most representatives initiated from apical cells or clusters of apical cells; plants diploid, meiosis occurring in conceptacles at gametogenesis; oogamous gametangia unilocular, sperm and eggs released from conceptacles prior to fertilization.

Key to the families

1. Branching dichotomous or pinnate; gas vesicles single or in pairs; mature apical cells four sided .... Fucaceae
1. Branching radial and alternate; gas vesicles in clusters; apical cells three sided .... Sargassaceae

Fucaceae

Branching dichotomous or pinnate; branches flattened or terete; gas vesicles intercalary, single or in pairs in branches; conceptacles in swollen, terminal receptacles; one to eight eggs per oogonium; 64–128 sperm per antheridium.

Key to the genera

1. Branching regularly dichotomous, gas vesicles mostly oppositely paired on flattened parts of branches .... Fucus
1. Branching dichotomous and pinnate, gas vesicles single in the midline of the branches .... Ascophyllum

Ascophyllum Stackhouse 1809, nom. cons.

Plants tough; branches linear and compressed, primary branching dichotomous, secondary branches irregularly pinnate; lacking distinct midribs or cryptostomata; gas vesicles single, intercalary in the main axes; receptacles formed pinnately from the axes on short stalks, deciduous; oogonia forming four eggs.

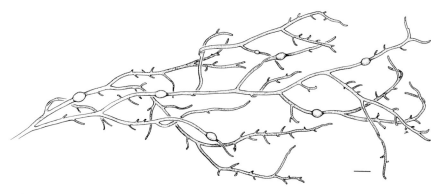

Figure 203. *Ascophyllum nodosum*, scale 1 cm.

Ascophyllum nodosum (Linnaeus) Le Jolis 1863, p. 96.
 Fucus nodosum Linnaeus 1753, p. 1628.
 Figure 203
 Plants 30–60 cm long, occasionally much larger; attached by discoid hold-fasts or often free floating or entangled with other plants; vesicles typically to 2 cm diameter, 2–3 cm long; main axes bearing pinnate rows of simple or forked, somewhat clavate, compressed branchlets 1–2 cm long; dioecious, with oval, yellowish receptacles single or in clusters of two to five on stalks to 2 cm long, replacing branchlets.
 Aziz and Humm 1962; Zaneveld and Willis 1976.
 On wreck at Salvo, N.C., just north of Cape Hatteras, and on mud in Beaufort harbor, N.C., July–August.
 Distribution: Canadian Arctic to Delaware, North Carolina, ?Bermuda, USSR to Portugal.
 Specimens of this alga have been occasionally collected in the drift along the North Carolina beaches for many years (Aziz and Humm 1962), but it was not known as an attached plant in North Carolina until reported by Zaneveld and Willis (1976) near the Virginia border. More recently, we have collected plants partially buried in the intertidal mud in Beaufort harbor, N.C. These plants are dichotomously branched, have gas vesicles, are sterile, and correspond to the free-floating ecad *mackaii* (Gibb 1957). Local collections vary from slender plants to 1.5 mm broad with vesicles 6 mm wide and 8 mm long, to coarse plants to 3 mm wide with vesicles 3 mm wide and 2.5 cm long.

Fucus Linnaeus 1753
Plants dichotomously branched, axes flattened, with conspicuous midrib, cryptostomata common in lamina; gas vesicles, if present, in lamina, usually in opposite pairs; monoecious or dioecious; conceptacles in terminal, often swollen, receptacles.

Fucus vesiculosus Linnaeus 1763, p. 1636.

Figure 204

Plants golden, olive, or dark brown; to 30–90 cm tall; branching dichotomous, sometimes proliferous below; branches 10–15 mm wide; vesicles (when present) opposite or, less commonly, single; cryptostomata common; receptacles broadly lanceolate to obovoid or forked; 1–3 cm long; dioecious.

Hoyt 1920; Taylor 1957, 1960; Kapraun 1984.

Intertidal and shallow subtidal on jetties and seawalls and other solid substrata in sounds and marshes from Beaufort, N.C, to as far south as Surf City, N.C., year-round.

Distribution: Canadian Arctic to Virginia, North Carolina, USSR to Portugal, Morocco.

Figure 204. *Fucus vesiculosus*, scale 1 cm.

Sargassaceae

Branching radial, more or less continuous from bases to apices; branches flat
with midribs, or transitional to cylindrical branches with macroscopic spinelike
projections, turbinate foliar organs or leaves with midribs, usually with cryptostomata and buoyant vesicles; conceptacles in ordinary branches or special receptacular branchlet systems; one egg formed per oogonium.

Sargassum C. Agardh 1820, nom. cons.
Plant erect, attached by a solid holdfast or by rhizoidal outgrowths of main axes,
or free floating; main axes terete, with determinate and indeterminate branches
bearing flat, broad to filiform, simple or occasionally forked, entire to serrate,
leaflike blades with midribs; stalked vesicles common; receptacular branches
developing in the axils of "leaves" or paniculate, generally cylindrical, less often
prismatic compressed or flattened; monoecious or dioecious.

One species occurs in the flora as attached plants and two as permanent members of the free floating offshore flora.

Key to the species

1. Plants free floating and sterile; cryptostomata absent or inconspicuous ... 2
1. Plants attached and often fertile; cryptostomata present ...... *S. filipendula*
2. Leaves linear, marginal teeth aculeate; vesicles apiculate or with terminal leaf
   .... *S. natans*
2. Leaves lanceolate, marginal teeth broad; vesicles not apiculate and without
   terminal leaf .... *S. fluitans*

Sargassum filipendula C. Agardh 1824, p. 300.
   Figures 205 and 206
   Plants 30–100(–200) cm tall, attached by conical, spreading, lobed holdfasts;
main axes smooth, sparingly forked, principal branches dominant; branchlets
bearing stalked leaves alternately; leaves serrate to subentire, 5–8(–15) mm wide,
3–12 cm long, linear lanceolate, simple (those near bases of plants sometimes
forked); midribs distinct, cryptostomata present; gas vesicles globose, 3–5 mm
diameter, stalked, the stalks about 5 mm long, sometimes apiculate; receptacles
simple, forked or sparsely, racemosely branched, axillary in the upper parts of
the plant, the raceme 0.2–0.5 times as long as the subtending leaf.

   Hoyt 1920; Williams 1948a, 1951; Taylor 1960; Schneider 1976; Schneider and
Searles 1979; Kapraun and Zechman 1982; Kapraun 1984; Searles 1987, 1988. As
S. vulgare sensu Johnson 1900. As *S. polyceratium sensu* Blair and Hall 1981.

   Growing on rocks and seawalls along the coast, in the sounds near inlets, common offshore from 14 to 55 m, throughout the area, year-round.

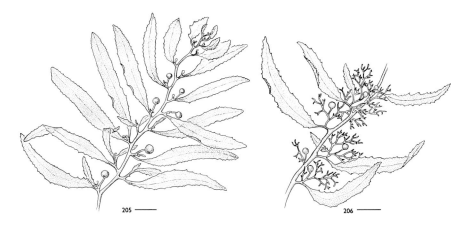

Figures 205, 206. *Sargassum filipendula*, scales 1 cm. 205. Sterile branch. 206. Branch with receptacles.

Distribution: Massachusetts to Virginia, North Carolina, South Carolina, Georgia, Bermuda, southern Florida, Gulf of Mexico, Caribbean, Brazil, West Africa.

S. filipendula var. *montagnei* (Bailey) Grunow 1916, p. 171.

Sargassum montagnei Bailey in Harvey 1852, p. 58, pl. 1A.

Figure 207

Stems long, slender, and smooth, or with a few minute papillae on younger axes; leaves linear, 3–6 mm wide, up to 15 cm long, simple, once or twice forked, or more often alternate to subpinnately branched, margins entire or very slightly dentate, cryptostomata few, small, scattered; gas vesicles globose to subpyriform, 2–3 mm diameter, tipped with single short spines or often with a leaf, length of pedicel 1.5–3 times diameter of vesicle, sometimes filiform, often compressed or with midribs; receptacle rachis filiform, smooth, sterile, with first- and sometimes second-order branches perpendicular, alternate, torulose, the ultimate division to 3 cm long.

Hoyt 1920; Blomquist and Pyron 1943; Williams 1948a, 1949, 1951; Taylor 1957, 1960; Schneider 1976.

Subtidal on coastal and estuarine jetties, and offshore in Onslow Bay from 15 to 40 m, year-round.

Distribution: Massachusetts to Virginia, North Carolina, South Carolina, Georgia, Bermuda, southern Florida, Gulf of Mexico.

There is great variation in the morphology of plants included in this species. The variety *montagnei* is traditionally recognized in the region but is just one of many variants. We have not seen the very long ultimate divisions of receptacles described for the variety; the longest in our specimens are 1 cm. There appear to be consistent differences between the shallow-water, coastal, and estuarine *Sargassum filipendula* plants in the Carolinas and the deep-water plants. These include morphological, phenological, and physiological differences. The blades of deep-water plants are broader than those in shallow water; deep-water plants persist at the end of the summer only as holdfasts, basal stalks, and a few leaves, some stalks forming swollen tuberlike structures that perhaps have a storage function; estuarine plants persist throughout the winter. Peckol and Ramus (1985) showed differences between deep-water and shallow-water plants in photosynthetic performance, pigment concentrations, and growth rates for acclimated plants growing under the same conditions. This is the most commonly collected species and the biomass-dominant species in Onslow Bay (Schneider and Searles 1979).

Sargassum fluitans Børgesen 1914, p. 66.
    Figure 208
    Plants pelagic, golden brown, without dominant axes; stems smooth or sparingly spinulose; leaves narrow to lanceolate, short stalked, broadly serrate, 2–6 cm long, 3–8 mm wide, cryptostomata absent or few and obscure; vesicles globose to ovoid, lacking an apiculus, 4–5 mm diameter, on 2–3 mm stalks, the stalks often winged, receptacles unknown.
    Blomquist and Pyron 1943; Chapman 1971; Kapraun 1984; Richardson 1987.
    Pelagic, present offshore and washed ashore primarily in summer.
    Distribution: Associated with *Sargassum natans*, it is common in the offshore waters of the western North Atlantic.

Sargassum natans (Linnaeus) Gaillon 1828, p. 355.
    Fucus natans Linnaeus 1753, p. 1160.
    Figure 209
    Plants pelagic, golden to dark brown, clumps 10–50 cm diameter, without dominant axes; stems smooth; leaves acutely linear, 2–5(–10) cm long, 2–4 mm wide, serrate, teeth slender, terete, their length 0.5–0.8 times leaf width, crypto-

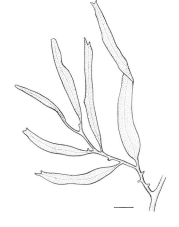

Figure 207. *Sargassum filipendula* var. *montagnei*, scale 0.5 cm.

Figure 208. *Sargassum fluitans*, scale 1 cm.
Figure 209. *Sargassum natans*, scale 0.5 cm.

stomata absent, midribs not prominent; vesicles 3–5 mm diameter, on stalks 3–5 mm long, apiculate or tipped with spines or small leaves; receptacles unknown.

Hoyt 1920; Blomquist and Pyron 1943; Taylor 1960; Kapraun 1984; Richardson 1986, 1987. As *Sargassum bacciferum*, Curtis 1876.

Pelagic, present offshore in all seasons and washed ashore in abundance in summer. This is the more common of the two pelagic species of *Sargassum*.

Distribution: As a pelagic species it has no defined range other than the North Atlantic and may be found floating offshore and cast up on beaches from Canada to Brazil and Europe.

Species Excluded

Sargassum hystrix var. *buxifolium* Chauvin
Sargassum linifolium (Turner) C. Agardh
Sargassum polyceratium Montagne
Sargassum pteropleuron Grunow

Blomquist and Pyron (1943) reported these four additional species of *Sargassum* as unattached plants in the drift on North Carolina beaches after a major storm, but these plants are normally attached, not pelagic, and were presumably carried north by currents from their places of origin; they have not been rediscovered in the flora.

RHODOPHYTA

Plants unicellular, colonial, pseudofilamentous, true filamentous, ligulate to broadly foliose, hollow tubular, discoid, or crustose; blades membranous to thick; crusts thin to thick; filamentous plants either simple or branched, many plants composed of aggregated filaments forming pseudoparenchymatous axes with one (uniaxial) or several (multiaxial) axes; few genera producing true parenchyma with cells dividing in more than one plane; texture varying from soft and gelatinous to wiry, cartilaginous, or stony with light to heavy calcification of calcite or aragonite, some impregnated forms jointed or articulated; plants living on or in plants (epiphytic, endophytic), animals (epizoic, endozoic), rocks (saxicolous), wood (lignicolous), sand or mud (terricolous), or shells (conchicolous); growth apical or intercalary, pit-connections forming between cells of most species; cells regularly to irregularly arranged, each with one (uninucleate) or more (multinucleate) nuclei and one to several variously shaped plastids with or without associated refractive pyrenoids; plants exhibiting an array of colors, predominantly reds, purples, and black, from dominant phycobiliproteins (pigments), phycoerythrin (red), and phycocyanin (blue), mostly masking the chlorophylls; some parasitic forms with pigment reduced or lacking; photosynthate stored in most species as floridean starch (an amylopectin-like substance); flagellated cells unknown; asexual reproduction by mono-, bi-, tetra-, or polyspores, akinetes, or vegetative propagules; many species fragmenting and reattaching by specialized hook-shaped tips, or becomingd entangled with other algae; sexual reproduction basically oogamous with non-motile spermatia fertilizing receptive carpogonia with hairlike extensions or trichogynes; fertilized carpogonia dividing directly into several carpospores, developing directly into carposporophytic filaments (gonimoblasts) which later bear carposporangia, or transferring the diploid nucleus to a near or distant receptive (auxiliary) cell which then develops gonimoblast filaments and ultimately carposporangia; some carposporophytes invested by loose involucral filaments or surrounded by highly organized vegetative filaments (pericarp); carposporangia releasing single, nonmotile carpospores; carpospores developing into free-living sporophytes similar (isomorphic) or dissimilar (heteromorphic) to the gametophytes; some species with direct development of sporophytes or gametophytes without the intervention of the other, therefore propagating without sexual reproduction.

Members of the Rhodophyta, or red algae, are mostly smaller than 1 m in length, and many are microscopic. They occur in the flora from the intertidal to a depth of 100 m. Offshore, *Gracilaria mammillaris* and *Botryocladia occidentalis* are two of the five most common and productive species on rock outcroppings. *Hypnea* and *Gracilaria* species are among the larger common red algae on jetties, seawalls, and pilings. Several genera, including many in the flora, have impregnations of the cellulosic cell walls by sulfated galactans, repeating units of galactose or its derivatives. Two of the familiar galactans include agar and carrageenan; the former was the basis of a commercial venture in the Carolinas during World War II.

RHODOPHYCEAE

This is the only class in the division. Two subclasses, the Bangiophycidae and Florideophycidae, are recognized in the Rhodophyta, although some authors (e.g., Dixon and Irvine 1977) distinguish them as classes. Both are represented in the flora. With the passage of time, the alleged distinctions between these taxa have become blurred. However, as seen in the general descriptions of the subclasses, the basic characteristics are divergent.

BANGIOPHYCIDAE

Plants unicellular, palmelloid colonies, pseudofilamentous, filamentous, to broadly foliose, saccate, or discoid with cells often embedded in a gelatinous matrix; endophytic and endozoic to epiphytic, epizoic, and saxicolous; growth intercalary, occasionally apical, pit-connections lacking in most; cells regularly to irregularly arranged, cylindrical, spherical, to ellipsoid, uninucleate with single, stellate, axial plastids or variously shaped parietal plastids with or without pyrenoids, few with several small plastids; flagellated cells unknown; asexual reproduction by monospores, akinetes, or the release of vegetative cells from a common sheath; sexual reproduction (where known) oogamous, with spermatia fertilizing carpogonia with trichogynes; fertilized carpogonia dividing into several carpospores, these producing alternate shell-boring, branched-filamentous "conchocelis" stages.

Key to the orders

1. Plants unicellular, palmelloid colonial, or pseudofilamentous with cells embedded somewhat loosely in a gelatinous matrix............Porphyridiales
1. Plants filamentous, foliose, saccate, or parenchymatous discs............ 2
2. Asexual reproduction by monosporangia cut off from undifferentiated vegetative cells by unequal curving or oblique divisions...... Compsopogonales
2. Asexual reproduction by monosporangia formed by transformation of vegetative cells, never by unequal divisions of them ....Bangiales

PORPHYRIDIALES

Plants unicellular to palmelloid colonies or pseudofilamentous, epiphytic to epizoic; pseudofilaments uniseriate to multiseriate, simple to irregularly branched; growth intercalary, pit-connections absent; cells regularly to irregularly arranged and separated by a thick gelatinous matrix, cylindrical, globose, to ellipsoid, uninucleate with single stellate plastid or several plastids with or without pyrenoids; flagellated cells unknown; asexual reproduction by monospores, akinetes, or the release of vegetative cells from a common sheath; sexual reproduction unknown.

Porphyridiaceae

Pyrenoids present and associated with plastids; reproduction by naked monospores produced by direct transformation of an undivided vegetative cell, or akinetes enclosed by thick membranes formed basipetally; other characteristics as for the order.

Key to the genera

1. Plants simple to sparingly branched, blue green to gray green . . . .
. . . . Chroodactylon
1. Plants repeatedly branched, red to purplish red and pink . . . . Stylonema

Chroodactylon Hansgirg 1885
Plants epiphytic, consisting of simple to irregularly and pseudodichotomously branched uniseriate gelatinous pseudofilaments attached by simple basal cells; cells narrowly to widely spaced, each with a single blue-green to gray-green stellate plastid with a single pyrenoid; monospores from transformed vegetative cells; akinetes covered with thick membranes transformed from vegetative cells.

Chroodactylon ornatum (C. Agardh) Basson 1979, p. 67, fig. 52.
   Conferva ornata C. Agardh 1824, p. 104.
   Figure 210
   Plants to 10 mm tall, generally much shorter; branching unilateral, pseudodichotomous to irregular, usually widely spaced; pseudofilaments 9–25 μm diameter, uniseriate throughout; cells oblong, oval, ellipsoid, or subquadrate, 3–8 μm diameter, 8–20 μm long; akinetes forming basipetally, subglobose to ellipsoid, 8–11 μm diameter, 14–15 μm long, with walls 2 μm thick, discharging laterally from the filaments.
   As Asterocytis ramosa, Williams 1948a.

Epiphytic on various small algae of intertidal mats on Cape Lookout jetty, and probably more widespread in shallow marine and brackish environments.

Distribution: North Carolina, southern Florida, Caribbean; elsewhere, widespread in temperate to tropical seas.

A culture study by Lewin and Robertson (1971) casts doubt as to whether both freshwater and marine entities of *Chroodactylon* (as *Asterocytis*) should be maintained. In light of salinity/culture studies on other members of the Bangiophycidae (see discussion for *Bangia*), we follow the logical approach of John et al. (1979) in accepting the single species, *C. ornatum.*

Stylonema Reinsch 1875

Plants epiphytic, consisting of simple to pseudodichotomously and irregularly branched, mostly uniseriate, gelatinous pseudofilaments attached by simple basal cells; cells narrowly to widely spaced, each with a single pink, red, to purplish red stellate plastid with a single pyrenoid; monospores from transformed vegetative cells.

Stylonema alsidii (Zanardini) Drew 1956, p. 72.

   Bangia alsidii Zanardini 1841, p. 471.

   Figure 211

   Plants to 6 mm tall but generally 1 mm or less, branching freely subdichotomous to irregular, rarely simple or subsimple; pseudofilaments 12–35 μm diameter, mostly uniseriate, multiseriate, or loosely and irregularly arranged below, to 50 μm diameter; cells globose, subglobose, crescent shaped, ellipsoid, cylindrical, to irregular, 6–13 μm diameter, 4–13 μm long; monospores dispersed by a breakdown in the gelatinous wall.

   Searles 1987, 1988. As *Goniotrichum alsidii*, Hoyt 1920; Williams 1948a; Humm 1952; Taylor 1957, 1960; Brauner 1975; Schneider 1976; Wiseman 1978; Kapraun 1980a.

   Common on various algae on jetties, seawalls, etc., throughout the area and to a depth of 40 m offshore, year-round.

   Distribution: North Carolina, South Carolina, Georgia, northeastern Florida to Gulf of Mexico, Caribbean, Brazil; elsewhere, widespread in temperate to tropical seas.

   For a clarification of the nomenclature of this species refer to Drew (1956) and Wynne (1985b).

COMPSOPOGONALES

Plants filamentous, crustose, membranous, to saccate; epiphytic, epizoic, endophytic, endozoic, or saxicolous; erect uniseriate to multiseriate filaments simple, prostrate uniseriate filaments much and irregularly branched, membranous blades and sacs normally entire or splitting at maturity; growth by intercalary divisions; cells with stellate axial or band- to bowl-shaped parietal plastids with or without pyrenoids, pit-connections lacking; flagellated cells unknown; asexual reproduction by monospores cut off from an undifferentiated vegetative cell by an unequal curving or oblique division; sexual reproduction reported but considered doubtful.

Erythropeltidaceae

Plants filamentous to membranous, erect and epiphytic, epizoic, or saxicolous, or prostrate and epiphytic to endophytic or endozoic; erect plants attached by single cells, single-layered to multilayered discs, or rhizoidal downgrowths from lower cells; sexual reproduction incompletely known, some with a shell-boring filamentous alternate stage lacking pit-connections (Heerebout 1968), known as a "pseudoconchocelis" generation (Garbary et al. 1980).

Figure 210. *Chroodactylon ornatum*, scale 20 μm.
Figure 211. *Stylonema alsidii*, scale 20 μm.

Key to the genera

1. Plants producing erect filaments or blades, often rising from a basal disc, plastids stellate axial 2
1. Plants consisting of prostrate filaments on or in other algae or animals, erect filaments lacking, plastids parietal 3
2. Erect axes remaining uniseriate below, occasionally expanding above ......
.... Erythrotrichia
2. Erect axes becoming single-layered broad blades, not uniseriate even below
.... Porphyropsis
3. Plants consisting of branched prostrate filaments with free ends ....
.... Erythrocladia
3. Plants consisting of orbicular discs with marginal growth by united border cells .... Sahlingia

Erythrocladia Rosenvinge 1909

Plants consisting of endophytic, epiphytic, endozoic, or epizoic prostrate branched filaments, pseudodichotomously to irregularly branched, forming loose, single-layered discs, some becoming irregularly multilayered centrally, or irregularly branched networks; growth apical, discs with marginal filaments free; pit-connections lacking; cells rectangular to irregular, with single band- or bowl-shaped plastids without distinct pyrenoids; monosporangia cut off from vegetative cells by oblique curving walls; sexual reproduction unknown.

Key to the species

1. Plants epiphytic or epizoic, filaments producing compact pseudoparenchymatous discs centrally, margins free .... E. irregularis
1. Plants mostly endophytic or endozoic, filaments producing an open network of cells .... E. endophloea

Erythrocladia endophloea Howe 1914, p. 81, pl. 30, figs 1–7.

Figure 212

Plants endophytic, endozoic, epi-endophytic, or epi-endozoic, minute, spreading to 2.5 mm, at first consisting of irregularly widespreading, radiating filaments, later becoming coalesced into pseudoparenchymatous areas centrally; cells in surface view variously shaped, 3–15 μm diameter, 8–40 μm long; monospores produced from older intercalary cells, ovoid to globose, 4–5 μm diameter.

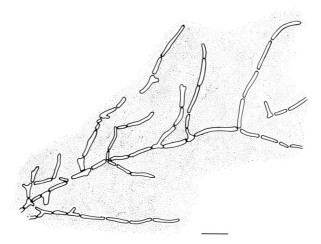

Figure 212. *Erythrocladia endophloea*, scale 25 μm.

As *E. recondita*, Howe and Hoyt 1916; Hoyt 1920. As *E. vagabunda*, Howe and Hoyt 1916; Hoyt 1920; Williams 1951.

In various hosts, in particular *Chondria* and *Dictyota*, offshore and intertidally from the Beaufort, N.C., area, June–August.

Distribution: Connecticut, North Carolina, South Carolina, Puerto Rico, Peru.

Heerebout (1968), basing his decision on the morphological variability shown in culture by other members of the family but lacking direct cultural proof, merged two plants previously described from North Carolina—*Erythrocladia recondita* Howe and Hoyt (1916) and *E. vagabunda* Howe and Hoyt (1916)—with *E. endophloea*. After studying our local material of these two species, we agree with his assessment.

Erythrocladia irregularis Rosenvinge 1909, p. 72, figs 4f–4h.

Figure 213

Plants epiphytic or epizoic, minute, to 100 μm across, consisting of compact to loose discs, single layered, coalesced centrally, some becoming irregularly multilayered, cells in surface view isodiametric, rectangular, oblong to irregular, 2–3 μm diameter, 2–7 μm long; monospores produced from somewhat central vegetative cells, globose to subglobose, 5 μm diameter.

On a variety of hosts, in particular *Polysiphonia*, from shallow and deep North Carolina subtidal habitats, June–August.

Distribution: North Carolina (new record), USSR to Portugal; elsewhere, widespread in temperate to tropical seas.

Although recent treatments found *Erythrocladia irregularis* and *E. subintegra* conspecific (Heerebout 1968; Garbary et al. 1981), Kornmann and Sahling (1985)

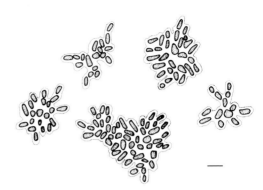

Figure 213.
Erythrocladia irregularis,
five crusts, scale 10 μm.

displayed the heterogeneity of the two species in culture, reassigning the latter species to *Sahlingia* Kornmann (1989). Both species can be found growing on the same host.

Erythrotrichia Areschoug 1850, nom. cons.
Plants epiphytic or saxicolous, consisting of erect simple filaments to single-layered blades arising from single- to multilayered suborbicular basal discs or simple expanded basal cells; erect filaments uniseriate, possibly becoming biseriate to multiseriate above with downgrowing rhizoids from lower cells; growth intercalary, pit-connections lacking; cells subquadrate rectangular to irregular, with single stellate plastids and distinct pyrenoids; monosporangia cut off from vegetative cells by oblique walls; spermatia produced in a similar manner to spores, smaller; carpogonia formed by a transformation of a vegetative cell; alternate generation a shell-boring "conchocelis" stage.

Key to the species

1. Plants with unicellular lobate bases or small, few-celled basal discs; erect filaments mostly uniseriate, narrow, 15–60 μm diameter .... *E. carnea*
1. Plants attached by hyaline rhizoidal downgrowths of several lower cells; mostly multiseriate, broad, to 185 μm diameter .... *E. vexillaris*

Erythrotrichia carnea (Dillwyn) J. Agardh 1883, p. 15, pl. 19.
   Conferva carnea Dillwyn 1807, p. 54, pl. 84.
   Figures 214 and 215
   Plants epiphytic, to 8 cm, generally less than 0.5 cm long; uniseriate below, occasionally becoming biseriate to multiseriate above, 15–60 μm diameter, arising from basal cells with lobed rhizoidal attachments or minute prostrate discs; cells of the erect filaments swollen, subquadrate rectangular to irregular, 15–20 μm diameter and 12–25(–32) μm long above, 9–13 μm diameter and 12–40 μm long near the base; monospores globose, 13–18 μm diameter.
   Hoyt 1920; Humm 1952; Taylor 1957, 1960; Chapman 1971; Brauner 1975; Kapraun 1980a; Hall and Eiseman 1981; Richardson 1987; Searles 1987, 1988. As *Bangia ciliaris sensu* Harvey 1858; Taylor 1957. As *E. ciliaris sensu* Williams 1948a; Kapraun 1980a.

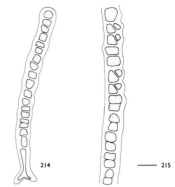

Figures 214, 215. *Erythrotrichia carnea*, scale 25 μm. 214. Germling with basal attachment. 215. Mature filament producing monospores.

Known as solitary epiphytes from deep offshore to intertidal habitats or as gregarious plants forming an abundant overgrowth on shallow subtidal populations of *Dictyota*, *Padina*, and other algae, throughout the area, most common in summer and fall months.

Distribution: Arctic to Virginia, North Carolina, South Carolina, Georgia, northeastern Florida to Gulf of Mexico, Caribbean, Brazil; elsewhere, widespread in temperate to tropical seas.

Wiseman (1978) observed that the South Carolina plants collected and reported by Harvey (1858) as *Bangia ciliaris* (=*Porphyrostomium ciliare* [Carmichael ex Harvey] Wynne) are best referred to *Erythrotrichia carnea*.

Erythrotrichia vexillaris (Montagne) Hamel 1929, p. 53.
   Porphyra vexillaris Montagne 1856, p. 450.
   Plants minute, to 185 μm diameter and 675 μm long, consisting of solitary filaments developing at first as uniseriate axes, becoming biseriate and then ligulate multiseriate blades by periclinal divisions of the initial row of cells; blades single layered, to twenty-four cells across, about 20 μm thick, tapering to the base or attenuate to truncate, attached by attenuate basal cells and hyaline rhizoidal downgrowths from lower cells; growth mostly terminal to subterminal, intercalary in older plants; blade cells 7–15 μm diameter, each containing single, stellate, parietal plastids.
   Williams 1948a, 1949; Taylor 1960. As *Porphyropsis vexillaris*, Heerebout 1968.
   Epiphytic on *Grateloupia*, *Rhodymenia*, and *Codium* from shallow subtidal of Cape Lookout jetty, summer–fall.
   Distribution: North Carolina, Caribbean.
   Heerebout (1968) placed this species in *Porphyropsis* on the basis of the attachment by hyaline rhizoidal downgrowths, a feature shared with *P. coccinea*, the type species. Although this attachment feature is unusual among *Erythrotrichia* species (Heerebout 1968), assigning *E. vexillaris* to *Porphyropsis* would necessitate a significant change in the developmental pattern assigned to that genus (South and Adams 1976).
   This plant has not been collected since it was reported as being common at

Cape Lookout jetty in 1947 (Williams 1948a), and voucher specimens have not been located.

Porphyropsis Rosenvinge 1909
Plants epiphytic, developing as cellular vesicles which rupture to form single-layered, foliose blades consisting of many rows of regularly arranged cells; blades attached to the substrate by hyaline rhizoidal downgrowths of several lower cells; growth marginal and intercalary, pit-connections lacking; each cell with single stellate to lobate plastids and single pyrenoids; monosporangia cut off vegetative cells by oblique walls.

Porphyropsis coccinea (J. Agardh ex Areschoug) Rosenvinge 1909, p. 69, figs 9–10.
  Porphyra coccinea J. Agardh ex Areschoug 1850, p. 407.
  Plants small, 2–5 cm long, gregarious, rosy red, blades oval to circular; attachments initially umbilicate, later the vesicles split, with several blades arising from the single basal attachment, secondarily affixed by numerous hyaline rhizoids developed from cells in the basal portions of each blade; margins initially involute, later flattening, at maturity undulate to crispate; blade cells variously shaped, 4–7 μm diameter, each with a single stellate to lobate, parietal plastid and single pyrenoid; monospores produced randomly over blade surfaces.
  Williams 1948a, 1949.
  Epiphytic on *Sargassum* from the shallow subtidal on Cape Lookout jetty, March–April.
  Distribution: Maritime Provinces, New Hampshire, New York, North Carolina, Faeroes, British Isles, Norway to Portugal, Japan, British Columbia to California.
  This plant has not been re-collected since it was reported as common at Cape Lookout in the spring of 1947 by Williams (1948a), and voucher specimens have not been located.

Sahlingia Kornmann 1989
Plants epiphytic or epizoic, consisting of prostrate, pseudodichotomously to irregularly branched filaments forming compact single-layered discs; growth marginal, outermost cells often bifurcate; pit-connections lacking; cells rectangular to irregular, with single band- or bowl-shaped plastids without distinct pyrenoids; monosporangia cut off somewhat central vegetative cells by oblique curving walls; sexual reproduction unknown.

Sahlingia subintegra (Rosenvinge) Kornmann 1989, p. 227, figs 1, 6–13.
  Erythrocladia subintegra Rosenvinge 1909, p. 73, figs 13–14.

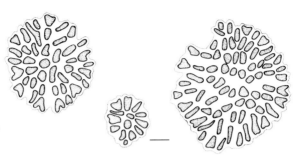

Figure 216. *Sahlingia subintegra*, three crusts, scale 10 µm.

Figure 216

Plants epiphytic or epizoic, minute (to 300 µm across), consisting of compact suborbicular discs, coalesced centrally, single layered; margin with coalesced bifurcate cells which continue the regular pattern of the disc; cells in surface view isodiametric, rectangular, oblong, to irregular, 2–5 µm diameter, (2–)7–13 µm long; monospores produced from older, nonmarginal disc cells, globose to subglobose, 6 µm diameter.

As *Erythrocladia subintegra*, Williams 1951; Humm 1952; Kapraun 1980a. As *E. irregularis* var. *subintegra*, Searles 1987, 1988.

Found on a variety of hosts, in particular *Polysiphonia*, from shallow and deep subtidal habitats, June–August in northern portions of the area, year-round in southern regions.

Distribution: Arctic, Connecticut, Virginia, North Carolina, Georgia, northeastern Florida, Caribbean, Brazil; elsewhere, widespread in polar to tropical seas.

Kornmann (1989) showed in culture that this species was not conspecific with *Erythrocladia irregularis* and could easily be separated by virtue of monospore size and integrity of the disc margin.

BANGIALES

Plants filamentous to membranous, erect and saxicolous, lignicolous, epiphytic, or endophytic in calcium carbonate substrates; erect plants unbranched, occasionally split, attached by basal discs formed by rhizoidal downgrowths from lower cells; cells with stellate axial plastids with central pyrenoids; flagellated cells lacking; asexual reproduction by monospores formed by transformation of vegetative cells, never by their unequal division; sexual reproduction known for some; carpogonia converted vegetative cells with trichogynes, fertilized by minute, colorless spermatia; fertilized cells dividing directly to form pigmented carpospores which produce the alternate shell-boring "conchocelis" stage; "conchocelis" stage pink to rosy red, comprised of uniseriate, highly branched microscopic filaments that penetrate calcium carbonate substrates, reproducing by conchospores (monospores); conchospores producing the erect generation or repeating the shell-boring stage; pit-connections found only in the penetrating stage.

Bangiaceae

Key to the genera

1. Plants filamentous to cylindrical .... Bangia
1. Plants broadly membranous .... Porphyra

Bangia Lyngbye 1819

Plants erect, initially uniseriate, becoming multiseriate, cylindrical; axes unbranched, occasionally tubular, expanded above, attached by expanded rhizoidal bases; growth intercalary, cells closely to widely spaced, encased in a firm gelatinous sheath, lacking pit-connections; cells cylindrical and shorter than broad, to quadrate and polyhedral, more or less in transverse tiers, each with stellate, axial plastid and central pyrenoid; monospores produced by direct transformation of surface vegetative cells, undivided or forming two to four spores per cell, on nonfertile or fertile plants; monoecious or dioecious, spermatia formed by repeated division of vegetative cells in three planes; carpogonia formed by direct transformation of vegetative cells, with or without obvious trichogynes; "carpospores" produced by direct repeated division in three planes of the fertilized carpogonium; alternate "conchocelis" stage with pit-connections.

Bangia atropurpurea (Roth) C. Agardh 1824, p. 76.

Conferva atropurpurea Roth 1806, p. 208, pl. 6.

Figure 217

Plants saxicolous or lignicolous, rosy red, purplish red, to brownish and black, to 20 cm, but generally 5–10 cm tall; when mature and multiseriate, 20–60 µm diameter below to 220 µm diameter above; axes unbranched, often irregularly constricted, occasionally tubular or torulose.

Wiseman and Schneider 1976; Kapraun 1980a; Richardson 1987. As *B. fuscopurpurea*, Harvey 1858; Hoyt 1920; Humm 1952; Taylor 1957, 1960; Chapman 1971.

On rock and woodwork throughout the area, often forming a distinct zone in the upper intertidal, winter–spring.

Distribution: Labrador to Delaware, North Carolina, South Carolina, Georgia, northeastern Florida to Gulf of Mexico, Caribbean, Brazil, Uruguay; elsewhere, widespread in temperate to subtropical seas.

Geesink (1973) has shown in a culture study that *Bangia fuscopurpurea* (Dillwyn) Lyngbye is conspecific with the freshwater *B. atropurpurea*. The marine form has been shown to alternate with a shell-boring "conchocelis" stage (Richardson and Dixon 1968; Geesink 1973), with the change precipitated by photoperiod (Richardson 1970) or temperature (Sommerfeld and Nichols 1973). Isolates from the Carolinas have not been studied in culture, and the species is not known from fresh water in our region.

Porphyra C. Agardh 1824, nom. cons.

Plants erect, single or double layered, with narrow to broad blades, usually undivided; margins entire, dentate, crenate, or undulate, blades solid to perforate, arising from basal discs of rhizoidal downgrowths from lower cells, with or without short stipes; growth intercalary, cells closely to widely spaced, encased in a firm, colorless gelatinous matrix, lacking pit-connections; cells subquadrate, polyhedral, elongate, oval, circular, to irregular, randomly arranged to packeted rosettes or tiers, each with one to two stellate axial plastids and central pyrenoids; monosporangia produced by direct transformation of undivided vegetative cells in laminal or marginal patches; monoecious or dioecious, spermatia formed by repeated division of vegetative cells in three planes, producing as many as 128 per cell, in mostly marginal patches; carpogonia formed by direct transformation of vegetative cells, with or without obvious trichogynes; carpospores produced by direct repeated division in three planes of the fertilized carpogonium, four to sixty-four per cell, in mostly marginal areas; alternate "conchocelis" stage with pit-connections, probably perennial.

Studies utilizing electron microscopy have recently advanced the understanding of the life history of *Porphyra* and conclusively proved that sexual reproduction occurs in at least one species (Hawkes 1978), a subject long disputed for this genus.

Figure 217. *Bangia atropurpurea*, scale 25 μm.

Key to the species

1. Plants small, margins serrulate .... *P. carolinensis*
1. Plants large, margins entire 2
2. Spermatangia in elongate patches, parallel to each other and blade axis.....
.... *P. leucosticta*
2. Spermatangia in irregular patches along the margin.......... *P. rosengurtii*

Porphyra carolinensis Coll et Cox 1977, p. 156, figs 1–8.
    Figure 218
    Plants to 4 cm tall and 2.5 cm wide, epiphytic, epizoic, or saxicolous, several blades arising from common basal holdfasts; blades rosy to pale purplish red, oval lanceolate to irregular, single layered, to 30 μm thick, entire to laciniate; margins serrulate, often folded; cells in surface view circular, oval, triangular, polyhedral, to irregular, 8–12 μm diameter, to 20 μm long, in younger parts in parallel rows, eventually forming rosettes, each with single stellate plastid and central pyrenoid; monosporangia marginal; monoecious, spermatia thirty-two per cell in four tiers, formed in marginal patches, not mixed with carpospores; carpospores sixteen per cell, formed in marginal areas.
    Coll and Cox 1977; Kapraun 1980a; Freshwater and Kapraun 1986; Richardson 1986, 1987. As *P. vulgaris sensu* Bailey 1848; Harvey 1958; Curtis 1867; Melvill 1875. As *P. umbilicalis* f. *laciniata sensu* Williams 1948a; 1949. As *P. umbilicalis sensu* ?Taylor 1960; ?Chapman 1971; Wiseman and Schneider 1976.
    On rocks, mussels, barnacles, and coarser algae from the upper intertidal of jetties and seawalls, most abundant in spring to early summer, but persisting year-round.
    Distribution: North Carolina (type locality), South Carolina, Georgia, ?northeastern Florida.
    This species does not adhere well to herbarium paper. Specimens from south of Georgia ascribed to *Porphyra umbilicalis* (L.) J. Agardh (Taylor 1960; Chapman 1971) need to be studied; they possibly represent, at least in part, southern records of *P. carolinensis*.

Porphyra leucosticta Thuret in Le Jolis 1863, p. 100
    Plants to 15 cm tall and 13 cm wide, epiphytic, epizoic, or saxicolous; one to a few blades arising from a basal or umbilicate holdfast, rarely stipitate; blades pink to purplish red, oblong or cordate, single layered, 25–50 μm thick, entire, rarely laciniate; margins entire, undulate, often folded; cells in surface view subquadrate to rectangular, 12–20 μm diameter, each with one lobed to stellate plastid and pyrenoid; monosporangia marginal; monoecious, spermatia to 64 or 128 per cell in eight tiers, 3–6 μm diameter, formed in elongate marginal or submar-

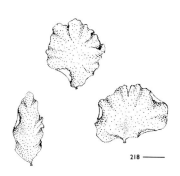

Figure 218. *Porphyra carolinensis*, scale 2 cm.
Figure 219. *Porphyra rosengurttii*, scale 1 cm.

ginal, colorless patches to 1.5 mm diameter and 1 cm long, patches parallel to each other and the blade axis; four to thirty-two carpospores per cell, 6–9 µm diameter, formed in elongate pink marginal bands.

Kapraun and Freshwater 1987; other records previously attributed to this species listed as *P. rosengurtii*.

From the intertidal in the Wrightsville Beach, N.C., area, November–March.

Distribution: Arctic, Newfoundland to New Jersey, ?Virginia, North Carolina, Brazil to Argentina, Falkland Islands, Norway to Portugal, Mediterranean, Black Sea, Azores.

At the time that Coll and Cox (1977) studied *Porphyra* in North Carolina, they could not attribute any specimens to *Porphyra leucosticta*, thus disagreeing with previous workers (e.g, Hoyt 1920; Williams 1948a). Recent work on the life history and chromosomes of the Carolina species of this genus (Kapraun and Freshwater 1987) has shown that *P. leucosticta* is included in our flora.

Porphyra rosengurtii Coll et Cox 1977, p. 157, figs 9–16.

Figure 219

Plants to 30 cm tall and 8 cm wide, epiphytic, epizoic, or saxicolous; one to a few blades arising from central or basal holdfasts; blades brownish and greenish to purplish red, reniform, oblong, sublanceolate, or cordate, single layered, 45–70 µm thick, entire to lanceolate; margins entire, undulate, often folded; cells in surface view oval to subquadrate and rectangular, 10–15 µm diameter, 15–25 µm long, each with one lobed to stellate plastid and pyrenoid; monosporangia marginal; monoecious, spermatia thirty-two to sixty-four per cell in two to four tiers, 5 µm diameter, formed in irregular marginal patches, not mixed with other reproductive cells; carpogonialike cells with long projections after fertilization producing four to eight carpospores per cell in one to two tiers, 9–11 µm diameter and 18–20 µm long, formed in elongate marginal patches.

Coll and Cox 1977; Kapraun 1980a; Kapraun and Luster 1980; Richardson 1986, 1987. As *P. leucosticta sensu* Hoyt 1920; ?Humm 1952; Williams 1948a, 1949; Stephenson and Stephenson 1952; Taylor 1960; ?Chapman 1971; Wiseman and Schneider 1976.

On rocks, oysters, and coarser algae from the intertidal of jetties, seawalls, and oyster beds from Beaufort, N.C., to Cumberland Island and possibly Marineland, November–May, occasionally remaining into July as dwarf plants.

Distribution: North Carolina (type locality), South Carolina, Georgia, ?northeastern Florida.

A detailed field and cultural study of this species in North Carolina was reported by Kapraun and Luster (1980). Specimens from Georgia (Chapman 1971) and Marineland (Humm 1952) listed as *Porphyra leucosticta* are potentially *P. rosengurtii* (Coll and Cox 1977). However, Kapraun and Freshwater (1987) recently reinstated *P. leucosticta* in North Carolina.

FLORIDEOPHYCIDAE

Plants filamentous and branched to membranous and pseudoparenchymatous, tubular and ligulate to broadly foliose, or crustose, with or without calcification; endophytic and endozoic to epiphytic, epizoic, and saxicolous; growth mostly apical or marginal, few with intercalary divisions, pit-connections present; cells regularly to irregularly arranged, polygonal, angular, subquadrate, rectangular, cylindrical, globose, to ellipsoid, mostly multinucleate with several variously shaped parietal plastids, a few with single plastids; a few with vegetative propagation by fragmentation or specialized propagules; most with alternate asexual and sexual generations, but some with direct development of sporophytes from sporophytes, gametophytes from gametophytes; motile cells unknown; asexual reproduction by monospores, bispores, tetraspores, polyspores, or paraspores; sexual reproduction oogamous, with nonmotile spermatia fertilizing carpogonia with extended trichogynes; fertilized carpogonia dividing directly into gonimoblast filaments (carposporophytes), or fertilized nucleus transferred from carpogonia to near or distant receptive auxiliary cells by connecting filaments (ooblasts), which then develop into carposporophytes; carposporophytes producing carposporangia from each vegetative cell or only terminally, each producing single carpospores; carposporophytes with or without surrounding vegetative filaments, loosely arranged (involucre) or compacted into a cortex (pericarp); carpospores producing free-living tetrasporophytes, or directly reverting to gametophytes.

Key to the orders

1. Auxiliary cells lacking... 2
1. Auxiliary cells present.. 6
2. Plants consisting of uniseriate, branched filaments .... Acrochaetiales
2. Plants not uniseriate, arranged into cortex and medulla 3

3. Alternation of isomorphic generations; carpogonia associated with sterile cell branches produced by supporting cells or chains of small nutritive cells produced after fertilization 4
3. Mostly alternation of heteromorphic generations; carpogonia not associated with sterile cell branches or chains of nutritive cells 5
4. Carposporophytes developing directly from fertilized carpogonia fused to hypogynous cells or long chains of small nutritive cells issued from fusion cells after fertilization .... Gelidiales
4. Carposporophytes developing from generative fusion cells composed of carpogonial and sterile cell branches produced by supporting cells .... Gracilariales
5. Growth uniaxial; vesicle cells present .... Bonnemaisoniales
5. Growth multiaxial; vesicle cells absent .... Nemaliales
6. Auxiliary cells produced prior to fertilization 7
6. Auxiliary cells produced from supporting cells after fertilization .... Ceramiales
7. Cell walls heavily impregnated with calcium carbonate; spores and gametes borne in conceptacles .... Corallinales
7. Plants rarely calcified, if so, calcium carbonate not impregnating cell walls; spores and gametes not borne in conceptacles 8
8. Auxiliary cells as terminal cells on two- or three-celled branches borne on supporting cells (procarpic) .... Rhodymeniales
8. Auxiliary cells terminal or intercalary in long or short, vegetative or specialized filaments (nonprocarpic), or as the supporting cells or other closely related cells (procarpic), but not in the above arrangement ...... Gigartinales

ACROCHAETIALES

Plants filamentous and saxicolous, epizoic, endozoic, primarily endophytic, epi-endophytic, or epiphytic, to 1 cm tall, arising from a persistent enlarged basal spore, a basal spore with accessory cells, a prostrate irregular filamentous system, filaments little to much branched, irregular, radial, lateral, to secund or opposite, uniseriate, branches ending abruptly or with unicellular hairs or multicellular hairlike extensions, the cells with one to several parietal, lobate, axial, stellate, or discoid plastids with none to several pyrenoids; asexual reproduction by mono-, bi-, tetra-, and polyspores, lateral, terminal, or opposite, sessile, or pedicellate, and globose, ovoid, to ellipsoid; sexual reproduction by spermatangia and carpogonia, monoecious or dioecious; spermatangia single or in clusters of two to several, terminal or lateral, sessile or on short pedicels; carpogonia flask shaped, simple, with extended trichogynes, sessile or pedicellate; carposporophyte developing directly from the fertilized carpogonium, naked, with short gonimoblasts and terminal or occasionally lateral or intercalary carposporangia each bearing a single carpospore.

This order is monotypic.

Acrochaetiaceae

This family has received a great deal of taxonomic attention, especially recently (Dixon and Irving 1977; Garbary 1979; Woelkerling 1983; Lee and Lee 1988). Several genera have been assigned to the family and variously segregated on the basis of plastid morphology and presence or absence of sexual reproduction. Because generic criteria can vary from one study to the next, and characteristics can differ according to location, season, and age, we agree with Dixon and Irvine (1977) that it is most suitable and least confusing to follow the one-genus concept proposed by Drew (1928, as *Rhodochorton*).

Audouinella Bory 1823, nom. cons.

In this genus we include *Acrochaetium*, *Colaconema*, and *Kylinia* of previous reports (Hoyt 1920; Williams 1948a, 1951; Searles and Schneider 1978). *Audouinella* is a persistent and sometimes obvious member of the epiphytic and epizoic flora from the area. Our treatment subscribes to the broad species concept proposed by Woelkerling (1971, 1973a, 1973b). Some species do not reproduce sexually and may actually be alternate stages of previously described gametophytes in the Nemaliales. For example, the gametophytic stage of *Scinaia complanata* has been shown to alternate with an acrochaetioid tetrasporophyte in Europe (van den Hoek and Cortel-Breeman 1970). A culture study of the life history of *Scinaia* in the southeastern United States is necessary to determine if it is linked with any of the *Audouinella* species listed below.

This treatment recognizes fifteen species in the flora; two of them, *Audouinella affinis* and *A. hoytii*, are considered endemic to the region. Acrochaetioids have been found that do not correspond to any of the taxa described below. Additional material and stages are needed for clarification. Specimens collected without obvious attachments and lacking sporangia are impossible to identify. It is best for the collector to leave these identified only as *Audouinella* sp.

Key to the species

1. Plant attached to the substrate by an obvious, persistent, single basal cell which may give rise to one or several accessory cells, an endophytic branched filament, or a prostrate filamentous system 2

1. Plant attached to the substrate by a multicellular prostrate system, the basal cell septate and persistent, or not obvious, obscured by divisions in the earliest stages 8

2. Plant minute, less than 250 μm tall . . . . *A. microscopica*

2. Plant larger, greater than 500 μm tall 3

3. Basal cells consistently globose or subglobose, obviously larger than the erect filaments they produce 4

3. Basal cells ovoid to panduriform, globose, or subglobose, the same size or only a few micrometers larger than the erect filaments they produce 7

4. Basal cell producing a single penetrating endophytic branched filament in addition to one or more erect filaments 5

4. Basal cell remaining undivided or occasionally producing a few accessory cells 6

5. Basal cell 12–15 μm diameter, erect filaments 8–13 μm diameter; dioecious, spermatangia forming corymbose clusters on lateral branches . . . .
. . . . *A. corymbifera*

5. Basal cell 7.5–12.5 μm diameter, erect filaments 3–5 μm diameter; monoecious, spermatangia paired, produced on opposite or whorled branches in a long series . . . . *A. ophioglossa*

6. Monosporangia 5–7.5 μm diameter, 11–15 μm long . . . . *A. hoytii*

6. Monosporangia 10–18 μm diameter, 18–27 μm long . . . . *A. affinis*

7. Basal cell remaining undivided and ovoid, elongate, to rectangular, occasionally with a flared disciform base . . . . *A. hallandica*

7. Basal cell, pyriform to panduriform, producing a prostrate filamentous system . . . . *A. dasyae* (in part)

8. Plants mostly endophytic or endozoic, upright axes less than 150 μm tall
. . . . *A. infestans*

8. Plants mostly epiphytic, epizoic, or saxicolous, upright axes greater than 200 μm tall 9

9. Plastids with more than one pyrenoid per cell . . . . *A. botryocarpa*

9. Plastids with only one pyrenoid per cell 10

10. Sporangia mostly borne singly or in pairs, not clustered at the base of lateral branches 11

10. Sporangia mostly borne in clusters of three or more on branched stalks at the base of lateral branches . . . . *A. daviesii*

11. Monosporangia 9–13 μm long 12

11. Monosporangia 15–40 μm long 14

12. Monosporangia 5–6 μm diameter; bisporangia, if present, 7–12 μm diameter, 12–18 μm long . . . . *A. bispora*

12. Monosporangia 6–10 μm diameter; bisporangia not present 13

13. Plants producing only single monosporangia, erect filaments to 1 mm tall arising from a uniseriate prostrate filament to small, compact pseudoparenchymatous disc .... *A. hypneae*
13. Plants producing single or two to three seriate monosporangia, erect filaments to 0.5 mm tall arising from a widespreading, much-branched, prostrate system, often with a persistent septate basal spore .... *A. densa*
14. Plastids stellate and parietal or axial, unicellular hairs common but not present on all plants, plants arising from an orbicular pseudoparenchymatous disc which is often later obscured by proliferating filaments ....
.... *A. secundata*
14. Plastids lobate and parietal or axial, unicellular hairs absent or rare, but terminal cells occasionally taper sharply to the tips, plants not arising from an orbicular pseudoparenchymatous disc in early stages 15
15. Prostrate system a widespreading mass of uniseriate filaments from an enlarged central pyriform to panduriform cell which is persistent but often obscured in older plants, uniseriate prostrate filaments giving rise to numerous erect filaments .... *A. dasyae* (in part)
15. Prostrate system a compact, often penetrating mass of uniseriate filaments, usually coalesced centrally to form a pseudoparenchymatous disc, basal spore not persistent .... *A. saviana*

Audouinella affinis (Howe et Hoyt) C. W. Schneider 1983, p. 4, figs 1a–1b.
  Acrochaetium affine Howe et Hoyt 1916, p. 118, pl. 15.
  Figures 220–224
  Plants epiphytic or epi-endophytic, to 4 mm tall, caespitose, arising from large globose to subglobose spores, spore walls persisting as basal cells, 14–26 μm diameter including walls 2.5–5 μm thick, half again as big as the cells which emanate from them; if embedded in host tissue, spores becoming vertically elongate to subpyriform or panduriform, if not, producing subcylindric obtuse or truncate foot extensions which penetrate to 24 μm, the extended spores 20–33 μm high; spore usually remaining simple, occasionally developing a few smaller accessory cells, rarely producing short, prostrate, irregularly branched filaments to five cells long; basal cell and occasionally accessory cells or basal filament cells producing one to four erect filaments, cells cylindrical, 5–14 μm diameter and 17–78 μm long, each with a parietal lobate plastid and one pyrenoid; for the most part, plants immediately branched from the distal end of the first cell of the erect filament, branching subdichotomous or subtrichotomous, appearing rigid below, more flexuous above, branching mostly subdichotomous to distinctly lateral, ultimate branches elongate virgate, 3–5.5 μm diameter, gradually tapering toward the apices, often terminating with incon-

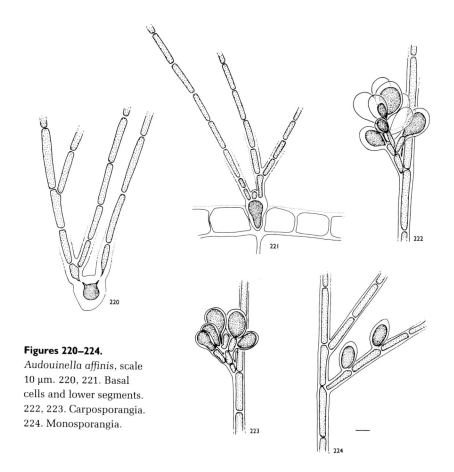

Figures 220–224.
Audouinella affinis, scale
10 μm. 220, 221. Basal
cells and lower segments.
222, 223. Carposporangia.
224. Monosporangia.

spicuous hairs; infrequent monosporangia lateral, secund, or, less commonly, terminal on upper portions of the plant, sessile or on one-celled pedicels, ovoid, obovoid, or oblong, 10–18 μm diameter, 18–27 μm long, occasionally on gameto-phytic plants; monoecious, spermatangia in proximity to sessile carpogonia, lateral or terminal, sessile or pedicellate, solitary or in groups of two or three; carposporophytes common, three to twelve spores, carpospores 8–25 μm diameter, 13–28 μm long.

Schneider 1983. As *Acrochaetium affine*, Howe and Hoyt 1916; Hoyt 1920. ?As *Audouinella alariae sensu* Kapraun 1980a.

This species is an obvious epiphyte of *Dictyota*, *Sargassum*, and other algae from deep (to 30 m) offshore and shallow subtidal habitats in the Beaufort and Wilmington, N.C., areas, at times turning the host plant red with a complete or marginal covering of fine red filaments, June–August.

Distribution: Endemic to North Carolina as currently known.

Hoyt (1920) listed several morphological characters useful in distinguishing *Audouinella affinis* from the previously described *A. hoytii*. Among these, the size of monosporangia seems to us to be the most reliable.

Audouinella bispora (Børgesen) Garbary 1979, p. 490.

   Chantransia bispora Børgesen 1910, p. 178, fig. 1.

   Figure 225

   Plants epi-endophytic, to 1.5 mm tall, arising from irregularly ramified, spreading prostrate systems which remain superficial or penetrate host tissues, prostrate cells cylindrical to irregular, to 17.5 μm in greatest dimension, all capable of producing an erect filament; erect filaments radially to irregularly branched and rebranched, infrequent below, more above, the branches sharply to narrowly angled from the axes which bear them, tapering toward the apices; cells cylindrical, 2.5–8 μm diameter, 10–32.5 μm long in the main axes, 1–3 μm diameter in the ultimate segments, occasionally ending in long multicellular, lightly pigmented hairlike extensions, unicellular hairs unknown, each cell containing a well-developed parietal plastid with a single large pyrenoid; monosporangia terminal or lateral on short branches to adaxial on longer axes, sessile or on one-celled pedicels, single or paired, narrow, ovoid to oblong, 5–6 μm diameter, 9.5–12.5 μm long, bisporangia situated as monosporangia, single or paired, broad, ovoid to ellipsoid, 7.5–11.5 μm diameter, 12.5–17.5 μm long; gametangia unknown.

   Schneider 1983; Searles 1987, 1988.

   Collected from 25 to 30 m offshore in Onslow Bay, May–June, and from 17 to 21 m on Gray's Reef, July–August.

   Distribution: North Carolina, Georgia, Virgin Islands.

Audouinella botryocarpa (Harvey) Woelkerling 1971, p. 37.

   Callithamnion botryocarpum Harvey 1855, p. 563.

   Figures 226 and 227

   Plants epiphytic, epi-endophytic, or saxicolous, to 1 cm tall, caespitose, arising from irregularly branched prostrate systems with entangled filaments free from one another or coalesced into pseudoparenchymatous discs, occasionally supplemented by corticating rhizoids, descending cells, or lower portions of erect filaments; base giving rise to several erect filaments, much and irregularly to unilaterally branched, sometimes terminating abruptly in multicellular hairlike extensions or gradually tapering toward the tips; cells of main axes cylindrical (8.5–)10–20(–30) μm diameter, 30–120 μm long, and 6–15 μm diameter in the ultimate segments, each cell containing a single parietal lobate plastid with one to several pyrenoids, unicellular hairs unknown; monosporangia terminal on short branches and adaxial, near the bases of branches, secund or scattered, sessile or on one-celled pedicels, single, paired, or rarely in groups of three to five, ovoid, 12–28 μm diameter, 18–43 μm long; tetrasporangia cruciate, with range of dimensions, shape and position similar to monosporangia but in general somewhat larger than monosporangia on the same plant; monoecious or dioecious,

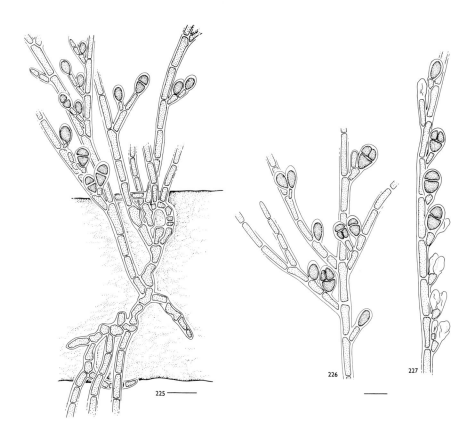

Figure 225. *Audouinella bispora*, with monosporangia and bisporangia, scale 20 μm.
Figures 226, 227. *Audouinella botryocarpa*, scale 25 μm. 226. Filament with monosporangia and tetrasporangia. 227. Sporangium walls remaining on filament after spore release.

spermatangia single to clustered on short lateral branches, sessile or pedicellate, ovoid to globose, to 6 μm diameter and long; carpogonia terminal on one- or two-celled pedicels usually near the axils of laterals, pedicel cells occasionally giving rise to a short, unbranched sterile branch shortly before or after fertilization, carposporophytes with numerous terminal ovoid carposporangia to 25 μm diameter, 35 μm long.

Searles and Schneider 1978; Schneider 1983.

Known only as an obvious epiphyte of *Codium decorticatum* and *C. fragile* from Pivers Island seawall and Radio Island jetty, Beaufort, N.C., but probably more widespread, July–March.

Distribution: North Carolina, Ireland, Netherlands, southern Africa, Australia, New Zealand.

Monosporangia are found in all collections, and tetrasporangia have been observed on December plants. Sexually reproductive plants are unknown in the flora.

Although reported from only a few disjunct regions, this species will probably prove to be much more widespread upon taxonomic investigation of related species (Guiry et al. 1987).

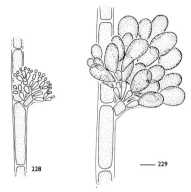

Figures 228, 229.
Audouinella corymbifera, scale 10 μm. 228. Spermatangia. 229. Carposporangia.

Audouinella corymbifera (Thuret in Le Jolis) Dixon 1976, p. 590.

Chantransia corymbifera Thuret in Le Jolis 1863, p. 107.

Figures 228 and 229

Plants epi-endophytic, 2–3 mm tall, arising from large, globose spores, spore walls persisting as basal cells, 12–15 μm diameter, which produce irregular, contorted, branched filaments that deeply penetrate host tissue with cells 8–13 μm diameter and 25–40 μm long; basal cells and occasional cells of the internal filament giving rise to erect branched axes, secund to alternate with branches more abundant above and not often rebranched; cells of erect filaments cylindrical, 7–16 μm diameter and 22–55 μm long, each cell containing a well-developed parietal lobed plastid with a single pyrenoid, unicellular hairs unknown; monosporangia terminal on short branches or lateral, near the bases of branches, sessile or on one-celled pedicels, single, ovoid, 8–10 μm diameter, 15–18 μm long, occasionally associated with gametangia; dioecious, spermatangia lateral on main axes, formed in a corymbose cluster on one- or two-celled pedicels, colorless, 4–5 μm diameter; carpogonia lateral on main axes, on one- or two-celled pedicels near the bases of branches, producing corymbose carposporophytes with terminal carposporangia, ovoid, 9–15 μm diameter, 14–18 μm long.

Schneider 1983. As *Acrochaetium corymbiferum*, Hoyt 1920. As *Acrochaetium bornetii*, Searles and Schneider 1978.

On *Dasya* from Beaufort, N.C., harbor, collected only once, May 1907.

Distribution: North Carolina, Bermuda, England, France, Mediterranean.

For a clarification of the correct epithet refer to Dixon and Irvine (1977).

Audouinella dasyae (Collins) Woelkerling 1973b, p. 545, figs 10–31.

Acrochaetium dasyae Collins 1906b, p. 191.

Figures 230–234

Plants epiphytic, to 3 mm tall, caespitose, arising from spores and the widespreading, uniseriate, prostrate filamentous axes they produce; spore walls persisting as elongated pyriform to panduriform basal cells, not greater in size than the cells they produce, rarely remaining simple, often becoming obscured by descending rhizoids and entangled prostrate filamentous axes; creeping prostrate system giving rise to several erect filaments, moderate to much and irregularly branched, often tapering to the tips, cells cylindrical, 6–12(–16) μm diameter and 25–70(–90) μm long in lower portions, 6–10 μm diameter and 15–60 μm long in the ultimate segments, each having a single parietal plastid with one pyrenoid, hairs unknown; monosporangia mostly single, occasionally in pairs, secund to scattered, sessile or on one-celled pedicels, ovoid, 7–12(–16) μm diameter, 16–27 μm long, borne on vegetative or gamete-producing plants; monoecious or dioecious, spermatangia terminal or lateral on specialized short lateral branched axes, globose to ovoid, 2–4 μm diameter, 3–5 μm long; carpogonia scattered, sessile or stalked, carposporophytes with numerous terminal and lateral ovoid carposporangia, 9–12 μm diameter, 16–24 μm long.

Kapraun 1980a, *pro parte*; Schneider 1983. As *Acrochaetium robustum*, Williams 1948a, 1951; Taylor 1960. As *Acrochaetium dasyae*, Aziz 1967.

Epiphytic on *Dasya*, *Sargassum*, and other hosts from shallow and deep subtidal environments in Onslow Bay, April–August.

Distribution: Prince Edward Island to Virginia, North Carolina, southern Florida, West Africa.

Plants in our area are often found with an early basal system of an ovoid to pyriform basal cell and one or two smaller accessory cells from which oblique upright filaments arise (see figure 231).

Figures 230–234.   *Audouinella dasyae*, scales 20 μm. 230–232. Filaments arising from bases with enlarged ovoid basal cells. 232–234. Monosporangia.

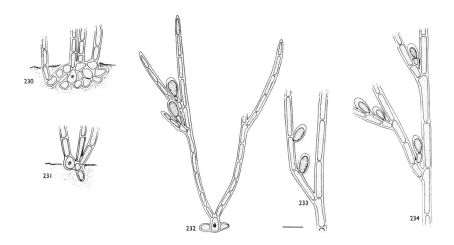

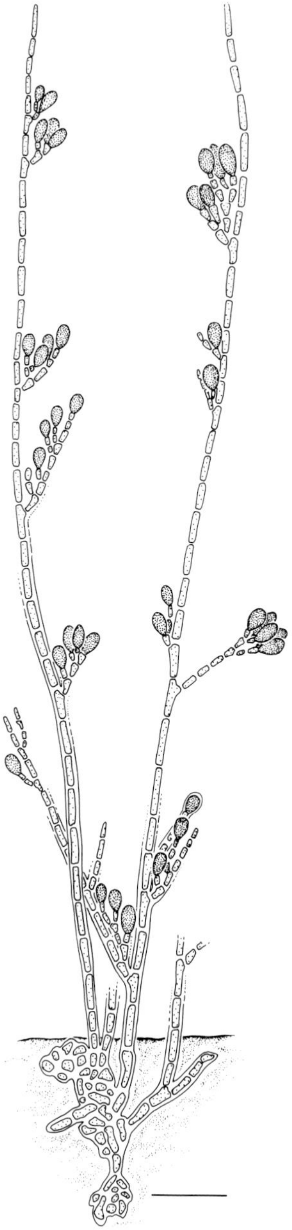

Figure 235. *Audouinella daviesii*, with monosporangia, scale 50 μm.

Audouinella daviesii (Dillwyn) Woelkerling 1971, p. 28, figs 7, 22.

Conferva daviesii Dillwyn 1809, p. 73, suppl. pl. F. Figure 235

Plants epiphytic, epi-endophytic, epizoic, or epi-endozoic, to 6 mm tall, caespitose, arising from spreading to penetrating branched filamentous systems, centrally coalesced into irregular pseudoparen-chymatous discs or entangled fungiform masses, original spores not persistent; erect filaments sparsely to freely and irregularly to unilaterally branched, branches often ending in long multicellular hairlike projections, 3–4 μm diameter and 60–120 μm long; cells of main axes cylindrical (6–)9–12 (–20) μm diameter, (8–)15–50(–70) μm long, and only slightly tapering to tips, 7.5–10 μm diameter where extensions not formed, each cell containing a single parietal lobate plastid with one pyrenoid; unicellular hairs unknown; monosporangia terminally clustered in groups of three or more on branched stalks or paired on one- or two-celled stalks adaxially on the lowermost cells of lateral branches, rarely single and scattered, ovoid, 7–13 μm diameter, 8–20 μm long; tetrasporangia cruciate, paired on unicellular pedicels, solitary or in groups of three, in similar positions to monosporangia, 13–22 μm diameter, 12–36 μm long; monoecious or dioecious, spermatangia borne terminally to laterally in small clusters on short lateral branches, ovoid to globose, to 4 μm diameter and 5 μm long; carpogonia terminal on one-celled branches, later displaced to a lateral position; carpo-sporophytes with terminal ovoid carposporangia, 9–18 μm diameter, 18–26 μm long.

Schneider 1983. As *A. hallandica sensu* Kapraun 1980a, *pro parte*.

Known only from the shallow subtidal of Masonboro Inlet jetty, Wrightsville Beach, N.C., January and April; the intertidal of Fort Macon jetty, June; and offshore from 30 m in Onslow Bay, June.

Distribution: Newfoundland to Virginia, North Carolina, Caribbean; elsewhere, widespread in temperate to tropical seas.

Only sporangiate plants have been found in the flora (Schneider 1983), and these agree in all respects with recently published descriptions (Woelkerling 1971, 1973b; Dixon and Irvine 1977). Multicellular hairlike projections were found only in the January collection and tetrasporangia only in April.

Audouinella densa (Drew) Garbary 1979, p. 490.
   Rhodochorton densum Drew 1928, p. 168, figs 17–24.
   Figures 236 and 237
   Plants epiphytic and epizoic, 0.2–0.3(–0.5) mm tall, caespitose, arising from irregularly branched prostrate systems, originally developed from spores with walls persisting around once-septate cells, filaments occasionally coalesced to form a pseudoparenchymatous disc, especially centrally, giving rise to several erect filaments, sparse to much branched, irregular, alternate, opposite, and secund, usually with moderate branching above, the branches sometimes tapering toward the tips; often ending in colorless unicellular hairs; cells of the upright filaments cylindrical to clavate, 4–8(–13) µm diameter and 7–18 µm long, each containing a single lobate to stellate plastid with a single pyrenoid; one, two, or three seriate monosporangia terminal, alternate secund, rarely opposite, sessile or on one-celled pedicels, single or in pairs, rarely clustered; monosporangia ovoid to ovoid-ellipsoid, 7–10 µm diameter, 10–13 µm long.
   Schneider 1983. As *Acrochaetium hummii* (*nomen nudum*), Aziz 1965. As

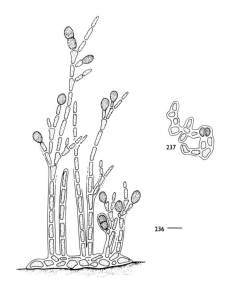

Figures 236, 237.
Audouinella densa, scale
10 µm. 236. Single and
seriate monosporangia.
237. Young germling with
original bi-cells.

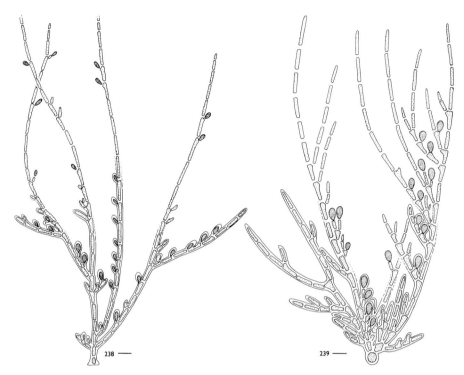

Figure 238. *Audouinella hallandica*, with monosporangia, scale 10 µm.
Figure 239. *Audouinella hoytii*, with monosporangia, scale 10 µm.

Acrochaetium densum, Stegenga and Vroman 1976. As *A. thurettii sensu* Kapraun 1980a.

Epiphytic on various animals and algae, in particular *Chaetomorpha aerea*, from the intertidal and shallow subtidal of Cape Lookout jetty, Fort Macon jetty, and the Pivers Island seawall, Beaufort, N.C., and Masonboro Inlet jetty at Wrightsville Beach, N.C., year-round. Also a single collection from 4.5 m in Port Royal Sound, S.C.

Distribution: North Carolina, South Carolina, Brazil, Netherlands, Japan, Kurile Islands, Korea, Pacific Mexico, California, Washington.

Germinating spores divide into two equal cells after attaching to a host (see figure 237). One or both of these gives rise to an erect and/or prostrate filament, occasionally one initial remaining undivided for some time. The prostrate system consists of either irregular filamentous branched axes or becomes compacted into a pseudoparenchymatous disc. For a discussion of the taxonomy of this species, refer to Schneider (1983).

Audouinella hallandica (Kylin) Woelkerling 1973a, p. 82, figs 1–4.
Chantransia hallandica Kylin 1906, p. 123, fig. 8.
Figure 238
Plants epiphytic, to 1 mm tall, arising from ovoid, obovate, subglobose, discoid, elongated, to rectangular spores, spore walls persisting around basal cells; if discoid, basal cell to 20 µm, otherwise 5–8 µm diameter, and the same size

or slightly larger than cells of the filaments it produces; basal cell remaining simple, producing one or two (to three) erect filaments, branching irregularly lateral to secund, not closely set, beginning in the lowermost portions of the plant, occasionally tapering toward the apices, hairs unknown; lower cells cylindrical, 3–7 μm diameter, 10–30 μm long, each containing a single parietal lobate plastid with one pyrenoid; monosporangia single or paired, lateral to secund, more rarely terminal, sessile or on one-celled pedicels, obovate to ovoid, 5–10 μm diameter, 8–11(–16) μm long; dioecious, spermatangia opposite on short lateral branches, rare; carpogonia sessile, rare.

Woelkerling 1973a; Schneider 1976; Kapraun 1980a, pro parte; Schneider 1983; Richardson 1987. As *Acrochaetium dufourii*, Hoyt 1920; Taylor 1960; Aziz 1965. As *Acrochaetium sargassi*, Taylor 1960; Aziz 1965.

Known as an epiphyte of *Dictyota* and *Sargassum* from shallow subtidal habitats near Beaufort and Wilmington, N.C., Cumberland Island jetty, and on a variety of red and brown algae from 17 to 35 m in Onslow Bay, June–November.

Distribution: North Carolina, South Carolina, Georgia, Bermuda, Sargasso Sea, southern Florida, Caribbean, Brazil, USSR to France, West Africa.

This taxon is clearly distinguishable when attached to a host by a broad disc of cell wall materials. The disc does not stain with aniline blue. The original spore wall remains around a simple elongated basal cell the same size or slightly larger than the cells of the erect filaments it produces.

A study by Stegenga and Borsje (1977) demonstrated that a plant provisionally determined as *Audouinella hallandica* is the gametophyte of a heteromorphic plant whose sporophyte was tentatively determined as *Acrochaetium polyblastum* (Rosenvinge) Børgesen.

Audouinella hoytii (Collins) C. W. Schneider 1983, p. 10, fig. 2j.
  Acrochaetium hoytii Collins 1908, p. 134.
  Figure 239
  Plants epiphytic, 0.2–1.3 mm tall, arising from persistent, large, globose spores, spore walls persisting around obvious basal cells, 9–15(–28) μm diameter including obvious walls 2–3 μm thick; basal cells superficial to slightly embedded in host tissue, elongating to 30 μm and obviously broader than the cells they produce, remaining simple, rarely forming one to a few accessory cells, producing one or two (to four) erect filaments; erect branching infrequent to frequent, often secund, with ultimate branches elongated, tapering to the apices for the most part, 2–3 μm diameter, hairs unknown; cells of the main axes cylindrical, 5–7 μm diameter, 10–20 μm long, each with a single parietal plastid and a large central pyrenoid; monosporangia secund or lateral on upper portions of the plant, sessile or on one-celled pedicels, oblong, 5–7.5 μm diameter and

11–15 μm long; monoecious, spermatangia lateral and terminal on short lateral branching systems; carpogonia sessile, lateral; carposporophytes developing as short branching systems, lateral on main axes or near the bases of branches, with four to sixteen terminal carpospores 4–9 μm diameter, 7–13.5 μm long.

Schneider 1983; Searles 1987, 1988. As *Acrochaetium hoytii*, Collins 1908; Hoyt 1920; Williams 1948a, 1951.

Known as an epiphyte of *Dictyota*, *Padina*, and other algae from shallow subtidal habitats in the Beaufort, N.C., area, June–October, and from 17 to 21 m on Gray's Reef, July.

Distribution: North Carolina (type locality), South Carolina, Georgia.

Although similar in morphology to *Audouinella alariae* (Jonsson) Woelkerling (1973b) from New England and Europe, the main axes, carposporangia, and monosporangia of *Audouinella hoytii* are only half the size of those in that species. *A. hoytii* has also been suggested as a possible alternate stage of *A. pectinata* (Kylin) Papenfuss (Woelkerling 1971).

Audouinella hypneae (Børgesen) Lawson et John 1982, p. 172.

Chantransia hypneae Børgesen 1909, p. 2, fig. 2.

Figure 240

Plants epiphytic or epi-endophytic, to 1 mm tall, arising from small, compact to spreading filamentous prostrate systems composed of one or more irregularly bent, branched filaments which can coalesce centrally into pseudoparenchymatous discs; each cell of the disc or ramified filaments capable of producing an erect filament; erect branches little to radially and much branched below, alternate, often pectinate secund above, cells cylindrical, 6–10 μm diameter and 16–22 μm long below, tapering to 4–7 μm diameter and 16–30 μm long in the ultimate segments, each containing a single parietal plastid with an obvious pyrenoid, occasionally bearing unicellular hyaline hairs to 60 μm long; monosporangia single, rarely paired, lateral in long secund series or rarely terminal, occasionally scattered, sessile or on one-celled pedicels, oblong ovoid, 6–10 μm diameter, 9–13 μm long; gametangia unknown.

Schneider 1983; Richardson 1987. As *Acrochaetium seriatum*, Williams 1948a; Aziz 1965; Chapman 1971. As *Acrochaetium hypneae*, Aziz 1965.

On a variety of plant and animal hosts from Beaufort, N.C., area jetties; on *Arthrocladia*, *Spatoglossum* and other hosts from 28 to 32 m offshore in Onslow Bay and Long Bay, July–September; and on *Hypnea* in intertidal of Cumberland Island jetty, August.

Distribution: North Carolina, South Carolina, Georgia, Bermuda, southern Florida, Caribbean, Brazil, East Indies, West Africa.

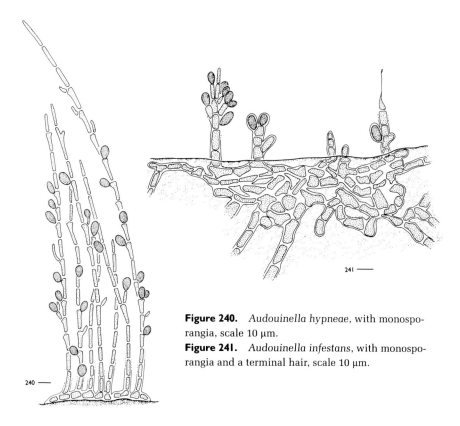

Figure 240. *Audouinella hypneae*, with monosporangia, scale 10 μm.

Figure 241. *Audouinella infestans*, with monosporangia and a terminal hair, scale 10 μm.

For a discussion of the taxonomy and variation in this species refer to Schneider (1983).

Audouinella infestans (Howe et Hoyt) Dixon 1976, p. 590.
  Acrochaetium infestans Howe et Hoyt 1916, p. 116, pl. 14.
  Figure 241

Plants endophytic or endozoic with emergent axes to 100 μm tall exclusive of hairs, internal axes consisting of tortuous, serpentine, or labyrinthine, irregularly branched filaments, sparse and straight for considerable distances, often crowded between host cells or compacted and subparenchymatous, cells irregular to elongate (2–)4–8(–13) μm diameter, 6–60 μm long; emergent axes to ten or more cells high but generally less, simple or sparingly and irregularly branched, occasionally bearing terminal hairs to 170 μm long, cells cylindrical to ellipsoid, 3–7 μm diameter, 6–30 μm long, containing a single parietal lobate plastid with a single pyrenoid; monosporangia single or paired, terminal or lateral, often on the adaxial surface, sessile or on one-celled pedicels, ovoid to ellipsoid, 4–6(–9) μm diameter and 6–15 μm long; gametangia unknown.

Schneider 1983. As *Acrochaetium infestans*, Howe and Hoyt 1916; Hoyt 1920. As *Kylinia infestans*, Taylor 1960. As *Colaconema infestans*, Woelkerling 1973a.

In various hydroids, *Dictyota*, and *Sargassum* from 30 m in Onslow Bay and shallow subtidal habitats near Beaufort and Wilmington, N.C., June–August.

Distribution: North Carolina (type locality), Bermuda, Sargasso Sea, British Isles, France, Mediterranean, Japan, Korea.

Audouinella microscopica (Nägeli) Woelkerling 1971, p. 33, figs 10, 23A.

Callithamnion microscopicum Nägeli in Kützing 1849, p. 640.

Figures 242–244

Plants minute epiphytes, 20–150(–220) μm tall exclusive of hairs, arising from subglobose, isodiametric, to cylindrical spores with or without thick spore walls, 5–15 μm diameter, close to the same size as the cells they produce; basal cells giving rise to one to six erect and/or lateral filaments, commonly arcuate, simple to much branched and secund, irregular, or opposite, branches short, tapering; cells barrel shaped, isodiametric, to cylindrical, 3–10 μm diameter, 3–11 μm long, having a single lobate to stellate plastid with one pyrenoid, often bearing hyaline hairs to 50 μm long; monosporangia terminal and secund to opposite, sessile or on one-celled pedicels, ovoid, 4–10(–15) μm diameter and 6–15(–22) μm long; tetrasporangia rare, tetrahedral, sessile, to 20 μm diameter and 30 μm long; dioecious, spermatangia scattered over erect filaments, sessile or on one-celled pedicels, ovoid, 2–4 μm long; carpogonia terminal to lateral, sessile; carposporophytes with six to fourteen terminal or lateral, spherical to ovoid carposporangia, 6–8 μm diameter and 7–10 μm long.

Kapraun 1980a; Schneider 1983. As *Acrochaetium parvulum*, Hoyt 1920. As *Acrochaetium compactum*, Williams 1948a. As *Acrochaetium trifilum*, Aziz 1965. As *Kylinia crassipes*, Taylor 1960.

Known as an epiphyte on a variety of intertidal and shallow subtidal seaweeds from the Beaufort and Wilmington, N.C., areas, year-round.

Distribution: Maritimes to Virginia, North Carolina, Bermuda, southern Florida, Gulf of Mexico, Caribbean, Brazil; elsewhere, widespread in polar to tropical seas.

Tetrasporangia have not been observed in the flora and only rarely elsewhere (Woelkerling 1972). For a discussion of the taxonomy and variation of this species refer to Schneider (1983).

Figures 242–244.
Audouinella microscopica, scale 10 μm.

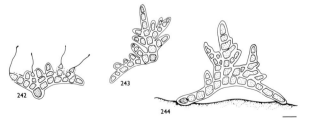

Audouinella ophioglossa C. W. Schneider 1983, p. 13, figs 3a–3p.

Figures 245 and 246

Plants epi-endophytic, 1–3.2 mm tall, arising from persistent, large, globose spores, spore walls persisting around obvious basal cells, 7.5–12.5 μm diameter, producing single unbranched or forked branched irregularly contorted filaments that deeply penetrate host tissue with cells 2–4 μm diameter, or remaining as unicellular bases; basal cells giving rise to one or two erect branched axes, secund to alternate and opposite, branching more common above; cells of the erect filaments cylindrical, 4–5 μm diameter, 45–55 μm long, in lower segments; each cell containing a single lobate to spiral, parietal plastid with a single inconspicuous pyrenoid; unicellular hairs terminal or lateral, to 2 mm, often associated with gametangial branches; monosporangia terminal on short branches or lateral, secund to alternate or opposite, sessile or on one-celled pedicels, single or in pairs, ovoid to ellipsoid, 6–12.5 μm diameter, 12.5–20 μm long, commonly associated with gametangia; monoecious, spermatangial branches in a whorl of four, opposite paired or, less commonly, one per node, in long series in the upper parts of the plant; spermatangia paired on one-celled pedicels, globose, 2.5 μm diameter; carpogonia lateral, opposite, or adjacent to monosporangia, occasionally forming in a whorl of spermatangial branches, producing only one carposporophyte per node, often in a long interrupted series; carposporangia terminal, globose, 10–15 μm diameter.

Schneider 1983; Searles 1987, 1988.

In *Dudresnaya crassa*, from 22 to 35 m offshore in Onslow Bay, June–September, and from 17 to 21 m on Gray's Reef, July–August.

Distribution: North Carolina (type locality), Georgia, Puerto Rico, and possibly wherever *Dudresnaya crassa* is found.

Along with *Audouinella corymbifera*, this is the only other species in the flora with a large basal cell remaining in the original spore wall that produces a penetrating endophytic filament (see figure 245). *A. ophioglossa* is distinguished by a smaller basal cell, narrower erect filaments, its monoecious habit, and the disposition of spermatangial branches.

Audouinella saviana (Meneghini) Woelkerling 1973b, p. 560, figs 56–60.

Callithamnion savianum Meneghini 1840b, p. 511.

Figures 247–249

Plants epiphytic to epi-endophytic, to 4 mm tall, arising from compact, mostly superficial prostrate systems of short, simple or branched filaments, free from each other to coalesced into irregular pseudoparenchymatous discs, the original spores not recognizable; prostrate filaments the same size or distinctly larger than erect ones; erect filaments with irregular or alternate to secund branching,

Figures 245, 246. *Audouinella ophioglossa*, scale 20 μm. 245. Filament with endophytic base and spermatangia. 246. Sporangia, some with released spores.

Figures 247–249. *Audouinella saviana*. 247. Basal system with erect filaments, scale 50 μm. 248, 249. Monosporangia and tetrasporangia, some with released spores, scale 25 μm.

Figures 250, 251. *Audouinella secundata*. 250. Filament with monosporangia and hairs, scale 25 μm. 251. Stellate plastid morphology, scale 10 μm.

cells of main axes and branches cylindrical, 7–14 μm diameter and 20–60 μm long, sometimes 4–6 μm diameter in the ultimate branches, unicellular hairs absent or rare, each cell containing a single parietal lobate plastid; monosporangia single or paired, adaxially secund to irregularly arranged, sessile or on one- to three-celled elongate pedicels, ovoid to ellipsoid, 10–15 μm diameter, 18–27 μm long; tetrasporangia cruciate, with shape and positions similar to monosporangia, often borne simultaneously, 17–24 μm diameter, 26–34 μm long; monoecious, spermatangia sparse on short branchlets near the base of branches, carpogonia sparse and solitary on lower segments near the base of branches, carpospores 9–13 μm diameter and 18–21 μm long.

Schneider 1976, 1983; Hall and Eiseman 1981; Richardson 1987. As *Acrochaetium thuretii*, Aziz 1965.

Usually epiphytic on *Codium* but also known on *Dictyopteris* and other coarse algae from shallow subtidal and deep offshore habitats in the Carolinas, May–October; from Cumberland Island jetty, August; and Haulover Cove, Indian River, Fla., year-round.

Distribution: Maritimes, Newfoundland, Massachusetts, Virginia, North Carolina, South Carolina, Georgia, northeastern Florida, Bermuda, Caribbean; elsewhere, widespread in temperate to tropical seas.

Audouinella secundata (Lyngbye) Dixon, 1976, p. 590.
Callithamnion daviesii var. *secundatum* Lyngbye 1819, p. 129, pl. 41, figs B4–B6.
Figures 250 and 251
Plants epiphytic or epizoic, to 3 mm tall, scattered to caespitose, arising from unistratose to bistratose, generally compact pseudoparenchymatous discs, the original spores not persistent; discs giving rise to several erect filaments, usually freely and irregularly branched with laterals often consisting of only one to five cells, with or without lateral and terminal unicellular hairs to 300 μm long; cells of main axes cylindrical 8–15(–20) μm diameter, 15–100 μm long, each with an axial or parietal stellate plastid and single central pyrenoid; monosporangia lateral or terminal on shorter lateral branches, sessile or pedicellate, single, paired, or rarely in groups of three, ovoid, (6–)10–20 μm diameter, (10–)15–26(–32) μm long; gametangia unknown.

Kapraun 1980a; Schneider 1983. As *Trentepohlia virgatula*, Johnson 1900. As *Acrochaetium virgatulum*, Williams 1948a, 1951. As *Kylinia virgatula*, Taylor 1957. As *Colaconema secundata*, Woelkerling 1973a.

Epiphytic on a variety of North Carolina intertidal algae, April–August, uncommon offshore.

Distribution: Arctic, Newfoundland to Virginia, North Carolina, Sargasso Sea,

USSR to Portugal, Mediterranean, West Africa, Canary Islands, southern Africa, Korea, Australia.

Although some authors maintain a specific distinction between *Audouinella secundata* and *A. virgatula* (Harvey) Dixon (e.g., Dixon and Irvine 1977; Kornmann and Sahling 1978), we have observed morphological plasticity in North Carolina plants (Schneider 1983) which supports the merger of the two species as suggested by Woelkerling (1973a, 1973b).

Easiest to distinguish in early developmental stages, the prostrate system develops from an orbicular, parenchymatous group of cells. This cluster may later proliferate as prostrate filamentous axes, obscuring the central disc and becoming more similar to mature stages of *Audouinella dasyae* and *A. saviana*.

NEMALIALES

Plants erect and/or prostrate, small and filamentous to large and pseudoparenchymatous; pseudoparenchymatous plants multiaxial with loose or compact aggregations of filaments, with or without calcification; cells uninucleate with one or more plastids and pyrenoids; asexual reproduction by tetraspores or polyspores; tetrasporophytes and gametophytes heteromorphic or isomorphic; monoecious or dioecious, spermatangia formed from surface vegetative cells or as terminal to subterminal clusters on lateral axes; carpogonial branches originating from cortex or terminal on lateral branches, with or without associated nutritive cells; carposporophyte developing directly from the fertilized carpogonium or the hypogynous cell, auxiliary cells lacking; carposporangia liberating single carpospores or four carpotetraspores.

An artificial grouping which contains diverse, well-differentiated families, the Nemaliales includes gelatinous forms as well as pseudoparenchymatous ones. Recent cultural studies have greatly increased our understanding of the life histories of many members of this order, which for many years was assumed to contain only haplontic organisms. The Acrochaetiaceae, Gelidiaceae, and Naccariaceae, until recently placed within the Nemaliales (Taylor 1960), are now recognized as orders.

Key to the families

1. Plants mostly soft, gelatinous; cortication loose and discontinuous . . . .
 . . . . Helminthocladiaceae
1. Plants firm; cortication compact and continuous . . . . Galaxauraceae

Helminthocladiaceae

Plants erect, soft and gelatinous to partly calcified; axes terete, multiaxial in structure, unbranched or dichotomously to irregularly branched; axial filaments bearing radiating cortical, branched, photosynthetic, lateral axes; pigmented cells with a single large parietal plastid, usually without a pyrenoid; asexual reproduction (where known) by monospores or tetraspores on microscopic acrochaetioid alternate stages (see *Audouinella*, p. 192); monoecious or dioecious, spermatangia borne in loose to dense clusters, terminal and occasionally subterminal on lateral axes; carpogonial branches one to eight celled, terminating on, or lateral to, vegetative axes; cystocarps immersed, not obvious to the unaided eye, involucres present or lacking; carposporangia terminal on the gonimoblasts, occasionally subterminal, intercalary, or in short series.

Helminthocladia J. Agardh 1851, nom. cons.

Plants soft, lax, irregularly to radially and dichotomously branched, one or more axes arising from small discoid holdfasts; axes terete, gelatinous, becoming firmer with age; medulla comprising a compact core of longitudinally arranged, enlarged, rectangular axial cells; cortex comprising few to several assimilatory dichotomously to subdichotomously branched lateral axes which issue perpendicularly from the central core, loosely to compactly arranged, terminal cells smaller, equal to, or greater than subterminal and other cortical cells, with or without terminal hairs; tetrasporophytes known only for some, only in culture; monoecious or dioecious, spermatangia clustered terminally, subterminally, or paniculately on lateral axes; carpogonial branches mostly three celled, but with as few as one to as many as five cells, borne laterally on cortical assimilatory filaments, carpogonia conical; gonimoblasts regularly branched, arising from the upper cell or both products of the initial division of the fertilized carpogonium; carposporangia terminal and/or subterminal; carposporophyte surrounded by few to many loosely arranged involucral filaments, rhizoidal downgrowths lacking.

Helminthocladia andersonii Searles et Lewis 1982, p. 166, figs 1–20.
  Figure 252
  Plants grayish pink to rosy red, to 5 cm tall; axes terete, 1.1 mm diameter near base, 0.3 mm diameter in ultimate branches; primary branching subdichotomous; secondary branches produced adventitiously around major axes; medullary filaments easily separable; terminal cells of assimilatory filaments smaller than subterminal cells; monoecious, spermatangial mother cells terminal or subterminal on assimilatory filaments; carpogonial filaments primarily three celled,

Figure 252. *Helmintho-cladia andersonii*, scale 0.5 cm.

but with as few as one to as many as five cells, lateral on supporting cell above the second dichotomy of an assimilatory branch; carpogonia conical; carpogonium dividing transversely, forming three cells after fertilization, each producing gonimoblast initials; supporting cell remaining smaller than adjacent vegetative cells; sterile involucral filaments arising from cells above and below the supporting cell and, if present, from first cell of the adjacent dichotomy, together forming a shallow cup; carposporophytes to 140 μm diameter and 90 μm long; carposporangia terminal and subterminal, to 10 μm diameter and 11 μm long; tetrasporophytes unknown.

Searles and Lewis 1982.

Known only from a single offshore outcropping at a depth of 27 m in Onslow Bay, June–August.

Distribution: Endemic to North Carolina as currently known.

This species exhibits several morphological characteristics which intergrade with those of the genus *Helminthora* J. Agardh (1852). Its reproductive features, however, are most closely aligned with *Helminthocladia*. For a clarification of these two genera, see Searles and Lewis (1982).

Galaxauraceae

Plants erect, bushy, and soft in texture or lightly calcified; axes terete to flattened, multiaxial, dichotomously branched; inner cortex pigmented or hyaline, filamentous or pseudoparenchymatous, of several simple or divided cell rows, extending to the outer cortex or beyond as multicellular pigmented hairs, the outer cortex consisting of hyaline or pigmented inflated cells, subquadrate to elongate or flattened in cross section, polyhedral and appearing parenchymatous in surface view, lightly or moderately calcified in some; pigmented cells usually with a single stellate plastid; asexual reproduction by monospores or tetraspores, in some produced on microscopic acrochaetioid alternate stages (see

Audouinella, p. 192), in others borne on slightly heteromorphic macroscopic alternate stages, and in others unknown; monoecious or dioecious, spermatangia formed in sori or scattered over the plant surface; carpogonial branches usually three celled, borne on the inner forks of the inner cortex, the hypogynous cell surrounded by a ring of nutritive cells; cystocarps embedded, pericarp of slender compacted filaments with a single ostiole to the plant surface; carposporangia seriate or terminal on the gonimoblast filaments.

Key to the genera

1. Plants lightly calcified, carposporangia terminal .... Galaxaura
1. Plants not calcified, carposporangia seriate .... Scinaia

Galaxaura Lamouroux 1812
Plants bushy, moderately soft to firm and wiry, dichotomously to pseudodichotomously branched, smooth or covered with pigmented multicellular hairs produced in the cortex; axes lightly to moderately pigmented, the latter often appearing segmented by the lack of carbonate at the dichotomies; medulla and inner cortex hyaline, outer cortex pigmented; outer cortical cells inflated, polyhedral in surface view, forming a pseudoparenchymatous epidermis, smooth in some, bearing one-celled subspinulose extensions, two- or three-celled special filaments, or unicellular hyaline hairs in others; tetrasporangia tetrapartite, sessile or pedicellate, lateral on specialized filaments; spermatangia formed in conceptacles, carpogonial branches usually three celled, formed in the inner cortex, associated with nutritive cells, after fertilization surrounded by a thin-walled ostiolate pericarp; carposporangia produced terminally on the gonimoblast filaments; gametophytes moderately dimorphic, at the microscopic and/or macroscopic level, from sporophytes.

Galaxaura obtusata (Ellis et Solander) Lamouroux 1816, p. 262.
  Corallina obtusata Ellis et Solander 1786, p. 113, table 22, fig. 2.
  Figure 253
  Plants erect, to 10 cm tall, forming nearly globose, bushy clusters arising from small discoid holdfasts; plants pink to rosy red, lightly calcified, opaque when dry, much and dichotomously branched; branches terete, jointed at the dichotomies, smooth or bearing abundant pigmented multicellular hairs; axes 1.5–4 mm diameter, the internodes one to four times as long as broad; apices obtuse; tetrasporangia scattered over the plant; tetrasporophytes having an inner cortical layer of contiguous enlarged hyaline cells, each outwardly supporting slender stalk cells which bear one or two smaller pigmented surface cells, polyhedral, contiguous in surface view, 27–45 μm diameter; cystocarps immersed

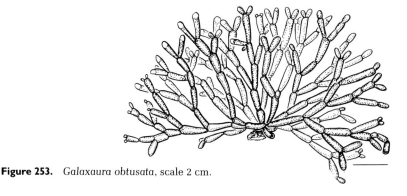

Figure 253. *Galaxaura obtusata*, scale 2 cm.

(not obvious macroscopically), somewhat irregular in shape, 160–530 μm diameter, carposporangia terminal, obovoid to ellipsoid, 14–27 μm diameter, (27–)34–50(–56) μm long; spermatangia ovoid, (2.5–)4–6 μm diameter, 5–9 μm long, formed in conceptacles 145–330 μm diameter; gametophytes having an inner cortical layer of contiguous enlarged hyaline cells outwardly bearing the surficial pigmented cells either directly or by means of short papillate outgrowths, surface cells similar to those on sporophytes; macroscopically, sporophytes and gametophytes indistinguishable.

Schneider and Searles 1973, 1979; Schneider 1976.

In our area, known only as one of the dominant species from Frying Pan Shoals, 65 km southeast of Cape Fear in Onslow Bay, at a depth of 23–32 m, May–November. Commonly associated with *Eucheuma isiforme* var. *denudatum*.

Distribution: North Carolina, Bermuda, southern Florida, Caribbean, Brazil, West Africa, Canary Islands, Indian Ocean, Japan, China.

Although most specimens are smooth, and this feature could be considered characteristic (Taylor 1960), we have found some plants with abundant multicellular pigmented hairs covering at least some of the axes. These hairs are unbranched and to 2 mm long and 15–20 μm diameter, with cells to 180 μm long.

Scinaia Bivona-Bernardi 1822

Plants bushy, soft, repeatedly dichotomously branched, arising from small discoid holdfasts; axes terete to slightly or completely flattened, slippery, with or without constrictions at the nodes and obvious axial strands; medulla composed of relatively few hyaline longitudinal filaments; inner cortex composed of loosely entwined dichotomous filaments, the outer two or three cells of which are pigmented; outer cortex composed of hyaline inflated cells, polyhedral in surface view, quadrate to elongate rectangular in section, forming a continuous pseudoparenchymatous epidermis, smooth in most, in some species interrupted by smaller, regularly spaced pigmented cells; monosporangia globose, scattered over the plant, formed between the epidermal cells; monoecious or dioecious, spermatangia globose, ovoid to pyriform in small sori, continuous or scattered over the plant, formed between the epidermal cells; carpogonial branches three celled, formed in the inner cortex, associated with nutritive cells, after fertiliza-

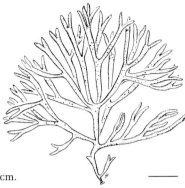

Figure 254. *Scinaia complanata*, scale 0.5 cm. ———

tion surrounded by a thin-walled ostiolate pericarp; carposporangia produced seriately on the gonimoblast filaments; tetrasporophytes (where known) small, uniseriate, branched filamentous, as in *Audouinella* (see p. 192).

Scinaia complanata (Collins) Cotton 1907, p. 260.
   Scinaia furcellata var. *complanata* Collins 1901, no. 836; 1906a, p. 110.
   Figure 254
   Plants rosy red, erect, 1–8 cm tall, 1–6 mm diameter, seven to nine times dichotomous at sexual maturity, arising from small discoid holdfasts; branches turgid, slippery, smooth, terete to slightly flattened, if the latter, especially below nodes; rarely constricted at the nodes, usually with an obvious axial strand of few (six to eight) to thirty longitudinal filaments; the internodes four to seven times as long as broad; apices obtuse, the ultimate segments fusiform, broader at the distal end; epidermal cells hyaline and polygonal, triangular, to subcircular, 12–44 µm diameter in greatest dimension, subquadrate to elongate rectangular in section, occasionally interrupted by smaller pigmented cells; subsurface layer composed of pyriform to subglobose heavily pigmented cells, loosely arranged; monoecious, cystocarps immersed, obvious macroscopically, 145–350 µm diameter, 130–220 µm tall, carposporangia narrowly obovate to ellipsoid, 4–5 µm diameter, 7.5–12.5 µm long, in a terminal series of two or three; spermatangia globose, 2.5–7.5 µm diameter, single or in patches of two (rarely, up to four) between epidermal cells, forming over broad portions of the plant; tetrasporophytes known in culture but not in the field.
   Schneider and Searles 1973; Schneider 1976; Kapraun 1980a; Searles 1981, 1987, 1988.
   Known occasionally from the shallow subtidal at Masonboro Inlet and commonly offshore from 18 to 47 m in Onslow Bay and from 17 to 21 m on Gray's Reef, June–September.
   Distribution: North Carolina, Georgia, Bermuda, southern Florida, Caribbean, Brazil, Mediterranean, Japan, Galapagos Islands, Pacific Mexico, Costa Rica.
   Plants from the region for the most part exhibit the characteristics described for *Scinaia complanata* var. *intermedia* Børgesen (1916). However, the value of recognizing this variety has been questioned owing to the broad range of intermediates described by Taylor (1942). For the most part, Onslow Bay and

Georgian specimens are terete, narrow, and show an obvious central strand, but broader or more complanate plants have been found. We therefore choose not to place our plants under the subspecific designation *intermedia*.

Van den Hoek and Cortel-Breeman (1970) have established a heteromorphic life history for this taxon in culture. Their gametophytic isolates from the Mediterranean alternate with an acrochaetioid tetrasporophyte that is capable of self-reproduction. Thus far, the alternate generation for gametophytes in our region has not been established either in culture or from field collections. However, because gametophytes are distinctly seasonal here, it is possible that the filamentous sporophyte is present as an overwintering stage.

GELIDIALES

Plants cartilaginous with erect and prostrate branching axes, flattened to terete; cells uninucleate with one or more plastids; each axis with a distinct apical cell, the uniaxial condition often obscured by subsequently produced longitudinal filaments, therefore appearing multiaxial; cortex single layered to multilayered with small pigmented cells, medulla of larger longitudinal, nonpigmented cells more or less traversed by rhizoid-like filaments (rhizines); tetrasporophytes and gametophytes isomorphic, or gametophytes lacking; dioecious, spermatangia produced in obvious hyaline sori below the apices; carpogonia sessile or on short branches, gonimoblasts arising directly from fertilized carpogonia which fuse with hypogynous cells, or from long nutritive filaments which issue from fusion cells after fertilization; auxiliary cells lacking.

Gelidiaceae

Plants exhibiting isomorphic alternation of generations; tetrasporangia scattered or in spatulate tips, cruciate or tetrahedral; cystocarps laminal, terminal, or more distal, often below branching axes, unilocular or bilocular; other characteristics as for the order.

Gelidium Lamouroux 1813, nom. cons.

Cystocarps biloculate, protruding on both surfaces of the blades; other characteristics as for the family.

The considerable taxonomic confusion in this genus has been reviewed recently (Santelices 1976; Dixon and Irvine 1977), and the literature for the *Gelidium* spp. in our flora contained evidence supporting a need for revision. Plants from our region have been assigned to six binomials, but we recognize only two.

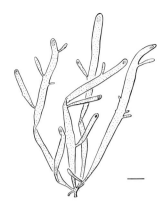

Figure 255. *Gelidium americanum*, scale 2 mm.

Key to the species

1. Plants 0.5–2(–4) cm tall, erect axes broadly flattened to ligulate, simple to sparingly branched, pinnate to palmate . . . . *G. americanum*
1. Plants 2–5(–15) cm tall, erect axes terete to somewhat compressed, smaller plants much branched and pinnate, larger plants irregularly branched but often with trifurcate tips . . . . *G. pusillum*

Gelidium americanum (W. R. Taylor) Santelices 1976, p. 173, figs 28–33.

Pterocladia americana W. R. Taylor 1943, p. 154, pl. IV, fig. 1.

Figure 255

Plants gregarious, spreading by terete to subterete multibranched stoloniferous axes to form dense turfs; erect axes 0.5–2(–4) cm tall, terete below, becoming flattened to 1 mm wide, often simple and ligulate or sparingly branched, alternate to pinnate, often palmate or fastigiate in regenerating areas, branch tips broadly acute, obtuse, or apiculate; internally, medulla with several axial filaments and numerous rhizines, cortex with one to a few layers of cells; sporangial sori forming in proximal regions of broadened simple blades or ultimate axes of branched plants, often forming in V-shaped rows covering almost the full width and breadth of the blade; tetrasporangia globose to ovoid, 30–45 μm in greatest dimension; cystocarps protruding on both surfaces, with one ostiole on each surface.

Kapraun 1980a. As *Pterocladia americana*, Taylor 1943, 1960; Chapman 1971. As *Gelidium coerulescens sensu* Hoyt 1920; Williams 1948a, 1951. As *G. pusillum sensu* Humm 1952; Schneider 1976; Wiseman and Schneider 1976. As *G. pusillum* var. *conchicola sensu* Williams 1948a.

Common year-round as a pinkish red, intertidal fuzz on oysters and other hard substrates of the Beaufort, N.C., town front and nearby jetties and seawalls; Fort Fisher outcropping; Fort Johnson and Bears Bluff, S.C.; and Marineland, Fla. Also, on subtidal shell and rock at Cape Lookout and less frequently from deep offshore rock outcroppings in the Carolinas and Georgia. In more exposed habitats such as the Fort Macon jetty, the habit is more robust and branched, the population more compact. In more protected areas such as the Pivers Island seawall in Beaufort, N.C., plants are simple and ligulate, forming less compact populations.

Distribution: North Carolina, South Carolina, Georgia, Bermuda, northeastern Florida to Gulf of Mexico, Caribbean, Brazil.

Gelidium pusillum (Stackhouse) Le Jolis 1863, p. 139.
   Fucus pusillus Stackhouse 1795, p. 16, pl. vi.
   Figures 256 and 257
   Plants gregarious, spreading by terete to subterete branched stoloniferous axes to form dense clumps or turfs; erect axes 2–5(–15) cm tall, mostly terete or subterete to compressed, especially in the upper portions of the plant, axes to 0.4 mm wide; smaller plants from simple to subpinnate and pinnate, larger plants tending to be more lax and subdichotomous to irregular, but both forms often with distinct broadened trifurcate tips to 0.5 mm wide; apices narrowly acute to acute; internally, medulla of tightly compact axial filaments and a few rhizines, mostly just below the cortex formed of one to a few cell layers; sporangial sori forming in terminal or lateral, distinctly swollen, branches, tetrasporangia ovoid to spherical, 25–38 μm in greatest dimension, occupying almost the full width and breadth of the swollen branch; cystocarps single, bilocular, embedded below the apex of a main axis or, more often, centrally in short, spindle-shaped branchlets.
   As *Gelidium crinale*, Johnson 1900; Hoyt 1920; Williams 1948a; Humm 1952; Taylor 1960; Wiseman and Schneider 1976; Kapraun 1980a; Richardson 1986, 1987. As *G. pulchellum*, Williams 1948a. As *G. corneum sensu* Curtis 1867; Williams 1948a; Wiseman 1978; Wiseman and Schneider 1976.
   Common on rigid intertidal substrates from the entire area, year-round.
   Distribution: Massachusetts to Virginia, North Carolina, South Carolina, Georgia, Bermuda, northeastern Florida to Gulf of Mexico, Caribbean, Brazil, Uruguay, Falkland Islands, British Isles to Portugal, Azores, Mediterranean, Black Sea, West Africa, China, California.
   Because of the great morphological variation and taxonomic confusion in this genus (Dixon 1961; Stewart 1968; Santelices 1976), we have chosen to follow the circumscription of *Gelidium pusillum* (including *G. crinale*) adopted by Dixon and Irvine (1977) for the British Isles. Following Dixon and Irvine, however, does cause some confusion with plants from our region. The terete *Gelidium* here, *G. pusillum*, includes none of the specimens previously assigned to *G. pusillum* by several workers (see citations for *G. americanum*), but rather contains all specimens previously known in the flora as *G. crinale*. The type of *G. pusillum* from England (Stackhouse 1795) is described as having cylindrical or slightly compressed erect axes (Dixon and Irvine 1977). This description does not fit the broadly flattened, strap-shaped plants from the Carolinas previously assigned to the species and presently named *G. americanum*. We have observed morphological variation in *G. pusillum* similar to that found in the British Isles and conclude that *G. pusillum* is the correct name for terete, slightly compressed specimens from our area. The epithet *pusillum*, however, belies the size of a majority

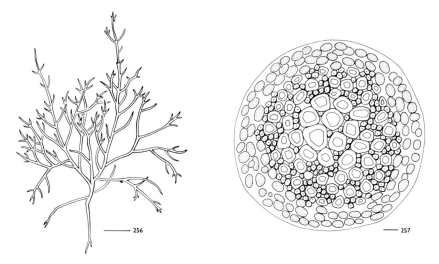

Figures 256, 257. *Gelidium pusillum.* 256. Habit, scale 0.5 cm. 257. Cross section, scale 10 µm.

of plants, which are generally larger than G. *americanum.* Dwarf populations of G. *pusillum* are difficult to distinguish from dwarf populations of G. *americanum;* therefore it is best to make identifications on mature, fertile plants.

In light of the British circumscription of *Gelidium pusillum,* western Atlantic records of both G. *crinale* and G. *pusillum* (e.g., Taylor 1960; Dawes 1974) warrant additional observation and clarification. A study of *Gelidium* in the Gulf of Mexico (Stewart and Norris 1981) has followed the Dixon and Irvine (1977) taxonomy. It is interesting that after considerable time, we have returned to the two species of *Gelidium* that were originally outlined by Hoyt (1920) for the Carolinas, separated by the same criteria, although with different names.

BONNEMAISONIALES

Plants erect or creeping, small, and filamentous, uniseriate to polysiphonous, to large, pseudoparenchymatous, and soft in texture, uniaxial; cells uninucleate with one or more plastids; with or without vesicle cells; tetrasporophytes and larger gametophytes heteromorphic, sporangia tetrahedral; monoecious or dioecious, spermatangia formed in superficial clusters spiraling around the axial row; carpogonial branches two to four celled, arising from basal cells of lateral axes, the hypogynous cells producing several small nutritive cells; gonimoblasts arising directly from the fertilized carpogonium or from the hypogynous cell; single carposporangia terminal, liberating one carpospore; pericarp lacking to well developed.

Key to the families

1. Plants with embedded carposporophytes, pericarps lacking . . . Naccariaceae
1. Plants with superficial carposporophytes, pericarps present . . . .
. . . . Bonnemaisoniaceae

Bonnemaisoniaceae

Plants much and bilaterally or radially branched, soft and gelatinous, attached by small discoid holdfasts or spreading stoloniferous axes with tufts of rhizoids; axial row producing short, determinate, lateral branches forming a compact cortex; tetrasporophytes (where known) uniseriate to polysiphonous branched filaments or creeping *Hymenoclonium*-type branched crusts; carpogonia fusing with fertile branch cells after fertilization, gonimoblasts develop to form superficial swollen cystocarps with pericarps.

Asparagopsis Montagne 1841
Gametophytes bushy, soft, alternately to radially branched, arising from stoloniferous, creeping axes and attached by tufts of rhizoids; erect axes terete to slightly compressed, slippery, distinct, and thicker than branches; branches rebranched and crowded; vesicle cells present; dioecious, spermatangia formed in superficial, dense sori on short, thickened branches off indeterminate axes; carpogonial branches two to four celled, formed in short branches off indeterminate axes; cystocarps on short stalks, ovoid and subpyriform to subglobose, with an irregular mass of carpospores at maturity.

Sporophytes ("*Falkenbergia*" stage) filamentous and prostrate with erect axes, much and irregularly branched, attached by ventrally issued, branched, multicellular holdfasts; axes terete, divided into three spirally produced pericentral cells, ecorticate; growth by transverse divisions of apical cells, trichoblasts lacking; branching alternate to irregular, directly from central buds off pericentral cells; sporangia tetrahedrally divided, sessile, produced from pericentral cells.

Asparagopsis taxiformis (Delile) Trevisan 1845, p. 45.
   Fucus taxiformis Delile 1813, pp. 151, 295, pl. 57, fig. 2.
Sporophytic stage:
   Falkenbergia hillebrandii (Bornet) Falkenberg 1901, p. 689.
      Polysiphonia hillebrandii Bornet in Ardissone 1883, p. 376.
   Figure 258
   Gametophytes purplish and brownish to rosy red, erect, 5–20 cm tall from basal stolons, with branches 2–5 cm long; axes and branches lax, slippery, terete, to 2 mm diameter, abundantly branched and rebranched to several orders, each axis and subaxis becoming narrowly pyramidal, lower portions naked; axial filament initially surrounded by three pericentral cells appearing tubular at maturity; pericentral cells dividing to form a continuous cortex of five to six layers in main axes, ultimate branchlets of only three uncorticated cells in cross section; dioecious, cystocarps borne on short determinate branches, swollen and appearing club shaped, including the stalk.

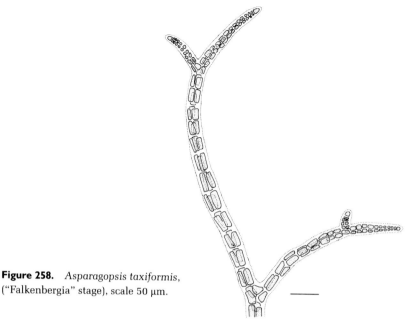

Figure 258. *Asparagopsis taxiformis,*
("Falkenbergia" stage), scale 50 μm.

Sporophytes rosy red, prostrate, and spreading to tangled, erect, to 2 cm tall, irregularly branched to few orders, main axes inconspicuous, attached by ventral multicellular, branched holdfasts; axes 30–80 μm diameter, with three pericentral cells surrounding slender axial filaments in 60 spirals, ecorticate, with prominent apical cells; segments irregular, 0.5–1.2 diameters long.

As *Falkenbergia hillebrandii,* Searles 1987, 1988.

Infrequent offshore from 32 to 34 m in the Frying Pan Shoals region southeast of Cape Fear, 17 to 21 m on Gray's Reef, and 30 to 35 m on Snapper Banks, July–August.

Distribution ("*Falkenbergia*" stage): North Carolina (new record), Georgia, Bermuda, southern Florida, Gulf of Mexico, Caribbean, Brazil.

Chihara (1961) found *Falkenbergia hillebrandii* to be the tetrasporophytic stage of *Asparagopsis taxiformis.* Thus far, gametophytes have not been found in the flora, despite the plant being widespread in the tropical and subtropical western Atlantic. All "*F. hillebrandii*" specimens collected in our area are vegetative, leading Searles (1987) to suggest that the species is maintained by fragmentation. Culture studies of local material are needed to clarify whether they maintain other reproductive capabilities.

Naccariaceae

Plants bushy, soft, and gelatinous, irregularly radially branched, attached by small discoid holdfasts; axial row producing short, determinate, lateral, deciduous axes forming a more or less densely compact cortex, loosely to compactly corticated below by rhizoidal downgrowths from the basal cells of lateral axes; tetrasporophytes (where known) small, uniseriate, branched filaments, as in *Audouinella* (p. 192); carpogonium fuses with hypogynous cell after fertiliza-

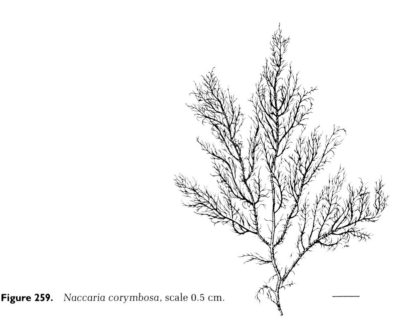

Figure 259. *Naccaria corymbosa*, scale 0.5 cm. ————

tion, gonimoblasts develop internally, forming obvious swollen cystocarps, pericarps lacking.

Naccaria Endlicher 1836
Plants bushy, soft, much and irregularly branched, arising from small discoid holdfasts; axes terete, gelatinous, tapering or occasionally nodose; growth by oblique divisions of the apical cell producing a single axial filament which bears short, determinate, lateral, more or less deciduous, photosynthetic axes; rhizoidal downgrowths from basal cells of lateral axes forming a pseudoparenchymatous cortication of the axial row below, becoming somewhat separated from the axial row with age; monoecious or dioecious, spermatangia formed in superficial clusters at the base of lateral axes, producing a spiral arrangement around the axial row; carpogonial branches two to four celled, cystocarps either causing a swelling of the branches or not, often subterminal.

Naccaria corymbosa J. Agardh 1899, p. 109.
Figure 259
Plants rosy red, erect, 2–15 cm tall, to 1 mm diameter in the main axes, 30–100 μm diameter in uncorticated axes, abundantly branched and rebranched to several orders; branches lax, delicate, gelatinous, terete, to somewhat nodose at times, irregularly branched; axial row of corticated axes composed of long cylindrical narrow cells, 4–6 μm diameter, 95–200 μm long, surrounded by enlarged isodiametric pericentral cells, 80–100 μm diameter; lateral axes sparingly branched, 27–43(–150) μm long, moniliform, cells subglobose, truncate oval, or cask shaped, to 13 μm diameter and long, mostly deciduous on the larger branches and main axes; presumably dioecious, cystocarps borne subterminally or in a median position, immersed in uncorticated fertile branches, constricted at the base, swollen to 120–130 μm diameter at the midpoint of the carposporo-

phyte, and not obvious to the unaided eye; carposporangia terminal, obovate, 10–13 μm diameter, 17–20 μm long; spermatangia unknown.

Searles and Leister 1980.

Known from 24 to 30 m offshore in Onslow Bay, July–August.

Distribution: North Carolina, Bermuda, Florida Keys.

CORALLINALES

Plants calcareous, crustose, and spreading, with lower surfaces wholly attached to the substrate, occasionally unattached, or parasitic, or upright and branching from basal crusts, multiaxial; branches rigid throughout (nonarticulated) or articulated with calcified segments (intergenicula) and noncalcified nodes (genicula), flattened to terete, branches dichotomous to oppositely pinnate and irregular; crustose forms organized into three layers: a lower layer (or hypothallus) of one to several compact filaments oriented parallel or perpendicular to the substrate, in some the parallel filaments are organized into decumbent arching tiers (coaxial), a middle layer (or perithallus) of filaments produced by the hypothallus roughly perpendicular to the substrate with or without large trichocytes (megacells), and an upper layer (or epithallus) of one or more layers of mostly compressed cover cells; some thin crusts with trichocytes in the hypothallus; one parasitic genus with hyaline filaments inside articulated genera producing obvious reproductive conceptacles on the surface; articulated plants organized in two layers: an inner layer (or medulla) of short to long parallel cells, mostly in arching tiers of similar-sized cells, and an outer layer (or cortex) of round to rectangular, densely pigmented cells obliquely ascending from the outer medulla; genicula of one or more tiers of long, thick-walled medullary cells, cortex often lacking; tetrasporophytes and gametophytes isomorphic; sporangia and gametangia immersed in elevated conceptacles either on crusts or laterally or terminally on erect branches, with one or more pores; sporangia zonately divided, occasionally bisporangiate, with or without paraphyses; dioecious, spermatangia formed on short, crowded filaments on floors and in some walls of male conceptacles; procarpic, carpogonial branches two celled, one (monocarpogonial) or more (polycarpogonial) borne on supporting cells which act as auxiliary cells after fertilization; gonimoblast filaments arising from a large fusion cell of many auxiliary cells, carposporangia large and terminal.

This order is probably the most difficult group of algae as a whole to investigate because many genera and species, particularly the crustose forms, are unidentifiable without decalcification and reproductive conceptacles. Books by Johansen (1981) and Woelkerling (1988) are invaluable aids to the study of these coralline red algae.

Corallinaceae

Of the thirteen genera recognized in the flora, four are articulated and nine are crustose. A genus with one questionable record, *Melobesia*, is included in the key.

Key to the genera

1. Plants with upright portion articulated (jointed) 2
1. Plants crustose or in part branched, but never articulated 5
2. Genicula of a single tier of elongated cells, intergenicula with medulla of uniform-sized tiers of cells, conceptacles axial 3
2. Genicula of more than one tier of cells, intergenicula with medulla of variable-sized tiers of cells, conceptacles lateral . . . . Amphiroa
3. Branching primarily dichotomous, occasionally irregular . . . . Jania
3. Branching primarily pinnate, occasionally in part dichotomous or irregular 4
4. Plants coarse, main axes mostly greater than 500 μm diameter; conceptacles formed terminally on pinnae and lacking any subtending branches . . . . Corallina
4. Plants delicate, main axes less than 500 μm diameter; tetrasporic and cystocarpic conceptacles axial, later becoming subtended by new branches issued from the fertile intergeniculum . . . . Haliptilon
5. Crusts epiphytic and thin, perithallus absent or generally less than five cells thick 6
5. Crusts saxicolous or unattached and thick, perithallus more than five cells thick at maturity 9
6. Tetrasporangial conceptacles each with several pores . . . . Melobesia
6. Tetrasporangial conceptacles each with a single pore 7
7. Crusts with hypothallial cells vertically elongated and palisadelike, secondary pit-connections present between adjacent cells . . . . Titanoderma
7. Crusts with horizontally elongated and spreading hypothallial cells, secondary pit-connections absent between adjacent cells, cell fusions present . . . 8
8. Crusts originating from central four-celled germinating elements, trichocytes terminal on hypothallial filament . . . . Fosliella
8. Crusts originating from central eight-celled germinating elements, trichocytes absent or intercalary in hypothallial filaments . . . . Pneophyllum
9. Plants unattached and branched . . . . Lithothamnion
9. Plants crustose without branched uprights 10
10. Tetrasporangial conceptacles each with several pores 11
10. Tetrasporangial conceptacles each with a single pore 13
11. Hypothallus coaxial . . . . Mesophyllum

11. Hypothallus noncoaxial 12
12. Conceptacle roofs to three cells thick; gonimoblasts marginal in conceptacles
.... Leptophytum
12. Conceptacle roofs four or more cells thick; gonimoblasts central in conceptacles .... Phymatolithon
13. Trichocytes vertical in perithallus, secondary pit-connections absent between adjacent cells, cell fusions present .... Neogoniolithon
13. Trichocytes lacking, secondary pit-connections present between adjacent
cells, cell fusions absent .... Lithophyllum

Articulated Corallines

Amphiroa Lamouroux 1812
Plants erect with some prostrate branches, repeatedly branched and bushy, articulated, calcified except at the nodes (genicula), with several uprights arising
from inconspicuous to conspicuous basal crusts, at times forming turfs lacking
basal crusts; branching basically dichotomous and in one plane, but usually irregular to some degree; calcified intergenicula terete to flattened, mostly three
or more times as long as broad; genicula of few to several tiers of medullary cells, rarely with one tier, cortex lacking; medulla of arching transverse
tiers of similar-length elongated cells, secondary pit-connections present, tier
length variable, patterned or long alternating with short tiers; cortex with few
to several layers, composed of small rounded cells radiating from sharply delineated medullary cells; tetrasporangia, or bisporangia in some, clustered in
dome-shaped conceptacles formed laterally on intergenicula, each conceptacle
with a single pore formed by degeneration of central cortical roof cells; dioecious, gametophytes generally more abundant than tetrasporophytes; spermatangia formed in low-relief, single-pored conceptacles with flat floors similar to
tetrasporangial conceptacles; mono- or polycarpogonial, carpogonial branches
two celled, borne on supporting cells which form the basal layer of female
conceptacles, later acting as auxiliary cells and fusing; carposporophytes borne
within singled-pored conceptacles, nearly all gonimoblast cells becoming carposporangia.

Although two species have been reported in the flora, only one is recognized
at present.

Amphiroa beauvoisii Lamouroux 1816, p. 299.
Figure 260
Plants erect, flabellate, saxicolous or conchicolous, heavily calcified, pink-
ish to rosy red, 2–6(–12) cm tall, with basal spreading crusts giving rise to sev-

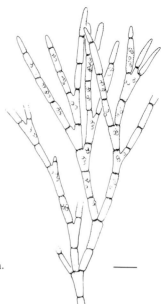

Figure 260. *Amphiroa beauvoisii*, scale 0.5 cm. ———

eral erect axes, some lower branches secondarily becoming prostrate and spreading; axes subterete to terete in lowermost portions, distinctly flattened above but occasionally subterete with increasing calcified layers in older specimens, segments 0.5–1.3(–1.7) mm wide, 0.5–1 mm thick, and 1.5–8(–10) mm long; axes dichotomously branched, mostly all in one plane, with some irregular and trichotomous branches; adventitious branches forming on older specimens and in wounded areas; lower intergenicula short rectangular to cuneate, upper intergenicula linear rectangular to cuneate, some forked, apices obtuse; medulla of three or four arching tiers of parallel compacted elongated cells, 7.5–12.5 µm diameter, (52–)75–105 µm long, then one tier of compacted shortened cells, 12–40 µm long; cortex 15–95 µm thick and composed of rounded angular to subquadrate, densely pigmented cells, 5–15 µm diameter, obliquely radiating from the medulla; genicula 42–77.5 µm long, comprised of two to five arching tiers of medullary cells and irregularly disposed cortical cells, calcification lacking on faces, loose and cracking at margins; zonate tetrasporangia and bisporangia obovoid, 17–25 µm diameter, 38–63 µm long, formed in raised, round to ellipsoid conceptacles scattered or crowded over the intergenicula, 270–850 µm diameter.

Schneider 1976; Kapraun 1980a. As *A. brasiliana*, Hoyt 1920; Williams 1948a, 1949, 1951; Taylor 1960. As *A. fragilissima sensu* Hoyt 1920.

Common offshore in Onslow Bay from 17 to 60 m, and rare to frequent from the shallow subtidal of area jetties and outcroppings, from Cape Lookout to Myrtle Beach, year-round.

Distribution: North Carolina, South Carolina, southern Florida, Caribbean, Brazil, Uruguay, Portugal, Azores, Mediterranean, southwest Indian Ocean, Gulf of California; probably more widespread in tropical and subtropical areas (pending taxonomic investigation of similar species).

Amphiroa beauvoisii is the only articulated coralline in the flora with more than one tier of medullary cells in the genicula and is easily recognizable by this anatomical trait. Inshore specimens of this taxon are mostly all large and robust. A continuum from narrow, spindly plants to thick, robust ones can be found throughout the offshore environment, similar to the variation reported in this highly variable species elsewhere (Norris and Johansen 1981). An examination of several offshore Carolina specimens reported by Hoyt (1920) as *Amphiroa fragilissima* show marked flattening of the axes and other features characteristic of *A. beauvoisii*. Numerous dredgings in Onslow Bay have failed to re-collect *A. fragilissima* while at the same time providing data that *A. beauvoisii* is common and variable in the area (Schneider 1976).

Tetrasporic plants of *Amphiroa beauvoisii* are common during the warmer months, but fertile gametophytes are as yet unknown from our area.

Questionable Record

Amphiroa fragilissima (Linnaeus) Lamouroux 1816, p. 298.
   Corallina fragilissima Linnaeus 1758, p. 806.
   ?Williams 1951; ?Humm 1952.

Williams's Carolina collections reported as *Amphiroa fragilissima* cannot be located. We have observed one of Humm's (1952) specimens of *A. fragilissima* plants from Marineland, Florida; it is possibly *Jania* but definitely not *Amphiroa* based upon vegetative anatomical features including a single tier of medullary cells in the genicula.

Corallina Linnaeus 1758

Plants erect, repeatedly branched, and bushy, articulated, calcified except at the nodes (genicula), with several uprights arising from conspicuous basal crusts; branching oppositely pinnate and basically in one plane; axial intergenicula terete to somewhat flattened, often cuneate, with or without marginal expansions, if present appearing lobed or winged; lateral nonaxial pinnae similar to axial intergenicula or narrower to long cylindrical; genicula of a single tier of long, thin, thick-walled medullary cells, cortex lacking; medulla of arching transverse tiers of similar length, elongated cells, secondary pit-connections lacking, tiers of similar length throughout; cortex with few to several layers, composed of small rounded cells radiating from sharply delineated medullary cells; intergenicula with conceptacles usually unbranched (lacking horns); several tetrasporangia clustered in expanded conceptacles formed terminally on axial or lateral intergenicula, each conceptacle with a single central pore; dioecious, spermatangia formed in single-pored terminal conceptacles with concave floors; carpogonial branches two celled, borne on supporting cells which form the basal layer of female conceptacles, later acting as auxiliary cells and fusing to form thin, broad conceptacle fertile floors; carposporophytes borne within single-pored terminal conceptacles, nearly all gonimoblast cells becoming carposporangia.

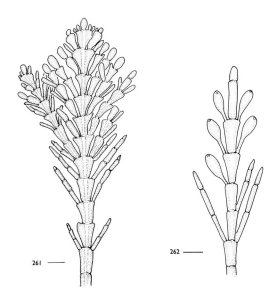

Figures 261, 262.
Corallina officinalis, scale
1 mm. 261. Vegetative
axis. 262. Axis with tetrasporangial conceptacles.

261 ——

262 ——

Corallina officinalis Linnaeus 1761, p. 539.
　Figures 261 and 262
　Plants erect, virgate to flabellate, saxicolous, heavily calcified, pinkish to dark rosy red, 2–8(–10) cm tall, with basal crusts giving rise to several erect axes; main axes terete to subterete in lowermost portions, distinctly flattened above, segments 0.2–1.7 mm wide, 0.2–1 mm thick, and 0.3–2 mm long; axes pinnately branched, mostly all in one plane, ultimate pinnae mostly cylindrical, some intergenicula palmately issuing up to five axes, apices obtuse; medulla of similar-length tiers of parallel compacted, elongated cells, 36–60 μm long; cortex 50–75 μm thick and composed of rounded, densely pigmented cells, 7–12.5 μm diameter, obliquely radiating from the medulla; genicula comprised of one tier of elongate medullary cells, 80–210 μm long centrally, cortex and calcification lacking; zonate tetrasporangia long obovoid, 32–48 μm diameter, 72–125 μm long, formed in ovoid, single-pored conceptacles, terminal on lateral, unbranched intergenicula.
　Williams 1951; Taylor 1960; Wiseman and Schneider 1976. As forma *vulgaris*, Williams 1948a.
　Common on the shallow outcroppings offshore of New River Inlet from 4 to 9 m, year-round, and less frequently from the shallow subtidal of Cape Lookout jetty, Myrtle Beach outcropping, and deep offshore reefs.
　Distribution: Arctic to Virginia, North Carolina, South Carolina, Caribbean to Argentina, USSR to Portugal, Mediterranean, South Africa, Central Pacific islands, Peru, Pacific Mexico.
　Only tetrasporic plants are known from the area.

Haliptilon (Decaisne) Lindley 1846
Plants erect, repeatedly branched, and bushy, articulated, calcified except at the nodes (genicula), with several uprights arising from conspicuous basal crusts;

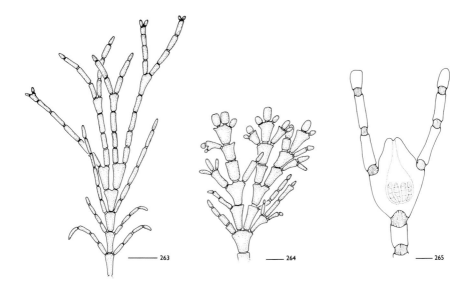

Figures 263–265. *Haliptilon cubense.* 263. Lax form, scale 0.5 cm. 264. Compact form, scale 0.5 mm. 265. Tetrasporangial conceptacle, scale 25 μm.

branching dichotomous to oppositely pinnate and basically in one plane; axial intergenicula terete to somewhat flattened, often narrowly cuneate; lateral non-axial pinnae similar to axial intergenicula or narrower; genicula of a single tier of long, thin, thick-walled medullary cells, cortex lacking; medulla of arching transverse tiers of similar-length elongated cells, secondary pit-connections lacking, tiers of similar length throughout; cortex with few to several layers composed of small, rounded cells radiating from sharply delineated medullary cells; intergenicula with conceptacles usually branched (with horns); few tetrasporangia in expanded conceptacles formed apically on axial intergenicula, each conceptacle with a single central pore; dioecious, spermatangia formed in single-pored terminal conceptacles with deep, narrow cavities; carpogonial branches two celled, borne on supporting cells which form the basal layer of female conceptacles, later acting as auxiliary cells and fusing to form thin, narrow conceptacle fertile floors; carposporophytes borne within single-pored apical conceptacles, nearly all gonimoblast cells becoming carposporangia.

Haliptilon cubense (Montagne ex Kützing) Garbary et Johansen 1982, p. 218.
   Jania cubensis Montagne ex Kützing 1849, p. 709.
   Figures 263–265
   Plants erect and creeping, flabellate, saxicolous or epiphytic, delicate, pinkish to rosy red, 1–3 cm tall, with basal crusts giving rise to several crowded erect axes, secondarily attached by adventitious calcareous pads; main axes terete to subterete, with or without broadening and flattening above, segments 50–400(–700) μm wide, 150–66 μm long; axes pinnately and dichotomously to irregularly branched, mostly all in one plane, ultimate pinnae mostly finer than the axes which produced them, occasionally tapering, apices obtuse; medulla

of similar-length tiers of parallel compacted, elongated cells, 36–78 μm long; cortex composed of rounded rectangular, densely pigmented cells, 5–7.5 μm diameter, obliquely radiating from the medulla; genicula comprised of one tier of elongate medullary cells, 80–210 μm long centrally, cortex and calcification lacking; zonate tetrasporangia long obovoid to arcuate, 20–70 μm diameter, 80–120 μm long, formed in urceolate, single-pored conceptacles, terminal on axial, branched (horned) intergenicula.

As *Corallina cubensis*, Hoyt 1920; Williams 1951; Taylor 1960; Schneider 1976; Kapraun 1980a.

Common offshore in Onslow Bay from 14 to 60 m, year-round, and infrequent on the shallow outcroppings offshore of New River Inlet.

Distribution: North Carolina, Bermuda, southern Florida, Gulf of Mexico, Caribbean, Brazil, Philippines.

Jania Lamouroux 1812

Plants erect, repeatedly branched, and bushy, articulated, calcified except at the nodes (genicula), with several uprights arising from conspicuous basal crusts; branching dichotomous with some irregular, basically in one plane; axial intergenicula terete to somewhat flattened, mostly cylindrical; genicula of a single tier of long, thin, thick-walled medullary cells, cortex lacking; medulla of arching transverse tiers of similar length, elongated cells, secondary pit-connections lacking, tiers of similar length throughout; cortex few to several layers composed of small rounded cells radiating from sharply delineated medullary cells; intergenicula with conceptacles usually branched (with horns); a few tetrasporangia in expanded conceptacles formed apically on axial intergenicula, each conceptacle with a single central pore; dioecious, spermatangia formed in single-pored terminal conceptacles with deep, narrow cavities; carpogonial branches two celled, borne on supporting cells which form the basal layer of female conceptacles, later acting as auxiliary cells and fusing to form thin, narrow conceptacle fertile floors; carposporophytes borne within single-pored apical conceptacles, nearly all gonimoblast cells become carposporangia.

Key to the species

1. Plants with axes greater than 125 μm diameter, habit corymbose, angle of branching mostly less than 30° .... *J. rubens*
1. Plants with axes less than 125 μm diameter, habit not corymbose, angle of branching mostly greater than 30° 2
2. Plants to 1 cm tall or less, axes 100 μm or less broad in median segments .... *J. capillacea*

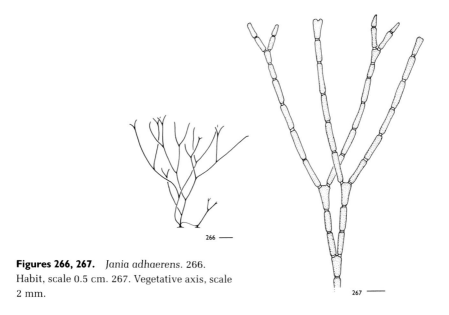

Figures 266, 267. *Jania adhaerens.* 266.
Habit, scale 0.5 cm. 267. Vegetative axis, scale
2 mm.

2. Plants greater than 1 cm tall, axes greater than 100 μm broad in median seg-
   ments . . . . *J. adhaerens*

Jania adhaerens Lamouroux 1816, p. 270.
   Figures 266 and 267
   Plants erect and creeping, flabellate, saxicolous or epiphytic, delicate, pinkish to rosy red, 1–3.5 cm tall, with basal crusts giving rise to several crowded
erect axes, secondarily attached by adventitious calcareous pads; main axes
terete to subterete, segments 90–200 μm diameter and 0.4–0.6(–1) mm long
below, axes tapering slightly above; axes dichotomously branched, angled at
30°–75°, mostly all in one plane, apices obtuse to conical; medulla of similar-
length tiers of parallel compacted, elongated cells, 60–75(–88) μm long; cortex composed of rounded rectangular, densely pigmented cells, 6–8 μm diameter, obliquely radiating from the medulla; genicula comprised of one tier of
elongate medullary cells, 65–100(–126) μm long centrally, cortex and calcification lacking; zonate tetrasporangia obovoid, to 70 μm diameter and 100 μm long,
formed in vasiform, single-pored conceptacles, terminal on axial intergenicula
with two branches (horns).
   Blomquist and Pyron 1943; Williams 1948a, 1951; Taylor 1960; Schneider
1976; Kapraun 1980a.
   Common offshore in Onslow Bay from 17 to 35 m, May–November, and infrequent on the shallow outcroppings offshore of New River Inlet and the intertidal
of Cape Lookout jetty, commonly growing with the similar *Haliptilon cubense.*
   Distribution: North Carolina, Bermuda, southern Florida, Gulf of Mexico,
Caribbean, Brazil; elsewhere, widespread in tropical and subtropical seas.

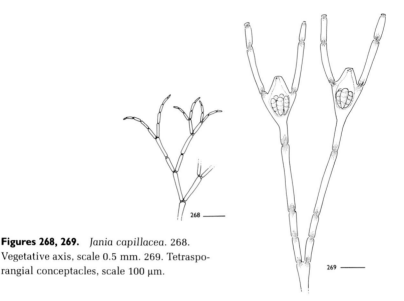

Figures 268, 269. *Jania capillacea*. 268. Vegetative axis, scale 0.5 mm. 269. Tetrasporangial conceptacles, scale 100 μm.

Jania capillacea Harvey 1853, p. 84.

Figures 268 and 269

Plants erect and creeping, flabellate, epiphytic or saxicolous, delicate and lightly calcified, pinkish to rosy red, 4–10 mm tall, with basal crusts giving rise to several erect axes, secondarily attached by adventitious calcareous pads; main axes terete, segments 45–100 μm wide and 300–600 μm long below, 30–60 μm diameter above; axes dichotomously, rarely trichotomously, branched, angled from 30° to 45°, mostly all in one plane, ultimate pinnae mostly tapering and somewhat recurved, apices truncate to conical and acute; medulla of similar-length tiers of parallel compacted, elongated cells, 26–42 μm long; cortex composed of rounded rectangular, densely pigmented cells, 5–8 μm diameter, obliquely radiating from the medulla; genicula comprising one tier of elongate medullary cells, 42–77 μm long centrally, cortex and calcification lacking; zonate tetrasporangia long obovoid, 30–38 μm diameter, 80–105 μm long, formed in swollen vasiform, single-pored conceptacles, terminal on axial intergenicula with two (up to four) branches (horns).

Williams 1951; Humm 1952. As *Corallina capillacea*, Hoyt 1920.

Infrequent offshore in deep Onslow Bay habitats and on shallow outcroppings offshore of New River Inlet, mostly on older stipes of *Sargassum filipendula*, May–August; intertidal on coquina at Marineland and on pelagic *Sargassum* spp. washed into coastal waters.

Distribution: North Carolina, Bermuda, northeastern Florida to Gulf of Mexico, Caribbean, Brazil; elsewhere, widespread in tropical and subtropical seas.

This species was merged with *Jania adhaerens* by Cribb (1983), but in our flora the two species are easily separable. *J. capillacea* has been found in the drift on pelagic *Sargassum* as far north as Virginia (Humm 1979).

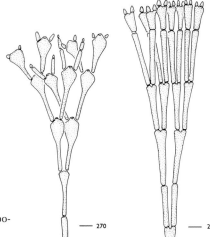

Figures 270, 271. *Jania rubens*, scales 0.5 mm. 270. Vegetative axis. 271. Tetrasporangial conceptacles.

Jania rubens (Linnaeus) Lamouroux 1816, p. 272, pl. IX, figs 6, 7.

  Corallina rubens Linnaeus 1758, p. 806.

  Figures 270 and 271

  Plants erect, branching systems straight to falcate, saxicolous, heavily calcified, pinkish to rosy red, 2–4(–8) cm tall, with discoid holdfasts giving rise to several erect axes, secondarily attached by adventitious calcareous pads; main axes terete to subterete, segments doliform near the base, becoming long cylindrical above and cuneate at the dichotomies, 125–280 μm diameter and 0.6–2.5 mm long in middle segments, only slightly tapering above; axes dichotomously branched, acutely angled from 10° to 20°, tightly packed and displaced from a single plane, becoming decidedly corymbose, apices acute, occasionally obtuse; medulla of similar-length tiers of parallel compacted, elongated cells, 80–146 μm long; cortex composed of rounded, densely pigmented cells, 7.5–12.5 μm diameter, several rows obliquely radiating from the medulla; genicula comprised of one tier of elongate medullary cells, 80–168 μm long centrally, cortex and calcification lacking; zonate tetrasporangia obovoid, 42–60 μm diameter, 100–110 μm long, formed in swollen vasiform, single-pored conceptacles, terminal on axial intergenicula with two (occasionally three) branches (horns); monoecious, cystocarpic conceptacles horned, spermatangial conceptacles terminal.

  Williams 1948a, 1949; Humm 1952; Taylor 1960.

  Common on lower intertidal and shallow subtidal rock of Cape Lookout jetty and intertidal coquina at Marineland, year-round.

  Distribution: North Carolina, Bermuda, northeastern Florida to Caribbean, Brazil to Argentina; elsewhere, widespread in warm temperate to subtropical seas.

Crustose Corallines

Measurements of the thick crustose species from Onslow Bay are taken from Suyemoto (1980).

Fosliella Howe 1920

Plants crustose, thin, lightly calcified, epiphytic or saxicolous, firmly affixed to the substrate over the entire lower surface by hypothallial cells; hypothallus a loosely to closely united single layer of filaments radiating from an original four-celled central element divided within the original spore wall, with frequent to occasional enlarged hair-producing trichocytes (megacells) formed terminally on filaments, although later outdistanced by the growing edge of the hypothallus, secondary pit-connections lacking, cell fusions occurring between adjacent filaments; perithallus often absent or incomplete, present at least near conceptacles, of a single layer; epithallus single layered, cells rounded in section; conceptacles slightly raised to conical and hemispherical, with single central pores.

Fosliella farinosa (Lamouroux) Howe 1920, p. 587.
  Melobesia farinosa Lamouroux 1816, p. 315, pl. 12, fig. 3.
  Figures 272–276

Crusts epiphytic, gregarious, lightly calcified, farinaceous, irregularly rimose from the center to the margins, pinkish red to whitish, 1–5 mm diameter and 17–23 μm thick in vegetative portions; hypothallus derived from central cruciate cell clusters, closely compact throughout, to loosely associated, to uniseriate and biseriate and creeping, composed of radially organized, dichotomously branched filaments, each branching system flabellate to irregular, cells variable in size, 7–18 μm diameter, 10–30 μm long; trichocytes terminal on hypothallial filaments, 12–30 μm diameter, 22–40 μm long, appearing ovoid to obovoid in surface view, producing elongate hyaline hairs; perithallus of occasional superficial cells derived from the dorsal side of hypothallial cells near the distal ends, compacted into complete layers at the conceptacles; cystocarpic and tetrasporic conceptacles 140–250 μm diameter, spermatangial conceptacles occasionally smaller; tetrasporangia ellipsoid, zonately divided, (20–)30–50 μm diameter, (40–)50–90 μm long.

Taylor 1957, 1960; Brauner 1975; Searles 1987, 1988. As *Fosliella farinosa* forma *callithamnioides*, Searles and Schneider 1978. As *Melobesia farinosa*, Hoyt 1920; Blomquist and Pyron 1943. As *M. farinosa* forma *callithamnioides*, Hoyt 1920.

Common on *Zostera*, *Sargassum*, and other coarse seaweeds in the shallow subtidal in the vicinity of Beaufort and Cape Lookout, N.C., and from the deep

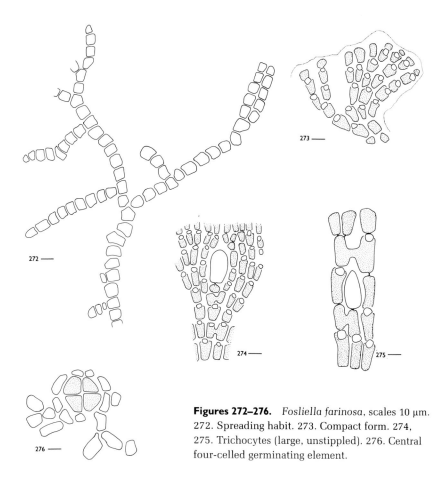

Figures 272–276. *Fosliella farinosa*, scales 10 μm.
272. Spreading habit. 273. Compact form. 274,
275. Trichocytes (large, unstippled). 276. Central
four-celled germinating element.

subtidal offshore in Onslow Bay, year-round, reaching peak size and abundance
from July–December; from 17 to 21 m on Gray's Reef, August.

Distribution: Quebec, Maritime Provinces, Massachusetts to North Carolina,
Georgia, Bermuda, southern Florida, Caribbean, Brazil; elsewhere, widespread
in temperate to tropical seas.

During a phenological study of the epiphytes on *Zostera* in the Beaufort, N.C.,
area, Brauner (1975) found specimens in the spring and fall to have a morphological gradient from the typical form to forma *callithamnioides* (see figure 272).
We have found a similar gradient offshore on a single host specimen, and thus
follow Brauner in not recognizing forma *callithamnioides*. This species is best
identified on a host with both young and older crusts, so that both the central
four-celled element and uniporate conceptacles can be observed.

Leptophytum Adey 1966

Plants crustose, thin to thick, heavily calcified, saxicolous, firmly affixed to the
substrate over the entire lower surface by hypothallial cells; hypothallus thick

and multilayered, parallel; perithallus of many tiers of similar-sized cells, thickening near conceptacles; trichocytes (megacells) and secondary pit-connections lacking, cell fusions occurring between adjacent filaments; epithallus single layered, cells rounded in section; asexual conceptacles elevated, circular to elongate, multiporate, sporangia zonately divided, individually covered by plugs in pores; gametangial conceptacles with single pores, gonimoblasts forming on periphery of conceptacles; conceptacle roofs thin.

Leptophytum species

Figures 277 and 278

Crusts saxicolous or conchicolous, conforming to the shape of the substrate, surface smooth, heavily calcified, 100–210 μm thick, firmly attached to the substrate over entire ventral surface; hypothallus parallel, four to six filaments thick, cells 4–7 μm diameter, 10–15 μm tall, with occasional cell fusions between adjacent filaments; perithallus tiered, composed of ovoid cells 3–6 μm diameter and 5–8(–10) μm tall, with abundant cell fusions between adjacent filaments; epithallus of one layer of rounded cells, to 6 μm diameter and 2 μm thick; trichocytes lacking; tetrasporangial conceptacles slightly raised, scattered over crusts, to 430 μm diameter, cavities 250–325 μm diameter and to 90 μm tall, ellipsoid in section, with individual pores and pore plugs for each tetrasporangium in the conceptacle; tetrasporangia ellipsoid, zonately divided, to 80 μm long; spermatangial conceptacles slightly raised, cavities to 100 μm diameter and 100 μm tall, flask shaped in section, uniporate; spermatangia formed on floor, roof, and walls of cavity.

Searles 1987, 1988.

Known from 17 to 21 m on Gray's Reef, August.

Distribution: Georgia.

Gray's Reef specimens are similar in vegetative characteristics to the arctic *Leptophytum foecundum* (Kjellman) Adey (1966), but differ in conceptacle dimensions (Searles 1988).

Lithophyllum Philippi 1837

Plants crustose, thick, heavily calcified, saxicolous, firmly affixed to the substrate over the entire lower surface by hypothallial, later perithallial, cells; hypothallus single layered, produced from an eight-celled central original germinating element, in time some plants cease growing at the margins and develop a secondary hypothallus of horizontal perithallial filaments, appearing as a coaxial hypothallus; perithallus of several tiers of similar-sized cells, thickening near conceptacles; trichocytes (megacells) lacking, secondary pit-connections occurring between adjacent filaments, cell fusions lacking; epithallus with one

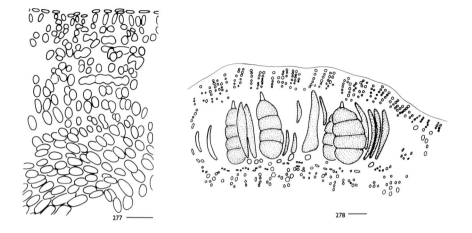

Figures 277, 278. *Leptophytum* species. 277. Cross section, scale 20 μm. 278. Section through tetrasporangial conceptacle, scale 25 μm (redrawn from Searles 1988).

to a few layers, cells rounded in section; conceptacles slightly elevated, circular to elongate, with single central pores, sporangia zonately divided.

Key to the species

1. Asexual conceptacles less than 300 μm across externally, cavities less than 200 μm diameter .... *L. intermedium*
1. Asexual conceptacles 300 μm or more across externally, cavities greater than 200 μm diameter .... *L. subtenellum*

Lithophyllum intermedium (Foslie) Foslie 1906, p. 23.
   Goniolithon intermedium Foslie 1901, p. 15.
   Crusts saxicolous, smooth, and conforming to the shape of the substrate, lacking protuberances, heavily calcified, rosy red, 0.3–1.3 mm thick, firmly attached to substrate over entire ventral surface; hypothallus nonparallel, initially a single layer of vertically elongated cells, 7.6–9.5 μm diameter, 7.6–19 μm tall, becoming several layered with age; perithallus multilayered, composed of elongate to rectangular cells, 5–8.5 μm diameter, 9.5–18.7 μm tall, with abundant secondary pit-connections; epithallus of one to three layers of rounded cells, 2.5–5 μm diameter, 2.5 μm tall; trichocytes lacking; tetrasporangial conceptacles slightly raised and scattered, 200–250 μm diameter, cavities 170–185 μm diameter and 75–90 μm tall, reniform in section, with single pores and central columnellas; tetrasporangia ellipsoid, zonately divided, 14–37.5 μm diameter, 40–75 μm long; female conceptacles unknown; spermatangial conceptacles flush with surface, occasionally slightly raised, 100–160 μm diameter, cavities 85–115 μm diameter and 30–50 μm high, flat bottomed in section, bases lined with spermatangial mother cells, spermatia elongate.
   Hoyt 1920; Williams 1948a, 1951; Taylor 1960; Suyemoto 1980.

Probably frequent deep offshore in Onslow Bay, year-round.

Distribution: North Carolina, Bermuda, southern Florida, Caribbean.

Vouchers for early reports of this species in North Carolina (Hoyt 1920; Williams 1948a, 1951) cannot be located, but a recent collection was made from 18 m in Onslow Bay. The description is based on measurements provided from that collection and isotypic material (Suyemoto 1980).

Lithophyllum subtenellum (Foslie) Foslie 1900b, p. 20.

Goniolithon subtenellum Foslie 1898, p. 11.

Crusts saxicolous or overgrowing other crustose corallines, smooth and conforming to the shape of the substrate; glossy, lacking protuberances, heavily calcified, purplish red, 0.1–2.2 mm thick, firmly attached to substrate over entire ventral surface; hypothallus nonparallel, composed of a single layer of vertically elongated cells, 5.7–19 μm diameter, 9.5–19 μm tall; perithallus multilayered, composed of squarish to short-rectangular cells, 7.6–13 μm diameter, 5.7–19 μm tall, with abundant secondary pit-connections; epithallus of one to two layers of rounded cells, 2.5–5 μm diameter, 2.5 μm tall; trichocytes lacking; tetrasporangial conceptacles slightly raised and scattered, 300–350 μm diameter, cavities 230–300 μm diameter and 70–120 μm tall, reniform in section, with single pores and central columnellas; tetrasporangia ellipsoid to ovoid, zonately divided, 31–56 μm diameter, 75–110 μm long; female conceptacles unknown; spermatangial conceptacles flush with surface, occasionally slightly raised, 150–200 μm diameter, cavities 110–155 μm diameter and 20–40 μm high, flat bottomed in section, bases lined with spermatangial mother cells, spermatia elongate.

New record for North Carolina (Suyemoto 1980).

Known offshore in Onslow Bay from 18 to 27 m, year-round.

Distribution: North Carolina, Atlantic and Mediterranean coasts of France, Algeria.

A comparison of North Carolina specimens with lectotype material confirmed the presence of this species in the flora (Suyemoto 1980).

Lithothamnion Heydrich 1897, nom. cons.

Plants crustose, thick, heavily calcified, saxicolous, and firmly affixed to the substrate over the entire lower surface by hypothallial cells, or unattached and free living; hypothallus multilayered, produced from an eight-celled central original germinating element; perithallus of several tiers of varying to similar-sized cells; trichocytes (megacells) and secondary pit-connections lacking, cell fusions occurring between adjacent filaments; epithallus mostly single layered, cells angular in section; conceptacles slightly elevated, circular to elongate, multiporate, sporangia zonately divided.

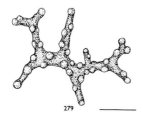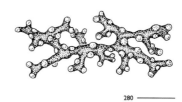

Figures 279, 280. *Lithothamnion occidentale*, scales 5 mm.

Lithothamnion occidentale (Foslie) Foslie 1908, p. 3.
   Lithothamnion fruticulosum var. *occidentale* Foslie 1904, p. 12.
   Figures 279 and 280
   Plants unattached, smooth, and glossy but with many protuberances, highly and irregularly branched, at times in one plane, prostrate with sediment, others highly three-dimensional, heavily calcified, pinkish to rosy red; axes terete to flattened, 1–2(–3) mm diameter, coalescing upon contact with other branches, the tips projecting; hypothallus quickly obscured by overgrowth of perithallus; perithallus multilayered, composed of swollen elongate to rectangular and occasionally irregular cells in filaments, 6–10 μm diameter, 10–32 μm tall, lacking secondary pit-connections but fusing to adjacent filaments, becoming subglobose and dark pigmented under the epithallus; perithallial filaments branched and spreading around axes; epithallus a single layer of compressed angular cells with distal extensions forming to the surface, 8–12 μm diameter, 5–7 μm tall, appearing polygonal in surface view; trichocytes lacking; tetrasporangial conceptacles inconspicuous, tetrasporangia ellipsoid, zonately divided, 75 μm diameter, 180 μm long.
   Schneider and Searles 1976.
   Common offshore from 15 to 45 m in Onslow Bay and Long Bay, unattached and entangled with other algae or lying on sand or rock, year-round.
   Distribution: North Carolina, South Carolina, southern Florida, Caribbean, Brazil.
   Only vegetative specimens of this species are known in the flora. Fragmented plants have been observed, demonstrating the likely form of propagation at the northern limit of distribution for this fragile species.

Species Excluded

Lithothamnion sejunctum Foslie
   Determined from North Carolina as *Lithothamnion sejunctum* by Hoyt (1920) with a query; additional crusts have since not been collected in the flora (Suyemoto 1980). The Hoyt specimen is male, with spermatangial mother cells lining the entire conceptacle cavity, including the roof, reminiscent of *Mesophyllum* and *Clathromorphum* (Adey and Johansen 1972). In *Lithothamnion*, spermatan-

gial mother cells form only at the cavity base (Johansen 1981); thus the specimen cannot be *Lithothamnion sejunctum* and it is removed from the flora. Until additional crusts are discovered with tetrasporangia and other necessary characteristics for accurate determination, Hoyt's specimen will remain undetermined.

Melobesia Lamouroux 1812

Questionable Record

Melobesia membranacea (Esper) Lamouroux 1816, p. 315.

    Corallina membranacea Esper 1806, pl. XII.

    No vouchers of this thin, epiphytic, crustose species can be located to verify reports by Williams (1948a, 1951; as *Epilithon membranaceum*). In an extensive survey of the epiphytes of *Zostera* near Cape Lookout, Brauner (1975) was careful to point out, despite common collections of *Fosliella*, *Pneophyllum* (as *Heteroderma*), and *Titanoderma* (as *Dermatolithon*), that he did not find *Melobesia*, which is easily distinguished from these other genera in that several pores form on each conceptacle (Chamberlain 1985). All of the specimens in our region previously labeled *Melobesia membranacea* have been found to be one of the above three thin crustose genera.

Mesophyllum Lemoine 1928

Plants crustose, thick, heavily calcified, saxicolous or epiphytic, firmly affixed to the substrate over the entire lower surface by hypothallial cells or loosely attached and becoming leafy; hypothallus thick and multilayered, parallel or coaxial; perithallus of a few tiers of similar-sized cells, thickening near conceptacles; trichocytes (megacells) and secondary pit-connections lacking, cell fusions occurring between adjacent filaments; epithallus (when present) single layered, cells rounded to rectangular in section; asexual conceptacles slightly elevated, circular to elongate, with multiple pores, sporangia zonately divided, individually covered by plugs in pores; gametangial conceptacles with single pores.

Mesophyllum floridanum (Foslie) Adey 1970, p. 24.

    Lithothamnium floridanum Foslie 1906, p. 11.

    Crusts saxicolous, conforming to the shape of the substrate, surface granular, with mammillate protuberances, heavily calcified, purplish red, 0.2–4.6 mm thick, firmly attached to the substrate over entire ventral surface; mammillae 1–4 mm tall, 1–2 mm diameter; hypothallus coaxial, 8–10(–15) filaments thick, cells 4.8–9.6 µm diameter, 12–22.8 µm tall, with cell fusions between adjacent filaments; perithallus tiered and folded, composed of elongate to rectangular cells, 4.8–6 µm diameter, 6–9.6 µm tall, with abundant cell fusions between adjacent filaments; epithallus of one layer of rounded cells, 5–7.5 µm diameter, 2.5–5 µm tall; trichocytes lacking; tetrasporangial conceptacles slightly raised, concentrated on mammillae and scattered over crusts, 400–600 µm diameter, cavities 290–360 µm diameter and 140–180 µm tall, ellipsoid in section, with individual pores and pore plugs for each tetrasporangium in the conceptacle; tetrasporangia ovoid to ellipsoid, zonately divided, 45–86 µm diameter, 108–150 µm long.

New record for North Carolina (Suyemoto 1980).

Known from 18 to 27 m offshore in Onslow Bay, year-round.

Distribution: North Carolina, southern Florida, Caribbean, ?West Africa.

Suyemoto (1980) found that Onslow Bay crusts agree well with lectotype material from Florida (Foslie 1906).

Questionable Record

Mesophyllum mesomorphum (Foslie) Adey 1970, p. 25.

Lithothamnium mesomorphum Foslie 1901, p. 5.

Vouchers for the tentative reports by Williams (1948a, 1949, 1951) (as *Lithothamnium mesomorphum* var. *ornatum* Foslie et Howe; the subspecific taxa were not transferred to *Mesophyllum* by Adey [1970]) for Cape Lookout jetty and offshore in Onslow Bay cannot be located, and a recent study of crustose corallines in North Carolina failed to re-collect this species (Suyemoto 1980).

Neogoniolithon Setchell et Mason 1943

Plants crustose, thick, heavily calcified, saxicolous, firmly affixed to the substrate over the entire lower surface by hypothallial cells; hypothallus thick and multilayered, parallel or coaxial, originally developed from an eight-celled central germinating element; perithallus of many tiers of similar-sized cells, thickening near conceptacles, secondary pit-connections lacking, cell fusions occurring between adjacent filaments; trichocytes (megacells) single or in vertical series within perithallial filaments; epithallus single layered, cells rounded in section; conceptacles slightly elevated, circular to elongate, with single central pores, sporangia zonately divided.

Key to the species

1. Trichocytes frequent, single or in vertical rows of two to three; asexual conceptacles greater than 200 µm across externally, cavities greater than 150 µm diameter .... *N. accretum*
1. Trichocytes rare, single; asexual conceptacles less than 200 µm across externally, cavities less than 120 µm diameter .... *N. caribaeum*

Neogoniolithon accretum (Foslie et Howe) Setchell et Mason 1943, p. 90.

Goniolithon accretum Foslie et Howe 1906, p. 131.

Crusts saxicolous, conforming to the shape of the substrate, surface rough, granular, heavily calcified, brownish red, 0.2–1.2 mm thick, firmly attached to the substrate over entire ventral surface; hypothallus parallel, ascending, non-coaxial, six to fifteen filaments thick, cells 5.7–9.5 μm diameter, 15.2–20.9 μm tall, with occasional cell fusions between adjacent filaments; perithallus tiered, composed of elongate to rectangular cells, 3.8–7.6 μm diameter, 5.7–9.5 μm tall, with abundant cell fusions between adjacent filaments; trichocytes frequent in perithallus, single or in vertical rows of two or three, 9.5–13.3 μm diameter, 10.4–14.2 μm tall; epithallus of one layer of rounded cells, 3.8–5 μm diameter, 2.5–3.8 μm tall; tetrasporangial conceptacles slightly raised, scattered over crusts, 290–380 μm diameter, cavities 170–220 μm diameter and 75–115 μm tall, short ellipsoid in section, with single pores; tetrasporangia ovoid to ellipsoid, once or zonately divided, 20–25 μm diameter, 55–72 μm long; monoecious, mature cystocarpic conceptacles raised, 350–400 μm diameter, cavities 265–300 μm diameter and 95–155 μm tall, ellipsoid in section, uniporate; carposporangia forming from a large discoid fusion cell at cavity base, developing in chains around the cavity perimeter; spermatangial conceptacles slightly raised, cavities 250–270 μm diameter and 40–50 μm tall, lenticular in section, uniporate, spermatangial mother cells dendroid, lining bases of cavities.

New record for North Carolina (Suyemoto 1980).

Known from 18 to 21 m offshore in Onslow Bay, year-round.

Distribution: North Carolina, southern Florida, Caribbean, Japan.

Suyemoto (1980) found that Onslow Bay specimens differed somewhat in measurements from isotype material from Florida and published reports for the Caribbean (Lemoine 1917), and were at variance with some reproductive characters reported by Masaki (1968) for Japanese crusts. She postulated that *Neogoniolithon accretum* was a widely distributed and variable species, perhaps divisible into distinct Atlantic and Pacific species.

Neogoniolithon caribaeum (Foslie) Adey 1970, p. 8.

Lithophyllum decipiens f. *caribaea* Foslie 1906, p. 18.

Crusts saxicolous, conforming to the shape of the substrate, surface rough, heavily calcified, pale purple, 0.1–0.5 mm thick, firmly attached to the substrate over entire ventral surface; hypothallus parallel, coaxial to noncoaxial, 6–10 filaments thick, cells 5.7 μm diameter and 11.4–15.2 μm tall, with cell fusions between adjacent filaments; perithallus tiered, composed of elongate to rectangular cells, 5.7 μm diameter and 5.7–7.6 μm tall, with abundant cell fusions between adjacent filaments; trichocytes rare, single, 7.5–10 μm diameter, 12.5–

15 µm tall; epithallus of one layer of rounded cells, 2.5–5 µm diameter, 2.5–5 µm tall; tetrasporangial conceptacles small, slightly raised, scattered over crusts, 150–180 µm diameter, cavities 85–105 µm diameter and 45–60 µm tall, short ellipsoid in section, with single pores, with occasional tetrasporangia released through simple breaks in the conceptacle roof; tetrasporangia ovoid to ellipsoid, zonately divided, 20–25 µm diameter, 38–60 µm long; monoecious, mature cystocarpic conceptacles slightly raised, 100–125 µm diameter, cavities 80–90 µm diameter and 25–40 µm tall, ellipsoid in section, uniporate; carposporangia forming from a large discoid fusion cell at cavity base, developing in chains around the cavity perimeter; spermatangial conceptacles not raised, cavities to 60 µm diameter and 33 µm tall, uniporate, spermatangial mother cells lining bases of cavities.

New record for North Carolina (Suyemoto 1980).

Known from 18 to 21 m offshore in Onslow Bay, year-round.

Distribution: North Carolina, Caribbean.

The Onslow Bay specimens are comparable to isolectotype material from the Virgin Islands (Suyemoto 1980).

Phymatolithon Foslie 1898
Plants crustose, thin to thick, heavily calcified, saxicolous, firmly affixed to the substrate over the entire lower surface by hypothallial cells; hypothallus thick and multilayered, parallel; perithallus of many tiers of similar-sized cells, thickening near conceptacles; trichocytes (megacells) and secondary pit-connections lacking, cell fusions occurring between adjacent filaments; epithallus single layered, cells rounded in section; asexual conceptacles slightly elevated, circular to elongate, multiporate, sporangia zonately divided, individually covered by plugs in pores; gametangial conceptacles with single pores, gonimoblasts scattered over conceptacle centers; conceptacle roofs thick.

Phymatolithon tenuissimum (Foslie) Adey 1970, p. 29.

   Lithothamnium tenuissimum Foslie 1900a, p. 20.

   Crusts saxicolous, conforming to the shape of the substrate, surface smooth, heavily calcified, purplish red, 0.2–0.6 mm thick, firmly attached to the substrate over entire ventral surface; hypothallus parallel, five to ten filaments thick, cells 4.5–8.1 µm diameter and 13.5–16.2 µm tall, with occasional cell fusions between adjacent filaments; perithallus tiered, composed of elongate to rectangular cells with cell elongation not limited to meristem, 6.3–7.2 µm diameter, 4.5–9 µm tall, with abundant cell fusions between adjacent filaments; epithallus of one layer of rounded cells, 2.5–5 µm diameter, 2.5–3.8 µm tall; trichocytes lacking; tetrasporangial conceptacles slightly raised, scattered over crusts, 180–200 µm diameter, cavities 150–170 μm diameter and 60–90 μm tall, ellipsoid in section, with individual pores and pore plugs for each tetrasporangium in the conceptacle; tetrasporangia ellipsoid, zonately divided; procarpic conceptacles slightly raised, 200–250 μm diameter, cavities 125–150 μm diameter and 60–90 μm tall, ellipsoid in section, uniporate.

New record for North Carolina and the western Atlantic (Suyemoto 1980).

Known from 18 m offshore in Onslow Bay, year-round.

Distribution: North Carolina, Morocco, West Africa, Canary Islands.

Onslow Bay specimens are in close agreement with isotype material from West Africa (Suyemoto 1980).

Pneophyllum Kützing 1843

Plants crustose, thin, lightly calcified, epiphytic, firmly affixed to the substrate over the entire lower surface by hypothallial cells; hypothallus a closely united single layer of filaments radiating from an original eight-celled central element divided within the original spore wall, trichocytes lacking or intercalary in hypothallial filaments, secondary pit-connections lacking, cell fusions occurring between adjacent filaments; perithallus of one to a few cell layers, thickening near conceptacles; epithallus single layered, cells rounded in section; conceptacles slightly raised, subhemispherical, with single central pores.

Key to the species

1. Plants lacking trichocytes; asexual conceptacles greater than 150 μm diameter, tetrasporangia greater than 25 μm diameter . . . . *P. lejolisii*
1. Plants with infrequent intercalary trichocytes; asexual conceptacles less than 150 μm diameter, tetrasporangia less than 25 μm diameter . . . .
. . . . *Pneophyllum* sp.

Pneophyllum lejolisii (Rosanoff) Chamberlain 1983, p. 359, figs 28–32.
Melobesia lejolisii Rosanoff 1866, p. 62.

Figures 281–284

Crusts epiphytic, gregarious, lightly calcified, pinkish red to whitish, 0.5–2 mm diameter and 15–30 μm thick in vegetative portions; hypothallus derived from central eight-celled element, closely compact and circular except at the margins of older crusts, composed of radially organized, dichotomously branched filaments, cells variable in size, 5–10 μm diameter, 5–10 μm long, squarish to short-rectangular and slightly broadened at the distal ends; trichocytes lacking; at margins, epithallus of occasional superficial cells derived from the dorsal side of hypothallial cells near the distal ends, in center of crusts, peri-

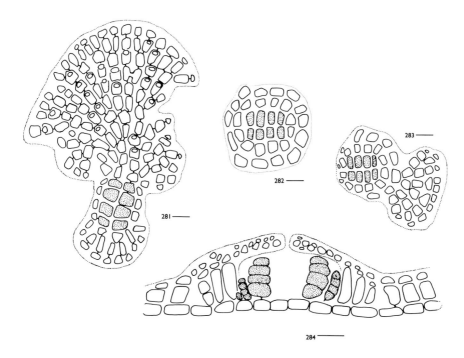

Figures 281–284. *Pneophyllum lejolisii*. 281–283. Compact discs with eight-celled germinating elements, scale 10 μm. 284. Section through tetrasporangia conceptacle, scale 20 μm.

thallus well developed by divisions of hypothallial cells parallel to the substrate, producing one to three perithallial layers, lower layer with cells 2–3 μm tall, second layer (when present) elongated, 17–20 μm tall, and upper layers similar to those at the base; conceptacles slightly elevated, circular, cystocarpic and tetrasporic conceptacles 150–250(–300) μm diameter, spermatangial conceptacles 75–100 μm diameter; tetrasporangia ellipsoid, arcuate, zonately divided, 25–50 μm diameter, 32–80 μm long.

As *Fosliella lejolisii*, Williams 1948; Humm 1952; Taylor 1960; Aziz and Humm 1962. As *Melobesia lejolisii*, Williams 1951. As *Heteroderma lejolisii*, Brauner 1975; Wiseman and Schneider 1976; Searles and Schneider 1978; Wiseman 1978; Kapraun 1980a.

Common on *Zostera*, *Sargassum*, and other coarse seaweeds in the shallow subtidal from Cape Lookout to Marineland, and on *Sargassum*, *Zonaria*, *Lobophora*, and others from deep offshore in Onslow Bay, year-round, reaching peak size and abundance, July–February.

Distribution: Newfoundland to Virginia, North Carolina, South Carolina, northeastern Florida to Gulf of Mexico, Caribbean, Brazil, Norway to Portugal, Mediterranean, West Africa, Red Sea, Indian Ocean.

Lack of agreement as to taxonomic placement of thin coralline crusts reflects a lively debate regarding generic concepts. Currently, the number of cells in the central element of germination discs and the presence or absence and position of trichocytes are used to separate *Pneophyllum* from *Fosliella* (Chamberlain 1983). These characters have been shown to remain constant under varied light and temperature regimes (Jones and Woelkerling 1984), as well as over a hypersalinity range (Harlin et al. 1985).

Pneophyllum species
Figures 285 and 286

Crusts epiphytic, gregarious, lightly calcified, pinkish red to whitish, to 3 mm diameter and two cells thick in vegetative portions; hypothallus derived from central eight-celled element, closely compact and circular except at the margins of older crusts, composed of radially organized, dichotomously branched filaments, cells variable in size, (3–)5–8 μm diameter, (5–)8–13(–15) μm long, rarely squarish, mostly short rectangular; trichocytes infrequent, intercalary in hypothallial filaments; at margins, epithallus of occasional superficial cells derived from the dorsal side of hypothallial cells near the distal ends; conceptacles elevated, hemispherical; cystocarpic conceptacles 110–140 μm diameter with cavities 54–70 μm diameter and 32–46 μm tall; spermatangial conceptacles with cylindrical snouts, 50–80 μm diameter with cavities 23–32 μm; tetrasporangia ellipsoid, slightly arcuate, zonately divided, 16–25 μm diameter, 32–40 μm long, in conceptacles 63–67 μm tall and 115–130 μm diameter, with cavities 54–63 μm diameter and 40–46 μm tall.

Searles 1987, 1988.

Known only as an epiphyte of *Rhodymenia pseudopalmata* from 17 to 21 m on Gray's Reef and 30 to 35 m on Snapper Banks, July–August.

Distribution: Georgia.

These Georgian plants are similar to *Pneophyllum confervicola* (Kützing) Chamberlain (1983) in reproductive dimensions, but they differ in the shape of hypothallial cells. To date, *P. confervicola* is known in the western Atlantic from Cuba (Vinogradova and Sosa 1977).

Figures 285, 286.
Pneophyllum species, scales 20 μm. 285. Cell fusions between adjacent filaments. 286. Trichocytes (stippled) (redrawn from Searles 1988).

285 ———

286 ———

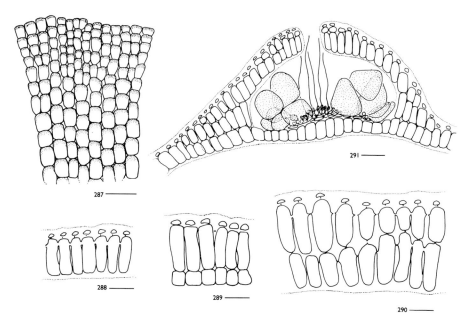

Figures 287–291. *Titanoderma pustulatum.* 287. Hypothallus at margin, scale 25 μm. 288–290. Sections through crusts, scale 20 μm. 291. Section through carposporangial conceptacle, scale 30 μm.

Titanoderma Nägeli 1858

Plants crustose, thin to thick, moderately to heavily calcified, epiphytic, firmly affixed to the substrate over the entire lower surface by hypothallial cells; hypothallus a closely united single tier of palisadelike cells, originally produced from an eight-celled central germinating element; perithallus of one to several tiers of similar-sized cells, thickening near conceptacles; trichocytes (megacells) lacking, secondary pit-connections occurring between adjacent filaments, cell fusions lacking; epithallus single layered, cells rounded in section; conceptacles slightly elevated, circular to elongate, with single central pores, sporangia once or zonately divided.

Titanoderma pustulatum (Lamouroux) Nägeli 1858, p. 532.
    Melobesia pustulata Lamouroux 1816, p. 315.
    Figures 287–291
    Crusts epiphytic, suborbicular to irregular, smooth, gregarious, and becoming confluent, moderately calcified, pinkish red to whitish at the margins, 2–10 mm diameter; hypothallus of a single layer of vertically elongated cells, 10–15 μm diameter, 12–55 μm tall; perithallus of one or two (up to eight) layers of squarish to rectangular cells, 10–15 μm diameter, 17–55 μm tall; epithallus of a single layer of lenticular cover cells; crusts becoming convex, to 350 μm thick

centrally; trichocytes lacking; conceptacles conspicuous, raised, and scattered, 150–500 μm diameter, spermatangial conceptacles only slightly elevated; tetrasporangia ellipsoid, zonately divided, 22–50(–70) μm diameter, 55–130 μm long.

As *Dermatolithon pustulatum*, Hoyt 1920; Brauner 1975; Kapraun 1980a. As *Lithophyllum pustulatum*, Taylor 1957, 1960.

Common on *Zostera*, *Corallina*, and other coarse seaweeds in the shallow subtidal of the sounds and near offshore outcroppings from Cape Lookout to Wilmington, N.C., year-round, reaching peak size and abundance from June–March.

Distribution: Maritime Provinces to Rhode Island, North Carolina, Bermuda, southern Florida, Gulf of Mexico, Caribbean, Brazil, Baltic Sea to British Isles and France, Azores, Mediterranean, Red Sea, Indian Ocean.

This species may include specimens reported in the Carolinas as *Titanoderma corallinae*.

Questionable Record

Titanoderma corallinae (P. Crouan et H. Crouan) Woelkerling, Chamberlain et Silva 1985, p. 333.

Melobesia corallinae P. Crouan et H. Crouan 1867, p. 150.

Although reported as an epiphyte exclusively on *Corallina officinalis* from the shallow subtidal of Cape Lookout and from the subtidal outcroppings offshore of New River Inlet, North Carolina (Williams 1948a, 1951; as *Lithophyllum corallinae*), all of the crusts found on host specimens from the southeastern United States fall within the limits of *Titanoderma pustulatum*. *T. corallinae* is similar to this species in most regards but is reported to be purplish gray and to contain only once-divided sporangia in wartlike conceptacles (Taylor 1960). Kylin (1944) reported that European specimens of *T. pustulatum* (as *Lithophyllum pustulatum*) have both bisporangia and zonate tetrasporangia, while those of *T. corallinae* (as *L. corallinae*) are strictly bisporangiate. By itself, then, the main character (division in the sporangium) seems taxonomically problematical. Aside from the report for the Carolinas, the next most southern limit of distribution for this species in the western Atlantic is Rhode Island, where year-round water temperatures are significantly lower. Without any voucher specimens of this crust or any mutually exclusive characteristics to differentiate the crusts on *Corallina* from those on *Zostera* or other coarse algae, it appears prudent to list *T. corallinae* as a questionable record for our area.

HILDENBRANDIALES

Hildenbrandiaceae

Hildenbrandia Nardo 1834, nom. cons.

Questionable Record

Hildenbrandia rubra (Sommerfelt) Meneghini 1841, p. 10.
  Verrucaria rubra Sommerfelt 1826, p. 140.
  As *H. prototypus*, Williams 1951; Taylor 1960.

Because no vouchers of this cold-water crustose species from the area have been located and it has not been re-collected since the original report for the Carolinas by Williams (1951), we question the existence of *Hildenbrandia rubra* in the flora.

GIGARTINALES

Plants erect and terete to ligulate and foliose, or prostrate and spreading to crustose, soft, and gelatinous, to fleshy, wiry, and cartilaginous; plants uniaxial and verticillate to pseudoparenchymatous, or multiaxial with loose or compact aggregations of filaments centrally and radial outer layers, or pseudoparenchymatous throughout, most without calcification, or crustose and comprised of spreading hypothallus and ascending perithallus, mostly calcified; tetrasporophytes and gametophytes heteromorphic or isomorphic, tetrasporangia cruciately or zonately divided, single on alternate filamentous stages, or scattered, in sori, or terminal or in chains in raised nemathecia on erect or crustose plants; monoecious or dioecious, spermatangia formed from outer cortical cells scattered over the plants or in discrete superficial to cryptlike sori; procarpic or nonprocarpic, carpogonial branches two to twenty-two celled, with one (monocarpogonial) or more (polycarpogonial) borne on supporting cells; auxiliary cells as supporting cells or cells cut off supporting cells, or intercalary cells in nearby or distant vegetative or specialized branches, if distant, with connecting filaments (ooblasts) from fertilized carpogonia or fusion cell; gonimoblasts issued inwardly or outwardly from auxiliary cells, junctions of connecting filaments and auxiliary cells, auxiliary cells and their proliferated cellular tissues, or small to large fusion cells, with or without involucres, most developing substantial pericarps with obvious ostioles; cystocarps sometimes embedded but more often projecting, scattered, and marginal, on papillae, spines, or marginal proliferations, carposporangia liberating one carpospore or four carpotetraspores.

We follow Kraft and Robbins (1985) in no longer accepting the Cryptonemiales and therefore include in the Gigartinales families formerly placed there, excluding the Corallinaceae (now recognized in the Corallinales).

Key to the families

1. Plants gelatinous or villose... 2
1. Plants not gelatinous or villose 5
2. Plants uniaxial, highly branched .... Dumontiaceae
2. Plants multiaxial, simple to lobate 3
3. Plants membranous 4
3. Plants thick gelatinous .... Gymnophloeaceae
4. Carpogonia and auxiliary cells borne in specialized ampullae ....
   .... Halymeniaceae (in part)
4. Carpogonia and auxiliary cells not borne in specialized ampullae ....
   .... Kallymeniaceae (in part)
5. Plants uniaxial 6
5. Plants multiaxial... 8
6. Tetrasporangia scattered over plants .... Cystocloniaceae
6. Tetrasporangia formed in discrete sori on ultimate branches 7
7. Plants alternately to irregularly, not pinnately, branched; gonimoblasts issued inwardly .... Hypneaceae
7. Plants repeated pinnately branched, ultimate branchlets often unilateral; gonimoblasts issued outwardly .... Plocamiaceae
8. Plants crustose 9
8. Plants not crustose 10
9. Plants in part calcified, crusts attached by rhizoids; tetrasporangia terminal or lateral on nemathecial filaments .... Peyssonneliaceae
9. Plants lacking calcification, crustose stages firmly attached without rhizoids; tetrasporangia intercalary on nemathecial filaments....................
   .... Petrocelidaceae (in part)
10. Cystocarps formed on papillae 11
10. Cystocarps formed on margins or blades........................... 12
11. Plants ligulate, pseudodichotomously branched ...Petrocelidaceae (in part)
11. Plants terete and radially alternately to suboppositely branched, or broadly foliose, irregularly to palmately and proliferously lobed ....
    .... Solieriaceae (in part)
12. Tetrasporangia zonately divided 13
12. Tetrasporangia cruciately divided................................ 14
13. Plants dichotomously branched, appearing pseudoparenchymatous in cross

Peyssonneliaceae

Plants crustose, the lower surfaces wholly or partly attached to the substrate, with or without unicellular or multicellular rhizoids; crusts calcified through-out, only in lower portions, or not at all; prostrate growth by radiating marginal rows of transversely dividing apical initials (basal layer) which later divide verti-cally to form single upper or lower perithallial cells; first perithallial cells giving rise to simple or branched erect filaments, together forming a loose to com-pact, upper and lower or upper only cortex (perithallus); tetrasporophytes and gametophytes isomorphic; sporangia formed among attenuated paraphyses in elevated or immersed nemathecia, cruciately divided; monoecious or dioecious, spermatangia in terminal sori, lateral or terminal on erect filaments in elevated to immersed nemathecia; procarpic and nonprocarpic, carpogonial branches two to six celled, auxiliary cell branches two to six celled, both borne laterally on erect filaments in nemathecia, connecting filaments often attaching to more than one auxiliary cell, forming small to sprawling fusion cells, short, chainlike gonimoblasts arising from one to several auxiliary cells, most cells developing into carposporangia bearing one carpospore.

Peyssonnelia Decaisne 1841
Crusts wholly or partly attached to the substrate by unicellular or multicellu-lar rhizoids issued from the lower surface, calcified throughout, hypobasally, or not at all, consisting of a single-layered hypothallus which generates above it an

erect perithallus of many layers; crusts conforming to the substrate or free from the substrate contour and planar, with or without concentric rings and/or radial lines in the perithallus; reproductive characteristics as for the family.

In order to best identify the individual species in this genus, both thin tangential sections and radial vertical sections are necessary. Reproductive crusts are very helpful in the identification process but are not often found.

Key to the species

1. Crusts attached by multicellular rhizoids, calcification throughout . . . . *P. polymorpha*
1. Crusts attached by unicellular rhizoids, calcification hypobasal only 2
2. Crusts loosely attached to the substrate, margins free, first cell of the perithallus giving rise exclusively to two erect filaments as viewed in radial section 3
2. Crusts firmly attached to the substrate over the entire lower surface, first cell of the perithallus giving rise to two or three erect filaments as viewed in radial section 5
3. Crusts lacking radial lines or concentric rings in the perithallus . . . . *P. simulans*
3. Crusts with radial lines or concentric rings in the perithallus 4
4. Crusts membranous and pliable, upper surface with uniform color when vegetative, showing faint radial lines and concentric rings . . . . *P. inamoena*
4. Crusts thin but not pliable, upper surface with greater pigmentation in the central region, showing conspicuous concentric rings . . . . *P. stoechas*
5. Crusts with a soft, fleshy texture in the perithallus; calcification appearing as a thin veneer over the lower surface; hypothallus composed of elongate rectangular cells in parallel rows . . . . *P. atlantica*
5. Crusts with a cartilaginous perithallus; calcification hypobasal in a layer nearly as thick as the crust proper; hypothallus composed of elongate, irregularly clavate cells in parallel rows . . . . *P. conchicola*

Peyssonnelia atlantica C. W. Schneider et Reading 1987, p. 176, figs 1–15.
Figures 292–294

Crusts conchicolous, purplish red, 2–5 cm broad, to 220 μm thick when sterile and 475 μm thick when fertile, round to irregular in outline, lacking conspicuous radial or concentric lines, conforming to and firmly attached over the entire lower surface by short, unicellular rhizoids; calcification hypobasal and extracellular below; hypothallus a distinct single layer composed of parallel rows of elongated rectangular cells, 6–13 μm diameter and 28–58 μm long as viewed from the ventral surface, irregularly elongate at the margins, some cells bear-

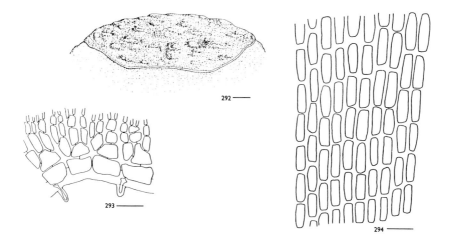

Figures 292–294. *Peyssonnelia atlantica.* 292. Habit, scale 0.5 cm. 293. Radial vertical section, scale 25 μm. 294. Hypothallial cells, scale 25 μm (from Schneider and Reading 1987).

ing a short, unicellular rhizoid at the anterior end of the cell, 6–10 μm diameter, to 43 μm long; hypothallial cells to 33 μm high, connected by a faint central line of cytoplasm which branches obliquely into the perithallial cells; upper wall of the hypothallial cells oblique, giving rise to the perithallial cells at angles of 60°–80° in lower portions; the first cell of the perithallus boot shaped to irregularly pentagonal and horizontally elongated, giving rise to two or three ascending filaments composed of three to seven cells, becoming erect above, apical cells 5–11 μm diameter, 6–16 μm tall, appearing as a loose cortex of irregularly rounded cells as viewed from the dorsal surface, to 12 μm in greatest dimension; tetrasporangia in immersed nemathecia, cruciately divided, 12–25 μm diameter, 40–65 μm tall, borne terminally on obvious supporting cells which sequentially regenerate new sporangia, paraphyses slender, simple to forked, cells 1.5–4 times taller than broad, shortening and expanding slightly at the apices, appearing club shaped, 80–265 μm long; monoecious, spermatangia and carpogonia borne in the same immersed nemathecia; spermatangia borne in dense ovoid to elongate clusters, terminal on occasional erect filaments, spermatangia irregularly rounded to angular, to 3 μm diameter; nonprocarpic, carpogonial branches two to four celled and borne laterally on erect filaments, auxiliary cell branches three to six celled and borne adjacent to and in similar positions to carpogonial branches, both stain darkly; carposporangia in chains of two to four from transversely divided gonimoblasts off sprawling fusion cells, globose to compressed, to 31 μm in greatest dimension.

Schneider and Reading 1987.

Known only from Cape Lookout jetty in the shallow subtidal, June.

Distribution: Endemic to North Carolina as currently known.

For a complete discussion of the morphology of this species, refer to Schneider and Reading (1987).

Peyssonnelia conchicola Piccone 1884, p. 317, pl. VII, figs 5–8.

Figures 295–297

Crusts saxicolous or conchicolous, purple to dull rosy red, to 3 cm broad, thin, to 220 μm when sterile and 400 μm when fertile, round to irregular in outline, lacking conspicuous radial or concentric lines, firmly attached over the entire lower surface by short, unicellular rhizoids; calcification hypobasal and extracellular below, hypothallus a distinct single layer composed of parallel rows of elongate irregular cells, 7–14 μm diameter and 22–44 μm long as viewed from the ventral surface, some cells bearing a unicellular rhizoid at the anterior end of the cell, 7–13 μm diameter, to 60 μm long; hypothallial cells to 28 μm high, connected by a faint central line of cytoplasm which branches obliquely into the perithallial cells; upper wall of the hypothallial cells oblique, giving rise to the perithallial cells at angles of 60°–75° in lower portions; the first cell of the perithallus irregularly pentagonal and horizontally elongate, giving rise mostly to three ascending filaments composed of three to six cells, becoming erect above, apical cells 5–11 μm diameter, 10–19 μm tall, appearing as a regular cortex of rounded irregular cells as viewed from the dorsal surface, to 13 μm in greatest dimension; tetrasporangia in immersed nemathecia, cruciately divided, 15–25 μm diameter, 32–53 μm tall, borne terminally on obvious supporting cells, paraphyses slender and mostly forked, cells 1.5 times taller than broad and tapering from bases to apices, 65–200 μm long; monoecious, spermatangia and carpogonia borne in the same nemathecium; spermatangia borne in dense ovoid clusters terminally on short, lateral branches of occasional erect filaments, spermatangia irregularly rounded to angular, to 4 μm diameter; procarpic, carpogonial branches three to six celled, borne laterally on erect filaments, auxiliary cell branches four to six celled, borne adjacent to and in similar posi-

Figures 295–297. *Peyssonnelia conchicola.* 295. Habit, scale 0.5 cm. 296. Radial vertical section, scale 25 μm. 297. Hypothallial cells, scale 25 μm (from Schneider and Reading 1987).

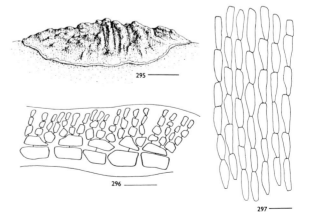

295 ——

296 ——

297 ——

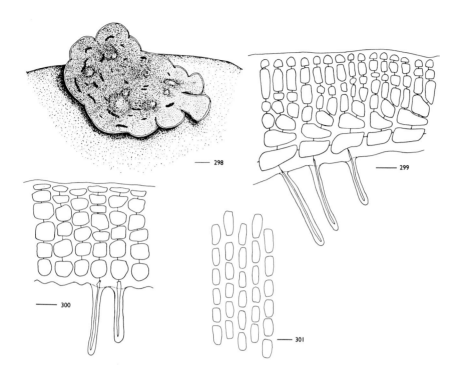

Figures 298–301. *Peyssonnelia inamoena.* 298. Habit, scale 0.5 cm. 299. Radial vertical section, scale 20 μm. 300. Tangential section, scale 20 μm. 301. Hypothallial cells, scale 20 μm (from Schneider and Reading 1987).

tions to carpogonial branches, both stain darkly; carposporangia in short chains from transversely divided gonimoblasts off connecting filaments or auxiliary cell complexes, globose to compressed, to 32 μm in greatest dimension.

Williams 1948a, 1949, 1951; Taylor 1960; Schneider and Reading 1987.

Verified only from off the New River Inlet in 5–8 m, but probably more widespread in subtidal of jetties, June.

Distribution: North Carolina, Virgin Islands, Red Sea, South Africa, Malaysia, ?northeast Australia, Japan, Pacific Mexico.

Peyssonnelia inamoena Pilger 1911, p. 311, figs 24–25.

Figures 298–301

Crusts saxicolous or conchicolous, membranous, pink to rosy red, 2–5 cm broad, thin, to 215 μm in sterile portions and to 385 μm when fertile, round to irregular and lobed in outline, faint radial and concentric lines present, very loosely attached over most of the lower surface by unicellular rhizoids, smooth plants conforming somewhat to the surface of the substrate but easily removed; calcification hypobasal and extracellular below, slightly greater than the height of the hypothallus and perithallus combined; hypothallus a distinct single layer composed of staggered, parallel rows of elongate, irregularly rectangular cells, 7–15 μm diameter and 18–38 μm long as viewed from the ventral surface, irregularly squarish to rectangular at the margins, many cells bearing a central unicellular rhizoid, 3–14 μm diameter, to 120 μm long; hypothallial cells to 26 μm high, connected by a faint central line of cytoplasm which branches obliquely into the perithallial cells; upper wall of the hypothallial cells oblique, giving rise to the perithallial cells at angles of 60°–80° in lower portions; the first cell of the perithallus boot shaped, giving rise to two ascending filaments composed of three to six cells, becoming erect above, apical cells 6–15 μm diameter, 3–12 μm tall, appearing as a cortex of irregularly rounded to angular cells as viewed from the dorsal surface, to 16 μm in greatest dimension; tetrasporangia in raised, deeply pigmented, confluent nemathecia forming concentric patterns, cruciately divided, 27–57 μm diameter, 70–100 μm tall, borne terminally on obvious supporting cells which sequentially regenerate new sporangia, paraphyses slender, simple to forked, 70–200 μm long, cells greatly elongated; ?dioecious, carposporic nemathecia raised and deeply pigmented as in tetrasporic nemathecia, but not as regularly confluent, forming small irregular patches not in concentric patterns; spermatangia borne in linear chains terminally on fascicles of erect filaments, spermatangia irregularly rounded to angular; nonprocarpic, carpogonial branches four to six celled and borne laterally on erect filaments, auxiliary cell branches four to six celled and borne adjacent to and in similar positions to carpogonial branches, both staining darkly; carposporangia in chains of two to four from transversely divided gonimoblasts, borne terminally on sterile gonimoblast cells off short fusion cells, globose to compressed, 105 μm in greatest dimension.

Schneider and Reading 1987. As *P. rubra sensu* Williams 1951; Taylor 1960; Schneider 1976; Wiseman and Schneider 1976; ?Kapraun 1980a; Blair and Hall 1981.

Known from 21 to 60 m offshore of Cape Lookout to Charleston, June–August.

Distribution: North Carolina, South Carolina, southern Florida, Caribbean, Brazil; elsewhere, widespread in warm temperate to tropical seas.

In a recent study (Schneider and Reading 1987), all of the local and several Caribbean specimens of *Peyssonnelia rubra* (Greville) J. Agardh were shown to lack the diagnostic cystoliths and multicellular rhizoids of that species and were found to conform best to *P. inamoena*. The status of *P. rubra* in the western Atlantic, except for Bermuda (Denizot 1968), is questionable. *P. inamoena* is easily distinguished with the unaided eye from other species in the flora by its nearly foliose habit, its loose attachment to the substrate, and its raised, darkly pigmented sporangial and gametangial nemathecia, which form patches or concentric lines on the perithallus.

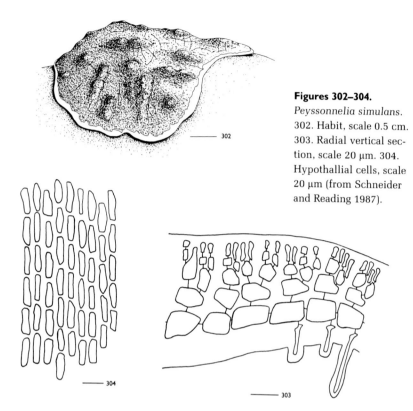

Figures 302–304.
Peyssonnelia simulans.
302. Habit, scale 0.5 cm.
303. Radial vertical section, scale 20 μm. 304. Hypothallial cells, scale 20 μm (from Schneider and Reading 1987).

Peyssonnelia simulans Weber-van Bosse in Børgesen, 1916, p. 142, figs 138, 140, 148.

Figures 302–304

Crusts saxicolous or conchicolous, pink to rosy red, 1–3 cm broad, thin, to 200 μm when sterile and 510 μm when fertile, round to irregular in outline, lacking conspicuous radial or concentric lines, new growth occasionally overgrowing older portions of the crust, conforming to but loosely attached over most of the lower surface by short, unicellular rhizoids; calcification hypobasal and extracellular below, approximating the height of the hypothallus and perithallus combined; hypothallus a distinct single layer composed of staggered, parallel rows of elongate, irregularly rectangular cells, 5–11 μm diameter and 20–36 μm long as viewed from the ventral surface, irregularly squarish at the margins, most cells bearing a unicellular rhizoid at the anterior end of the cell, 10–13 μm diameter, to 65 μm long; hypothallial cells to 45 μm high, with grossly high apical (marginal) cells—to 40 μm tall prior to division, nearly twice the height of other marginal hypothallial cells—connected by a dark-staining central line of cytoplasm which branches obliquely into the perithallial cells; upper wall of the hypothallial cells oblique, giving rise to the perithallial cells at angles of 60°–70° in lower portions; the first cell of the perithallus boot shaped and giving rise to two ascending filaments composed of three to six cells, becoming erect above, apical cells 6–13 μm diameter, 9–22 μm tall, appearing as a compact cortex of angular isodiametric cells as viewed from the dorsal surface, to 16 μm in greatest dimension; tetrasporangia in immersed nemathecia, cruciately divided, 12–30 μm diameter, 40–63(–80) μm tall, borne laterally on erect filaments on obvious supporting cells which sequentially regenerate new sporangia, paraphyses slender, simple or forked, cells 1.5 times taller than broad, and tapering from bases to apices, 80–120 μm long; ?dioecious, spermatangia unknown; carposporic nemathecia to 200 μm thick, carposporangia in chains of two to three from transversely divided gonimoblasts, globose to compressed, 100 μm in greatest dimension.

Schneider and Reading 1987.

Known from the shallow subtidal on Cape Lookout jetty, July.

Distribution: North Carolina, Gulf coast of Florida, Virgin Islands, Brazil, Tanzania.

Peyssonnelia stoechas Boudouresque et Denizot 1975, p. 53, figs 97–106.

Figures 305–308

Crusts saxicolous or conchicolous, rosy red, to 1–4 cm broad, delicate, thin, to 150 μm when sterile, circular to irregular in outline, bearing conspicuous concentric rings, loosely attached over most of the lower surface by long, unicellular rhizoids; plants barely conforming to the surface of the substrate, being most attached in the darker central region of the crust, the edges planar to curled upward, plants easily removed; calcification hypobasal and extracellular below, greater than twice the height of the hypothallus and perithallus combined; hypothallus a distinct single layer composed of staggered parallel rows of elongated hexagonal cells, 7–18 μm diameter and 20–45 μm long as viewed from the ventral surface, irregularly squarish to rectangular at the margins, many cells bearing a central unicellular rhizoid produced from an anterior division near the margin, 4–12 μm diameter, greater than 230 μm long; hypothallial cells to 28 μm high, connected by a faint central line of cytoplasm which branches obliquely into the perithallial cells; upper wall of the hypothallial cells oblique, giving rise to the perithallial cells at angles of 60°–80° in lower portions; the first cell of the perithallus boot shaped and giving rise to two ascending filaments composed of three to five cells, becoming erect above, apical cells 7–18 μm diameter, 7–20 μm tall, appearing as a cortex of angular, isodiametric cells as viewed from the dorsal surface, to 17 μm in greatest dimension; reproduction unknown.

Schneider and Reading 1987.

Known only from 29 m in Onslow Bay, July.

Distribution: North Carolina, Mediterranean France.

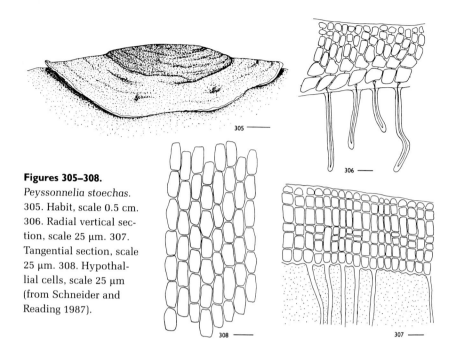

Figures 305–308.
Peyssonnelia stoechas.
305. Habit, scale 0.5 cm.
306. Radial vertical section, scale 25 μm. 307.
Tangential section, scale 25 μm. 308. Hypothallial cells, scale 25 μm
(from Schneider and Reading 1987).

Questionable Record

Peyssonnelia polymorpha (Zanardini) Schmitz 1879, p. 264.

   Nardoa polymorpha Zanardini 1844, p. 37.

   Voucher specimens for the reports of this species (Williams 1948a, 1949, 1951) cannot be located, and a study of recent *Peyssonnelia* collections (Schneider and Reading 1987) did not verify its existence here. However, because *P. polymorpha* has certain characteristics not shared by any of the other species known in the flora, including multicellular rhizoids and perithallial calcification, it is unlikely that Williams erred in his determination. We retain it as a questionable record.

Dumontiaceae

Plants soft and gelatinous or firm and fleshy, erect or prostrate, simple or irregularly to alternately and pinnately branched or irregularly lobed, arising from small discoid holdfasts, usually of somewhat firmer consistency than upper portions, some with alternating crustose stages; upright plants terete to strongly compressed, ligulate to foliose; axes uniaxial, at least in young stages, some later appearing multiaxial; axial cells each producing four radiating branched, fasciculate, mostly corymbose filaments at right angles to the axis, outermost cells deeply pigmented and forming loose to compact surfaces; medulla filamentous, often with descending rhizoids from basal cells of lateral axes, axial rows often obscured below the tips or hollow centered; tetrasporophytes and gametophytes isomorphic or heteromorphic; sporangia (where known) cruciately or zonately divided, formed singly or in intercalary chains from inner cortical cells and

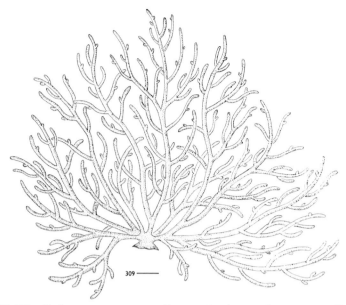

Figures 309–311. *Dudresnaya crassa.* 309. Tetrasporic plant, scale 1 cm. 310. Female plant, scale 0.5 cm. 311. Axis with apex overtopped with cortical filaments, scale 20 μm.

scattered over upper portions of plants or in discrete laminal or marginal sori, or formed in alternating crusts; monoecious or dioecious, spermatangia subglobose, formed from outer cortical cells in widespread superficial sori; nonprocarpic, carpogonial branches four to twenty-two celled and borne laterally on inner cortical cells; fertilized carpogonium fusing with, or emitting a short connecting filament to, a nearby cell in the carpogonial branch which in turn issues one or more connecting filaments; connecting filaments often seeking additional auxiliary cells after contact with first; auxiliary cells obvious, intercalary or terminal on long specialized branches, issuing gonimoblasts directly or from connecting filaments near union with auxiliary cells; carposporophytes with or without special involucral filaments, nearly all gonimoblast cells becoming carposporangia; cystocarps remaining embedded, scattered over the blades, without cortical ostioles.

Dudresnaya P. Crouan et H. Crouan 1835, nom. cons.
Plants gelatinous, some so lightly gelatinous as to appear bushy villose, others embedded in copious mucilage, appearing vermiform; erect, terete to compressed, and sparingly to greatly and repeatedly radially branched; uniaxial, obscure to obvious exserted apical cells transversely divided, axial rows obvious near tips, often obscured below by descending rhizoids, axial cells each producing four (occasionally three) determinate or indeterminate branches; determinate branches alternately, oppositely, irregularly, or dichotomously branched, mostly loosely arranged and embedded in a light to copious gelatinous matrix, outermost cells lightly to densely pigmented; indeterminate branches replacing determinate branches, ultimately repeating the branching pattern of the primary

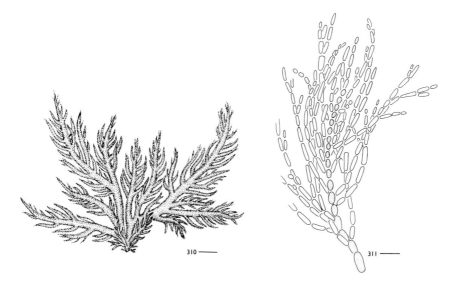

310 ——   311 ——

axial row; tetrasporophytes and gametophytes isomorphic or slightly heteromorphic; sporangia (where known) formed singly on the determinate filaments,
zonately divided; monoecious or dioecious, spermatangia as for the family; lateral carpogonial branches five to nine (occasionally to as many as nineteen)
celled, proximal on determinate branches, distant from obvious, intercalary auxiliary cells in nine- to fourteen-celled branches; carpogonia connecting with one
or more cells in their fertile branches, fusion cells issuing one or more connecting filaments; gonimoblasts issued directly from auxiliary cells or from connecting filaments near the union with auxiliary cells; carposporophytes remaining embedded, involucral filaments rudimentary or absent and cortical pores
lacking.

Key to the species

1. Plants lacking internal crystals in axial cells 2
1. Plants with distinct crystals in axial cells . . . . *D. puertoricensis*
2. Plants dioecious; apical cell obscured by overtopping lateral branches . . . . . .
. . . . *D. crassa*
2. Plants monoecious; apical cell exserted and obvious . . . . *D. georgiana*

Dudresnaya crassa Howe 1905, p. 572, pls XXVII, XXIX, figs 12–26.
    Figures 309–311
    Plants saxicolous or conchicolous, soft and gelatinous, pinkish or brownish
to rosy red, 5–25 cm tall, attached by small, somewhat fleshy discoid holdfasts; axes terete, irregularly radially branched, apical cells obscure; main axes
2–10 mm diameter, tapering to ultimate branches 1–1.5 mm diameter, apices
acute to rounded; axial row obvious throughout but surrounded by attenuate rhizoidal filaments from inner cortical cells; axial cells 50–100(–165) µm diameter, each giving rise centrally or distally to four radiating, slightly tapering,
pseudodichotomously to alternately branched cortical assimilatory filaments,
basal cells with as many as four secondary lateral filaments; outer cortical cells

deeply pigmented, elongate, and cylindrical, 3–5 μm diameter; tetrasporophytes more vermiform than gametophytes and invested with many fewer fine ultimate branches; sporangia obliquely zonate, cylindrical to long obovoid, 10–12.5 μm diameter, 37–53 μm long, borne terminally on tips and lateral axes of cortical filaments; dioecious and dimorphic, spermatangial plants smaller and less robust than carpogonial plants; spermatangia cylindrical, pyriform, or obovoid, 2–3 μm diameter, forming hemispherical to subglobose clusters terminally on lateral axes of cortical filaments; carpogonial branches six to ten celled, borne at sites of branching in the inner cortex; auxiliary cells intercalary, below with six to thirteen enlarged, densely staining cells, while above branches reduced or resembling vegetative branches, auxiliary cells rectangular to subglobose, 8–13 μm diameter; gonimoblast initials issued laterally from bulges at the junction of the connecting filaments and auxiliary cells or from the auxiliary cells directly, producing subglobose gonimolobes, 60–320 μm diameter, lacking involucral filaments; carposporangia ovoid to ellipsoid and irregular, 7.5–15 μm diameter.

Schneider and Searles 1973; Schneider 1976; Searles 1987, 1988.

Infrequent but at times locally abundant, offshore on rock and coral heads from 21 to 40 m between Cape Lookout and New River Inlet in Onslow Bay, and from 17 to 21 m on Gray's Reef, June–September.

Distribution: North Carolina, Bermuda, Georgia, southern Florida, Caribbean.

The reproductive morphology of *Dudresnaya crassa* was reported in detail by Taylor (1950). In our flora, large mature specimens can easily be separated into cystocarpic and tetrasporic plants. Male plants are smaller and infrequent. Cystocarpic plants are bushy and copiously branched with a marked gradation between massive major axes and fine ultimate branches (see figure 310). In contrast, tetrasporic specimens have few attenuate ultimate branches and are mostly comprised of several vermiform branches not much reduced in size from the primary axes (see figure 309). Although the plant is described as dioecious elsewhere, Norris and Bucher (1982) reported monoecious specimens from Belize.

Dudresnaya crassa is a frequent host for the epi-endophyte *Audouinella ophioglossa*.

Dudresnaya georgiana Searles 1983, p. 309, figs 1–18.

Figure 312

Plants saxicolous or conchicolous, soft and villose, pinkish red, 1–4 cm tall, attached by small discoid holdfasts; axes indistinct, forming villose tufts or terete, irregularly radially branched main axes to 1.6 mm diameter, tapering to acute apices with exserted apical cells; axial row obvious in younger axes where lightly invested with rhizoidal filaments from inner cortical cells; rhizoidal filaments well developed in older portions, axial cells obscured; axial

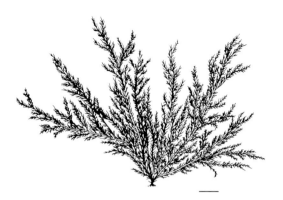

Figure 3I2. *Dudresnaya georgiana*, scale 2 mm (redrawn from Searles 1983).

cells to 105 μm diameter, each giving rise distally to four radiating, slightly tapering, pseudodichotomously to laterally branched cortical assimilatory filaments to 0.6 mm long, basal cells with as many as five secondary lateral filaments; inner cortical cells elongate to cylindrical, outer cortical cells deeply pigmented, cylindrical to ellipsoid, 3–7 μm diameter; tetrasporophytes unknown; monoecious, spermatangia subglobose, 1.5–2.5 μm diameter, forming laterally on determinate cortical filaments; carpogonial branches with eight cells (sometimes as few as six or as many as nine), the proximal one or two cells of branches often bearing short lateral filaments; auxiliary cell branches ten to twelve (rarely to fourteen) celled, with one or two proximal cells often bearing one to two sterile lateral filaments, middle cells enlarged and densely staining, above cells small and cylindrical, resembling vegetative branches; auxiliary cells third or fourth from branch base, enlarged, subglobose, 5–6 μm diameter; gonimoblast initials issued laterally from bulges at the junction of the connecting filaments and auxiliary cells, producing subglobose gonimolobes to 120 μm diameter, lacking involucral filaments; carposporangia ovoid to irregular, 5–9 μm diameter.

Searles 1983, 1987, 1988.

Infrequent from 17 to 21 m on Gray's Reef, June–August.

Distribution: Endemic to Georgia as currently known.

Dudresnaya puertoricensis Searles et Ballantine 1986, p. 389, figs 1–24.

Figures 313–315

Plants saxicolous or epiphytic, soft and gelatinous, villose and occasionally annulate above, firmer and more cylindrical below, pinkish red, to 8 cm tall, attached by small discoid holdfasts; axes terete, irregularly radially branched; main axes to 1.8 mm diameter, tapering to acute apices with exserted apical cells; axial row lightly invested with attenuate rhizoidal filaments from inner cortical cells in younger portions, heavily invested by rhizoids and enlarged proximal cells of assimilatory filaments in older portions; axial cells 75–240 μm diameter, forming up to four elongate hexagonal to irregular-shaped crystals, each cell giving rise distally to three to five (primarily four) radiating, laterally to pseudodichotomously branched cortical assimilatory filaments to 0.7 mm long, basal cells with one (sometimes two) secondary lateral filament; inner cortical cells elongate to cylindrical, often bearing distinct lateral, sessile or pedicellate, globose to pyriform cells; outer cortical cells deeply pigmented, ellipsoid, 4–

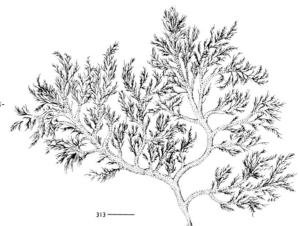

Figures 313–315. *Dudresnaya puertoricensis.* 313. Habit, scale 0.5 cm. 314. Axis with exerted apical cell, scale 10 μm. 315. Axial filament with crystals, scale 10 μm (redrawn from Searles and Ballantine 1986).

313 ⸺

7 μm diameter, terminal hair cells common; tetrasporophytes unknown; monoecious, spermatangia globose or ovoid, 3 μm diameter, forming terminally or subterminally on determinate cortical filaments; carpogonial branches with seven to nine cells, proximal cells of branches occasionally bearing short lateral filaments; auxiliary cell branches with ten to twelve (to fourteen) cells, with proximal and occasional median cells bearing sterile lateral filaments, the three to five lower to median cells enlarged and densely staining, cells above small and cylindrical, resembling vegetative branches; auxiliary cells second to fifth from branch base, enlarged, subglobose, 5–7 μm diameter; gonimoblast initials issued laterally from junction of the connecting filaments and auxiliary cells, producing subglobose gonimolobes to 140 μm diameter, with few or usually no involucral filaments; carposporangia obovoid, pyriform to angular, 5–9 μm diameter.

Searles and Ballantine 1986; Searles 1987, 1988.

Infrequent from 17 to 21 m on Gray's Reef, July–August.

Distribution: Georgia, Puerto Rico.

Gymnophloeaceae

Plants soft and gelatinous, firm and fleshy, or lightly calcified, erect or peltate, simple to branched or irregularly lobed and bullate, arising from small discoid holdfasts, usually of firmer consistency than upper portions; upright plants terete to strongly compressed or bladelike; axes multiaxial, cortex composed of radiating branched, fasciculate, mostly corymbose filaments at right angles to the axis, with or without gland cells, outer cells deeply pigmented; medulla composed of loosely arranged parallel hyaline filaments; tetrasporophytes and gametophytes isomorphic or heteromorphic; sporangia (where known) cruciately or zonately divided, formed from inner cortical cells scattered over upper portions of plants, or formed singly on filamentous "acrochaetioid" prostrate filaments or crusts; monoecious or dioecious, spermatangia subglobose, formed from outer cortical cells in superficial sori; nonprocarpic, carpogonial branches two to five celled and borne laterally on inner cortical cells; fusion cells lack-

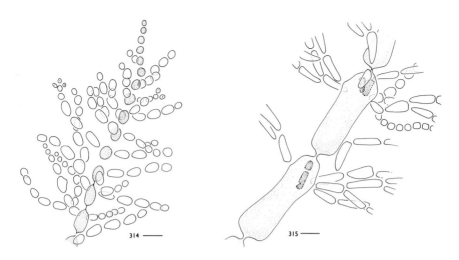

ing, auxiliary cells obvious, intercalary, issuing gonimoblasts directly or from the junction of connecting filaments and auxiliary cells, with or without associated nutritive filaments; gonimoblasts developing outwardly, becoming carposporophytes with or without special involucral filaments, nearly all gonimoblast cells becoming carposporangia; cystocarps remaining embedded, scattered over the blades, with or without cortical ostioles.

Predaea G. De Toni 1936

Plants gelatinous, erect or peltate, compressed, and simple to greatly lobate and bullate; anatomy as for the family, with or without gland cells; gametophytes and tetrasporophytes (known only in culture) heteromorphic, tetrasporophytes composed of uniseriate branched prostrate filaments, often compacted into single-layered crusts; sporangia formed singly, cruciately, zonately, or irregularly divided; monoecious or dioecious, spermatangia as for the family; lateral carpogonial branches two or three celled, mostly distant from obvious intercalary auxiliary cells with associated nutritive filaments on adjacent contiguous cells; in some, supporting cells unusually act as auxiliary cells, though diploidization not effected by associated carpogonia; one to three gonimolobes issued terminally from auxiliary cells or laterally from union of connecting filaments and auxiliary cells, carposporophytes remaining embedded, involucral filaments and cortical pores lacking.

A great deal of attention has recently been given to this genus (Kraft and Abbott 1971; Kraft 1984; Ganesan and Lemus 1975). Both species in the flora are widely distributed, with most data derived from specimens collected at great distances from the type localities.

Key to the species

1. Plants multilobate and bullate in more than one plane, erect, to 30 cm tall; outer cortical cells elongate cylindrical; gland cells lacking ... *P. feldmannii*
1. Plants broad, flat blades to 8 cm tall; outer cortical cells moniliform; gland cells present . . . . *P. masonii*

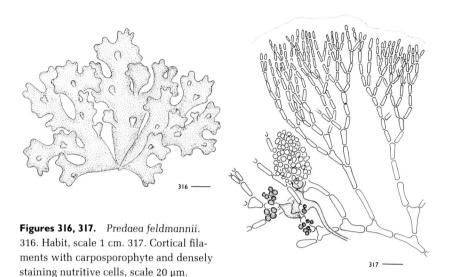

Figures 316, 317. *Predaea feldmannii.*
316. Habit, scale 1 cm. 317. Cortical filaments with carposporophyte and densely
staining nutritive cells, scale 20 μm.

Predaea feldmannii Børgesen 1950, p. 7, figs 1–3.

Figures 316 and 317

Plants saxicolous or conchicolous, gelatinous, pinkish to rosy red, 2–30 cm
tall, attached by small, somewhat fleshy discoid holdfasts; axes subcylindrical
to flattened, irregularly lobed in more than one plane, to 0.5–6 cm wide and
bearing few to many bullate projections; blades to 1 cm thick, medulla of thin-
walled axial filaments 2–3 μm diameter, giving rise to radiating, pseudodichotomously branched and corymbose cortical assimilatory filaments; cortical cells
deeply pigmented, elongate, and cylindrical, 4–5 μm diameter, tapering slightly
to the apices; unbranched rhizoidal filaments produced adventitiously on cortical and medullary cells; dioecious and dimorphic, spermatangial plants with
more lobes and bullations than carpogonial plants; spermatangia subglobose, 2–
3 μm diameter, one or two formed terminally on outer cortical cells; carpogonial
branches three celled, borne at sites of branching in the inner cortex; auxiliary
cells densely staining, large, broadly ellipsoid, 10–13(–30) μm diameter, 12.5–
25(–35) μm long, and adjacent cells above and below bearing four to six branched
chains of dense-staining, globose nutritive cells to six cells long and 5–6 μm
diameter; gonimoblast initials issued laterally from bulges at the junction of the
connecting filaments and auxiliary cells, producing two to four ovoid to pyriform gonimolobes 40–80 μm diameter and 55–190 μm long; carposporangia
ovoid to irregular, 7.5–15 μm diameter.

Schneider and Searles 1975; Schneider 1976; Searles 1987, 1988.

Locally common offshore on rock and coral heads from 18 to 55 m in Onslow
Bay and from 17 to 21 m on Gray's Reef, June–September.

Distribution: North Carolina, Georgia, Caribbean, Ghana, Saint Helena.

Culture studies of Venezuelan specimens have provided the first report of the tetrasporangial phase (Lemus and Ganesan 1977). Cruciate, zonate, and irregular sporangia were developed on small, spreading, branched filaments similar to the genus *Audouinella*. These, as well as male plants, are unknown in the flora.

Predaea masonii (Setchell et Gardner) G. De Toni 1936, p. [5] (pages unnumbered).

Clarionea masonii Setchell et Gardner 1930, p. 174.

Figures 318–320

Plants epizoic or saxicolous, gelatinous, rosy red, 2–12 cm tall, attached by small, somewhat fleshy discoid holdfasts; axes compressed into simple to few-lobed broad blades to 9 cm wide, with or without bullate projections, with or without faint central and lateral veins; blades to 8 mm thick, medulla of thin-walled axial filaments, 2.5–4 µm diameter, giving rise to radiating, pseudo-dichotomously branched cortical assimilatory filaments; cortical cells deeply pigmented, 2.5–5 µm diameter, with occasional obvious globose gland cells, 8–28 µm diameter, innermost cortical cells forked cylindrical, outermost cortical cells pyriform to globose, tapering to the apices; dioecious, spermatangial plants unknown; carpogonial branches two celled, borne at sites of branching in the inner cortex; auxiliary cells densely staining, large, subglobose to ovoid, 8–10 (–30) µm diameter, to 15 µm long, and adjacent cells above and below bearing five to nine branched chains of dense-staining, globose nutritive cells to seven cells long and 2.5–4 µm diameter; gonimoblast initials issued laterally from bulges at the junction of the connecting filaments and auxiliary cells, producing

Figures 318–320. *Predaea masonii.* 318. Habit, scale 0.5 cm. 319. Cortical filaments with auxiliary cell and densely staining nutritive filaments, scale 20 µm. 320. Carposporophyte, scale 20 µm.

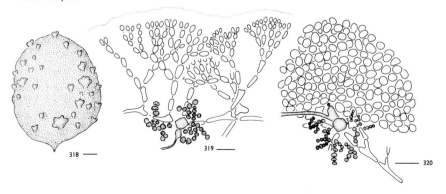

a single hemispherical gonimolobe 50–200 μm diameter and 50–100 μm long; carposporangia angular, ovoid, to subglobose, 6–8(–15) μm diameter.

Schneider and Searles 1973, 1975; Schneider 1976; Wiseman and Schneider 1976.

Infrequent offshore from 20 to 50 m in Onslow Bay and Long Bay, June–August. Although it is not known what substrate this species grows on at the deeper dredge sites, it is found by divers growing exclusively on alcyonarians at 32 m or less.

Distribution: North Carolina, Caribbean, Brazil, Ghana, Pacific Mexico, Gulf of California, California.

Although this species is at times a simple, broad blade in our flora, occasionally it produces surface bullations that give it a markedly different appearance. Plants in the southeastern United States do not have a faint vein system, but they do have obvious gland cells, two-celled carpogonial branches, and pyriform to moniliform outer cortical cells. Fresh or liquid preserved specimens of *Predaea masonii* from the type locality, Clarion Island, are needed for a critical comparison of plants from the Atlantic and Pacific oceans.

Sebdeniaceae

Plants soft and elastic to firm, erect, branched, arising from small discoid holdfasts; plants terete to strongly compressed to bladelike; axes multiaxial, cortex composed of few to several layers of rounded to elongate cells, decreasing in size from inner to outer cortex; outer cortical cells small, deeply pigmented, and radially elongated; medulla comprising the greater part of the plant, centrally composed of loosely arranged parallel to intertwined hyaline filaments, grading outward to many-pointed stellate ganglioid cells connecting to the cortex; tetrasporophytes and gametophytes isomorphic; sporangia cruciately divided, formed laterally from outer cortical cells, scattered throughout the plants; monoecious or dioecious, spermatangia pyriform to subglobose, formed from outer cortical cells in large superficial sori; nonprocarpic, carpogonial branches three or four celled, borne laterally on middle cells of the cortex; auxiliary cells obvious, intercalary, issuing gonimoblasts directly from auxiliary cells, with or without associated nutritive filaments; gonimoblasts developing outwardly, becoming carposporophytes without special involucral filaments, nearly all gonimoblast cells becoming carposporangia; cystocarps remaining embedded, scattered over the blades, with ostioles in slightly raised to conical pericarps produced after fertilization by periclinal development of the cortex above carposporophytes.

The family contains a single genus.

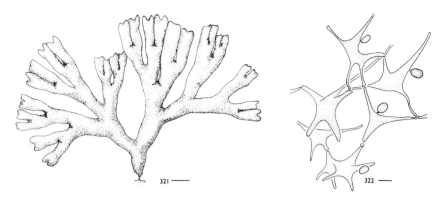

Figures 321, 322. *Sebdenia flabellata.* 321. Habit, scale 1 cm. 322. Medullary ganglia with glandlike cells, scale 25 μm.

Sebdenia Berthold 1882

Sebdenia flabellata (J. Agardh) Parkinson 1980, p. 12.

    Isymenia flabellata J. Agardh 1899, p. 62.

    Figures 321 and 322

    Plants saxicolous, soft to turgid and elastic, dark rosy to brownish and pur-plish red, 4–30 cm tall, attached by broad, somewhat fleshy discoid holdfasts; axes subcylindrical, occasionally flattened at the nodes, repeatedly dichoto-mously branched to several orders in one plane, the first dichotomy 0.5–2.5(–4) cm above the base, others spaced at distances of 0.7–3.5 cm, becoming flabel-late; axes 0.3–1 cm wide, to 2.3 cm wide below the nodes and bearing occasional small marginal proliferations; axes arising and quickly broadening from short cuneate to attenuate bases, little if at all tapering to obtuse to short-forked apices; inner medulla of thin-walled, loosely arranged axial filaments, 3–10 μm diame-ter, and long-armed stellate ganglia with centers 18–40 μm diameter, giving rise to outer ganglioid, loosely arranged medullary cells with centers 12–25 μm diameter, these often pit-connected to spherical to pyriform and ellipsoid, dark-staining, glandlike cells; cortex of four to seven layers, inner cells large, 20–35 μm diameter, lightly pigmented, irregularly lobed, and loosely arranged; outer cortical cells deeply pigmented, thick walled, elongate, and cylindrical, 3–6(–9) μm diameter and 3–12.5 μm long, appearing subpolygonal in surface view; tetrasporangia cruciately divided, subglobose to ellipsoid, 13–20(–27) μm diameter, 19–30 μm long; monoecious or dioecious, spermatangia pyriform, 2–2.5 μm diameter, one or two formed terminally on outer cortical cells, forming large, dense, circumscribing sori below the apices; carpogonial branches three or four celled, borne laterally in the middle cortex; auxiliary cells and surround-ing vegetative cells darkly staining, the latter producing small-chained nutritive filaments after fertilization; gonimoblast initials issued terminally from auxil-iary cells, carposporophytes pedicellate on bases of nutritive cells and their fila-ments, appearing reniform in surface view, 150–200 μm diameter, lacking in-volucral filaments but with four to seven layered pericarps of flattened cortical cells, becoming slightly raised to conical, with small ostioles; carposporangia subspherical to irregular, 12–20 μm diameter.

As *S. polydactyla*, Schneider 1976; Searles 1981, 1987, 1988. As *Halymenia polydactyla*, Schneider and Searles 1975.

Known from frequent scattered plants offshore on rock from 21 to 55 m in Onslow Bay, May–September, and from 17 to 21 m on Gray's Reef, June–September.

Distribution: North Carolina, Georgia, southern Florida, Caribbean; elsewhere, widespread in tropical and subtropical seas.

Confusion exists concerning the taxonomy of similar terete *Halymenia* and *Sebdenia* species in the western Atlantic and Caribbean. After observing the syntypes of *Isymenia flabellata* (the basionym of *H. agardhii* De Toni) in J. Agardh's Herbarium (Lund), we agree with Parkinson (1980) that they are *Sebdenia*, and not *Halymenia*. Because the epithet *flabellata* was not previously occupied in *Sebdenia* as it was in *Halymenia* (Parkinson 1980), *S. flabellata* displaces the names used here previously: *H. polydactyla* Børgesen (Schneider and Searles 1975) and *S. polydactyla* (Børgesen) Balakrishnan (Searles 1981). Additionally, specimens in the flora identified as *H. agardhii*, no longer a valid taxon, are referred to *H. trigona*.

Sebdenia flabellata has been confused with the vegetatively similar *Halymenia trigona* (as *H. agardhii*, Schneider and Searles 1975; Norris and Bucher 1976). In fact, the two plants illustrated for *H. agardhii* by Taylor (1960, pl. 51, figs 1–2) could represent both species. In the flora, *S. flabellata* is more robust and darker pigmented than *H. trigona* and has branches in only one plane, unlike the latter. Furthermore, many of the larger medullary ganglia of *S. flabellata* have unique darker-staining, glandlike cells associated with them (see figure 322). Balakrishnan (1961, as *S. polydactyla*) considered these glandlike cells to be rhizoidal initials, although we cannot find any evidence for this in our specimens. Reproductively, the two species are substantially different; for instance, auxiliary cells are intercalary in vegetative cortical cell filaments in *Sebdenia*, while those of *Halymenia* are found in specialized accessory ampullae.

Halymeniaceae

Plants soft and gelatinous to firm and fleshy or cartilaginous, erect or crustose and papillate, simple or alternately to dichotomously and pinnately branched or irregularly lobed, arising from small to large discoid holdfasts; upright plants terete to strongly compressed, ligulate to foliose, with or without papillate outgrowths and midribs; axes multiaxial, cortex composed of radiating simple to branched files of small cells at right angles to the axis, outermost cells deeply pigmented and forming compact surfaces; medulla filamentous and surrounded by refractive stellate cells with long filamentous arms; tetrasporophytes and gametophytes isomorphic; sporangia cruciately divided and scattered over upper portions of plants or restricted to proliferations, occasionally in raised laminal sori; monoecious or dioecious, spermatangia formed from outer cortical cells, scattered or in sori; nonprocarpic, carpogonial branches two celled and borne laterally on innermost cortical cells in specialized accessory branch systems (ampullae); fertilized carpogonium or fusion of carpogonium and hypogynous cell issuing one or more connecting filaments; connecting filaments often seeking additional auxiliary cells after contact with first; auxiliary cells obvious, intercalary in lower portions of specialized ampullae, issuing gonimoblasts outwardly; carposporophytes with or without involucral filaments or pericarps, nearly all gonimoblast cells becoming carposporangia; cystocarps remaining embedded or slightly raised, scattered over the blades or in fertile proliferations, with or without ostioles.

Key to the genera

1. Plants with midribs, at least in lower portions .... Cryptonemia
1. Plants lacking midribs 2
2. Plants broadly foliose, simple to marginally lobed or divided, or terete and pseudodichotomously branched; tetrasporangia scattered over the blades ...
.... Halymenia
2. Plants narrowly ligulate, bearing marginal pinnae; tetrasporangia grouped in branches .... Grateloupia

Cryptonemia J. Agardh 1842
Plants erect, firm, irregularly to palmately lobed or dichotomously to somewhat irregularly branched, with or without obvious midribs in lower parts of the blades; blades flattened and ligulate, mostly above terete stipes, with one or more axes arising from small to massive discoid holdfasts, with or without marginal proliferations; anatomy as for the family; tetrasporangia developed from cortical cells scattered over the blades or restricted to marginal bladelets; dioecious, spermatangia in large, irregular, superficial sori; carpogonial branches in ampullae, two celled; carposporophytes immersed, scattered, nearly all gonimoblast cells becoming carposporangia, slightly elevating one surface of blades by thickening of the cortex, with ostioles and few involucral filaments.

Cryptonemia luxurians (C. Agardh) J. Agardh 1851, p. 228.

    Sphaerococcus lactuca var. *luxurians* C. Agardh 1822, p. 232.

    Figure 323

    Plants saxicolous, new growth membranous and pinkish red, older portions firm, leathery, and dark rosy red, 5–25 cm tall, attached by small to large discoid holdfasts giving rise to one or more simple to branched erect axes; axes broadly ligulate above brief to long stipes which extend into the attenuate blade bases as midribs reaching about half the total distance, blades pseudodichotomously to alternately branched, 1.5–2.5 cm wide, with few to many marginal, proliferous, obovoid to subcircular bladelets, at times concentrated on the truncate to rounded apices, but occasionally found covering margins throughout as well, often several blades issued from extended older terete stipes; margins undulate and entire to erose; blades 115–300 μm thick in vegetative portions, medulla comprising much of the blade width and composed of densely arranged narrow filaments 2–5 μm diameter surrounded by stellate cells with bodies 7–23 μm diameter, giving rise to a few-layered cortex that is progressively smaller from medulla outward, surface cells 3–5 μm diameter, irregularly angular to rounded in surface view; tetrasporangia obovoid to ellipsoid, cruciately divided, 7–10 μm diameter, 10–13 μm long, densely formed in marginal bladelets near the apices; dioecious, spermatangia in patches on marginal bladelets, short ellipsoid, 2–3 μm diameter; carposporophytes immersed in marginal bladelets or blade tips, globose, 100–150 μm diameter, with basal placentas, involucral filaments, and small cortical ostioles; carposporangia subquadrate to irregularly angled, 7.5–10 μm diameter.

    Schneider and Searles 1973; Schneider 1976.

    Common offshore from 17 to 40 m in Onslow Bay, obvious May–December, but present winter–spring as holdfasts and denuded short stipes; reaches peak size, abundance, and reproductive maturity in late summer.

    Distribution: North Carolina, Bermuda, southern Florida, Caribbean, Brazil, West Africa.

    This species is rarely collected without heavy encrustations of bryozoans, crustose corallines, or other plants and animals, at least in lower (older) portions.

Grateloupia C. Agardh 1822, nom. cons. prop.

Plants erect and soft, main axes pseudodichotomously to somewhat irregularly branched, these beset with few to numerous compressed, attenuate, pinnate branches; blades narrowly to broadly ligulate and foliose, with one or more axes arising from small to massive discoid holdfasts; anatomy as for the family; tetrasporangia developed from outer cortical cells, scattered over the blades

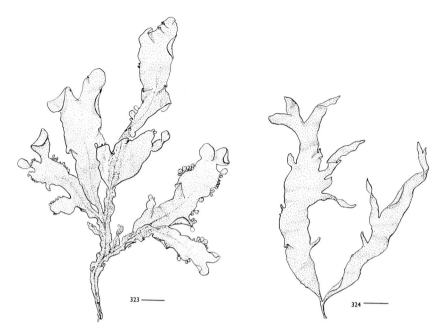

Figure 323. *Cryptonemia luxurians*, scale 1 cm.
Figure 324. *Grateloupia cunefolia*, scale 1 cm.

or in sori; monoecious or dioecious, spermatangia in small superficial sori or scattered; carpogonial branches in ampullae and two celled; carposporophytes small, scattered, and immersed, nearly all gonimoblast cells becoming carposporangia, slightly elevating one surface of blades by extensive thickening of the cortex, with ostioles and few involucral filaments.

Key to the species

1. Plants linear, with numerous marginal pinnate branches 2
1. Plants narrowly foliose, mostly lacking marginal pinnae, appearing asymmetric to lunate . . . . *G. cunefolia*
2. Plants with main axes 200 μm or less thick; monoecious with cystocarps greater than 150 μm diameter . . . . *G. filicina*
2. Plants with main axes thicker than 200 μm; dioecious with cystocarps to 150 μm diameter . . . . *G. gibbesii*

Grateloupia cunefolia J. Agardh 1849, p. 85.
   Figure 324
   Plants saxicolous, purplish red to greenish black, 4–40 cm tall, foliose, short stipitate, attached by small discoid holdfasts; blades asymmetrical, simple, and linear lanceolate to crescent shaped or irregularly to palmately dividing into narrower strips, tapering to the cuneate bases; margins smooth to dentate or erose, undulate, with few narrow to lanceolate proliferations, apices acute; main axes 1–5 cm wide, 160–250 μm thick, medulla composed of loosely arranged entangled filaments 1–3 μm diameter surrounded by stellate cells with bodies to 5–

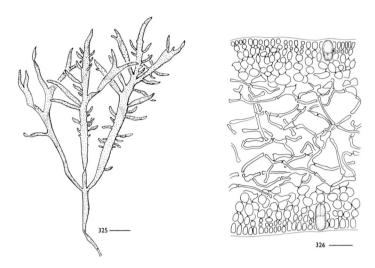

Figures 325, 326. *Grateloupia filicina.* 325. Habit, scale 0.25 cm. 326. Cross section through tetrasporangial blade, scale 20 μm.

10 μm diameter, giving rise to a three- to seven-layered cortex, the cortical cells only slightly diminishing from medulla outward, surface cells 2–4 μm diameter, irregularly rounded in surface view; tetrasporangia ellipsoid to ovoid, cruciately divided, 8–19 μm diameter, 26–35 μm long, formed in the inner cortex of blades; carposporophytes immersed in laminae, with few involucral filaments, carposporangia subglobose to elongate and irregular, 8–13 μm diameter; cystocarps globose, to 130 μm diameter, ostioles inconspicuous.

Humm 1952.

Known only from intertidal coquinoid outcroppings at Marineland, year-round.

Distribution: Northeastern Florida to Caribbean, Brazil, Uruguay.

Grateloupia cunefolia from Marineland has very few marginal proliferations and is greater than 1 cm wide, easily distinguishing it from the other two species in the flora.

Grateloupia filicina (Lamouroux) C. Agardh 1822, p. 223.

Delesseria filicina Lamouroux 1813, p. 125.

Figures 325 and 326

Plants saxicolous, firm, purplish and tannish to rosy red, or greenish purple to purplish black, 2–10(–45) cm tall, attached by small discoid holdfasts, giving rise to one or more branched erect axes on short stipes; axes flattened to subterete, simple to di- and trichotomously branched when young, becoming markedly pinnate with age, rarely proliferating from the laminae, main axes 0.5–2.5(–5) mm wide, short to greatly extended pinnae 0.3–1 mm wide, lower ones occasionally rebranched; blades 150–200 μm thick, medulla composed of loosely arranged, mostly longitudinal filaments 3–4(–7) μm diameter and surrounded by small stellate cells, giving rise to a multilayered cortex, progressively smaller from medulla outward, surface cells 3–5 μm diameter, irregularly angular to rounded in surface view; tetrasporangia ellipsoid, cruciately divided, 11–27 μm diameter, 25–47 μm long, often clustered, formed in the inner cortex of marginal pinnae; monoecious, spermatangia in scattered patches over the blades, formed from outer cortical cells; carposporophytes immersed in upper portions of blades and throughout pinnae, with few to several involucral filaments, carposporangia obovoid to irregular, 7.5–15(–23) μm diameter; cystocarps compressed globose, 150–260 μm diameter, ostioles inconspicuous or lacking.

Hoyt 1920; Williams 1948a, 1949, 1951; Humm 1952; Stephenson and Stephenson 1952, pro parte; Taylor 1960; Chapman 1971; Wiseman and Schneider 1976; Kapraun 1980a; Richardson 1987.

Common, at times in dense patches, on high intertidal to shallow subtidal rock of Cape Lookout jetty, April–November, and infrequently to frequently on other jetties and outcroppings throughout the area, year-round.

Distribution: North Carolina, South Carolina, Georgia, northeastern Florida to Caribbean, Brazil; elsewhere, widespread in warm temperate to tropical seas.

Highly variable within a short vertical distance on Cape Lookout, *Grateloupia filicina* has a wide range of branching patterns of the main axes from simple to many times dichotomous and trichotomous, and color variation from green to dark purplish red. Plants in the high intertidal are large, subterete, and dark purplish red. In the lower intertidal and shallow subtidal, plants are smaller, greatly flattened, and rosy red unless constantly bombarded with suspended sand particles, in which case a long, slender green form is common. Dense patches of this species trap and hold sediment, occasionally becoming inundated on the jetty.

Grateloupia gibbesii Harvey 1853, p. 199, pl. XXVI.
Figure 327
Plants saxicolous, greenish to blackish purple, 3–60 cm tall, attached by small discoid holdfasts; main axes simple and linear lanceolate to alternately and subdichotomously ligulate, most tapering to the base, marginally pinnate with narrow to ligulate proliferations, apices acute; main axes 0.3–5 cm wide, short to greatly extended pinnae to 9 cm long and 2–3 mm wide, these often bearing marginal or, uncommonly, ligulate, linear to irregular pinnules; blades 240–950 μm thick, medulla composed of loosely arranged entangled filaments 1–4 μm diameter, surrounded by stellate cells with bodies to 5 μm diameter, giving rise to a three- to eight-layered cortex, the cortical cells only slightly diminishing from medulla outward, surface cells 2–4 μm diameter, irregularly rounded in surface view; tetrasporangia ellipsoid to obovoid, cruciately divided, (7–)10–15(–18) μm diameter, (11–)28–35(–45) μm long, formed in the inner cortex of marginal pin-

nae; dioecious, spermatangia in scattered sori over the blades, formed from outer cortical cells and 2–3 µm diameter; carposporophytes immersed in blades, with several involucral filaments, carposporangia subglobose to irregular, 9–16 µm diameter; cystocarps globose to 150 µm diameter, ostioles inconspicuous.

Harvey 1853; Melvill 1875; Hoyt 1920; Taylor 1960; Wiseman and Schneider 1976. As *G. filicina sensu* Stephenson and Stephenson 1952, *pro parte*.

Known only from the sea jetties at Charleston and Winyah Bay and seawalls at the western tip of Sullivan's Island where the intracoastal waterway leads into Charleston Harbor, year-round.

Distribution: South Carolina (type locality), southern Florida, Caribbean.

Although placed in synonymy with *Grateloupia lanceolata* J. Agardh by Ardré and Gayral (1961), and with *G. doryophora* (Montagne) Howe by Dawson et al. (1964), we agree with Wiseman (1966), who kept *G. gibbesii* separate from these taxa on the basis of a number of differences in structure and measurements. Wiseman (unpublished data) reports greenish-purple as well as blackish-purple specimens from turbid waters at the type locality on Sullivan's Island and finds most specimens to be 20 cm or less, significantly smaller than the upper dimensions of height reported by Harvey (1853) and Hoyt (1920).

Halymenia C. Agardh 1817, nom. cons. prop.
Plants erect, lax, and mucilaginous, foliose and simple to irregularly lobed or mostly terete and pseudodichotomously to irregularly and marginally branched, with or without short terete stipes, one or more blades arising from small discoid holdfasts, with or without marginal proliferations; medullary filaments often making bridges between stellate ganglia or inner cortical cells on either side of the blade, other anatomy as for the family; tetrasporangia developed from cortical cells, scattered over the blades; monoecious, spermatangia in small to large irregular, superficial sori; carpogonial branches in ampullae, two celled; carposporophytes small, immersed, and scattered, nearly all gonimoblast cells becoming carposporangia, slightly elevating one or both surfaces of blades, with few to several involucral filaments, with or without ostioles.

Key to the species

1. Plants terete and greatly branched .... *H. trigona*
1. Plants narrowly to broadly foliose, simple to lobate or marginally branched
   2
2. Plants marginally branched to several orders.................. *H. floresia*
2. Plants simple to lobate or dissected, but not regularly branched.......... 3
3. Stipes present and obvious, blade texture leathery........ *H. bermudensis*
3. Stipes lacking, blade texture soft...................................... 4

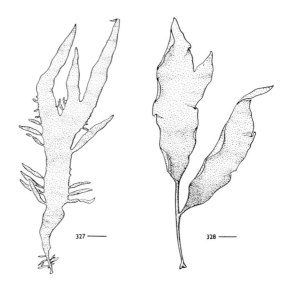

Figure 327. *Grateloupia gibbesii*, scale 0.5 cm.
Figure 328. *Halymenia bermudensis*, scale 1 cm.

327 ——

328 ——

4. Plants less than 4 cm tall and 2 cm broad, blades to 50 µm thick . . . .
. . . . *H. hancockii*
4. Plants greater than 4 cm tall and 2 cm broad, blades thicker than 50 µm
. . . . *H. floridana*

Halymenia bermudensis Collins et Howe 1916, p. 169.
Figure 328
Plants saxicolous, firm, foliose, purplish to rosy red, 5–30 cm tall, attached by discoid holdfasts giving rise to distinct, thick, simple to branched stipes 2–30 mm long, sometimes becoming subrhizomatous and spreading; blades cordate, obovate to suborbicular, simple to lobed, margins plain to undulate, entire to dentate; main blades 4–30 mm wide, 60–180 µm thick, apices rounded obtuse; medulla composed of mostly transverse bridging filaments 3–8 µm diameter, usually interspersed in older portions with coarser interconnecting filaments radiating from obvious stellate ganglia, 15–65 µm diameter, with three to ten arms; subcortex with one to three layers in older portions, cells 13–25 µm diameter; cortex with one to three layers, outermost 5–10 µm diameter, subglobose to irregularly angular in surface view; tetrasporangia subglobose to ellipsoid, cruciately divided, 7–13 µm diameter, 12–20 µm long, scattered in the outer cortex.
Schneider 1975a, 1976.
Uncommon offshore from 25 to 50 m in Onslow Bay, June–August.
Distribution: North Carolina, Bermuda, southern Florida, Gulf of Mexico, Caribbean, Brazil.
Only tetrasporic plants are known in the flora.

Halymenia floresia (Clemente y Rubio) C. Agardh 1817, p. 19.
Fucus floresius Clemente y Rubio 1807, p. 312.
Figure 329
Plants saxicolous, gelatinous, pinkish tan to rosy red, 10–40 cm tall, to 30 cm wide, attached by small discoid holdfasts giving rise to one or more highly

Figure 329. *Halymenia floresia*, scale 2 cm.

branched, erect axes on short stipes; main axes distinctly flattened, 0.5–5 cm wide, marginally pinnate to several orders with slender linear branches, subterete to flattened, 0.2–1 cm wide, each order successively smaller, occasionally proliferous on the blades, apices acute; blades 120–400 μm thick, medulla composed of loosely arranged filaments 3–10 μm diameter, grading into occasional stellate cells; cortex of one or two layers, cells 5–8 μm diameter, elongate in section and irregular in surface view; tetrasporangia subglobose to short ellipsoid, irregularly cruciately divided, 11–18 μm diameter, 13–30 μm long, scattered in the outer cortex; dioecious, carposporophytes immersed throughout axes, on basal pedicels with few involucral filaments, carposporangia subglobose to obovoid and irregular, 7.5–12.5 μm diameter; cystocarps compressed globose to reniform in section, 150–230 μm diameter, not projecting, ostioles lacking.

Hoyt 1920; Taylor 1928, 1960; Schneider 1976.

Infrequent offshore from 30 to 35 m in Onslow Bay, June–August.

Distribution: North Carolina, Bermuda, southern Florida, Gulf of Mexico, Caribbean, Brazil, Portugal, Spain, Mediterranean, Canary Islands.

Plants in the flora range from typical *Halymenia floresia* with narrow main axes and long, linear pinnae to those with broader axes and short, dentate pinnae. These latter specimens are similar to *H. pseudofloresia* Collins and Howe (1916), but are much smaller than those typifying the taxon. Because we have found specimens intermediate between the two extremes, it would be difficult to recognize *H. pseudofloresia* as a specific entity in the flora. A comparison of type material of these two species would be helpful in distinguishing differences, if, in fact, they exist.

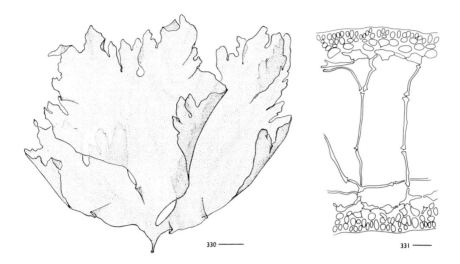

Figures 330, 331. *Halymenia floridana.* 330. Habit, scale 2 cm.
331. Cross section, scale 25 μm.

Halymenia floridana J. Agardh 1894, p. 59.

Figures 330 and 331

Plants saxicolous or conchicolous, gelatinous and foliose, pinkish to rosy red, often in part discolored by brown or green endophytes, 5–35(–60) cm tall, 4–20 cm wide, attached by discoid holdfasts giving rise to one or more erect blades with cuneate bases on short stipes; blades ovate to suborbicular, simple and entire to multilobate and greatly erose and laciniate, apices rounded; blades 60–550 μm thick, medulla composed of loosely to densely arranged, mostly longitudinal but with some bridging, filaments 3–6 μm diameter, radiating from few to many conspicuous large stellate ganglia in the inner cortex, 15–30 μm diameter; subcortex one or two layered; cortex of one or two fairly uniform, densely pigmented, cell layers, cells 5–10 μm diameter, irregularly rounded in surface view, flattened to rounded triangular and long ellipsoid in section, embedded in conspicuous surface jelly; tetrasporangia subglobose to ellipsoid, cruciately divided, 8–13 μm diameter, 12–16 μm long, scattered in the outer cortex; dioecious, spermatangia in large, spreading, superficial sori, formed from outer cortical cells, 2–3 μm diameter; carposporophytes in dense to scattered patches, immersed, with few to several involucral filaments, carposporangia irregularly rounded, 4–8 μm diameter; cystocarps subglobose to reniform, 100–250 μm diameter, slightly swelling both plant surfaces, ostioles lacking.

Hoyt 1920; Blomquist and Pyron 1943; Williams 1951; Taylor 1960; Chapman

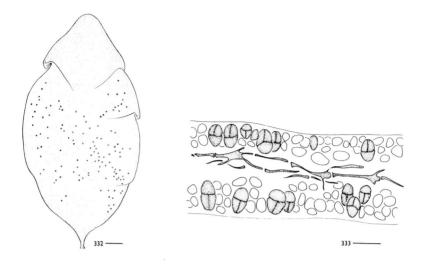

Figures 332, 333. *Halymenia hancockii.* 332. Habit of cystocarpic plant, scale 2 mm. 333. Cross section of tetrasporic blade, scale 20 μm.

1971; Schneider 1976; Kapraun 1980a; Searles 1981, 1987, 1988. As *H. gelinaria*, Collins and Howe 1916; Hoyt 1920; Williams 1951; Taylor 1960; Kapraun 1980a. As *H. vinacea sensu* Blair and Hall, 1981.

Common offshore from 17 to 55 m in Onslow Bay, May–November, and from 17 to 21 m on Gray's Reef, June–August; uncommon offshore in South Carolina.

Distribution: North Carolina, South Carolina, Georgia, southern Florida, Gulf of Mexico, Caribbean, Brazil.

The distinctions used to separate *Halymenia floridana* and *H. gelinaria* Collins and Howe (1916) do not hold upon observation of many specimens. Blade thickness, orientation of the medullary filaments, appearance of the outer cortex in section, and number of stellate ganglia are variable, and each character shows a continuum between extremes. We have therefore been unable to find any characteristic we can reliably associate with different species, and we retain all of the specimens under the older epithet. A specimen attributed to *H. vinacea* Howe and Taylor (1931) from South Carolina (Blair and Hall 1981) is also *H. floridana*.

Older specimens of *Halymenia floridana* often appear brown or greenish brown due to massive and nearly ubiquitous infestations of *Onslowia endophytica*. These plants are often heavily epiphytized as well, most commonly with *Antithamnionella flagellata*.

Halymenia hancockii W. R. Taylor 1942, p. 98, pl. 3 fig. 6; pl. 14.

Figures 332 and 333

Plants saxicolous, firm, small bladed, pinkish to purplish and rosy red, 0.7–3.5(–10) cm tall, 3–20 mm wide, attached by small discoid holdfasts giving rise to erect axes on stipes 1–3 mm in length; blades oblanceolate to nearly lanceolate, simple to infrequently bifurcate, margins entire, occasionally with marginal or laminal proliferations similar to the main blades, apices rounded; blades 35–50 μm thick, medulla composed of few loosely arranged, longitudinal fila-

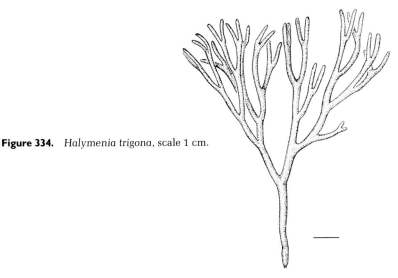

Figure 334. *Halymenia trigona*, scale 1 cm.

ments 3–8 μm diameter, connected to inconspicuous three- or four-armed sub-cortical cells, conspicuous stellate ganglia absent; cortex single layered, cells 8–10 μm diameter, rounded angular and compact in surface view; tetrasporangia ellipsoid to obovoid, cruciately divided, 7–13 μm diameter, 12–16 μm long, scattered between cortical cells; ?dioecious, spermatangia unknown; carposporo-phytes scattered over lower portions of blades, immersed, with few involucral filaments, carposporangia irregularly angled and rounded, 5–7 μm diameter; cystocarps circular in surface view, compressed globose to hemispherical in section, 50–280 μm diameter, greatly swelling both plant surfaces, ostioles lacking.

Schneider 1975a, 1976; Wiseman and Schneider 1976; Searles 1987, 1988.

Infrequent, only in localized populations offshore from 32 to 50 m in Raleigh Bay, Onslow Bay, and Long Bay, June–December, and from 17 to 21 m on Gray's Reef, August.

Distribution: North Carolina, South Carolina, Georgia, southern Florida, Atlantic Colombia.

Southeastern United States specimens of this species are entire and much smaller than the 10 cm plants reported from the type locality, Colombia (Taylor 1942), but sexual specimens assure the identification (Schneider 1975a).

Halymenia trigona (Clemente y Rubio) C. Agardh 1822, p. 211.

Fucus trigonus Clemente y Rubio 1807, p. 318.

Figure 334

Plants saxicolous, gelatinous, pinkish and brownish to light rosy red, 5–20 cm tall, attached by broad discoid holdfasts giving rise to branched erect axes, stipes lacking; axes terete to somewhat flattened, repeatedly dichotomously branched to several orders, usually not all in the same plane, the first dichotomy 1.5–2 cm above the base, others spaced at distances of 0.5–1.5 cm below, lengthening to 3 cm above, sinuses rounded, overall habit becoming flabellate; main axes 2–6 mm wide, gradually tapering from bases to apices, ultimate axes 1–2 mm wide, apices rounded acute; medulla composed of loosely arranged, mostly longitudinal filaments 2–5 μm diameter and occasional stellate cells; cortex of a single layer, cells pyriform to clavate, 5–7.5 μm diameter, irregularly rounded to pyriform and ovoid in surface view; tetrasporangia subglobose to ellipsoid, cruciately divided, 10–15 μm diameter, 15–20 μm long, scattered in the outer cortex; dioecious, carposporophytes immersed in median portions of plants, with few to several involucral filaments, carposporangia obovoid to irregular, 5–15(–23) μm diameter; cystocarps compressed globose, 70–180 μm diameter, not projecting, ostioles inconspicuous.

As *H. agardhii*, Hoyt 1920; Williams 1951; Taylor 1960; Schneider and Searles 1975; Schneider 1976; Searles 1987, 1988.

Uncommon offshore from 15 to 40 m in Onslow Bay and from 17 to 21 m on Gray's Reef, June–November.

Distribution: North Carolina, Georgia, Bermuda, southern Florida, Gulf of Mexico, Caribbean, Brazil, Spain, Mediterranean, Morocco; elsewhere, probably widespread in tropical and subtropical seas (identified as *H. agardhii*); however, some records are undoubtedly misapplied *Sebdenia flabellata*.

Because Parkinson (1980) pointed out that *Halymenia agardhii* was in actuality *Sebdenia flabellata*, terete specimens of *Halymenia* from the southeastern United States, Bermuda, and the Caribbean are now referred to *H. trigona* (Millar 1990). Codomier (1974) and Millar (1990) have found that all of the terete species of *Halymenia* worldwide are conspecific with this species originally described from Atlantic Spain. *H. trigona* can easily be distinguished from the similar *S. flabellata*, which bears obvious glandlike cells on inner medullary stellate ganglia (see figure 322). *H. trigona* does not produce glandlike cells (Schneider and Searles 1975). Male plants are unknown in the flora.

Kallymeniaceae

Plants soft and gelatinous to firm and fleshy, erect, simple to irregularly divided or dichotomously to subpinnately branched, arising from small to large discoid holdfasts, or parasitic and pulvinate with irregular lobes; axes strongly compressed, ligulate to foliose; axes multiaxial, cortex with one to a few layers, forming compact surfaces; medulla thick, filamentous, some surrounded by refractive stellate cells with long filamentous arms, or pseudoparenchymatous and interspersed with rhizoidal filaments; tetrasporophytes and gametophytes isomorphic or heteromorphic; sporangia cruciately divided, scattered over plants or restricted to upper portions of fronds, or formed singly on branched, prostrate *Hymenoclonium*-like phase; dioecious, spermatangia formed from outer cortical cells and scattered or in small patches; nonprocarpic with connecting filaments, or procarpic, carpogonial branches with three cells, monocarpogonial or polycarpogonial, borne on supporting cells which in some act as auxiliary cells, in others auxiliary cells are distant but homologous to supporting cells of carpogonial branches; connecting filaments often seeking additional auxiliary cells after contact with first; auxiliary cells issuing gonimoblasts outwardly; carposporophytes with cortical pericarps, nearly all gonimoblast cells become carposporangia; cystocarps protruding, scattered over the blades or in fertile areas, with or without ostioles.

Key to the genera

1. Plants ligulate, branched, and firm . . . . Cirrulicarpus
1. Plants foliose, simple to lobed, and perforated . . . . Kallymenia

Cirrulicarpus Tokida et Masaki 1956
Plants erect, firm to cartilaginous, pseudodichotomously to polychotomously branched; blades flattened and narrowly to broadly ligulate, axes often increasing from bases to apices, with or without short stipes arising from small discoid holdfasts, with or without marginal proliferations; medulla of densely intertwined hyaline filaments of varying sizes, with or without stellate ganglioid connections, giving rise to several cortical layers of successively smaller cells; plants isomorphic, tetrasporangia (where present) developed terminally from surface cortical cells or intercalary in subcortical layer, cruciately divided, scattered over the blades; dioecious, spermatangia in large, irregular, superficial sori; mono- or polycarpogonial, carpogonial branches three celled, with one to five per supporting cell which bears sterile subsidiary cells; nonprocarpic, carposporophytes immersed, scattered or clustered and restricted to outer portions, outermost gonimoblast cells become carposporangia, separated by vegetative filaments; cystocarps slightly elevating one or both surfaces of blades by thickening of the cortex, without ostioles.

Cirrulicarpus carolinensis Hansen 1977a, p. 1, figs 1–63.
    Figures 335 and 336
    Plants saxicolous and epizoic, new growth fleshy and pinkish red, older portions firm, leathery to cartilaginous and dark rosy to purplish red, 2–16 cm tall, attached by small to large discoid holdfasts, giving rise to one or more branched erect axes on short stipes; axes broadly ligulate, pseudodichotomously branched, lower branches narrowly cuneate, newest growth broadly obovate, to 2 cm wide, margins undulate and entire to erose, apices obtuse; blades 115–400 μm thick, medulla comprising much of the blade width, composed of densely entangled narrow filaments 2–3 μm diameter surrounded by stellate cells with bodies 10–40 μm diameter, giving rise to a one- or two-layered subcortex with subglobose cells 12–40 μm diameter, and a one- or two-layered

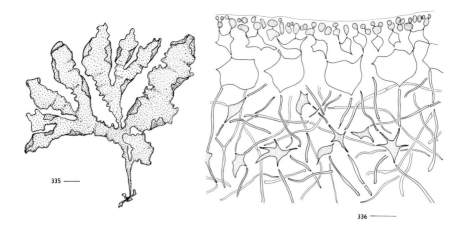

Figures 335, 336. *Cirrulicarpus carolinensis.* 335. Habit, scale 1 cm. 336. Partial cross section of blade, scale 20 µm.

cortex with subglobose cells 4–8 µm diameter, irregularly angular to rounded in surface view; tetrasporangia unknown; dioecious, spermatangia in scattered patches, subglobose to short ellipsoid, 2–3 µm diameter; one or two carpogonial branches, with or without subsidiary cells, on supporting cells; carposporophytes diffuse with gonimoblasts interspersed with vegetative tissue, immersed in upper portions of plants, globose to ellipsoid, to 3 mm diameter, with occasional concentric rings of carposporangia, involucral filaments and ostioles lacking; carposporangia subglobose to irregularly angled, 8–12 µm diameter.

Hansen 1977a, 1977b. As *Cirrulicarpus* sp., Schneider 1976. As *Cryptonemia crenulata sensu* Hoyt 1920; Blomquist and Pyron 1943; Williams 1951; Taylor 1960.

Frequent on rocks and sea squirts from 4 to 9 m off New River Inlet at Mile Hammock Rock (type locality), year-round, and infrequent in localized populations offshore from 21 to 24 m in Onslow Bay, June–September.

Distribution: Endemic to North Carolina and southern Florida.

A detailed morphological and reproductive investigation by Hansen (1977a) showed that specimens from North Carolina previously assigned to *Cryptonemia crenulata* J. Agardh (Hoyt 1920; Williams 1951) were *Cirrulicarpus*, a genus with two other species from Japan and Australia (Hansen 1977b). *Cirrulicarpus carolinensis* is unique in that it has no true auxiliary cell and produces gonimoblasts directly on vegetative filaments. Furthermore, Hansen (1977a) discovered a direct-type life history in culture without tetrasporophytes. Carpospores germinated into filamentous *Hymenoclonium*-like prostrate systems, which in time developed into multilayered crusts with uprights. The erect axes became multiaxial gametophytes similar to those found in the field.

Kallymenia J. Agardh 1842
Plants erect, broadly foliose, and lobate, membranous to firm, occasionally perforate or spinose, with or without short stipes arising from small discoid hold-

fasts, with or without laminal or marginal proliferations; medulla of loosely arranged hyaline filaments of varying sizes surrounded by stellate ganglioid cells, giving rise to two to four cortical layers of successively decreasing cells; plants isomorphic, tetrasporangia developed from inner or middle cortical cells, cruciately divided, scattered in the outer cortex of blades; dioecious, spermatangia in large, irregular, superficial sori; mono- or polycarpogonial, carpogonial branches three celled and borne on supporting cells with subsidiary cells; nonprocarpic, carposporophytes immersed, scattered, nearly all gonimoblast cells becoming carposporangia, separated by vegetative filaments; cystocarps slightly elevating one or both surfaces of blades by thickening of the cortex, without ostioles.

Kallymenia westii Ganesan 1976, p. 169, figs 1–12, 14, 16–21.
Figure 337
Plants saxicolous, gelatinous, and foliose, pinkish to rosy red, 5–20(–30) cm tall and wide, attached by small discoid holdfasts; blades suborbicular to reniform, simple to lobate, perforate, margins smooth to erose, apices rounded; perforations variable in size, with small and large intermixed, oval to circular in shape, becoming irregular in older plants; blades 200–500 μm thick, medulla comprising much of the blade thickness, composed of a few mostly transverse bridging filaments 4–6 μm diameter, surrounded by a subcortex of inflated, thin-walled, transversely elongate cells, 15–62 μm diameter, becoming conspicuously stellate with projecting arms, giving rise to a single-layered cortex (sometimes with two layers) with subglobose cells 3–12 μm diameter, irregularly rounded in surface view; tetrasporangia subglobose and obovoid to ellipsoid, irregularly cruciately divided, 10–15 μm diameter, 12–25 μm long, scattered in the

Figure 337. *Kallymenia westii*, scale 1 cm.

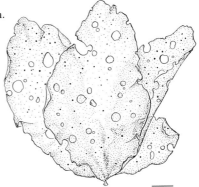

outer cortex; dioecious, spermatangia in large superficial sori formed in rosettes around outer cortical cells, globose to subglobose, 1–2.5 μm diameter; polycarpogonial, two to four (up to eight) carpogonial branches with subsidiary cells on large supporting cells; nonprocarpic, auxiliary cells surrounded by eight to fifteen globose to elongate subsidiary cells, forming a large fusion cell after fertilization; carposporophytes in scattered clusters, immersed, globose, 0.5–1 mm diameter, involucral filaments and ostioles lacking.

As *K. perforata sensu* Schneider and Searles 1973; Schneider 1976.

Infrequent offshore from 26 to 45 m in Onslow Bay, June–August.

Distribution: North Carolina, Venezuela.

A small fragment of a carpogonial specimen from Onslow Bay has verified that the North Carolina specimens previously assigned to the monocarpogonial *Kallymenia perforata* J. Agardh (Schneider and Searles 1973) are polycarpogonial and conform to *K. westii*. In his description of *K. westii* from Venezuela, Ganesan (1976) suggested that all records of *K. perforata* from the Caribbean based on vegetative specimens might represent his newly described species, with *K. perforata* being restricted to the Indian and Pacific oceans. Along with differences in the number of carpogonial branches per supporting cell, *K. perforata* has lobate carpogonial and subsidiary cells, while those of *K. westii* are not lobate (Ganesan 1976). The descriptions of tetrasporangia and spermatangia from North Carolina are the first for the species.

Plocamiaceae

Plants erect and firm, flattened to membranous and repeatedly pinnately branched, bushy, arising from small to large discoid holdfasts, or parasitic and pulvinate, terete and pectinately branched; plants uniaxial, pseudoparenchymatous throughout, apical cells distinct, ultimate branches pectinate; cortex compact, generally few layered, composed of small, deeply pigmented cells enclosing the medulla of progressively larger, isodiametric, hyaline cells, largest centrally; tetrasporophytes and gametophytes isomorphic; sporangia zonately divided, formed below the surface among cortical cells of specialized branchlets; spermatangia formed in superficial sori on ultimate branchlets; procarpic, carpogonial branches three celled, borne laterally on supporting cells which function as auxiliary cells, forming fusion cells after fertilization, gonimoblasts radiating outwardly in rows, outer cells developing as seriate carposporangia; cystocarps without involucres or ostioles, marginal and projecting from the plant surface or laminal in special branchlets and appearing stalked at maturity.

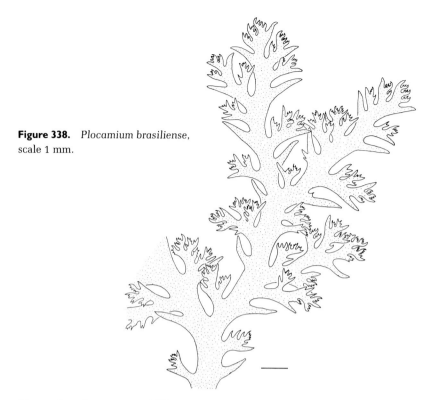

Figure 338. *Plocamium brasiliense*, scale 1 mm.

Plocamium Lamouroux 1813, nom. cons.

Plants erect and firm, flattened to membranous and repeatedly pinnately branched, sympodial, bushy, arising from small to large discoid holdfasts and spreading by secondarily attached prostrate branches with rhizoidal pads; plants uniaxial, pseudoparenchymatous throughout, apical cells distinct, branching alternately distichous, at least the final two orders pectinate; anatomical and reproductive characteristics as for the family.

Plocamium brasiliense (Greville) Howe et Taylor 1931, p. 14, figs 7–8.

   Thamnophora brasiliensis Greville in Saint-Hilaire 1833, p. 448.

   Figure 338

   Plants saxicolous and epiphytic, erect, 2–5(–10) cm tall, light rosy to purplish red, attached by small discoid holdfasts, entangled and spreading, secondarily attached by rhizoidal pads issued from tips and margins of prostrate stoloniferous branches and branches entwined around other algae; axes flattened, 0.7–1(–1.5) mm wide centrally, 130–330 μm thick, with or without inconspicuous midribs in tapering lower portions but microscopically showing at apices, axes two to four times alternately pinnate, branches near apices falcate secund curving toward axes, apices acute; branches regular to irregular in alternate pairs of threes, proximal members of the pairs falcate to long deltoid, simple, entire to few toothed on the outer margins, 0.5–1(–1.5) mm diameter at bases, 1–2(–4) mm long, distal members short to long, rebranched in the manner of main axes, occasionally undeveloped or subsimple, ultimately beset with short, rostrate to acute, secund, simple or similarly rebranched pinnae; pseudo-parenchymatous medulla composed of mixed-sized cells, largest 100–250 µm diameter, below obscuring the axial row and showing through the cortical surface, abruptly producing a one-celled inner and then outer cortex with mixed-sized cells 8–35 µm diameter in greatest dimension and subglobose to irregular in surface view; tetrasporangia ovoid, zonate, 31–52 µm diameter, 57–76 µm long, in paired series in simple to furcate and bifurcate, obtuse ultimate branchlets; dioecious, spermatangia in superficial sori on ultimate branchlets, 3.5–5 µm diameter; carposporangia pyriform, 27–42 µm diameter, 42–55 µm long, cystocarps globose, 0.6–0.7 mm diameter, without ostioles, scattered throughout the plant.

Schneider and Searles 1973; Schneider 1976; Wiseman and Schneider 1976; Blair and Hall 1981.

Infrequent offshore from 23 to 50 m, southeast of Cape Fear in the Frying Pan Shoals area to off Charleston, June–September.

Distribution: North Carolina, South Carolina, Netherlands Antilles, Venezuela, Brazil.

Only vegetative plants are known in our flora.

Sarcodiaceae

Plants firm, wiry to cartilaginous, or soft and flexible, erect, branched or lobed, one or more blades arising from small to large discoid to crustose holdfasts; plants terete to strongly flattened, slender to broadly ligulate and foliose, with or without marginal proliferations; axes multiaxial, cortex composed of radiating files of progressively smaller cells at right angles to the axis, outer rows deeply pigmented; medulla composed of axially elongated, hyaline, thin-walled pseudoparenchymatous or filamentous cells; tetrasporophytes and gametophytes isomorphic; sporangia (where known) laterally produced from inner to outer cortical cells, zonately divided, in scattered sori or raised nemathecia; dioecious, spermatangia formed in widespread to subterminal superficial raised sori; procarpic and unicarpogonial or nonprocarpic and polycarpogonial; carpogonial branches three or four celled, borne on large supporting cells in the inner cortex which in some genera later function as auxiliary cells, others with more distant auxiliary cells; gonimoblasts issued outwardly, carposporophytes developing from large, basal, irregular to reticulate fusion cells or from auxiliary cells producing small-celled central vegetative tissue, carposporangia developing from nearly all gonimoblast cells or from the most distal two to four cells; cystocarps strongly projecting, marginal, scattered over the blades or in specialized papillae, single or grouped, with thick pericarps and obvious ostioles.

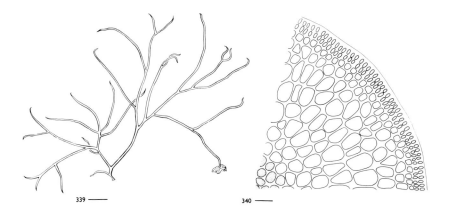

Figures 339, 340. *Trematocarpus papenfussii*. 339. Habit, scale 1 cm. 340. Partial cross section, scale 25 μm.

Trematocarpus Kützing 1843

Plants erect to prostrate and spreading in dense turfs, wiry to cartilaginous, repeatedly dichotomously branched, becoming fastigiate; axes terete to slightly flattened above, slender to narrowly ligulate; medulla loosely to densely filamentous, remaining anatomical characteristics as for the family; tetrasporangia formed in scattered nemathecial sori, zonately divided; gonimoblasts issued from uppermost cells of large, irregular, basal fusion cells, the thin, compact fascicles producing terminal carposporangia; cystocarps scattered throughout the plants, single or clustered, protruding, hemispherical to globose, with thick pericarps and obvious ostioles.

Trematocarpus papenfussii Searles 1972, p. 21, fig. 3.

Figures 339 and 340

Plants saxicolous, wiry and cartilaginous, light brownish to rosy red, erect to 22 cm tall, firmly attached by crustose holdfasts, each giving rise to several highly branched upright axes and becoming crowded; axes terete, some slightly compressed above, 0.3–0.5(–0.7) mm diameter below, tapering to 2.5–3 mm diameter near the rounded apices, sparsely and suboppositely branched below, dichotomously to pseudodichotomously branched to several orders above, fastigiately organized; multiaxial, medulla of compact, thin-walled, axially elongated parallel cells, appearing pseudoparenchymatous in transverse section, 14–30 μm diameter, 160–225 μm long; inner cortex of axially elongate cells, 20–26 μm diameter, 60–90 μm long, outer cortex of radially elongate cells, 3–8 μm diameter, 8–15 μm long, elongate and irregularly rounded in surface view; tetrasporangia and gametangia unknown.

Searles 1972; Schneider 1976.

Infrequent offshore from 23 to 50 m, southeast of Cape Fear in the Frying Pan Shoals area, May–December.

Distribution: Endemic to North Carolina as currently known.

Rhizophyllidaceae

Contarinia Zanardini 1843

Questionable Record

Contarinia magdae Weber-van Bosse in Børgesen 1916, p. 128, fig. 137.

Williams (1951) reported this species in the flora based on material rarely collected offshore in Onslow Bay. Unfortunately, this record cannot be verified with voucher, or more recent, specimens. If the Williams specimens were tetrasporic, it is doubtful that they could be confused with *Peyssonnelia*. Nevertheless, macroscopic features of *Contarinia magdae* are shared with *P. atlantica* Schneider and Reading (1987). Both crusts are firmly affixed to the substrate, and calcification is hypobasal to the fleshy, thick perithallus. Until its existence in our area can be proven, this species will remain a questionable member of the flora.

Cystocloniaceae

Plants erect or prostrate and firm, wiry to thick cartilaginous, alternately to irregularly radially branched, often marginally proliferous, arising from small discoid or fibrous holdfasts, some later becoming secondarily attached and spreading; axes terete to flattened, slender to foliose, uniaxial, pseudoparenchymatous with persistent axial rows, some with rhizoids investing the axial row; cortex compact, few layered, composed of small, deeply pigmented cells enclosing the medulla of progressively larger, isodiametric, hyaline cells of parallel filaments, outer layer complete or incomplete and forming rosettes around subsurface cortical layer; outer surface with or without delicate hyaline hairs; tetrasporophytes and gametophytes isomorphic; sporangia zonately divided, formed among cortical cells scattered throughout, in sori, or restricted to proliferations; monoecious or dioecious, spermatangia formed in outer cortex, scattered or in discrete sori, or restricted to ultimate branchlets; procarpic, carpogonial branches three (rarely four) celled, borne on inner cortical cells; cells of associated branches or daughter cells of supporting cells functioning as auxiliary cells, issuing one gonimoblast which radiates inwardly or outwardly, with large fusion cells or small celled, basal or central zones bearing gonimoblast filaments, in some genera some gonimoblasts becoming sterile threads which separate fertile filaments, outer gonimoblast cells developing as seriate carposporangia; carposporophytes enclosed by multilayered pericarps without ostioles; cystocarps formed on blades, margins, or specialized papillae or branchlets, greatly projecting from plant surfaces.

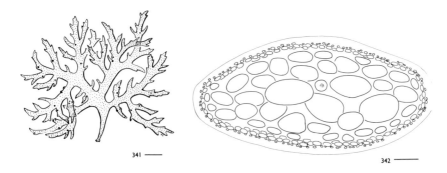

Figures 341, 342. *Craspedocarpus humilis.* 341. Habit, scale 0.5 cm. 342. Cross section, scale 100 μm.

Craspedocarpus Schmitz in Schmitz et Hauptfleisch 1896

Plants erect, narrowly to broadly ligulate, pseudodichotomously to irregularly branched to several orders, arising from small discoid holdfasts, occasionally having lower axes secondarily attached by adventitious marginal holdfasts; main axes flattened throughout, with occasional to frequent flattened to terete marginal proliferations, indistinct nerves usually present at least in young portions of axes, but often throughout; anatomy as for the family, with distinct cortical rosettes and often with abundant hair cells; tetrasporangia formed from outer cortical (rosette) cells, enlarging inwardly, scattered over the axes or restricted to marginal proliferations, in one formed in laminal sori; monoecious or dioecious, spermatangia scattered over the axes, formed from rosette cells; carpogonial branches three (rarely four) celled, borne on inner cortical cells along margins of main axes and proliferations; carposporophytes with small-celled nutritive tissue and gonimoblast initially developing inwardly, at maturity with central large-celled tissues issuing carposporangia in chains, interspersed with nutritive filaments and elongate sterile filaments which connect with the pericarps, without involucral filaments; cystocarps prominent and protuberant, scattered on margins or restricted to proliferations, occasionally on blades, ostioles inconspicuous or lacking.

Craspedocarpus humilis C. W. Schneider 1988, p. 2, figs 1–18.

Figures 341 and 342

Plants erect and lax, saxicolous or conchicolous, dark rosy red, 2–3.5 cm tall; axes flattened, 2–3 mm diameter in main axes, tapering to 1 mm in lesser branches, 300–450 μm thick, alternately and suboppositely branched to four orders from the margins, appearing pinnate, often arcuate, with marginal linear-lanceolate proliferations in lower portions, apices acute; axial row obvious in cross section, cells 7–20 μm diameter, to 500 μm long, surrounded by a medulla of large, hyaline, transversely compressed cells to 200 μm wide, decreasing in size from center outward to a three-layered, densely pigmented cortex, outermost cells forming an incomplete rosettelike layer around subsurface cortical cells, rosette cells occasionally bearing short unicellular hairs to 500 μm long; rosette cells 5–30 μm diameter, subglobose to highly irregular in surface view, subsurface cells subglobose to irregularly ellipsoid in surface view; tetrasporangia ellipsoid to obovoid and arcuate, zonately divided, 28–43 μm diameter, 50–75 μm long, in extensive irregular sori covering much of the main axes and branches, occasionally restricted in small distinct sori in youngest portions of the axes; monoecious, spermatangia globose to obovoid, formed on rosette cells in small rosettes throughout, 2–3 μm diameter; carpogonial branches three celled, carposporangia subglobose to obovoid, 17–30 μm diameter, formed in terminal chains on gonimoblasts surrounding large-celled vegetative tissues; cystocarps marginal, less often on laminae, hemispherical with globose carposporophytes, 425–625 μm diameter, surrounded by stretched cortical cells and producing thin pericarps without ostioles.

Schneider 1988.

Infrequent offshore from 35 to 40 m in Onslow Bay, June–August.

Distribution: Endemic to North Carolina as currently known.

A detailed discussion of the reproductive and vegetative morphology of this taxon is given by Schneider (1988).

Wurdemanniaceae

Wurdemannia Harvey 1853

Questionable Record

Wurdemannia miniata (Duby) Feldmann et Hamel 1934, p. 544.

Gigartina miniata Duby 1830, p. 953.

As *W. setacea*, Blomquist and Pyron 1943; Williams 1948a; Searles and Schneider 1978.

Vouchers of this species from the flora have not been located, nor has it been collected since Williams (1948a) reported it as infrequent on the seaward side of Cape Lookout jetty in fall and summer.

Solieriaceae

Plants firm and cartilaginous but flexible, erect, branched or lobed, arising from small to large discoid to crustose holdfasts, or parasitic and pulvinate with several lobes; plants terete to strongly flattened, slender to broadly ligulate and foliose, with or without marginal and laminal proliferations; axes multiaxial or uniaxial, cortex composed of radiating files of progressively smaller cells at right angles to the axis, outer rows deeply pigmented; medulla composed of loosely to densely arranged, axially elongated, hyaline filamentous cells; tetrasporophytes and gametophytes isomorphic; sporangia laterally produced from outer cortical cells, zonately divided, scattered throughout the plants; dioecious, spermatangia formed in widespread to subterminal superficial raised sori; nonprocarpic, carpogonial branches three (rarely four) celled and borne on large supporting cells in the inner cortex; auxiliary cells scattered and in positions similar to carpogonial branches, cells surrounding auxiliary cells dividing prior to gonimoblast initiation, single gonimoblast initials issued inwardly into the medulla; carposporophytes developing from large central fusion cells or from small-celled central vegetative tissue, carposporangia developing from nearly all or only distal gonimoblast cells; cystocarps embedded to swelling one or both plant surfaces, marginal, scattered over the blades or in specialized papillae, with pericarps and obvious ostioles.

Key to the genera

1. Plants broadly foliose, lobate . . . . Meristotheca
1. Plants terete to flattened and ligulate, branched 2
2. Plants with spines or spurlike branches 3
2. Plants lacking spines or spurlike branches 4
3. Axes terete; cystocarps with central fusion cells . . . . Eucheuma
3. Axes flattened; cystocarps with multicellular sterile central tissues . . . .
. . . . Meristiella
4. Axes terete 5
4. Axes flattened 6
5. Axes to 1 mm diameter; cystocarps with central fusion cells . . . . Solieria
5. Axes to 5 mm diameter; cystocarps with multicellular sterile central tissues
. . . . Agardhiella (in part)
6. Plants dichotomously branched . . . . Sarcodiotheca
6. Plants alternately to oppositely pinnately branched . . . . Agardhiella (in part)

Agardhiella Schmitz in Schmitz et Hauptfleisch 1896
Plants firm, erect, terete to flattened, radially alternately to pinnately branched, with or without adventitious branches arising from cartilaginous discoid holdfasts; multiaxial, medulla of densely intertwined hyaline filaments giving rise to inner cortical cells and progressively smaller, anticlinally developed, pigmented outer cortical cells; tetrasporangia and spermatangial sori as for the family; carpogonial branches three (sometimes four) celled, with reflexed trichogynes; gonimoblasts forming toward the medulla from the inner cortex throughout the blades, carposporophytes with large, diffuse, small- and large-celled central vegetative tissue, surrounded by fertile gonimoblast filaments with short

Figure 343. *Agardhiella ramosissima.* Habit, scale 0.5 cm.

distal chains of three or four carposporangia interspersed with isolated sterile strands of involucral tissue; cystocarps embedded, swelling one blade surface or marginal; thick pericarps with filamentous involucral linings and obvious ostioles.

Key to the species

1. Plants flattened, pinnately branched .... *A. ramosissima*
1. Plants terete, radially to subpinnately branched .... *A. subulata*

Agardhiella ramosissima (Harvey) Kylin 1932, p. 17.
  Chrysymenia ramosissima Harvey 1853, p. 190, pl. XXXB.
  Figures 343 and 344
  Plants saxicolous, firm, fleshy, flattened, light brownish to rosy red, 5–20(–30) cm tall, attached by cartilaginous discoid holdfasts; main axes markedly compressed, 2–6(–15) mm diameter, alternately to oppositely pinnately branched to three or four orders, with or without adventitious branches, smaller ultimate branches less flattened to terete, tapering to bases and acute apices; medulla of loosely arranged filaments, 2.5–5 μm diameter, giving rise to the radially organized inner cortex of three to four layers of hyaline, radially elongate to angular,

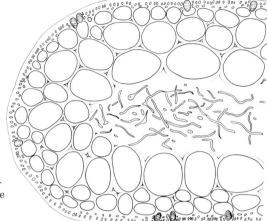

Figures 344.
Agardhiella ramosissima.
Partial cross section, scale
75 μm.

thick-walled cells, 50–150 μm diameter, tapering to the outer one- to two-celled cortex of densely pigmented, elongate cells 7.5–12.5 μm diameter; tetrasporangia in densely scattered patches in the outer cortex throughout except at tips and on older main axes, irregularly obovoid to ellipsoid, zonately divided, 30–50 μm diameter, 47–78 μm long; dioecious, carposporangia formed in terminal chains of three to four on short gonimoblasts surrounding large, diffuse central vegetative tissues; carposporangia subglobose to irregularly ellipsoid, 17–38 μm diameter, 20–45 μm long; cystocarps globose, 1–1.7 mm diameter, with thick pericarps and obvious ostioles.

Taylor 1942, 1960. As *Rhabdonia ramosissima*, Hoyt 1920. As *Neoagardhiella ramosissima*, Schneider and Searles 1973; Schneider 1976.

Locally frequent offshore from 19 to 40 m in Onslow Bay, May–August.

Distribution: North Carolina, Bermuda, southern Florida, Gulf of Mexico, Caribbean, Brazil.

Agardhiella subulata (C. Agardh) Kraft et Wynne 1979, p. 329.

Sphaerococcus subulatus C. Agardh 1822, p. 328.

Figure 345

Plants saxicolous, firm, fleshy to cartilaginous, terete, purplish, greenish, and brownish to rosy red, 7–45 cm tall, attached by cartilaginous discoid holdfasts; axes slender to greatly thickened, 0.5–5 mm diameter, radially and alternately branched to several orders, occasionally subpinnate or secund, with or without adventitious branches, branches basally constricted or tapering to the bases, tips acute to rounded; medulla of densely tangled filaments, 2.5–7.5 μm diameter,

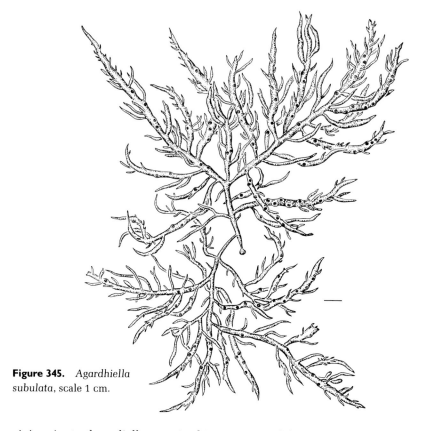

Figure 345. *Agardhiella subulata*, scale 1 cm.

giving rise to the radially organized inner cortex of three or four layers of hyaline, radially elongate to angular, thick-walled cells, 60–250 μm diameter, tapering to the outer one- to two-celled cortex of densely pigmented, elongate cells 5–10 μm diameter; tetrasporangia scattered in the outer cortex throughout except at tips and on older main axes, obovoid to ellipsoid, zonately divided, 22–44 μm diameter, 47–78 μm long; dioecious, spermatangia formed in scattered patches throughout, ellipsoid, 6–7 μm diameter, 14–17 μm long; carposporangia formed in terminal chains of three or four on short gonimoblasts surrounding large, diffuse central vegetative tissues; carposporangia irregularly ellipsoid, 15–30 μm diameter, 34–47 μm long; cystocarps globose, 1–2 mm diameter, with thick pericarps and obvious ostioles.

Gabrielson and Hommersand 1982b. As *Rhabdonia baileyi*, Bailey 1851. As *Rhabdonia tenera*, Melvill 1875; Johnson 1900. As *Agardhiella tenera*, Hoyt 1920, *pro parte*; Williams 1948a, 1951; Taylor 1957, 1960, *pro parte*. As *Neoagardhiella baileyi*, Schneider 1976; Wiseman and Schneider 1976; Kapraun 1980a. As *Solieria chordalis sensu* Harvey 1853; Curtis 1867. As *Eucheuma gelidium sensu* Hoyt 1920, *pro parte*.

Common on intertidal rock of area jetties throughout the area, year-round, but reaching peak size and abundance April–July, and infrequent offshore from 29 to 45 m in Onslow Bay, June.

Distribution: Massachusetts to Virginia, North Carolina, South Carolina, northeastern Florida to Gulf of Mexico, Caribbean, Brazil.

Gabrielson (1985) provided a clarification of the current generic and specific names of this species with a confused nomenclatural history. A thorough morphological study of this species, including specimens from the Carolinas, was made by Gabrielson and Hommersand (1982b). Throughout its range *Agardhiella subulata* is pleiomorphic both between and among populations. It ranges from thick, cartilaginous, and sparingly branched to thin or thick, soft, and copiously branched. Slender plants can be confused with *Solieria filiformis*, which is restricted to offshore habitats; however, *A. subulata* is rare offshore and common intertidally. Cystocarpic specimens are useful in distinguishing these two species.

Eucheuma J. Agardh 1847
Plants firm, erect, terete to flattened, radially to bilaterally branched, with or without papillae and/or spines, arising from cartilaginous discoid holdfasts, or prostrate and secondarily attached by rhizoidal bundles issued from branches; multiaxial, medulla of thin, densely intertwined hyaline filaments, or large cells interspersed with rhizoidal filaments, giving rise to inner cortical cells and progressively smaller, anticlinally developed, pigmented outer cortical cells; tetrasporangia and spermatangial sori as for the family; carpogonial branches three celled, with reflexed trichogynes; gonimoblasts forming toward the medulla from the inner cortex throughout the blades, carposporophytes with large, regular, central fusion cells surrounded by short, fertile gonimoblast filaments with terminal carposporangia interspersed with isolated sterile filaments connected to involucral tissue; cystocarps embedded, swelling one blade surface or marginal; thick pericarps with filamentous involucral linings and obvious ostioles.

Eucheuma isiforme var. *denudatum* Cheney 1988, p. 214, fig. 7.
Figures 346 and 347
Plants saxicolous and conchicolous, cartilaginous, terete, dark rosy red, erect, 16–30(–50) cm tall, attached by cartilaginous discoid holdfasts or spreading by secondary attachments from prostrate branches; axes 2–5(–8) mm diameter, smooth to nodulose, radially alternately to suboppositely branched to two to four orders, branches not basally constricted, but tapering to bases and acute tips, ultimate branches infrequent, 1–5 cm long; short, acute spines few or lacking; medulla of densely tangled filaments, 3–6 µm diameter, giving rise to the radially organized inner cortex of two or three layers of hyaline, ellipsoid and subglobose to stellate, thin-walled cells, 12–60 µm diameter, decreasing in size to the outer layer of densely pigmented ellipsoid cortical cells 5–7 µm diame-

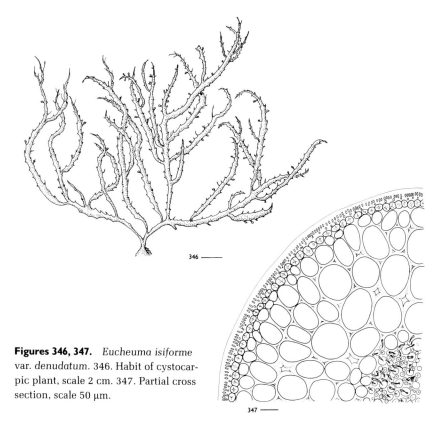

Figures 346, 347. *Eucheuma isiforme* var. *denudatum*. 346. Habit of cystocarpic plant, scale 2 cm. 347. Partial cross section, scale 50 μm.

ter; tetrasporangia scattered in the outer cortex throughout except at tips and on older main axes, obovoid to ellipsoid, zonately divided, 12–20 μm diameter, 35–55 μm long; dioecious, carpogonial branches forming in nemathecia in upper one-third of papillate outgrowths; carposporangia formed in terminal chains of two on short gonimoblasts surrounding large central fusion cells; carposporangia obovoid to ellipsoid, 9–14 μm diameter, 15–26 μm long; cystocarps single, paired, or clustered, immersed in the tips of papillae, thus appearing stalked, 0.5–1.8 mm diameter, with thick pericarps and obvious ostioles.

Gabrielson 1983; Cheney 1988. As *Eucheuma gelidium sensu* Hoyt 1920, *pro parte*. As *E. isiforme*, Schneider and Searles 1973, 1979; Schneider 1976.

Locally frequent offshore from 21 to 40 m, southeast of Cape Fear in the Frying Pan Shoals area, June–September.

Distribution: North Carolina, southern Florida, Gulf of Mexico.

Recent morphological studies of *Eucheuma isiforme*, including specimens from North Carolina, have greatly increased our understanding of the infrafamilial relationships of this taxon (Gabrielson 1983). Although specimens of var. *denudatum* from the west coast of Florida are totally devoid of spines

(Cheney 1988), spines are always present on plants from North Carolina and are therefore useful in helping distinguish this species from other terete species with filamentous medullas, such as *Agardhiella subulata*. Interestingly, although Cheney (1988) described his new variety as "conspicuously spineless," he cited Carolina specimens under the entity, and in a key for the varieties distinguishes var. *denudatum* as having "few or no spines."

Meristiella Cheney in Gabrielson et Cheney 1987
Plants firm, erect, terete to flattened, alternately to suboppositely branched, mostly in one plane, with ultimate short spikelike or spinelike branches, with or without adventitious branches, axes arising from discoid holdfasts; multiaxial, medulla of densely intertwined hyaline filaments giving rise to hyaline inner cortical cells and progressively smaller, anticlinally developed, pigmented outer cortical cells; tetrasporangia and spermatangial sori as for the family; carpogonial branches three celled, with reflexed trichogynes; gonimoblasts forming outwardly from the inner cortex throughout the blades, carposporophytes with large, diffuse, small- and large-celled central vegetative tissue surrounded by fertile gonimoblast filaments with short distal chains of two or three carposporangia interspersed with isolated sterile strands of gonimoblast tissue; cystocarps embedded, greatly swelling one blade surface or marginal, often with one or more associated spines of varying lengths; thick pericarps with filamentous involucral linings and obvious ostioles.

Meristiella gelidium (J. Agardh) Cheney et Gabrielson in Gabrielson et Cheney 1987, p. 483, figs 1–4, 7–26.
   Sphaerococcus gelidium J. Agardh 1841, p. 17.
   Figures 348 and 349
   Plants saxicolous, firm, fleshy, flattened, rosy red, 10–18 cm tall, attached by cartilaginous discoid holdfasts; main axes markedly compressed, 3–7 mm diameter, mostly percurrent, gradually tapering to acute apices, alternately to suboppositely branched to four or five orders, with few or no adventitious branches, ultimate spinelike to spikelike branches flattened to terete, 4–36 mm long, often dense, causing margins to appear subentire to spinulose proliferous; medulla of loosely arranged filaments, 4–8 μm diameter, giving rise to the radially organized inner cortex of three to five layers of hyaline, radially elongate to angular, thick-walled cells, 80–180 μm diameter, decreasing in size to the outer two- to four-celled cortex of densely pigmented elongate cells 5–10 μm diameter; tetrasporangia in densely scattered patches in the outer cortex throughout except at tips and on older main axes, irregularly obovoid or ovoid to ellipsoid, zonately divided, 10–20 μm diameter, 34–51 μm long; dioecious,

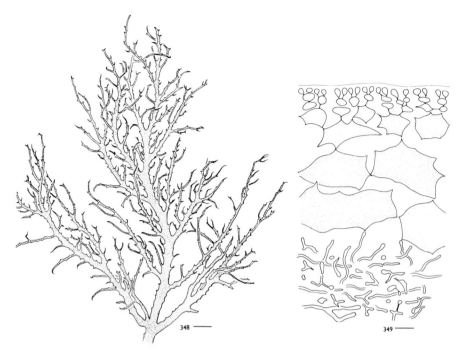

Figures 348, 349. *Meristiella gelidium*. 348. Habit, scale 1 cm. 349. Partial cross section, scale 20 µm.

carposporangia formed in terminal chains of two or three on short gonimoblasts surrounding large, diffuse central vegetative tissues; carposporangia subglobose to irregularly ellipsoid, 22–26 µm diameter, 30–38 µm long; cystocarps globose, 0.8–1 mm diameter, with thick pericarps, few to several, short to long spines, and obvious ostioles.

Gabrielson and Cheney 1987. As *Eucheuma gelidium*, ?Hoyt 1920.

Infrequent, southeast of Cape Fear in Onslow Bay from 30 to 32 m, July–August.

Distribution: North Carolina, Bermuda, southern Florida, Gulf of Mexico, Caribbean, Brazil.

All of the Hoyt specimens labeled *Eucheuma gelidium* that we have discovered are either *Agardhiella subulata* or *Eucheuma isiforme* var. *denudatum* (Searles and Schneider 1978). Our recent collections of robust reproductive plants in Onslow Bay confirm the occurrence of *Meristiella gelidium* in the Carolinas.

Meristotheca J. Agardh 1872

Plants firm, erect, broadly ligulate to foliose, at first simple, but becoming irregularly to palmately and proliferously lobed or divided, with or without papillate outgrowths of the margins and blades, arising from cartilaginous discoid holdfasts; multiaxial, medulla of loosely arranged, intertwined hyaline filaments giving rise to inner stellate cortical cells and progressively smaller, anticlinally developed, pigmented outer cortical cells; tetrasporangia and spermatangial sori as for the family; carpogonial branches three celled with reflexed trichogynes, in some associated with sterile cells on supporting cells; gonimoblasts forming mostly on short to long papillae throughout the blades, carposporophytes with a large, small-celled central vegetative tissue surrounded by fertile gonimoblast filaments with distal carposporangia periodically separated by filamentous incursions of involucral tissue lining the pericarps; cystocarps embedded, occasionally appearing stalked from positions in papillae, swelling one or both surfaces; thick pericarps with obvious ostioles.

Meristotheca floridana Kylin 1932, p. 29, figs 6a–6c.
  Figures 350 and 351
  Plants saxicolous, firm, fleshy, foliose, light to dark rosy red, 10–52 cm tall, attached by cartilaginous discoid holdfasts; axes greatly flattened above short stipes, 300–700 µm thick, 5–25 cm diameter, simple to irregularly divided and palmately lobed, with or without laminal and marginal papillate outgrowths, with or without marginal proliferations; medulla of loosely arranged filaments 2.5–12.5 µm diameter, giving rise to the radially organized inner cortex of anastomosing, transversely elongate, irregular to stellate cells, 12.5–50 µm diameter, 37.5–100 µm long, decreasing in size to the two-layered outer cortex with sur-

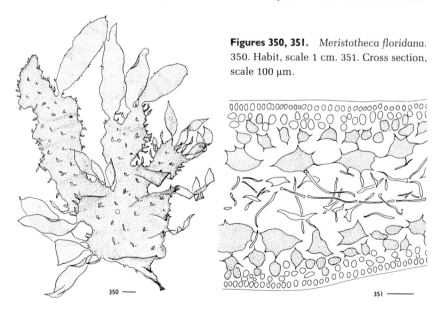

Figures 350, 351. *Meristotheca floridana.* 350. Habit, scale 1 cm. 351. Cross section, scale 100 µm.

350 ——

351 ———

face cells 5–7.5 μm diameter, cortical cells irregularly rounded in surface view; tetrasporangia borne on plants lacking papillae, ellipsoid, zonate to irregularly cruciate, 12.5–18 μm diameter, 22–25(–40) μm long, scattered in the outer cortex; dioecious, spermatangia produced from outermost cortical cells in superficial sori over all but the most basal regions on nonpapillate plants, giving the surfaces mottled appearances, spermatangia 1–3 μm diameter; carposporophytes forming in small papillae, carposporangia ellipsoid to subglobose, 7.5–18 μm diameter, 7.5–25 μm long; cystocarps hemispherical, protruding on only one surface, with obvious ostioles, single or paired, to 0.7–2 mm diameter, the tips of papillae often appearing as thorns on sessile or stalked cystocarps.

Taylor 1960; Schneider 1976; Kapraun 1980a. As *M. duchassaingii*, Hoyt 1920; Williams 1951.

Common offshore in Onslow Bay, infrequent in Raleigh Bay, from 15 to 55 m, May–December, reaching peak size and abundance June–September; on the shallow reef at 6.5 m off New River Inlet, and occasionally subtidally on the Masonboro Inlet jetty, summer.

Distribution: North Carolina, Bermuda, southern Florida, Caribbean.

Both zonate and irregularly cruciate tetrasporangia can be found on the same specimen.

Sarcodiotheca Kylin 1932

Plants firm, erect, narrowly or broadly ligulate and dichotomously to pseudo-dichotomously branched, or terete and radially alternately branched, with or without marginal proliferations or adventitious branches, arising from cartilaginous discoid holdfasts; multiaxial, medulla of loosely to densely arranged hyaline filaments giving rise to inner thin-walled cortical cells and progressively smaller, anticlinally developed, pigmented outer cortical cells; tetrasporangia and spermatangial sori as for the family; carpogonial branches three (occasionally four) celled, with reflexed trichogynes; gonimoblasts forming toward the medulla from the inner cortex throughout the blades, carposporophytes with compact, regular, small- and large-celled central vegetative tissues surrounded by fertile gonimoblast filaments with distal branched chains to as many as eight carposporangia long, these chains interspersed with isolated sterile strands of involucral tissue; cystocarps embedded, swelling one blade surface or marginal; thick pericarps with filamentous involucral linings and obvious ostioles.

Sarcodiotheca divaricata W. R. Taylor 1945, p. 223, pl. 74, fig. 1.

Figures 352 and 353

Plants saxicolous, firm, fleshy, ligulate, dull purplish to rosy red, 7–15 cm tall, attached by cartilaginous discoid holdfasts, occasionally secondarily attached

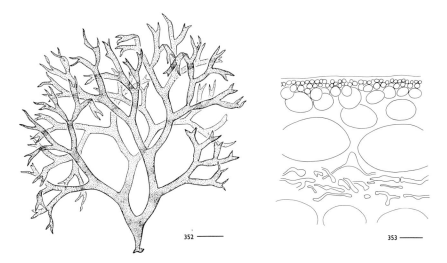

Figures 352, 353. *Sarcodiotheca divaricata.* 352. Habit, scale 1 cm. 353. Partial cross section, scale 50 μm.

by rhizoidal outgrowths from tips and margins; axes greatly flattened above short terete stipes, 300–650 μm thick, 2–7 mm diameter, repeatedly and widely dichotomously branched to several orders, with rounded sinuses and divaricate, acute tips; medulla of loosely arranged filaments 2.5–20 μm diameter, giving rise to the radially organized inner cortex of two to three layers of hyaline, transversely elongate to subglobose or stellate, anastomosing, thin-walled cells, 12–60 μm diameter, tapering to the outer cortical layer of densely pigmented, thick-walled cells 5–20 μm diameter, cortical cells angular in surface view; tetrasporangia obovoid to ellipsoid, zonately divided, 24–29 μm diameter, 32–39 μm long, in scattered patches in the outer cortex; spermatangia unknown; cystocarps subglobose to pyriform, marginal, occasionally appearing minutely pedicellate, with obvious ostioles.

Schneider and Searles 1976; Schneider 1976; Wiseman and Schneider 1976. Infrequent offshore from 35 to 45 m in Onslow Bay and Long Bay, June–August.

Distribution: North Carolina, South Carolina, southern Florida, Galapagos Islands.

Only tetrasporic specimens are known from the Atlantic Ocean.

Solieria J. Agardh 1842

Plants firm, erect, terete to slightly compressed, radially alternately to pinnately branched, with or without adventitious branches, arising from cartilaginous discoid holdfasts; multiaxial, medulla of loosely arranged hyaline filaments giving rise to inner thin-walled cortical cells and progressively smaller, anticlinally developed, pigmented outer cortical cells; tetrasporangia and spermatangial sori as for the family; carpogonial branches three (sometimes four) celled, with reflexed trichogynes; gonimoblasts forming toward the medulla from the inner cortex throughout the blades, carposporophytes with large, regular, central fusion

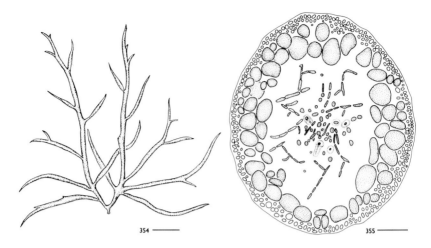

Figures 354, 355. *Solieria filiformis.* 354. Habit of young plant, scale 0.5 cm. 355. Cross section, scale 0.25 mm.

cells, the contents diminished at cystocarp maturity, surrounded by short fertile gonimoblast filaments with terminal carposporangia interspersed with isolated sterile filaments connected to involucral tissue; cystocarps embedded, swelling one blade surface or marginal; thick pericarps with filamentous involucral linings and obvious ostioles.

Solieria filiformis (Kützing) Gabrielson 1985, p. 278, figs 1b, 1d, 2.
   Euhymenia filiformis Kützing 1863, p. 15.
   Figures 354 and 355
   Plants saxicolous, firm, fleshy, terete, light to dark rosy red, 5–24 cm tall, attached by fibrous discoid holdfasts; axes slender, 0.5–1 mm diameter, similar sized throughout except at tips, pseudodichotomously branched below, alternately branched above, branches basally constricted, tips acute; medulla of loosely arranged filaments and rhizoids 2–18 μm diameter, giving rise to the radially organized inner cortex of three or four layers of hyaline, subglobose cells, 60–160 μm diameter, tapering to the outer one- to two-celled cortex of densely pigmented, subglobose to ellipsoid cells 4–10 μm diameter, 9–15 μm long, cortical cells irregularly rounded to angular in surface view; tetrasporangia scattered in the outer cortex throughout except at tips and on older main axes, obovoid to ellipsoid, zonately divided, 15–35 μm diameter, 30–55 μm long; dioecious, spermatangia formed in scattered patches throughout, obovoid to ellipsoid, 2 μm diameter, 3 μm long; carposporangia formed terminally on short gonimoblasts surrounding large central fusion cells; carposporangia obovoid and subglobose to ellipsoid, 10–15 μm diameter, 10–47 μm long; cystocarps globose, 0.5–1 mm diameter, with thick pericarps and obvious ostioles.
   Gabrielson and Hommersand 1982a, 1982b; Searles 1987, 1988. As *A. tenera* sensu Hoyt 1920, pro parte; Taylor 1960, pro parte. As *Solieria tenera*, Schneider and Searles 1973; Schneider 1976; Wiseman and Schneider 1976.

Common offshore from 14 to 48 m in Onslow Bay and Long Bay, and from 17 to 21 m on Gray's Reef, June–January, reaching peak size August–September.

Distribution: North Carolina, South Carolina, Georgia, southern Florida, Gulf of Mexico, Caribbean, Brazil, West Africa.

Gabrielson (1985) provided a clarification of the current generic and specific names of this species, which has a confused nomenclatural history. A thorough morphological study of this species, including specimens from the Carolinas, was made by Gabrielson and Hommersand (1982a).

Hypneaceae

Plants erect or prostrate and firm, wiry to cartilaginous, alternately to irregularly radially branched, becoming bushy, arising from small discoid to crustose holdfasts, some later becoming secondarily attached and spreading, or parasitic and pulvinate with several lobes; axes terete to flattened, slender, uniaxial, pseudoparenchymatous with persistent axial rows; cortex compact, few layered, composed of small, deeply pigmented cells enclosing the medulla of progressively larger, isodiametric, hyaline cells, largest centrally; outer surface with or without delicate hyaline hairs; tetrasporophytes and gametophytes isomorphic; sporangia zonately divided, formed among cortical cells in swollen sori on short ultimate branchlets; dioecious, spermatangia formed in superficial, slightly swollen sori in positions similar to tetrasporangia; procarpic, carpogonial branches three celled, borne laterally on cortical cells; daughter cells of supporting cells functioning as auxiliary cells, fusion cells lacking, issuing one gonimoblast which radiates inward, some gonimoblast cells becoming elongated threads which connect to pericarps all around the carposporophytes, outer gonimoblast cells developing as seriate carposporangia; carposporophytes enclosed by multilayered pericarps with or without ostioles; cystocarps on ultimate branchlets greatly projecting from plant surfaces.

Hypnea Lamouroux 1813

Plants tufted, entangled, and bushy, virgate, or spreading; nonparasitic; axes more or less covered with short linear spinulose and stellate branchlets; other characteristics as for the family.

More study of this taxonomically difficult tropical genus is needed, especially on a worldwide basis.

Key to the species

Hypnea cervicornis J. Agardh 1851, p. 451.

Figure 356

Plants saxicolous, firm, and cartilaginous, light brownish to rosy red or bleached, 2–15 cm tall, initially attached by small discoid holdfasts giving rise to one or more greatly branched erect axes, soon becoming secondarily attached by adventitious rhizoidal pads issued from the ventral surfaces of decumbent axes and branches, becoming bushy and tufted; axes terete, 0.4–1 mm diameter in upper portions, main axes widely pseudodichotomously branched below, alternate and cervicorn above, loosely covered with spirally arranged simple filiform to branched ultimate branchlets, apices acute with small apical cells; axial row obvious in cross section, surrounded by a medulla of large, thick-walled, hyaline cells, 100–320 µm diameter, giving rise to a two-layered, densely pigmented cortex; cortical cells 7.5–25 µm diameter, irregularly angular to rounded in surface view; tetrasporangia ellipsoid, zonately divided, 17–23 µm diameter, 38–49 µm long, in raised sori on subulate to ovate, simple or forked branchlets on fertile axes; dioecious, carposporangia irregularly subglobose, 5–12.5 µm diameter; cystocarps hemispherical, 0.7–0.9 mm diameter, pericarps of several radiating layers of cortical cells, protruding, with ostioles.

Schneider and Searles 1976; Schneider 1976.

Rare, known from a single collection from 27 m offshore in Onslow Bay, November.

Distribution: North Carolina, Bermuda, southern Florida, Caribbean, Gulf of Mexico, Brazil; elsewhere, widespread in tropical and subtropical seas.

Questions have been raised as to whether this species is conspecific with the equally widespread *Hypnea spinella* (C. Agardh) Kützing (Lawson and John 1982; Cribb 1983).

Hypnea musciformis (Wulfen) Lamouroux 1813, p. 131.
    Fucus musciformis Wulfen in Jacquin 1789, p. 154, pl. 14, fig. 3.
    Figures 357–359

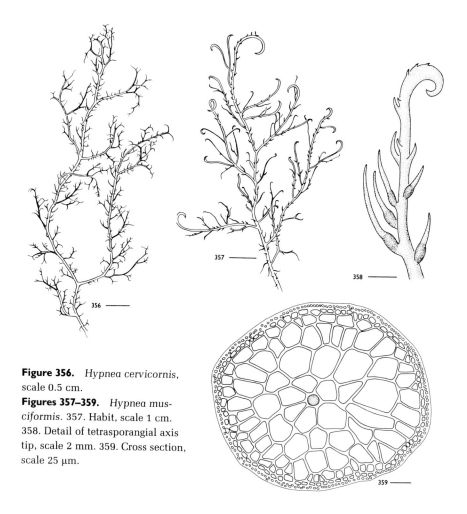

Figure 356. *Hypnea cervicornis,*
scale 0.5 cm.
Figures 357–359. *Hypnea musciformis.* 357. Habit, scale 1 cm.
358. Detail of tetrasporangial axis
tip, scale 2 mm. 359. Cross section,
scale 25 μm.

Plants saxicolous, conchicolous, and epiphytic, firm and cartilaginous, yellowish green to greenish, purplish, and brownish red, 4–50 cm tall, attached
by small discoid holdfasts giving rise to one or more branched erect axes, becoming bushy and greatly entangled by crosier tips and secondarily issued
rhizoids; axes terete, 0.5–2 mm diameter, main axes sparingly to repeatedly
alternately branched, loosely to densely covered with simple to branched
branchlets, secund on lesser branches, apices acute with small apical cells; tips
of axes elongate, mostly unbranched and with few short spines, a few to several tips broadening and hooking into large crosiers, these often bearing short,
spurlike branchlets; axial row not usually obvious in cross section, surrounded
by a medulla of large, thick-walled, hyaline cells, 40–280 μm diameter, giving
rise to a two-layered, densely pigmented cortex; cortical cells 7–17.5 μm diameter, irregularly angular to rounded in surface view; tetrasporangia ellipsoid to
obovoid, zonately divided, 22–33 μm diameter, 42–84 μm long, in raised sori
girdling the bases of simple, often extended, ultimate branchlets; dioecious,

carposporophytes forming in positions similar to tetrasporangial sori on female gametophytes, carposporangia pyriform to subglobose, 19–30 µm diameter, 19–42.5 µm long; cystocarps mostly clustered on ultimate or single on penultimate branches, sessile, globose, 0.3–1 mm diameter, often among spurlike branchlets, pericarps of several radiating layers of cortical cells, protruding, with ostioles.

Harvey 1853; Curtis 1867; Melvill 1875; Johnson 1900; Hoyt 1920; Williams 1948a, 1949, 1951; Humm 1952; Taylor 1960; Chapman 1971; Brauner 1975; Schneider 1976; Wiseman and Schneider 1976; Kapraun 1980a; Richardson 1987.

Common on intertidal rocks, seawalls, and pilings throughout the area, as well as free floating or attached to shells in protected sounds, year-round; infrequent from 6.5 m on the shallow reef off New River Inlet, summer; and locally abundant offshore from 21–26 m in Onslow Bay, July–September.

Distribution: Massachusetts to Virginia, North Carolina, South Carolina, Georgia, northeastern Florida to Gulf of Mexico, Caribbean, Bermuda, Brazil, Uruguay, France to Portugal; elsewhere, widespread in tropical and subtropical seas.

This species shows great morphological plasticity between and among populations. Some plants even lack crosier tips and can at first be confused with *Hypnea valentiae*. The latter species, however, can be distinguished by the presence of unique stellate ultimate branchlets or, if these are lacking, by the much denser branching near the tips.

Hypnea valentiae (Turner) Montagne 1840 [1839–1842], p. 161.
Fucus valentiae Turner 1809, p. 17, pl. 78.
Figures 360 and 361
Plants saxicolous, firm, and cartilaginous, purplish and brownish red to black, 4–12(–40) cm tall, attached by small discoid holdfasts giving rise to one or more branched erect axes, becoming bushy; axes terete, 0.4–1.5 mm diameter, main axes repeatedly alternately branched, different-length axes intermixed, densely covered with simple spinelike to cornute branchlets to 2.2 mm long, at times interspersed with sessile or peltate forked to three- to six-pointed stellate branchlets, apices acute with small apical cells; axial row obvious in cross section, surrounded by a medulla of large, thick-walled, hyaline cells, 50–280 µm diameter, giving rise to a two-layered, densely pigmented cortex; cortical cells 7.5–12.5 µm diameter, irregularly angular to rounded in surface view; tetrasporangia ellipsoid, zonately divided, 17–23 µm diameter, 38–49 µm long, in raised sori girdling the bases of simple ultimate branchlets; dioecious, carposporophytes forming in positions similar to tetrasporangial sori on female gametophytes, carposporangia irregularly subglobose to pyriform, 5–10 µm diameter; cystocarps mostly clustered on simple branchlets, urceolate to subglobose, 325–

Figures 360, 361. *Hypnea valentiae*. 360. Habit of cystocarpic plant, scale 2 mm. 361. Detail of cystocarpic apex, scale 1 mm.

500 μm diameter, pericarps of several radiating layers of cortical cells, protruding, with ostioles.

As *H. cornuta*, Williams 1951; Taylor 1960; Kapraun 1980a.

Infrequent to frequent on intertidal rock of Radio Island, Fort Macon, and Masonboro Inlet jetties; free floating or attached to shells in protected sounds near Wrightsville Beach and Atlantic Beach, spring–summer; and rare from 6.5 m on the shallow reef off New River Inlet, summer.

Distribution: North Carolina, Bermuda, southern Florida, Gulf of Mexico, Caribbean, Brazil; elsewhere, widespread in tropical and subtropical seas.

As pointed out by Wynne (1986), previous workers found *Hypnea valentiae* synonymous with plants called *H. cornuta* (Lamouroux ex Kützing) J. Agardh in the western Atlantic. Unfortunately, the diagnostic stellate branchlets of this taxon are not always present, and this can cause confusion with those specimens of *Hypnea musciformis* lacking crosier tips. However, the tips of *H. musciformis* axes are mostly unbranched and attenuate with few short, spinulose branchlets, whereas those of *H. valentiae* are densely ramified throughout. These species can be found growing together on many jetties in our area.

Hypnea volubilis Searles in C. W. Schneider et Searles 1976, p. 53, figs 3–7, 9–10.

Figures 362 and 363

Plants epiphytic and saxicolous, firm, rosy red, prostrate, and spreading, to 4–6 cm tall, attached by rhizoidal pads issued from margins, blades, and tips, attaching and entwining around other algae; axes flattened, 0.4–0.9(–2) mm diameter, 400–750 μm thick, main axes sparingly subdichotomously and alter-

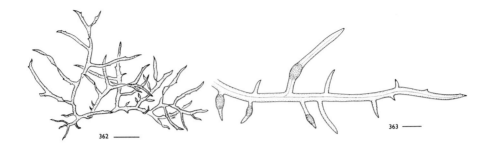

Figures 362, 363. *Hypnea volubilis.* 362. Habit, scale 0.5 cm. 363. Detail of tetrasporangial axis, scale 150 μm.

nately to irregularly branched from the margins, sparsely producing simple marginal subulate branchlets 1–6 mm long, often secund on lesser branches, apices acute with obvious apical cells; axial row obvious in cross section and whole mount, surrounded by a medulla of large hyaline cells, 42–180 μm diameter, giving rise to a single-layered, densely pigmented cortex, naked or with occasional unicellular hairs to 400 μm long; cortical cells 6–48 μm diameter, isodiametric to rounded in surface view; tetrasporangia ellipsoid, zonately divided, 17–24 μm diameter, 25–42 μm long, in slightly raised sori girdling branches and simple ultimate branchlets in median positions.

Schneider and Searles 1976; Schneider 1976; Wiseman and Schneider 1976. Commonly entwined with other algae offshore from 23 to 45 m in Onslow Bay and Long Bay, year-round, but reaching peak size and abundance June–September.

Distribution: North Carolina (type locality), South Carolina, southern Florida, Gulf of California.

Phyllophoraceae

Plants firm, wiry to cartilaginous, erect, simple to branched or lobed, one or more blades arising from small to large discoid to crustose holdfasts, or compact crusts with marginal growth firmly attached over the entire ventral surface without rhizoids, or small tuberous to lobate parasites on larger red algae; upright plants terete to strongly flattened, slender to broadly ligulate, without papillate outgrowths but often proliferous; axes multiaxial, cortex composed of radiating files of small cells at right angles to the axis, outer cells deeply pigmented; medulla composed of large, compact, pseudoparenchymatous, hyaline, thick-walled cells; tetrasporophytes and gametophytes isomorphic or heteromorphic, some directly reproducing the same generation; sporangia (where known) formed in simple chains of three or more from inner cortical cells or directly from carposporangia, cruciately divided, associated with others in swollen sori or nemathecia; monoecious or dioecious, spermatangia cylindrical, formed in widespread sori that are superficial or partially embedded to deep and pitlike; procarpic, carpogonial branches three celled, with or without an associated sterile branch, borne on large supporting cells in the inner cortex which later function as auxiliary cells; fusion cells lacking, auxiliary cells issuing more than one slender, branched gonimoblast inward into the medulla, developing carposporophytes without special involucral filaments, gonimoblasts bearing seriate carposporangia interrupted by attenuate sterile filaments; cystocarps remaining embedded, often swelling one or both plant surfaces, marginal, scattered over the blades or in proliferations, with or without small ostioles.

Key to the genera

1. Plants with narrow, terete, wiry axes; collected in the intertidal . . . .
   . . . . Gymnogongrus
1. Plants broadly ligulate; collected in the deep subtidal offshore . . . .
   . . . . Petroglossum

Gymnogongrus Martius 1828
Plants dichotomously and somewhat irregularly branched to several orders, terete to flattened, with or without marginal or laminal proliferations, often pleiomorphic; anatomy as for the family; dioecious, spermatangia in superficial sori; carposporophytes immersed, elevating one or both plant surfaces by extensive thickening of the cortex, lacking ostioles; carposporangia seriate, terminal on fertile gonimoblasts in between sterile filaments; tetrasporangia developed by cruciate division of carposporangia (carpotetrasporangia) in raised, wartlike nemathecia on female gametophytes of certain species; free-standing tetrasporophytes as yet unknown for species without carpotetrasporangia; other reproductive characteristics as for the family.

Gymnogongrus griffithsiae (Turner) Martius 1828, p. 27.
   Fucus griffithsii Turner 1808, p. 80.
   Figures 364 and 365
   Plants saxicolous, wiry, and cartilaginous, dark brownish, greenish, or purplish red to black, 1–6 cm tall, firmly attached by spreading, calluslike discoid holdfasts or pads, each giving rise to several highly branched upright axes and becoming crowded and turflike; axes terete, somewhat compressed at the tips, 0.3–0.8 mm diameter, slightly tapering to acute apices, dichotomously and irregularly pinnately branched to several orders, occasionally becoming corymbose or appearing polychotomous; multiaxial, medulla of compact, thick-walled, axially elongated cells to 30 μm diameter in cross section, cortex of radiating series of smaller, radially elongate, anastomosing cells, inner cells to 8 μm

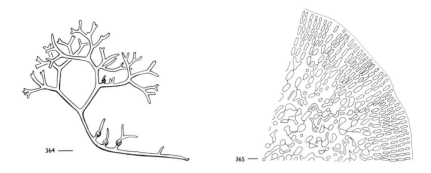

Figures 364, 365. *Gymnogongrus griffithsiae.* 364. Habit of nemathecial plant, scale 2 mm. 365. Partial cross section, scale 10 μm.

in greatest dimension, decreasing in size to the outermost elongate cells, to 2 μm diameter, cortical cells irregularly rounded in surface view; carpotetrasporangia ellipsoid to ovoid, often imperfectly divided, (7–)15–20 μm diameter, (10–)22.5–42.5 μm long, formed in unbranched chains on outwardly extended fertile gonimoblasts forming greatly raised, wartlike, spreading nemathecia, 1–3 mm long, mostly at dichotomies on female gametophytes and eventually completely surrounding axes; male gametophytes unknown.

Hoyt 1920; Williams 1948a, 1949; Humm 1952; Taylor 1957, 1960; Chapman 1971; Wiseman and Schneider 1976; Kapraun 1980a; Richardson 1987. As *Actinococcus aggregatus*, Hoyt 1920.

Common, forming extensive mats trapping sand on intertidal and shallow subtidal (to 6 m deep) rock of jetties, breakwaters, and seawalls near Beaufort and Wilmington, N.C., Georgetown, S.C., and Cumberland Island, Ga.; on natural outcroppings at Fort Fisher, N.C., Myrtle Beach, S.C., and Marineland, Fla.; and as occasional free-floating tufts in bays and sounds; year-round.

Distribution: Massachusetts, Rhode Island, Virginia, North Carolina, South Carolina, northeastern Florida to Caribbean, Brazil, Uruguay, Azores, British Isles, France to Portugal, Mediterranean, Salvage Islands, Canary Islands.

The carpotetrasporic nemathecia of this and other species of *Gymnogongrus*, originally thought to be parasites, were placed in the genus *Actinococcus* (Kützing 1843) until it was established that they were part of the life history of *Gymnogongrus* (Gregory 1934).

Petroglossum Hollenberg 1943

Plants dichotomously and somewhat irregularly branched to several orders, appearing flabellate; blades flattened and ligulate above short, terete stipes or greatly narrowed bases, with or without marginal or laminal proliferations; one or more axes arising from spreading discoid to crustose holdfasts or somewhat stoloniferous bases; medulla as for the family, cortex with only a few layers of small, radiating, highly pigmented cells; tetrasporangia (where known) devel-

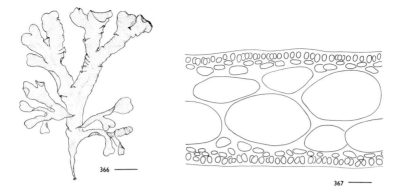

Figures 366, 367. *Petroglossum undulatum.* 366. Habit, scale 1 cm. 367. Cross section, scale 20 μm.

oped from cortical cells in simple chains of three to five producing raised hemispherical to ellipsoid nemathecia, median on blades; monoecious or dioecious, spermatangia in superficial, slightly raised sori; carposporophytes immersed, elevating one or both surfaces of blades or marginal proliferations by extensive thickening of the cortex, lacking involucral filaments and ostioles; carposporangia clustered on fertile gonimoblasts located between sterile, elongate filaments.

Petroglossum undulatum C. W. Schneider in C. W. Schneider et Searles 1976, p. 57, figs 11–16.

Figures 366 and 367

Plants saxicolous or conchicolous, firm and cartilaginous, dark brownish to rosy red, 4–11(–15) cm tall, attached by small to large discoid holdfasts giving rise to one or more greatly branched erect axes, becoming bushy and flabellate; axes ligulate above brief stipes, slightly broadening from the bases to the bifurcate obtuse apices, 5–8(–17) mm wide in upper portions, repeatedly dichotomously branched, with occasional adventitious branching and laminal—rarely marginal—cordate proliferations below; margins undulate and crenate to erose except at extreme bases; blades 115–300 μm thick in vegetative portions; medulla of compact, thick-walled, transversely elongate (in cross section) cells, distinctly different sizes intermixed, to 200 μm diameter in greatest dimension, giving rise to a single-layered cortex, cortical cells irregularly angular to rounded in surface view, 4–5 μm diameter; tetrasporangia unknown; monoecious, spermatangia formed in round to irregular superficial sori, usually in upper portions of blades, not flush with margins; procarps unknown, carposporophytes forming on blades near the margins, carposporangia irregularly subglobose, 5–10 μm diameter; cystocarps globose or hemispherical, pericarps of several radiating layers of cortical cells, protruding on one or both surfaces, without ostioles but rupturing at maturity, one to a few per blade, to 750 μm diameter.

Schneider and Searles 1976; Schneider 1976; Wiseman and Schneider 1976.

Common offshore from 23 to 54 m in Onslow Bay and Long Bay, year-round.

Distribution: North Carolina (type locality), South Carolina.

Locally obvious on the middle and outer continental shelf of the Carolinas, this species is usually heavily epiphytized with crustose coralline algae, hydrozoans, and bryozoans. *Petroglossum undulatum* recorded from south of Cape Canaveral (Eiseman 1979) has recently been assigned to *Gracilaria* ?*cuneata* (Hanisak and Blair 1988).

Gigartinaceae

Plants firm, wiry to cartilaginous, erect, simple to branched or lobed, one or more blades arising from small to large discoid to crustose holdfasts, some later becoming secondarily attached and spreading; axes terete to strongly flattened, slender to broadly foliose, with or without papillate outgrowths; axes multiaxial, cortex composed of radiating files of small cells at right angles to the axis, innermost cells loose and appearing reticulate, outer cells compact and deeply pigmented; medulla composed of loosely to densely organized filaments mostly parallel to the axis, in transverse section appearing reticulate from secondary pit-connections; tetrasporophytes and gametophytes isomorphic; sporangia formed in simple or branched chains of four or more from inner cortical cells or specialized accessory filaments produced from the medulla, cruciately divided, associated with others in swollen sori or nemathecia; dioecious, spermatangia globose, formed in widespread superficial to partially embedded sori; procarpic, carpogonial branches three celled, borne on large supporting cells in the inner cortex that later function as auxiliary cells; fusion cells lacking, auxiliary cells issuing more than one slender, branched gonimoblast inward into the medulla, developing carposporophytes with or without special involucral filaments, most gonimoblast cells becoming carposporangia; cystocarps remaining embedded but swelling one plant surface, scattered over the blades or in papillate projections, lacking ostioles, rupturing when mature.

Gigartina Stackhouse 1809

Plants simple and foliose or irregularly to pseudodichotomously branched, terete to flattened, with sparse to dense papillate outgrowths, often pleiomorphic; anatomy as for the family; tetrasporangia developed from inner cortical cells in circular to spreading immersed sori, elevating portions of papillae or the blades proper; carposporophytes with an obvious dense ring of sterile involucral filaments, cystocarps forming only on papillae; other reproductive characteristics as for the family.

Recently, species of *Gigartina* with heteromorphic alternation of generations and other shared characteristics have been removed to the former subgenus,

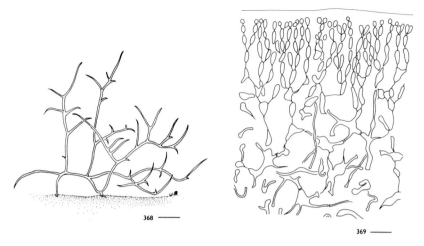

Figures 368, 369. *Gigartina acicularis.* 368. Habit, scale 1 cm. 369. Partial cross section, scale 20 μm.

Mastocarpus, now elevated to generic status and placed in the Petrocelidaceae (Guiry et al. 1984). The isomorphic members, such as the one below, were retained in the larger, Pacific-based genus, *Gigartina*.

Gigartina acicularis (Wulfen) Lamouroux 1813, p. 136.
   Fucus acicularis Wulfen 1803, p. 63.
   Figures 368 and 369
   Plants saxicolous, wiry, and cartilaginous, dark purplish red to black, 2–10 cm tall, initially attached by small discoid holdfasts, giving rise to one or more sparsely branched, erect to prostrate, curved axes, soon becoming secondarily attached by adventitious rhizoidal pads issued from the ventral surfaces of decumbent axes and branches, spreading and becoming entangled; axes terete with occasional flattening in upper portions, 0.5–1(–2) mm diameter, irregularly to alternately and pseudodichotomously branched below, above with mostly short alternate to secund recurved branches, rarely pinnate, tapering to acute apices and less markedly to bases; medulla of loosely to densely tangled filaments, 2.5–15 μm diameter, giving rise to the radially organized inner cortex with anastomosing, stellate cells, 5–17.5 μm diameter, tapering to an outermost cortex with surface cells 3–5(–7) μm diameter, cortical cells round in surface view; tetrasporangia ellipsoid, cruciately divided, 15–18 μm diameter, 22–40 μm long, embedded in loose sori in the inner cortex, causing a slight swelling of the lower branch surface; carposporophytes forming in small papillae, carposporangia ellipsoid, 13–18 μm diameter, 18–23 μm long; cystocarps hemispherical, protruding on only one surface, without obvious ostioles, single or in short series, median in short ultimate, often recurved, branchlets, to 1 mm diameter.
   Blomquist and Humm 1946; Williams 1948a, 1949, 1951; Taylor 1960; Kapraun 1980a.
   Common from 1 to 6 m on subtidal rock of Cape Lookout jetty, June–October;

rarely from 6.5 m on the shallow reef off New River Inlet, summer; and occasionally from the drift on Shackleford Bank and free living in the bays behind Wrightsville Beach.

Distribution: North Carolina, Bermuda, southern Florida, Caribbean, Brazil, Uruguay, British Isles, France to Portugal, Mediterranean, Black Sea, Azores, West Africa, Canary Islands, India, Japan.

Most of the abundant summer collections from Cape Lookout are sterile or tetrasporic and are found associated with several macroscopic algae, including *Sargassum*, *Grateloupia*, and *Gelidium*. *Gigartina* has not been located on protected jetties such as those on Shackleford Bank and Radio Island near Beaufort, N.C., and is thus limited to open, shallow, exposed ocean environments. Occasional plants detached from such areas are known to sustain themselves in protected bays with free-living plants of *Gracilaria* (Kapraun 1980a).

Petrocelidaceae

Plants cartilaginous, erect, branched, and ligulate to foliose, one or more blades arising from small to large discoid to crustose holdfasts, or compact crusts with marginal growth, firmly attached over the entire ventral surface without rhizoids; upright plants strongly flattened, slender to broadly foliose, with papillate outgrowths; axes multiaxial, cortex composed of radiating files of small cells at right angles to the axis, innermost cells loose and appearing reticulate, outer cells compact and deeply pigmented; medulla composed of loosely organized filaments mostly parallel to axes, in transverse section appearing reticulate from secondary pit-connections; crusts with a prostrate, spreading hypothallus which gives rise above to branched, ultimately erect, loosely to compactly associated filaments composing the perithallus, and below to a solid layer of short descending filaments affixing to rock; tetrasporophytes and gametophytes heteromorphic, some directly reproducing the same generation; sporangia intercalary, formed singly on perithallial filaments, cruciately divided, associated with others in nemathecia; dioecious, less commonly monoecious; spermatangia cylindrical, formed in widespread superficial to partially embedded sori; procarpic, carpogonial branches three celled, bearing a sterile branch from the second or third cell, borne on large supporting cells in the inner cortex which later function as auxiliary cells; fusion cells lacking, auxiliary cells issuing more than one slender, branched gonimoblast outwardly into the cortex, developing carposporophytes without special involucral filaments, most gonimoblast cells becoming carposporangia; cystocarps remaining embedded in papillate projections, with obscure lateral ostioles.

Mastocarpus Kützing 1843

Erect plants ligulate to foliose, flattened throughout, pseudodichotomously to palmately branched, with sparse to dense papillate outgrowths, often pleiomorphic; anatomy as for the family; crusts as for the family; tetrasporangia single within perithallial filaments of crusts; carposporophytes lacking involucral filaments, cystocarps forming only in marginal and laminal papillae; other reproductive characteristics as for the family.

Mastocarpus stellatus (Stackhouse in Withering) Guiry in Guiry, West, Kim, et Masuda 1984, p. 56.

Fucus stellatus Stackhouse in Withering 1796, p. 99.

Plants saxicolous, firm, and cartilaginous, dark brownish to purplish red and black, 5(–17) cm tall, attached by small, discoid overlapping holdfasts giving rise to one or more branched erect axes; axes flattened throughout from above short stipes, expanded above, 2–10 mm wide, widely pseudodichotomously to palmately or irregularly branched, the slightly thickened margins inrolled and the blades becoming contorted and twisted, papillae dense over both blades and, less commonly, over margins, ligulate to cylindrical; medulla of loosely to densely tangled filaments 2–12 μm diameter, giving rise to the radially organized inner cortex with anastomosing, stellate cells, 7.5–20 μm diameter, tapering and elongating to an outermost cortex with surface cells 2–6 μm diameter, cortical cells irregular to round in surface view; tetrasporangia in "*Petrocelis*" crustose alternate stage; gametophytes dioecious, spermatangia superficial, terminal on short, branched filaments from outer cortical cells of papillae; carposporophytes forming in rounded papillae, carposporangia subglobose to ellipsoid, 7.5–12(–20) μm diameter; cystocarps globose with small lateral ostioles, to 1 mm diameter.

As *Gigartina stellata*, Schneider and Searles 1973; Schneider 1976.

Rare, known from a single hand-collected cystocarpic specimen on rock from 30 m offshore in Onslow Bay, September.

Distribution: Newfoundland, Maritime Provinces to Connecticut, North Carolina, Iceland, Faeroes, British Isles, USSR to Portugal, Baltic Sea.

The record of this cold-water species in the Carolinas is most unusual, as it is found continuously in the western Atlantic from Canada to Long Island Sound. These northern populations are basically intertidal and reach to a few meters subtidally. Clearly, the specimen collected from 30 m in Onslow Bay is anomalous, and the species is not a regular member of our flora. Although many general collections have subsequently been made at the same site where *Mastocarpus stellatus* was found in 1972, few dives have been made in fall when bottom temperatures cool and the species might more likely be present.

Petrocelis cruenta, the tetrasporophyte of *Mastocarpus stellatus* in some European locations (West et al. 1977), is unknown in the flora, along with all other species of the genus. Temperate North American plants generally have a direct development of "gametophyte" from "carpospore" (Edelstein et al. 1974); however, Guiry and West (1983) found that some carpospores from Maine produced *P. cruenta* crusts in culture.

GRACILARIALES

Plants erect and firm, wiry to cartilaginous, branched, often bushy, as one or more blades arising from small to large, discoid to crustose holdfasts, or small, colorless parasites found on larger members of the order; calcification lacking; axes terete to flattened, slender to broadly ligulate, uniaxial but mostly obscured below and appearing multiaxial, pseudoparenchymatous throughout; cortex compact, few layers to multilayered, composed of small, deeply pigmented cells enclosing the medulla of progressively larger, isodiametric, hyaline cells, largest centrally; outer surface with or without delicate hyaline hairs; tetrasporophytes and gametophytes isomorphic; sporangia cruciately divided, formed obliquely below the surface on intercalary cortical cells, scattered, in sori or in specialized branchlets; spermatangia formed in superficial sori or small saucer-shaped to flask-shaped depressions among surface cells; carpogonial branches two to three celled and borne laterally on intercalary cortical cells; sterile cell branches produced by supporting cells fusing with carpogonium and other branch cells after fertilization, forming large generative fusion cells, auxiliary cells lacking; gonimoblasts radiating outwardly in rows, outer cells developing as seriate or clustered carposporangia; carposporophytes enclosed by multilayered pericarps with distinct ostioles, occasionally connected by thin, tubular nutritive cells, some lacking inner pericarps; cystocarps laminal or marginal, projecting from the plant surface.

The order contains a single family (see Fredericq and Hommersand 1989a).

Gracilariaceae

Key to the genera

1. Spermatangia formed in saucer-shaped to flask-shaped depressions; carposporophytes connected to pericarps by thin, tubular nutritive cells . . . .
. . . . Gracilaria
1. Spermatangia formed in superficial sori in the outer cortex; carposporophytes lacking connective tubular nutritive cells . . . . Gracilariopsis

Gracilaria Greville 1830, nom. cons.

Plants simple or dichotomously, alternately, irregularly, to proliferously branched, arising from small to large discoid holdfasts; cortex of two to six radiating layers; tetrasporangia formed obliquely from intercalary cells in cortical filaments and scattered among cortical cells over the entire plant or various sections; dioecious, spermatangia formed singly from outer cortical cells in saucer-shaped to flask-shaped depressions in the outer cortex; carpogonial branches two celled, supporting cells formed obliquely from intercalary cells in cortical filaments, ultimately bearing two sterile cell branches which fuse with fertilized carpogonia to form generative fusion cells that produce gonimoblasts directly; cystocarps ostiolate, hemispherical, having large central cellular placentas, with thin, tubular nutritive cells between the carposporophyte and pericarp; carposporangia organized in terminal clusters or irregular chains on gonimoblast filaments.

This genus is represented by an extremely variable group of species (Taylor 1960), and the distinctions between some taxa remain dubious. Recent studies of plants found in the western Atlantic north of the Carolinas and previously ascribed to *Gracilaria foliifera* (Forsskål) Børgesen and *G. verrucosa* have shown these plants to be distinct from their European counterparts and both members of a highly plastic species, *G. tikvahiae* McLachlan (McLachlan et al. 1977; Bird et al. 1977; Edelstein et al. 1978; McLachlan 1979). We have found that local specimens of *G. foliifera* conform with *G. tikvahiae* and are highly pleiomorphic (see figures 377–380), but are distinct from plants we maintain as *G. verrucosa* in the flora (figure 381). On a worldwide basis, further clarifications of other species are necessary, including some of the species listed below.

Key to the species

1. Plants composed of narrow to broad blades, compressed 2
1. Plants composed of terete axes, sparingly to much branched 4
2. Plants sparingly branched, largest segments exceeding 1 cm broad below branch nodes . . . . *G. curtissiae*
2. Plants much branched, largest segments less than 1 cm broad below branch nodes 3
3. Branch apices acute . . . . *G. tikvahiae*
3. Branch apices obtuse to emarginate . . . . *G. mammillaris*
4. Plants sparingly branched, axes 2–4 mm diameter in upper portions . . . .
. . . . *G. cylindrica*
4. Plants much branched, axes 0.5–2 mm diameter in upper portions 5
5. Tetrasporangia scattered over main axes and branches; at least some spermatangia formed in deep pits . . . . *G. verrucosa*

5. Tetrasporangia borne in nemathecial swellings above the constricted bases of ultimate branches; all spermatangia formed in shallow, concave sori . . . .
. . . . *G. blodgettii*

Gracilaria blodgettii Harvey 1853, p. 111.

Figures 370 and 371

Plants conchicolous or saxicolous, erect to 20 cm, bushy, purplish, pinkish, or rosy red, attached by small discoid holdfasts; axes terete, to 1(–2) mm diameter below, freely alternately radially to irregularly branched throughout to four or five orders, ultimate branches gradually tapering to acute apices, generally short and spindle shaped, 0.2–0.5(–1) mm diameter and, as most other branches, markedly constricted at their points of attachment; medulla composed of large, hyaline, thin-walled cells to 600(–1200) µm diameter, abruptly contiguous with a one- or two-layered inner cortex of small cells and a single-layered outer cortex of rounded-angular smaller cells, 7.5–25 µm diameter in surface view; tetrasporangia globose to ellipsoid, cruciately divided, 20–37.5 µm diameter, 30–38(–50) µm long, borne in cortical nemathecial swellings above the constricted bases of ultimate branches; spermatangia formed in numerous small, concave sori scattered in the outer cortex; carposporophytes with occasional tubular nutritive cells connecting to pericarps; carposporangia ellipsoid, rounded angular to globose, to 38 µm long; cystocarps abundant, prominent, mammillate, to 1 mm diameter.

Schneider and Searles 1973; Schneider 1976; Kapraun 1980a; Reading and Schneider 1986; Searles 1987, 1988. As *G. confervoides sensu* Hoyt 1920, pro parte.

Known occasionally from near-shore ledges off Wilmington and commonly from 16 to 35 m offshore in Onslow Bay, year-round, achieving maximum development in summer and fall; and from 17 to 21 m on Gray's Reef, July–August.

Distribution: North Carolina, Georgia, southern Florida, Gulf of Mexico, Caribbean, Brazil, ?Japan, ?China.

At least one of the offshore reef specimens attributed to *Gracilaria verrucosa* (as *G. confervoides* (L.) Greville) by Hoyt (1920) is a male specimen of *G. blodgettii*. Florida specimens of this taxon were introduced into the sounds near Beaufort, N.C., in April 1944 during the time of the short-lived agar industry here; however, those plants probably did not survive the inshore winter conditions (Causey et al. 1946). At present, *G. blodgettii* has not been found attached on local jetties or on shells in sounds. The offshore plants are finer than offshore *G. verrucosa* plants, which are likewise constricted, but inshore specimens of *G. verrucosa* are macroscopically indistinguishable from *G. blodgettii*. The two taxa can only be separated reliably when fertile.

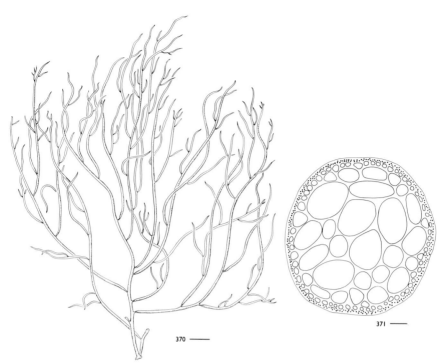

Figures 370, 371. *Gracilaria blodgettii*. 370. Habit, scale 2 cm. 371. Cross section, scale 100 μm.

A report on Japanese specimens attributed to *Gracilaria blodgettii* is at variance with some characters attributed here to the taxon. Ohmi (1958) found tetrasporangia densely scattered over the greater part of the frond and illustrated antheridia in much deeper, more elongate crypts than the shallow depressions of Carolina plants (Schneider and Searles 1973, fig. 2B; Reading and Schneider 1986, fig. 2). Ohmi found his specimens to be similar also to *G. cylindrica*; however, he tentatively placed them in *G. blodgettii* on the basis of vegetative morphological characters because the plant is "extremely variable in external morphology." Our specimens nevertheless bear the short, spindle-shaped branchlets described by Harvey (1853) for his type specimen from Florida, not pictured by Ohmi (1958) for Asian specimens. Despite the fact that Harvey's specimens were vegetative, our plants conform to those from Florida and the Caribbean and we continue to name our reproductive plants *G. blodgettii*. Reproductive specimens from the type locality, Key West, would be of great value in clarifying this taxon.

Gracilaria curtissiae J. Agardh 1885, p. 61.
 Figure 372
 Plants saxicolous or conchicolous, erect to 40 cm, bright rosy red, attached by small discoid holdfasts; axes foliaceous, membranous, 0.5–1 mm thick, dichotomously to polychotomously branched; segments ligulate to lanceolate, somewhat contracted at the bases, 1–3.5 cm broad, 3–22 cm long; medulla composed of slightly compressed to globular hyaline, thin-walled cells to 600 μm diameter,

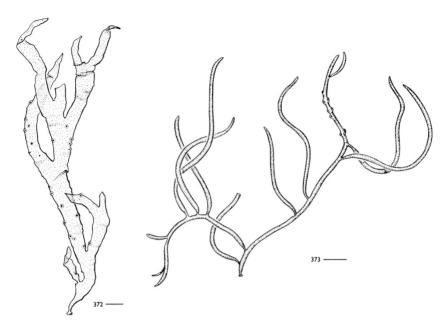

Figure 372. *Gracilaria curtissiae*, scale 1 cm.
Figure 373. *Gracilaria cylindrica*, scale 1 cm.

abruptly contiguous with a two-layered inner cortex of small compressed cells and a one- to three-layered outer cortex of elongate cells, 7.5–13 μm diameter in surface view; carposporophytes with numerous radial tubular nutritive cells connecting to pericarps; carposporangia ellipsoid, obovoid to globose, to 38 μm long; cystocarps abundant, prominent on margins and laminae, hemispherical and slightly apiculate, to 1.2 mm diameter.

Schneider and Searles 1975; Schneider 1976.

Known from 27 to 40 m offshore in Onslow Bay, June–November.

Distribution: North Carolina, southern Florida, Caribbean, Brazil.

Williams (1951) stated that some of the offshore material of *Gracilaria foliifera* (Forsskål) Børgesen could have passed for *G. curtissiae*; however, that report has not been included in the flora due to lack of specimens and the known abundance of a similar taxon offshore, *G. mammillaris* (Schneider and Searles 1979). *G. curtissiae* is uncommon in North Carolina (the only known site in the flora), and only cystocarpic specimens have been found.

Gracilaria cylindrica Børgesen 1920, p. 375, figs 364, 365.
    Figure 373
    Plants saxicolous or conchicolous, erect to 38 cm, rosy red, attached by small discoid holdfasts, axes terete, tapering to the holdfasts, but 2–4 mm diameter in middle sections and above, sparsely and radially, alternately to unilaterally branched to two or three orders; branches not significantly different in size from main axes, arcuate for the most part, sharply constricted to pedicellate or tapered at their points of attachment, slightly tapering to acute apices or not

tapering with blunt apices; medulla composed of large hyaline or lightly pigmented cells, 200–525 µm diameter and surrounded by one or two layers of smaller pigmented, elongate cortical cells, 5–7.5 µm diameter in surface view; tetrasporangia ellipsoid to subglobose, cruciately divided, 20–40 µm diameter, 40–60 µm long, scattered or grouped in irregular patches in the cortex; carposporophytes with few tubular nutritive cells connecting to pericarps; carposporangia obovoid, ellipsoid, to subglobose, to 40 µm long; cystocarps scattered or localized, mammillate, to 2 mm diameter.

Schneider 1975a, 1976.

Known only from 27 to 35 m offshore in Onslow Bay, June–December. Male plants are unknown in the flora.

Distribution: North Carolina, southern Florida, Caribbean, Brazil, ?Japan.

Gracilaria mammillaris (Montagne) Howe 1918, p. 515.

Rhodymenia mammillaris Montagne 1842a, p. 242.

Figures 374–376

Plants saxicolous, flabellate, erect to 14 cm, most commonly 7–9 cm tall, dark purplish red to dull rosy red, attached by small discoid holdfasts; axes ligulate, (3–)5–7(–10) mm diameter, arising from short, terete, simple or branched stalks; blades regularly to irregularly dichotomously branched to four or more orders, commonly with cuneate bases and obtuse to emarginate apices, to 250 µm

Figures 374–376. *Gracilaria mammillaris*. 374, 375. Habits, scale 1 cm. 376. Cross section, scale 25 µm.

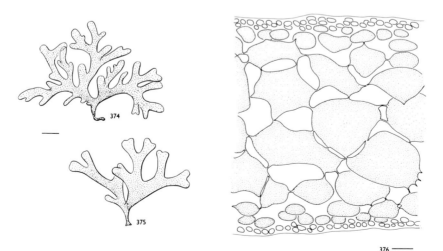

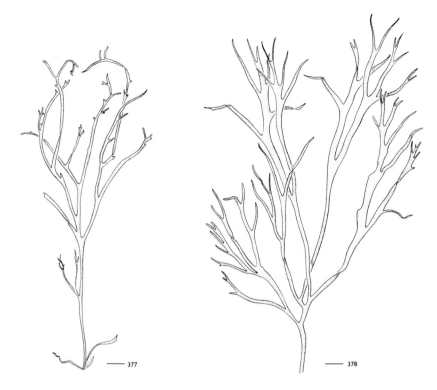

—— 377

—— 378

thick; medulla composed of large, globular to occasionally compressed, lightly pigmented, thick-walled cells 50–170 μm diameter, contiguous with a one-, sometimes two-layered cortex of much smaller quadrate to angular, densely pigmented cells 5–17.5 μm diameter in surface view; tetrasporangia globose, occasionally ellipsoid, cruciately divided, 25–35 μm diameter, scattered in the slightly modified cortical layers; spermatangia formed in numerous small, well-defined, shallow, concave sori scattered over the blades, usually separated by anticlinally elongate cells; carposporophytes with few to several tubular nutritive cells connecting to pericarps, carposporangia globose, subglobose, subquadrate, angular, to irregular, to 30 μm long; cystocarps scarce to abundant, prominent, mammillate, to 1.3 mm diameter.

Schneider 1975c, 1976; Wiseman and Schneider 1976; Schneider and Searles 1979; Searles 1987, 1988.

Common from 14 to 60 m in Onslow Bay and Long Bay, year-round, reaching peak biomass of 17 g/m² during June in a large central area of Onslow Bay; rare from 17 to 21 m on Gray's Reef, July.

Distribution: North Carolina, South Carolina, Georgia, Bermuda, southern Florida, Gulf of Mexico, Caribbean, Brazil, Galapagos Islands, Pacific Mexico, California (see Schneider 1975b).

This species is the most commonly collected *Gracilaria* offshore and is often the dominant member of the standing crop of the red seaweeds there (Schneider and Searles 1979). Inshore records of Blomquist and Humm (1946) and offshore records of an atypical form (Schneider and Searles 1973; also refer to Schneider 1975c) are now ascribed to *G. tikvahiae*.

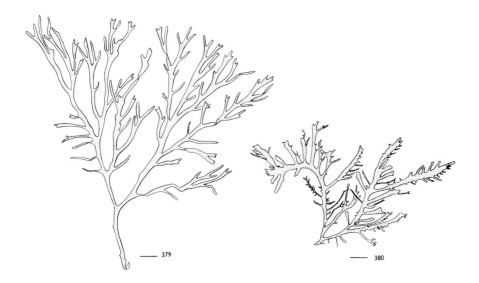

Figures 377–380. *Gracilaria tikvahiae,*
specimens showing morphological variation, scales 1 cm.

Gracilaria tikvahiae McLachlan 1979, p. 19, fig. 1.

Figures 377–380

Plants saxicolous, conchicolous, gregarious, flabellate to fastigiate, erect to
37 cm, black, greenish and purplish red, to dull rosy red, olive brown, yel-
low green, or light green, attached by small discoid holdfasts; axes, except at
the extreme base, compressed to markedly flattened, 0.2–1 mm thick, 1–15 mm
wide, irregularly di-, tri-, to polychotomously and alternately branched in single
plants from one to several orders, often bearing short proliferous branches
on lower internodes; internodal segments broadening toward their distal ends,
apices acute; medulla composed of large, somewhat compressed, lightly pigmented cells, (70–)90–240(–270) μm in greatest dimension, surrounded by a
one- or two-layered cortex of smaller, rounded to angular, densely pigmented
cells, 5–12.5 μm diameter in surface view; tetrasporangia obovoid to ellipsoid
and subglobose, cruciately divided, 10–35 μm diameter, (17–)22–37(–45) μm
long, scattered in the cortex throughout or concentrated in upper portions of the
plant; spermatangia formed in well-defined, shallow, concave sori scattered over
the blades; carposporophytes with a few to several tubular nutritive cells connecting to pericarp or lacking; carposporangia globose, obovoid, and ellipsoid,
15–33(–40) μm in greatest dimension; cystocarps abundant and prominent over
the faces and margins of the blades, hemispherical, to 1 mm diameter.

Richardson 1987. As *G. multipartita sensu* Harvey 1853; Curtis 1867; Mel-
vill 1875; Hoyt 1920. As *G. multipartita* var. *angustissima sensu* Harvey 1853;
Hoyt 1920. As *G. foliifera sensu* Williams 1948a, 1951; Humm 1952; Kim and
Humm 1965; Chapman 1971; Schneider 1976; Wiseman and Schneider 1976;

Kapraun 1980a. As *G. foliifera* var. *angustissima sensu* Williams 1948a, 1949; Wiseman and Schneider 1976. As *G. lacinulata sensu* Kim and Humm 1965. As *G. mammillaris sensu* Williams 1948a, 1949; "atypical form" *sensu* Schneider and Searles 1973; Schneider 1975c.

Known intertidally from jetties throughout the area and also from shallow subtidal and estuarine habitats on rocks and shells, or free living, year-round, achieving maximum development during the warmest months; known from 14 to 48 m offshore in Onslow Bay, April–December, and from 19 m in offshore Georgia, March.

Distribution: Maritime Provinces to Virginia, North Carolina, Georgia, northeastern Florida to Gulf of Mexico, Caribbean, and probably south to Brazil (currently recorded as *G. foliifera* [Oliveira F. 1977]).

Despite lacking specimens for comparison, McLachlan (1979) suspected that plants south of New Jersey known as *Gracilaria foliifera* (Forsskål) Børgesen were probably members of the same morphologically variable species of the Canadian Maritimes and New England that he described as *G. tikvahiae*. Vegetatively, *G. tikvahiae* cannot be distinguished from European *G. foliifera*. However, cystocarpic specimens from the area compare perfectly with the description and illustrations of *G. tikvahiae*, not *G. foliifera* (McLachlan 1979).

Kim and Humm (1965) claimed to be able to distinguish on morphological grounds between *Gracilaria tikvahiae* (as *G. foliifera*) and *G. lacinulata* Vahl. After examining over 100 specimens in the flora, we find a high degree of morphological variation from narrow, nearly terete forms to broad, complanate ones. They are impossible to segregate either at the specific or subspecific levels. Offshore, deep-water specimens of *G. tikvahiae* previously recorded as both *G. foliifera* (Schneider 1976) and *G. mammillaris* "atypical form" (Schneider and Searles 1973, fig. 4B; Schneider 1975c) are now believed to be different expressions of the morphological variability shown in inshore populations. Although offshore cystocarpic specimens compare suitably with *G. tikvahiae*, no males have been discovered.

Populations of *Gracilaria tikvahiae* from Canada have received a great deal of attention to their biochemistry, genetics, life history, and ecology (refer to McLachlan 1979). In the Carolinas, Kim and Humm (1965) have studied growth rates and characterized the properties of the agar extracted from certain populations of this species.

Figure 381. *Gracilaria verrucosa*, scale 1 cm.

Gracilaria verrucosa (Hudson) Papenfuss 1950, p. 195.

    Fucus verrucosus Hudson 1762, p. 470.

    Figure 381

    Plants conchicolous, saxicolous, erect to 60 cm, or free living, bushy, dull to dark brownish, purplish, greenish, grayish, and rosy red to light pinkish red, attached by small discoid holdfasts; axes terete throughout, 0.5–2(–3) mm diameter, freely alternately and pseudodichotomously, radially to irregularly branched to four orders, usually with numerous short proliferous branches, the main axes often percurrent; branches gradually tapering toward acute apices, and tapered to somewhat constricted at their points of attachment; medulla composed of large, hyaline, thin-walled cells to 650 μm diameter, abruptly contiguous with a two- or three-layered cortex of small, pigmented, rounded-angular cells 5–12.5 μm diameter in surface view, occasional surface cells producing deciduous fine hairs; tetrasporangia ovoid to subglobose, cruciately divided, 20–38 μm diameter, 27–45 μm long, sparse to numerous, scattered over the main axes and branches; spermatangia formed in small, shallow to mostly deep pits to 50 μm diameter in the cortex of younger parts of the plant; carposporophytes with few to numerous tubular nutritive cells connecting to pericarps, carposporangia obovoid to subglobose, to 38 μm long; cystocarps abundant, prominent, hemispherical, 0.3–1 mm diameter.

Taylor 1960; Schneider 1976; Wiseman and Schneider 1976; Kapraun 1980a; Reading and Schneider 1986; Richardson 1986. As *G. confervoides*, Harvey 1853; Johnson 1900; Hoyt 1920, *pro parte*; Humm 1942, 1944; Causey et al. 1946.

Known frequently from the shallow subtidal on jetties near Beaufort and Wilmington, N.C., and on shells or free living in area sounds, tidal pools, and brackish-water estuaries, summer–fall, and infrequently from 23 to 30 m offshore in Onslow Bay, August.

Distribution: Virginia, North Carolina, South Carolina, Georgia, Bermuda, Gulf of Mexico, Caribbean, Brazil, USSR to Portugal, and possibly more widespread.

Free-living sterile specimens of *Gracilaria verrucosa* (as *G. confervoides* (L.) Greville) formed the basis of a short-lived agar industry in Beaufort, N.C., during World War II (Humm 1942, 1944; Causey et al. 1946), and the species is still prolific in local sounds during the summer and fall. Attached plants are also common during this period and are often fertile. Inshore plants grow intermixed with the similar and less common *Gracilariopsis lemaneiformis*, which is easily distinguished when fertile. *Gracilaria verrucosa* could also be confused with *G. blodgettii* from offshore (see p. 320).

Plants in the flora compare favorably in all respects with descriptions of *Gracilaria verrucosa* from the British Isles, the type locality (Dixon and Irvine 1977). They do not compare, however, with plants collected from New England and Canada previously placed under this binomial and now recognized as nearly terete forms of *G. tikvahiae*. Therefore, aside from the record of Humm (1979) from Virginia, records to the north of Cape Hatteras require verification.

Spermatangia have been given greater taxonomic weight in the genus *Gracilaria* since Yamamoto (1975, 1978) organized the species on the basis of several male conceptacle types. In a recent study of *G. verrucosa* from North Carolina (Reading and Schneider 1986), a single male specimen was shown to have a continuum from deep cryptlike conceptacles (Verrucosa type) to shallow bowl-shaped conceptacles (Textorii type). More species in the genus need to be assessed for this characteristic to establish whether or not conceptacle type is constant, as for *G. blodgettii* (Reading and Schneider 1986), or variable as for this taxon.

Gracilariopsis Dawson 1949

Plants erect and terete, alternately, irregularly, to proliferously branched, arising from small discoid holdfasts; cortex of few to several radiating layers; tetrasporangia formed obliquely from intercalary cells in cortical filaments, scattered over the entire plant; dioecious, spermatangia formed singly from outer cortical cells in superficial sori; carpogonial branches two celled, supporting cells formed obliquely from intercalary cells in cortical filaments, ultimately bearing two sterile two- or three-celled branches which fuse with fertilized carpogonia to form generative fusion cells that produce gonimoblasts directly; cystocarps ostiolate, hemispherical, having large central cellular placentas, without tubular nutritive cells between the carposporophyte and pericarp, gametophytic cells of inner pericarp becoming dark-staining nutritive tissue, gonimoblast filaments fused to the cystocarp floor by small conjunctor cells; carposporangia basically organized in straight terminal chains.

Fredericq and Hommersand (1989b) recently reestablished this genus on the basis of cytological and anatomical studies of *Gracilariopsis lemaneiformis* from the Pacific Ocean.

Gracilariopsis lemaneiformis (Bory) Dawson, Acleto, et Foldvik 1964, p. 59, pl. 56, fig. A.

   Gigartina lemanaeformis Bory 1828, p. 151.

   Figure 382

Plants saxicolous or conchicolous, erect to 0.5(–2) m, purplish and brownish to rosy red, attached by small discoid holdfasts, often attached secondarily by short, prostrate, rhizomelike branches issued variously on the axes; axes terete, 0.5–2 mm diameter, often sparsely but occasionally freely branched to a few orders, branching irregular, radial, and often short proliferous, main axes often percurrent; branches gradually tapering to acute apices, somewhat tapered but not sharply constricted at their points of attachment; medulla composed of large, hyaline, thin-walled cells surrounded by four to six layers of much smaller pigmented cortical cells, 2.5–7.5 μm diameter in surface view; tetrasporangia ellipsoid to subglobose, cruciately divided, 25–27.5 μm diameter, 32.5–47.5 μm long, scattered in the outer cortex; spermatangia formed in open, nondepressed, superficial hyaline sori, contiguous over large portions of the outer cortex; carposporophytes lacking radial trabecular filaments, carposporangia ellipsoid, obovoid to globose, 12.5–25 μm diameter, 20–38 μm long; cystocarps sparse to abundant, prominent, hemispherical, 0.8–1.8 mm diameter.

   As *Gracilariopsis sjoestedtii*, Dawson 1953. As *Gracilaria sjoestedtii*, Taylor 1960; Schneider 1976; Wiseman and Schneider 1976; Kapraun 1980a. As *G. confervoides* var. *longissimus sensu* Harvey 1853.

Known infrequently from the shallow subtidal on jetties and on shells in sounds, mixed with and less common than *Gracilaria verrucosa* near Wilmington and Beaufort, N.C., spring and summer; infrequent offshore in Onslow Bay from 24 to 50 m, June–August; and frequent on shells and pebbles in the shallow subtidal and lower intertidal of the natural outcropping near Fort Fisher, year-round.

Distribution: North Carolina, South Carolina, southern Florida, Gulf of Mexico, Caribbean, Brazil, Indian Ocean, Central Pacific islands, China, southern British Columbia to Chile.

Abbott (1983) found *Gracilariopsis lemaneiformis* (as *Gracilaria lemaneiformis*) synonymous with *Gracilaria sjoestedtii*, the name under which plants from our flora had previously been reported. Although there has been speculation that Atlantic specimens reported as *G. lemaneiformis/sjoestedtii* are incorrect identifications (Bird and Oliveira F. 1986), Dawson (1953) reported plants from North Carolina and the Caribbean with superficial male sori. Unfortunately, male plants of *Gracilariopsis* and *Gracilaria* are uncommon.

Gracilariopsis lemaneiformis possibly represented a portion of the *Gracilaria verrucosa* (as *G. confervoides*) collected during the 1940s in the sounds near Beaufort and Morehead City, N.C., for use in the wartime agar industry (Humm 1942, 1944; Dawson 1953; Causey et al. 1946). The two can only be separated as fertile gametophytes.

Figure 382.
*Gracilariopsis
lemaneiformis,*
scale 1 cm.

RHODYMENIALES

Plants erect and terete to ligulate and foliose, or prostrate and spreading to crustose, soft and gelatinous to fleshy and cartilaginous; plants multiaxial and pseudoparenchymatous or multiaxial with loose or compact aggregations of filaments centrally and radial outer layers, or with central hollow mucilage-filled cavities, some with both solid axes and hollow branches, some with dark-staining gland cells; tetrasporophytes and gametophytes isomorphic, tetrasporangia cruciately or tetrahedrally divided, terminal or intercalary in cortical filaments, scattered, in flush to depressed sori or in chains in raised nemathecia, polysporangia produced in some; dioecious, spermatangia formed from outer cortical cells, scattered or in discrete spreading or girdling superficial sori; procarpic, procarps scattered or in nemathecia, carpogonial branches three or four celled, one or more two- or three-celled auxiliary cell branches forming on the same supporting cells, with short connecting filaments (ooblasts) or connecting cells; gonimoblasts issued outwardly from small to large fusion cells, with or without a reticulum of attenuate attachment filaments (*tela arachnoidea*) between carposporophytes and pericarps; pericarps with or without beaks (rostrate) and with obvious ostioles; cystocarps projecting and scattered or marginal, carposporangia liberating one carpospore.

Key to the families

1. Plants solid or in part hollow; if hollow, lacking attenuate medullary filaments; sporangia cruciately divided . . . . Rhodymeniaceae
1. Plants always in part hollow, with attenuate longitudinal medullary filaments at least in youngest portions; sporangia tetrahedrally divided 2
2. Hollow central cavities with single-layered traversing septa; tetrasporangia intercalary, scattered; carpogonial branches four celled, only terminal gonimoblast cells becoming carposporangia . . . . Champiaceae
2. Septa lacking or present as few-layered partitions at branch insertions; tetrasporangia terminal in depressed sori; carpogonial branches three celled, nearly all gonimoblast cells becoming carposporangia . . . . . . Lomentariaceae

Rhodymeniaceae

Plants soft and gelatinous to firm and fleshy, erect, irregularly, pinnately, or palmately to dichotomously branched, arising from small to large discoid holdfasts, or parasitic and tuberculate to pulvinate with irregular lobes; axes terete to strongly compressed, ligulate to foliose, some beset with mucilage-filled bulbous vesicles; structure multiaxial, cortex of one to few complete to incomplete layers or composed of loosely to closely associated, dichotomously branched filaments; medulla compact and pseudoparenchymatous to hollow and gelatinous, with or without traversing medullary filaments; tetrasporophytes and gametophytes isomorphic; sporangia cruciately divided, intercalary or terminal in cortical filaments, scattered over plants or restricted to subapical sori; dioecious, spermatangia formed from outer cortical cells, scattered or in small patches; procarpic, three- to four-celled carpogonial branches borne on a supporting cell also bearing a two- or three-celled auxiliary cell branch, connecting cell present in some; auxiliary cells issuing gonimoblasts outwardly; carposporophytes with cortical pericarps with or without a reticulum of attenuate attachment filaments (*tela arachnoidea*) between them, most gonimoblast cells becoming carposporangia; cystocarps protruding, scattered over the blades or branches, with ostioles.

Key to the genera

1. Plants hollow throughout or in part hollow 2
1. Plants solid throughout 3
2. Plants with short to long solid axes, bearing terminal and lateral hollow, bladderlike vesicles.. Botryocladia
2. Plants with hollow axes, at least in part, bladders lacking .... Chrysymenia
3. Plants peltate, dorsiventrally arranged .... Halichrysis
3. Plants otherwise 4
4. Axes basically dichotomously branched 5
4. Axes palmately, irregularly, or pinnately divided...................... 8
5. Plants leathery, tetrasporangia in subterminal sori .... Rhodymenia
5. Plants soft, tetrasporangia scattered or in sori scattered over the blades ... 6
6. Main axes to 2.5 mm wide 7
6. Main axes 5–55 mm broad .... Agardhinula (in part)
7. Plants irregularly dichotomously divided, tetrasporangia scattered over the main blades....................................... Gloioderma (in part)
7. Plants regularly dichotomously branched, tetrasporangia in sori scattered over the blades .... Leptofauchea
8. Axes narrow, pinnately or radially branched .... Gloioderma (in part)
8. Axes broadly ligulate to foliose, irregularly to palmately branched ....
.... Agardhinula (in part)

Agardhinula De Toni 1897

Plants erect, soft, irregularly to di- and trichotomously branched, appearing palmate to flabellate; blades broadly flattened throughout, with cuneate bases arising from small discoid holdfasts; medulla of several layers of large cells interspersed with markedly smaller ones, giving rise to one to three layers of anticlinally arranged cortical cells, outer layer not completely covering subsurface cortical layer; tetrasporangia developed from surface cortical cells, cruciately

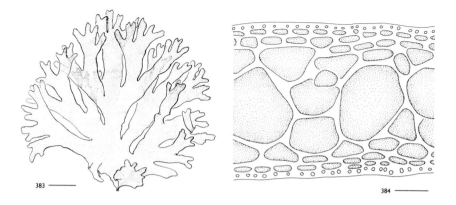

Figures 383, 384. *Agardhinula browneae.* 383. Habit, scale 1 cm. 384. Cross section, scale 100 μm.

divided, in mostly large confluent, reticulate sori over the blades, giving the plant a mottled appearance; dioecious, spermatangia in widespread irregular, superficial sori; procarpic, carposporophytes scattered over blades, nearly all gonimoblast cells becoming a hemispherical to spherical mass of carposporangia, attached to flat base by a few gonimoblast filaments, connecting *tela arachnoidea* lacking; cystocarps prominently elevating one surface of blade by thickening of the cortex, with inconspicuous ostioles.

Agardhinula browneae (J. Agardh) De Toni 1897, p. 64.
  Callophyllis browneae J. Agardh 1885, p. 36.
  Figures 383 and 384
  Plants saxicolous, firm gelatinous, pinkish and brownish to rosy red, 10–20 cm tall, attached by discoid holdfasts, giving rise to one or more palmate to flabellate erect blades; axes broadly ligulate, to 2.5 cm broad below nodes, regularly to irregularly dichotomously to trichotomously branched, lower axes broader than above, sinuses narrow and rounded, bases cuneate, margins entire, occasionally proliferous near the base, apices rounded acute, obtuse to truncate; blades 350–625 μm thick, medulla comprising much of the blade thickness, composed of large and intermixed smaller cells 35–370 μm in greatest dimension in cross section, giving rise to a one- to three-layered cortex with irregular to subglobose cells 4–10 μm diameter, irregularly rounded to circular in surface view; tetrasporangia ellipsoid, cruciately divided, 12.5–22.5 μm diameter, 17.5–37.5 μm long; spermatangia ellipsoid to globose, 1–2 μm diameter; cystocarps scattered over blades and margins, occasionally confluent, hemispherical, to 1 mm diameter, ostioles inconspicuous; carposporangia subglobose to irregularly angled, 10–15 μm diameter.
  Hoyt 1920; Williams 1951; Taylor 1928, 1960; Schneider and Searles 1973; Schneider 1976; Wiseman and Schneider 1976.
  Frequent and occasionally abundant in localized populations, offshore from 17 to 55 m in Onslow Bay and Long Bay, June–November.

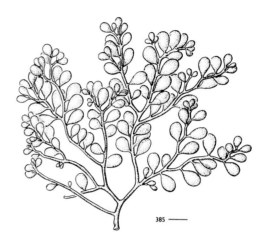

Figures 385–387.
Botryocladia occidentalis.
385. Habit of robust plant,
scale 1 cm. 386. Habit
of lax deep-water plant,
scale 1 cm. 387. Inner
medullary cells bearing
gland cells, scale 25 µm.

385 ——

Distribution: North Carolina, South Carolina, southern Florida.

At times from June to September *Agardhinula* is the dominant species on the wreck of the *Suloide* (Buoy WR-13; 34°32′48″N, 76°53′43″W; depth- top 13 m, bottom 21.5 m), but usually is found as isolated plants at most sites.

Botryocladia (J. Agardh) Kylin 1931, nom. cons.

Plants erect, simple or irregularly to pseudodichotomously branched, radially to bilaterally beset with few to many soft, stipitate, obpyriform to obovoid vesicles; terete axes slender, solid, cartilaginous, arising from small to massive discoid holdfasts; vesicles with hollow mucilage-filled central cavity, medulla of one to a few layers of large, mostly colorless cells, some bearing single or clustered dark-staining gland cells projecting into the cavity, cortex of a continuous or discontinuous layer of small pigmented cells; tetrasporangia developed from subsurface cortical cells, cruciately divided, scattered over the vesicles; monoecious or dioecious, spermatangia in small, irregular, superficial sori on vesicles; procarpic, carposporophytes scattered over vesicles, nearly all gonimoblast cells becoming carposporangia, attached to base by a few gonimoblast filaments, connecting *tela arachnoidea* lacking; cystocarps slightly elevating the vesicle surface by thickening of the cortex, with ostioles.

Key to the species

1. Vesicles with one or two gland cells on medullary cells, plants with multiple vesicles and richly branched .... *B. occidentalis*
1. Vesicles with more than two gland cells on medullary cells, plants with few vesicles and sparsely branched if at all 2
2. Outer cortication on vesicles not covering entire surface of subsurface cortical cells that obscure the cells of the medulla; gland-bearing cells unmodified .... *B. pyriformis*
2. Outer cortication on vesicles restricted to rosette formation above subsurface cortical cells and between anticlinal walls of exposed cells of the medulla; gland-bearing cells occasionally stellate .... *B. wynnei*

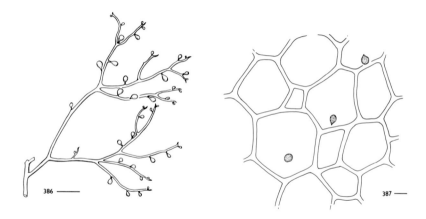

Botryocladia occidentalis (Børgesen) Kylin 1931, p. 18.

Chrysymenia uvaria var. *occidentalis* Børgesen 1920, p. 403, fig. 388.

Figures 385–387

Plants saxicolous, rosy red, attached by massive discoid holdfasts giving rise to one or more erect, terete, cartilaginous, alternately to pseudodichotomously branched axes, 2–20(–25) cm tall, radially to bilaterally beset with few to many mucilage-filled vesicles; axes solid, pseudoparenchymatous, 0.5–1.5 mm diameter; vesicles obpyriform to obovoid and subglobose, short stipitate, 2–8 mm diameter, 3–11 mm long; medullary cells polygonal, 50–150 μm diameter, several producing one (to two noncontiguous) globose to ellipsoid gland cells toward the inner vesicle cavity, 7.5–20 μm diameter; outer cortex with irregular to subglobose anastomosing cells 3–15 μm diameter forming a near-continuous layer over subsurface layer of intermediate-sized cells, many subsurface cells not completely covered by outer cortex; tetrasporangia unknown; cystocarps few per vesicle, to 1.5 mm diameter, ostioles present; carposporangia rounded and irregularly angled, 12–25 μm diameter.

Taylor 1960; Chapman 1971; Schneider 1976; Wiseman and Schneider 1976; Schneider and Searles 1979; Searles 1981, 1987, 1988. As *Chrysymenia uvaria*, Hoyt 1920; Taylor 1928; Blomquist and Pyron 1943. As *B. uvaria*, Williams 1951.

Common and abundant offshore from 14 to 48 m in Onslow Bay and Long Bay, year-round; from deep water off Sapelo Island and from Gray's Reef, March–September; inshore, rarely collected on Fort Macon jetty during the summer.

Distribution: North Carolina, South Carolina, Georgia, Bermuda, southern Florida, Gulf of Mexico, Caribbean, Brazil.

Along with *Gracilaria mammillaris*, *Botryocladia occidentalis* is one of the two most commonly dredged red algae offshore in Onslow Bay, having by far the greatest year-round biomass of any of the red algae (Schneider and Searles 1979). We have estimated its biomass at as much as 41 g/m^2 during peak production in June. Only female reproductive characteristics have been observed for this species.

Some deep-water plants contain sparse, small bladders, giving the appearance of a different species (see figure 386). This form, although sometimes found at lesser depths, is usually found at depths of 40 m or greater.

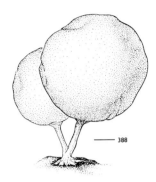

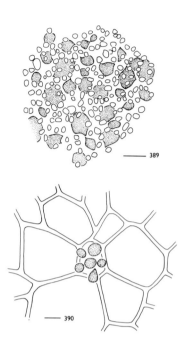

Figures 388–390.
Botryocladia pyriformis.
388. Habit, scale 2 mm.
389. Surface cortication,
scale 20 μm. 390. Inner
medullary cells bearing
gland cells, scale 25 μm.

Botryocladia pyriformis (Børgesen) Kylin 1931, p. 18.

    Chrysymenia pyriformis Børgesen 1910, p. 187, figs 8–9.

    Figures 388–390

    Plants saxicolous, pinkish red, attached by small to large discoid holdfasts, giving rise to one or two terete, wiry, simple or branched axes, 1–5 cm tall, radially beset with few mucilage-filled vesicles, or with one or two stipitate vesicles arising directly from small discoid holdfasts; vesicles obpyriform, short stipitate, 3–30(–40) mm diameter, 4–40(–50) mm long; medullary cells rounded polygonal, 30–138 μm diameter, with large and occasional smaller cells interspersed, some of the smaller cells producing a cluster of four to eight central, pyriform to clavate gland cells toward the inner vesicle cavity, 5–25 μm diameter; outer cortex with rounded-angular to subglobose cells 3–8 μm diameter forming an incomplete layer over a nearly complete subsurface layer of intermediate-sized cells; tetrasporangia ovoid, to 34 μm diameter and 44 μm long; dioecious, spermatangia clustered on outer cortical cells, 1–2.5 μm diameter; cystocarps globose, several per vesicle, to 1 mm diameter, ostioles present; carposporangia subglobose to irregularly angled, 7.5–22.5 μm diameter.

    Schneider and Searles 1973; Schneider 1976; Wiseman and Schneider 1976.

    Infrequent offshore in Onslow Bay and Long Bay from 18 to 55 m, June–September.

    Distribution: North Carolina, South Carolina, Bermuda, southern Florida, Caribbean, Brazil.

    In this area, only the large-vesicled form of *Botryocladia pyriformis* is found, a form at some variance with that originally described (Taylor 1960; Schneider

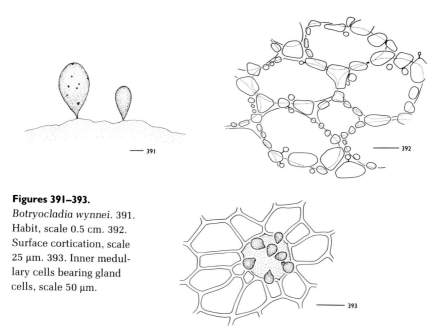

Figures 391–393.
Botryocladia wynnei. 391.
Habit, scale 0.5 cm. 392.
Surface cortication, scale
25 µm. 393. Inner medul-
lary cells bearing gland
cells, scale 50 µm.

and Searles 1973) for stalked plants from the Caribbean (Børgesen 1910). Because only vesicle size is at variance, however, we are retaining our plants as
B. pyriformis until more significant differences can be delineated. Only male and
female gametophytes of this species are known.

Botryocladia wynnei Ballantine 1985, p. 199, figs 1–5.
    Figures 391–393
    Plants saxicolous, pinkish to rosy red, 6–35 mm tall, attached by small dis-
coid holdfasts giving rise to short simple or branched stalks bearing one to five
vesicles; vesicles pyriform and becoming obovoid at maturity, short stipitate, 3–
14 mm diameter, 6–35 mm long; medullary cells rounded, elongate polygonal,
30–90 µm diameter, with occasional modified stellate or unmodified cells producing generally three to six (although as few as one to as many as seventeen),
central, pyriform to ovoid gland cells toward the inner vesicle cavity, 15–33 µm
diameter, 17–26 µm long; inner cortex of intermediate-sized cells wedged in between upper medullary cell surfaces, outer cortex with irregular to subcircular cells 7–14 µm diameter forming an incomplete layer over anticlinal walls of
medullary cells and intermediate-sized cells themselves in rosette fashion, occasionally covering gland-bearing cells completely; tetrasporangia globose, cruciately divided, 18–30 µm diameter; monoecious, spermatangia clustered on outer
cortical cells, 1–3 µm diameter; cystocarps several per vesicle, 440–670 µm
diameter, ostioles present; carposporangia subglobose to irregularly angled, 7.5–
22.5 µm diameter.
    Searles 1987, 1988.

Infrequent from 30 to 35 m in Onslow Bay, June; and 17 to 21 m on Gray's Reef, June–August.

Distribution: North Carolina (new record), Georgia, Puerto Rico.

Botryocladia wynnei can easily be distinguished from the other large-vesicled species, *B. pyriformis*, by its incomplete, yet regular, surface cortication between medullary cells on the vesicles, and occasional stellate medullary cells bearing gland cells. Plants from the area are occasionally twice as large (to 35 mm tall) as those described initially from Puerto Rico (to 14 mm) but conform in all other characteristics (Ballantine 1985).

Chrysymenia J. Agardh 1842

Plants erect, soft, and tubular to foliose, irregularly to alternately branched or split, branches radial to marginal; axes terete and mucilage filled throughout, appearing saccate to vesiculate, to broadly foliose and locally hollow but not saccate, arising from small discoid holdfasts; medulla hollow and with or without a few filaments traversing the mucilage-filled cavity; inner cortex of large compacted cells, these occasionally bearing gland cells toward the mucilage-filled cavity, giving rise to one to three cortical layers of anticlinally arranged cortical cells, outer layer completely to incompletely covering subsurface layer; tetrasporangia developed from cortical cells, cruciately divided, and scattered over the plants; dioecious, spermatangia in isolated irregular superficial sori; procarpic, carposporophytes scattered over plants, nearly all gonimoblast cells becoming a hemispherical to globose mass of carposporangia, attached to flat base by a few gonimoblast filaments, connecting *tela arachnoidea* lacking; cystocarps prominently elevating one surface of blade by thickening of the cortex.

Key to the species

1. Plants flattened, ligulate to foliose .... *C. agardhii*
1. Plants terete and tubular .... *C. enteromorpha*

Chrysymenia agardhii Harvey 1853, p. 189, pl. XXXa.

Figures 394–397

Plants saxicolous, firm gelatinous, membranous, brownish to rosy red, 10–20 cm tall, blades arising from short stipes and cuneate bases attached by small discoid holdfasts; axes broadly ligulate to foliose, 1–5 cm broad, simple to pseudodichotomously to palmately branched, axes broadest centrally, tapering to bases and rounded acute to obtuse apices, sinuses narrow and rounded, margins smooth to erose dentate, undulate and often proliferous, at times appearing irregularly pinnate; blades 0.7–1.1 mm thick, central region mostly hollow and mucilage filled, uncommonly traversed by few to several medullary filaments,

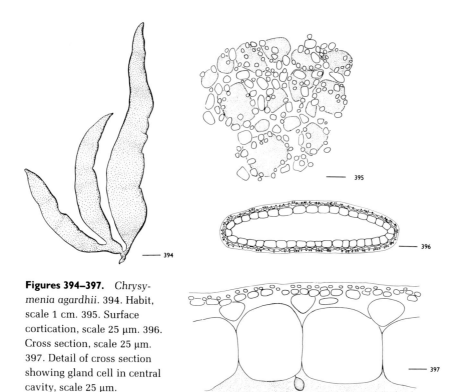

Figures 394–397. Chrysymenia agardhii. 394. Habit, scale 1 cm. 395. Surface cortication, scale 25 μm. 396. Cross section, scale 25 μm. 397. Detail of cross section showing gland cell in central cavity, scale 25 μm.

5–15(–20) μm diameter; one or two inner cortical layers lining the cavity and composed of large, hyaline, rounded polygonal to transversely elongate cells, 170–320 μm diameter, some bearing a single hemispherical gland cell toward the cavity, giving rise to a two- or three-layered outer cortex with radially elongate to subglobose cells 4–5 μm diameter, ovoid and ellipsoid to circular in surface view, incompletely covering subsurface cortical cells; cystocarps scattered over blades, hemispherical, ostioles inconspicuous.

Hoyt 1920; Taylor 1928; Schneider 1976. As Cryptarachne agardhii, Taylor 1960.

Infrequent offshore from 17 to 50 m in Onslow Bay, June–August.

Distribution: North Carolina, Bermuda, southern Florida, Caribbean.

Simple blades of this species could be confused with Halymenia floridana; however, Chrysymenia agardhii does not contain stellate ganglia in the inner cortex and bears occasional gland cells toward the central cavity, as typical of the genus. Only vegetative plants have been collected in the flora.

Chrysymenia enteromorpha Harvey 1853, p. 187.

Figures 398–401

Plants saxicolous or conchicolous, gelatinous, cylindrical, and highly branched, yellowish and pinkish to rosy red, iridescent, 5–30 cm tall, axes arising from short stipes and cuneate bases attached by small discoid holdfasts; axes

Figures 398–40l.
*Chrysymenia enteromor-
pha.* 398. Habit of large
plant, scale 1 cm. 399.
Habit of small plant, scale
1 cm. 400. Surface corti-
cation, scale 20 μm. 401.
Inner medullary cells
bearing gland cells, scale
25 μm.

cylindrical, mostly terete with slight flattening in some lower axes, 4–10 mm
diameter below, tapering to 1–2 mm in ultimate branches, alternately to irregularly and radially branched to several orders, although at times appearing distichous; axes broadest centrally, markedly constricted at the bases, and slightly
to greatly tapering to obtuse or rounded acute apices; central region of axes hollow and mucilage filled, cavities lined with one to two inner cortical layers of
large, hyaline, rounded to axially elongate cells, 100–250 μm long, some bearing
three to eighteen obovate to pyriform gland cells toward the cavity, giving rise to
a two-layered outer cortex with obovate to subglobose cells, 5–10 μm diameter,
irregularly rounded to pyriform in surface view, forming an incomplete network
over the anticlinal walls of inner cortical cells in young plants, becoming somewhat more corticated with maturity, but not completely covering the large inner
cortical cells, except those bearing gland cells; constrictions with a layer of large
inner cortical cells covered by smaller ones, these occasionally bearing clusters
of gland cells; tetrasporangia scattered over the axes, occasionally contiguous,
short ovate to globose, cruciately divided, 27.5–40 μm diameter, 35–45 μm long;
spermatangia in irregular sori, produced from outer cortical cells, 1 μm diameter; cystocarps scattered over blades, occasionally confluent, blunt conical, to
780 μm diameter, ostioles conspicuous; carposporophytes globose to reniform,
to 600 μm diameter, of several lobes, carposporangia subglobose to irregularly
angled, 15–20 μm diameter.

Hoyt 1920; Taylor 1928, 1960; Williams 1951; Schneider 1976; Blair and Hall
1981.

Frequent offshore in Onslow Bay and infrequent in Long Bay from 19 to 45 m,
May–September.

Distribution: North Carolina, South Carolina, Georgia, Bermuda, southern
Florida, Gulf of Mexico, Caribbean, Brazil, West Africa.

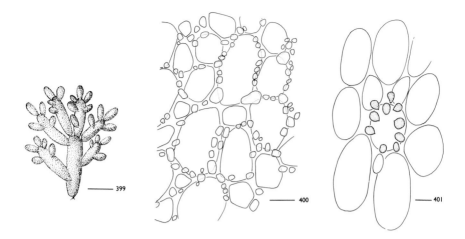

Very young plants bear superficial resemblance to *Botryocladia* but can be distinguished because gland-bearing cells are completely corticated in *Chrysymenia*.

Gloioderma J. Agardh 1851

Plants prostrate and erect, gelatinous, tubular to foliose, pinnately and irregularly to alternately or dichotomously branched, branches radial to marginal, arising from small discoid holdfasts; medulla pseudoparenchymatous, comprised of one to several layers of large, hyaline, rounded to ellipsoid cells; inner cortex of intermediate-sized cells forming an anastomosing network parallel to the surface, outer cortex consisting of anticlinally produced, dichotomously branched, loose, gelatin-enclosed filaments; branches often fusing with other portions of the plant; tetrasporangia developed terminally on inner cortical cells or on lateral branches of cortical filaments, cruciately divided, scattered over the plants; dioecious, spermatangia (where known) formed on outer cortical cells in extensive superficial sori; procarpic, carpogonial branches three celled; carposporophytes scattered over plants, nearly all gonimoblast cells becoming a globose mass of carposporangia, attached to flat base by a few gonimoblast filaments, connecting to pericarp with a *tela arachnoidea*; cystocarps prominently elevating one surface or margin of blade by thickening of the cortex, with or without overtopping hornlike projections, with conspicuous ostioles.

R. E. Norris (in manuscript) proposes that *Gloioderma* is synonymous with *Gloiocladia* J. Agardh 1842 and will transfer all species of *Gloioderma* to the earlier taxon.

All of the three species known in the flora were originally described from North Carolina specimens.

Key to the species

1. Plants flattened to ligulate, branching lateral 2
1. Plants terete, branching radial . . . . G. *rubrisporum*
2. Axes 1–2 mm diameter, branching pinnate . . . . G. *atlanticum*
2. Axes 2–5 mm diameter, branching dichotomous . . . . G. *blomquistii*

Glioderma atlanticum Searles 1972, p. 23, figs 1b, 4.

Figure 402

Plants erect and in part prostrate, saxicolous and epiphytic, gelatinous, complanate, and highly branched, pinkish to rosy red, 1–5 cm tall, primarily attached by discoid holdfasts, secondarily attached by rhizoidal outgrowths of tips and margins; main axes 1–2 mm diameter, at times slightly tapering to the branch bases, pinnately branched to three or four orders, ultimate branches alternate to subopposite, 0.5–2.5 mm long, apices rounded acute; medulla composed of four to six layers of closely packed, axially elongate polygonal cells, 75–240 µm long; inner cortex composed of a network of stretched, interconnected, axially elongate cells; outer cortex of anticlinal dichotomous filaments with obovate to subglobose and ellipsoid cells, 3–4 µm diameter, 5–10 µm long, filaments becoming parallel to the axis at surface; tetrasporangia ellipsoid to subglobose, 27–37.5 µm diameter, 35–50 µm long, scattered over the blades; spermatangia 1–2 µm diameter, forming in widespread sori over the blades; cystocarps scattered over margins and blades, globose to mammillate, to 1.1 mm diameter, with or without one to three hornlike, overtopping projections, ostioles conspicuous; carposporangia subglobose to irregularly angled, 12.5–20 µm diameter.

Searles 1972; Schneider and Searles 1975; Schneider 1976; Wiseman and Schneider 1976.

Infrequent offshore from 24 to 54 m in Onslow Bay and Long Bay, June–November.

Distribution: North Carolina, South Carolina, Bahamas, Puerto Rico.

Florida specimens attributed to this taxon from deep water south of Cape Canaveral (Eiseman 1979) are now placed in *Glioderma rubrisporum* (Searles 1984b).

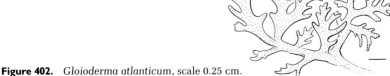

Figure 402.   *Glioderma atlanticum,* scale 0.25 cm.

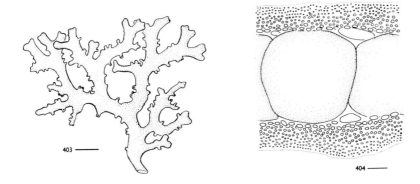

Figures 403, 404. *Glioderma blomquistii*. 403. Habit of cystocarpic plant, scale 0.5 cm. 404. Cross section, scale 50 μm (redrawn from Schneider and Searles 1975).

Glioderma blomquistii Searles in C. W. Schneider et Searles 1975, p. 87, figs 6–9.

Figures 403 and 404

Plants erect, saxicolous, gelatinous, ligulate, and branched, pinkish red, 2–4 cm tall, attached by discoid holdfasts; main axes 2–5 mm diameter and of similar width throughout, 240–320 μm thick, dichotomously and irregularly branched to three or four orders, overlapping and repeatedly fused to one another, apices truncate; medulla composed of a single layer of squarish cells in section, 140–290 μm in diameter; inner cortex composed of a network of stretched, interconnected, axially elongated cells; outer cortex of anticlinal dichotomous filaments with ellipsoid to subglobose cells, 4–6 μm diameter, 8–10 μm long, irregularly rounded and loosely organized in surface view; tetrasporangia ellipsoid, 19–20 μm diameter, 30–34 μm long, scattered over the blades; spermatangia unknown; cystocarps globose and marginal, sessile or short pedicellate, occasionally contiguous, 0.7–1.5 mm diameter, with or without one to three hornlike, short to overtopping projections, ostioles conspicuous; carposporangia ellipsoid to globose, 17.5–35 μm diameter, 17.5–40 μm long.

Schneider and Searles 1975; Schneider 1976.

Rare offshore from 39 to 45 m in Onslow Bay, June–September.

Distribution: North Carolina, southern Florida.

In Florida this species is known from 38 to 43 m, south of Cape Canaveral in March and November (Eiseman 1979).

Glioderma rubrisporum Searles 1984b, p. 217, figs 1–5.

Figure 405

Plants erect and in part prostrate, saxicolous, gelatinous, terete, and highly branched, rosy red, 1–2.5 cm tall, primarily attached by discoid holdfasts, secondarily attached by rhizoidal outgrowths of branch tips; main axes 0.4–1 mm diameter and of similar width throughout, radially branched to three or

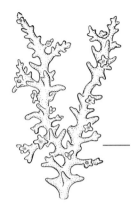

Figure 405. *Gloioderma rubrisporum*, Cystocarpic plant, scale 2 mm (redrawn from Searles 1987b).

four orders, ultimate branches alternately arranged, to 1(–1.8) mm long, apices rounded acute; medulla composed of several layers of closely packed, axially elongated polygonal cells, 60–100 μm diameter, 120–200 μm long; inner cortex composed of a network of stretched, interconnected, axially elongated cells; outer cortex of anticlinal dichotomous filaments with obovate to subglobose cells, to 2 μm diameter, 3–8 μm long terminally, with some intercalary cortical cells that are larger, subglobose, to 10 μm diameter and 14 μm long, cortical filaments widely spaced and loosely organized in surface view; tetrasporangia ellipsoid, 20–24 μm diameter, 30–34 μm long, scattered over the blades; spermatangia unknown; cystocarps scattered over branches, 450–900 μm diameter, with two to five hornlike, overtopping projections, ostioles conspicuous; carposporophytes globose, to 600 μm diameter, carposporangia subglobose to elongate and irregularly angled, 10–14 μm diameter, to 40 μm long.

Searles 1984b.

Rare offshore from 30 to 33 m in Onslow Bay, May–August.

Distribution: North Carolina, southern Florida.

In Florida this species is only known from deep water (30 to 37 m) south of Cape Canaveral (Eiseman 1979, as *Gloioderma atlanticum*).

Halichrysis (J. Agardh) Schmitz 1889

Plants prostrate, soft, lubricous, foliose, dorsiventrally arranged; axes flattened and irregularly to subdichotomously branched or palmately lobed, initially arising from small to large discoid, peltate, or marginal holdfasts, secondarily attached by rhizoidal haptera issued from the margins and blades; medulla pseudoparenchymatous, comprising few layers of large, hyaline, polygonal cells, some with smaller cells interspersed; inner and outer cortex of progressively smaller layers of cells, dorsal surfaces having far greater development than incomplete ventral layers; tetrasporangia developed from cortical cells, cruciately divided, scattered over one or both plant surfaces; procarpic, carposporophytes scattered on dorsal and ventral surfaces or restricted to one, most gonimoblast cells becoming a hemispherical to globose mass of carposporangia, attached to the base by pedicels of gonimoblast filaments, connecting *tela arachnoidea* lack-

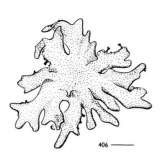

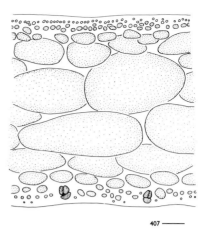

Figures 406, 407. *Halichrysis peltata.* 406. Habit, scale 0.5 cm. 407. Cross section of tetrasporic plant, scale 20 μm (redrawn from Schneider 1975a).

ing; cystocarps prominently elevating one surface of blade by thickening of the cortex, with conspicuous ostioles.

Halichrysis peltata (W. R. Taylor) P. Huvé et H. Huvé 1977, p. 106.
    Fauchea peltata W. R. Taylor 1942, p. 113, pl. 3 fig. 9; pl. 16 figs 1–5.
    Figures 406 and 407

Plants saxicolous, prostrate, dorsiventrally organized, firm gelatinous, brownish to rosy red, 1–9 cm wide, initially attached by short stalks to peltate or marginal holdfasts, secondarily attached by haptera issued from downturned margins or the ventral surfaces of the blades; axes initially lobate, becoming branched ligulate, 3–10 mm broad, irregularly to subdichotomously branched, branches overlapping and fused to one another, margins undulate and curled ventrally, occasionally proliferous, apices obtuse; blades 160–480 μm thick, medulla comprising much of the blade thickness, composed of large cells 80–175 μm diameter, giving rise to a three- or four-layered cortex with irregular to subglobose outer cortical cells 3–5.5 μm diameter, irregularly rounded to circular in surface view; dorsal surface cortical cells widely and regularly spaced, ventral surface cortication irregular; outer dorsal cortex with occasional packets of obvious translucent cells; tetrasporangia ellipsoid to subglobose, cruciately divided, 10–15 μm diameter, 17.5–20 μm long, sparsely to densely scattered over ventral surfaces; spermatangia unknown; cystocarps scattered over both blade surfaces, occasionally confluent, mammillate with extended necks, to 0.9 mm diameter, ostioles conspicuous; carposporangia subglobose to irregularly angled, 12.5–17.5 μm diameter.

As *Weberella peltata*, Schneider 1975a, 1976.

Infrequent offshore from 22 to 45 m in Onslow Bay, May–December.

Distribution: North Carolina, southern Florida, Gulf of Mexico, Caribbean, Brazil.

Leptofauchea Kylin 1931

Plants erect, soft, and lubricous, ligulate; axes flattened and irregularly to dichotomously branched, arising from small discoid holdfasts; medulla pseudoparenchymatous, comprised of few layers of large, hyaline, polygonal cells; cortex of a single layer of smaller cells; tetrasporangia developed terminally or in chains in raised nemathecia on blades, cruciately divided; procarpic, carposporophytes restricted to margins, nearly all gonimoblast cells becoming a hemispherical to globose mass of carposporangia, with connecting *tela arachnoidea* to the pericarp; cystocarps prominently elevating margins of blades by thickening of the cortex, with conspicuous ostioles.

Leptofauchea brasiliensis Joly 1957, p. 130, text fig. 2, pl. 12 fig. 7; pl. 19 fig. 4.
    Figures 408 and 409
    Plants saxicolous, erect, soft, and lubricous, rosy red, 1.5–4 cm tall, attached by short stalks to small discoid holdfasts, axes narrow-ligulate, (2–)3–6 mm broad, irregularly to dichotomously branched, the nodes often constricted, terminal blades mostly becoming subspatulate, margins smooth, apices obtuse; blades 160–290 µm thick, medulla comprising much of the blade thickness, composed of a single layer of large cells 60–140 µm diameter, giving rise to a single-layered cortex with transversely elongate to subglobose outer cortical cells 5–10 µm diameter, irregularly rounded to angular in surface view; tetrasporangia ellipsoid to rounded rectangular, cruciately divided, 10–21 µm diameter, 10–25(–32) µm long, in raised nemathecial sori, terminally or in chains of three or four on proliferous cortical cells; gametangia unknown.
    Schneider and Searles 1976; Schneider 1976.
    Rare, several specimens from a single collection offshore from 50 m in Onslow Bay, June.
    Distribution: North Carolina, western Florida, Brazil.

Figures 408, 409.
Leptofauchea brasiliensis, habits of tetrasporic plants, scale 0.5 cm.

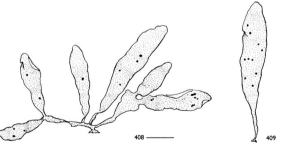

408 ————   409

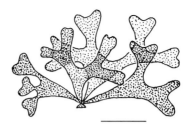

Figure 410. *Rhodymenia divaricata*, scale 1 cm.

Rhodymenia Greville 1830, nom. cons.

Plants erect or prostrate and spreading, firm, ligulate to foliose; axes flattened and irregularly to repeatedly dichotomously branched, with occasional apical or marginal proliferous branches, some simple, with or without perforations, arising from small to large discoid holdfasts, sessile or stalked, many spreading by secondary attachments on cylindrical stoloniferous branches; medulla pseudoparenchymatous, comprised of few to several layers of large, hyaline, axially elongated polygonal cells; cortex of one to five layers of progressively smaller cells; tetrasporangia developed in the outer cortex in distinct subterminal to terminal sori or individually scattered over the blades, cruciately divided; dioecious, procarpic, three- to four-celled carpogonial branches, carposporophytes scattered or restricted to apices or margins, nearly all gonimoblast cells becoming a hemispherical to globose mass of carposporangia, *tela arachnoidea* lacking; cystocarps prominently elevated, mammillate to hemispherical, with conspicuous ostioles.

Key to the species

1. Branches diverging at angles of 80°–100°; tetrasporangia in sori scattered over the blades; plants restricted to deep offshore waters . . . .*R. divaricata*
1. Majority of branches diverging at angles of 40°–65°, and always less than 80°; tetrasporangia in obvious subapical sori; plants common in shallow subtidal, less frequent offshore . . . . *R. pseudopalmata*

Rhodymenia divaricata Dawson 1941, p. 141, pl. 23 fig. 31.

Figure 410

Plants saxicolous, erect and prostrate, firm, light rosy red, 4–5 cm tall, with or without minute stalks, attached by small to large discoid holdfasts, axes cuneate above holdfast and ligulate to broadening above nodes, 1.5–3 mm broad, regularly dichotomously branched, branch angles 80°–100°, margins smooth, apices rounded obtuse; blades 100–250 μm thick, medulla comprising much of the blade thickness, composed of few to several layers of axially elongated polygonal cells 90–180 μm in diameter, giving rise to a two- or three-layered cortex with ellipsoid to irregularly angular, axially elongated outer cortical cells, 5–7.5 μm diameter, 7.5–17 μm long in surface view; tetrasporangia ellipsoid, cruciately divided, 12–20 μm diameter, 20–25 μm long, intercalary in cortical filaments, in irregular yellowish-pink sori scattered over the blades; ?dioecious, spermatangia

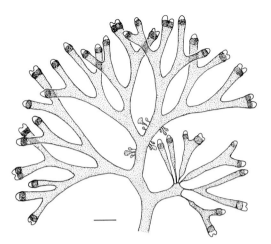

Figure 4II. *Rhodymenia pseudopalmata,* tetrasporic plant, scale 1 cm.

unknown; cystocarps scattered over the blades, hemispherical, to 0.8 mm diameter, ostioles conspicuous; carpospores forming a hemispherical mass, darkly pigmented, rounded polygonal to ellipsoid, 7–18 μm diameter.

Schneider and Searles 1976; Schneider 1976; Wiseman and Schneider 1976; Blair and Hall 1981.

Uncommon offshore from 40 to 54 m in Onslow Bay and Long Bay, and off Charleston, S.C., June–October.

Distribution: North Carolina, South Carolina, southern Florida, Gulf of California, Galapagos Islands.

Rhodymenia pseudopalmata (Lamouroux) Silva 1952, p. 265.

Fucus pseudopalmatus Lamouroux 1805, p. 29.

Figure 411

Plants saxicolous, erect, firm to cartilaginous, light to dark rosy red, 4–12(–18) cm tall, lax to rigidly flabellate, with short stalks 0.3–0.7 mm diameter, attached to small or spreading discoid holdfasts, axes cuneate above short stalks, becoming narrowly to broadly ligulate above, many subspatulate, 3–5 mm broad, regularly dichotomously branched, branch angles (25°–)40°–65°(–75)°, occasionally irregular to palmate, margins smooth and occasionally proliferous, especially at wound sites, apices obtuse or tapering to subacute; blades 100–250 μm thick, medulla comprising much of the blade thickness, composed of axially elongated cells 45–88 μm in diameter, giving rise to a two- or three-layered cortex with irregularly ellipsoid, axially elongated outer cortical cells, 5–8(–12) μm diameter and 9–14 μm long in surface view, outer cells not completely covering subsurface cortical cells; tetrasporangia obovoid to ellipsoid, cruciately divided, 15–25 μm diameter, 25–40 μm long, intercalary in cortical filaments, in raised, darkly pigmented, subapical, transversely ellipsoid to subcircular sori on one or both surfaces, becoming lighter pigmented and deteriorating upon spore release; dioecious, spermatangia forming in lightly pigmented subapical, transversely ellipsoid to subcircular sori, spermatangia paired on mother cells, 2–3 µm diameter; carpogonial branches three celled; cystocarps mostly developing on margins, few laminal, occasionally contiguous, globose or ovoid to mammillate, to 1 mm diameter, ostioles conspicuous; carpospores darkly pigmented, rounded polygonal to ellipsoid and irregular, 5–20 µm diameter.

Taylor 1960; Schneider 1976; Wiseman and Schneider 1976; Kapraun 1980a; Searles 1981, 1987, 1988. As *R. palmetta*, Hoyt 1920; Williams 1948a, 1949, 1951; Stephenson and Stephenson 1952. As *R. pseudopalmata* var. *caroliniana*, Taylor 1960; Chapman 1971; Wiseman and Schneider 1976; Richardson 1987. As *R. palmata sensu* Bailey 1848. As *R. occidentalis sensu* Blair and Hall 1981. As *Leptofauchea rhodymenioides sensu* Blair and Hall 1981.

Common offshore from 14 to 45 m in Onslow Bay and Long Bay and from 17 to 22 m off Georgia, occasional in Raleigh Bay, May–December, and in subtidal of Cape Lookout, Radio Island, Masonboro Inlet, and Cumberland Island jetties and occasional pilings, floating docks, and retaining walls throughout the area, year-round.

Distribution: North Carolina, South Carolina, Georgia, southern Florida, Brazil, British Isles to West Africa, Mediterranean, Azores, Canary Islands.

Within the flora, specimens of *Rhodymenia pseudopalmata* exhibit a continuum of morphology from the typical narrow form to the larger, broader var. *caroliniana* described from Pawleys Island, S.C. (Taylor 1960, p. 485), and typical of shallow-water specimens known at that time from the Carolinas. Therefore we find it unnecessary to recognize separate varieties that are distinct only at the extremes of the range of variation.

Plants collected in the deepest waters are more lax and narrow and exhibit longer internodes than most shallow-water specimens. An examination of specimens from South Carolina attributed to *Rhodymenia occidentalis* Børgesen (Blair and Hall 1981) has shown them to be lax, deep-water specimens of *R. pseudopalmata* with none of the polychotomous branching typical of that species (Taylor 1960). In fact, we have found similar lax specimens not only in deep water but also in shaded specimens from the shallow subtidal of Beaufort, N.C., area jetties. It appears that light intensity or quality plays a role in the morphological development of *R. pseudopalmata*.

It is possible that *Rhodymenia occidentalis* is not a distinct species at all, but rather a variant of the morphologically plastic *R. pseudopalmata*. We have seen Børgesen's type material and suggest this possibility, but as *R. occidentalis* is only known from vegetative specimens, its generic placement must remain tenuous.

Champiaceae

Plants soft and lubricous, prostrate or erect, radially to distichously organized, irregularly or alternately branched, arising from small discoid holdfasts; axes terete to compressed, hollow except in basal stalks in some; structure multi-axial, cortex of few to several complete to incomplete layers; medulla hollow and gelatinous, with transverse single-cell-layered septa at regular intervals; tetrasporophytes and gametophytes isomorphic; sporangia tetrahedrally divided, some producing polysporangia, intercalary in cortical filaments, scattered over plants; dioecious, spermatangia formed from outer cortical cells, in girdling sori throughout plants; procarpic, four-celled carpogonial branches borne on a supporting cell also bearing one or two two-celled auxiliary cell branches, connecting filament short; fusion cells issuing gonimoblasts outwardly; carposporophytes with cortical pericarps, outer gonimoblast cells becoming carposporangia; cystocarps protruding, scattered over the blades or branches, with ostioles.

Champia Desvaux 1809

Plants erect to prostrate, soft, alternately or suboppositely to irregularly branched; blades terete to flattened, arising from small discoid holdfasts; medulla hollow and mucilage filled, traversed by single-layered septa of inner cortical cells at regular intervals, appearing as constrictions in outer surface, cavities lined with inner cortical cells larger than those to the exterior, giving rise to one or two layers of anticlinally arranged outer cortical cells, outer layer not completely covering subsurface cortical layer; tetrasporangia intercalary in cortical filaments, tetrahedrally divided, scattered over the blades; dioecious, spermatangia in girdling regular, superficial sori, spermatangial mother cells producing two or three spermatangia; procarpic, carposporophytes scattered over blades, outer gonimoblast cells becoming an ellipsoid to globose mass of carposporangia; cystocarps prominently elevating one surface of blade by thickening of the cortex, with conspicuous ostioles.

Champia parvula (C. Agardh) Harvey 1853, p. 76.
   Chondria parvula C. Agardh 1824, p. 207.
   Figure 412
   Plants epiphytic, saxicolous, and lignicolous, firm gelatinous, single or tufted, brownish, greenish, or pinkish to rosy red, 2–10 cm tall, attached by small discoid holdfasts giving rise to one or more erect, highly branched axes; axes terete to slightly compressed, 0.5–2 mm diameter, generally broadest in middle portions, alternately to oppositely and irregularly radially branched, at times mostly distichous, branches occasionally anastomosing, apices rounded obtuse; axes

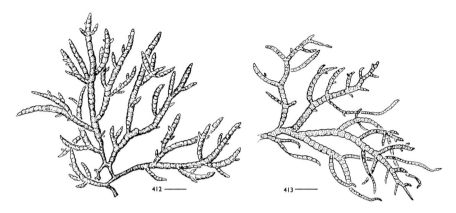

Figure 412. *Champia parvula*, scale 0.5 cm.
Figure 413. *Champia parvula* var. *prostrata*, scale 0.5 cm.

markedly constricted at regular intervals in positions where medullary cavity is traversed by cellular septa, segments doliform, one to two times as long as broad, ultimate segments dome shaped; medulla hollow and mucilage filled, divided by septa of cortical cells one cell layer thick; cavities lined with regularly spaced longitudinal, attenuate filaments visible through the cortical layers, occasionally bearing obpyriform gland cells, giving rise to axially elongated, rectangular to ovoid inner cortical cells, 22–38 μm in diameter, 50–130 μm long, and a single-layered outer cortex with irregularly rounded to subglobose cells 7–12.5 μm diameter, sparsely forming between the anticlinal walls of subsurface cortical cells; tetrasporangia globose, scattered, tetrahedrally divided, 50–100 μm diameter; spermatangia in spreading sori, occasionally on terminal segments, ellipsoid to circular, 1–2 μm diameter; cystocarps scattered over segments, single, paired or clustered, ovoid and occasionally rostrate, to 0.9 mm diameter, ostioles conspicuous; carposporangia obpyriform to irregularly angled, 40–50 μm diameter.

C. parvula var. *prostrata* L. Williams 1951, p. 155.

Figure 413

Plants caespitose and spreading, epiphytic or epizoic, axes conspicuously flattened and laterally branched, bearing numerous secondary attachments from axes and tips; tetrasporangia forming in extensive patches in major axes, obovoid to globose, 50–90 μm diameter; cystocarps marginal, sessile or short stalked, 0.7–0.8 mm diameter, scarcely rostrate.

Curtis 1867; Hoyt 1920; Williams 1951; Taylor 1960; Chapman 1971; Brauner 1975; Schneider 1976; Kapraun 1980a. Var. *prostrata*, Williams 1951; Schneider 1976; Searles 1981, 1987, 1988.

The typical form is common and occasionally abundant in the shallow subtidal of area jetties and pilings, and in sounds on *Zostera*, shells, and other seaweeds, year-round, reaching peak size and abundance during spring and summer; offshore, the typical form is infrequent from 17 to 27 m in Onslow

Bay, June–September. The prostrate variety is restricted to deep offshore waters throughout the region, being collected as a common epiphyte of other seaweeds from 15 to 40 m, May–December.

Distribution: Maritimes, Massachusetts to Virginia, North Carolina, South Carolina, Georgia, Bermuda, southern Florida, Gulf of Mexico, Caribbean, Venezuela, Brazil, British Isles, France to Portugal; elsewhere, widespread in temperate to tropical seas. Var. *prostrata* reported only from North Carolina to Georgia.

Lomentariaceae

Plants soft and lubricous, prostrate or erect, radially to distichously organized, irregularly or alternately to dichotomously branched, arising from small discoid holdfasts; axes terete to strongly compressed and ligulate, hollow except in basal stalks in some; structure multiaxial, cortex of few to several complete to incomplete layers; medulla with longitudinally arranged, attenuate, loosely organized filaments at least distally, in some appearing hollow proximally, with transverse multilayered septa at branch nodes or lacking septa throughout; tetrasporophytes and gametophytes isomorphic; sporangia tetrahedrally divided, terminal in outer cortical filaments, forming in depressed sori over plants, with filaments of sterile cells within sori, associated with openings in the outer cortex; dioecious, spermatangia formed from outer cortical cells, in sori on ultimate branches; procarpic, three-celled carpogonial branches borne on a supporting cell also bearing one or two two-celled auxiliary cell branches, with or without small connecting cells; fusion cells issuing gonimoblasts outwardly; carposporophytes with cortical pericarps, nearly all gonimoblast cells becoming carposporangia; cystocarps protruding, scattered over the blades or branches, with ostioles.

Lomentaria Lyngbye 1819

Plants erect to prostrate, soft, alternately or oppositely to irregularly branched; blades terete to compressed, arising from small discoid holdfasts, secondarily attached by rhizoidal outgrowths from branches and tips; medulla with attenuate, loosely arranged, and longitudinally organized filaments, to hollow, mucilage filled, and traversed by multilayered septa of inner cortical cells at constrictions of branch bases, medulla or cavities lined with inner cortical cells larger than those to the exterior, giving rise to one or two layers of anticlinally arranged outer cortical cells, outer layer not completely covering subsurface cortical layer; tetrasporangia terminal on outer cortical filaments, tetrahedrally divided, in depressed sori scattered over the blades, associated with pores in the outer cortex; dioecious, spermatangia in superficial sori in ultimate branches, spermatangial

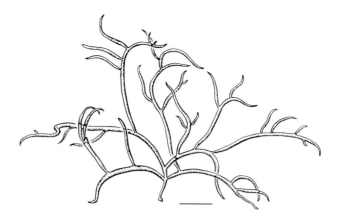

Figure 414. *Lomentaria baileyana*, scale 0.5 cm.

mother cells producing two or three spermatangia; procarpic, carposporophytes scattered over blades, nearly all gonimoblast cells becoming an ellipsoid to globose mass of carposporangia; cystocarps prominently elevating one surface of blade by thickening of the cortex, with conspicuous ostioles.

Key to the species

1. Axes terete, branching irregular, branches curved . . . . *L. baileyana*
1. Axes flattened, branching regularly pinnate, branches straight . . . .
. . . . *L. orcadensis*

Lomentaria baileyana (Harvey) Farlow 1876, p. 698.
    Chylocladia baileyana Harvey 1853, p. 185, pl. XXc fig. 1.
    Figure 414
    Plants epiphytic or saxicolous, firm gelatinous, single or tufted, yellowish, pinkish, to rosy red, 1–8(–12) cm long, initially attached by small discoid holdfasts giving rise to a single, sparsely to highly branched axis, spreading by reattachments from recurved branch tips; axes terete, 0.1–1 mm diameter, irregularly radially branched, portions occasionally secund, branches distinctly tapering to both ends, recurved, ultimate branches to 1 cm long, apices tapering, obtuse; medulla hollow and mucilage filled with few attenuate longitudinal filaments lining the cavities, 2–3 μm diameter; branch insertions with septa of several cortical cell layers traversing the medulla, the axes otherwise remaining undivided; medullary filaments connecting to a network of axially elongate inner cortical cells 25–40 μm diameter and 50–115 μm long, occasionally bearing subglobose gland cells toward the cavities; two-layered outer cortex with a subsurface of axially elongate to irregular cells and a surface layer of irregularly rounded to subcircular cells, 7–13 μm diameter, forming an incomplete layer between anticlinal walls of subsurface cortical layer; tetrasporangia globose to short ellipsoid, in depressed sori in upper branches, tetrahedrally divided, 30–68 μm diameter at maturity; spermatangia in raised sori on ultimate branches, ellipsoid to circular, 1–2 μm diameter; cystocarps scattered over seg-

ments, usually sparse, occasionally confluent, ovoid, to 0.6 mm diameter, ostioles conspicuous; carposporangia obpyriform and subglobose to ellipsoid, 17–23 μm diameter.

Williams 1948a, 1949, 1951; Taylor 1957, 1960; Chapman 1971; Brauner 1975; Schneider 1976; Wiseman and Schneider 1976; Kapraun 1980a; Searles 1981, 1987, 1988. As *Chylocladia baileyana*, Harvey 1853; Curtis 1867. As *C. baileyana* var. *valida*, Melvill 1875. As *L. uncinata*, Hoyt 1920.

Abundant below the low-water mark on area jetties, pilings, and in sounds on buoys, shells, and other seaweeds, year-round, reaching peak size and abundance during spring and summer; frequent offshore from 15 to 40 m in North Carolina and Georgia on rocks or as an epiphyte on other seaweeds, May–August.

Distribution: Maritimes, New Hampshire to New Jersey, Virginia, North Carolina, South Carolina, Georgia, Bermuda, southern Florida, Gulf of Mexico, Caribbean, Pacific Mexico, Galapagos Islands.

In deep Onslow Bay waters, some plants of this species are in part flattened and exhibit subopposite branching reminiscent of *Lomentaria clavellosa* (Turner) Gaillon (see Irvine 1983), a species known from New Hampshire to Connecticut in the western Atlantic. *L. clavellosa*, however, can be distinguished by its much more pronounced triangular-shaped habit and subapical tetrasporangial sori.

Lomentaria orcadensis (Harvey) Collins ex W. R. Taylor 1937, p. 309.

Chrysymenia orcadensis Harvey 1849, p. 100.

Figures 415 and 416

Plants prostrate or erect, saxicolous, firm gelatinous, caespitose, pinkish to dark rosy red, 2–3(–8) cm tall, initially attached by small discoid holdfasts giving rise to a single highly branched axis becoming triangular in outline, spreading by reattachments from prostrate stolons and flattened branches; axes distinctly flattened, 0.5–3.5 mm diameter, often naked below, oppositely to alternately pinnate to one or two orders above, segments pad shaped to lanceolate, ultimate branches to 1–6 mm long, apices obtuse; medulla vacuous, mucilage filled, lined with few longitudinal filaments, 2–5(–12) μm diameter, compacting into a plug of traversing cortical cell layers at branch insertions, the axes otherwise remaining undivided; cavity lined with a network of stretched anastomosing cells connected to medullary filaments giving rise to a layer of axially elongated inner cortical cells, 12–75 μm diameter, and a single-layered outer cortex with irregularly rounded to subcircular cells, (3–)5–12 μm diameter, forming a sparse to near-complete network over subsurface cells between anticlinal walls of subsurface cells; tetrasporangia ellipsoid to subglobose, in depressed sori in

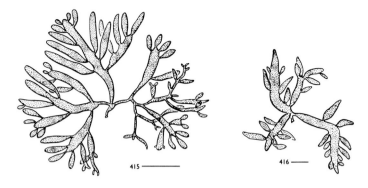

Figures 415, 416. *Lomentaria orcadensis*. 415. Habit, scale 0.5 cm. 416. Habit, scale 2 mm.

upper branches, tetrahedrally divided, 35–55 μm diameter at maturity; gametangia unknown.

Schneider 1975a, 1976. As *Lomentaria rosea*, Hoyt 1920; Williams 1948a.

Rare in the shallow subtidal of Cape Lookout jetty, June, and offshore from 21 to 23 m off Cape Lookout in Onslow Bay, May–June.

Distribution: Maritimes to Connecticut, North Carolina, Iceland, British Isles, Faeroes to Portugal.

North Carolina plants display much less outer cortical development than those observed from New England, which show nearly complete outer cortical layers. Onslow Bay is the known southern limit of distribution for this species in the western Atlantic.

CERAMIALES

Plants erect and/or prostrate, small and filamentous to large and pseudoparenchymatous; uniaxial from apical cells or initials, monopodial or sympodial, radially, bilaterally, or dorsiventrally organized, with or without coverings of monosiphonous or polysiphonous determinate branches, and with or without partial to heavy cortication, without calcification; uniseriate, polysiphonous to ligulate and foliose, with greater or lesser branching; cells uninucleate or multinucleate with several plastids lacking pyrenoids; asexual reproduction by mono-, tetra-, poly-, or paraspores; vegetative propagation by fragmentation or propagules; tetrasporangia produced singly in clusters or sori, or organized into distinct whorled series on specialized determinate pseudolaterals (stichidia); tetrasporophytes and gametophytes isomorphic, some with direct development of one generation; monoecious or dioecious, spermatangia in branched clusters, compact to spreading superficial sori, stichidia, or flat apical sori; two- to four-celled carpogonial branches, auxiliary cells formed on supporting cells after fertilization; carpogonial branches fusing directly or indirectly with auxiliary cells after fertilization; carposporophytes developing directly from the auxiliary cells or small to large fusion cell complexes, with or without involucral filaments or pericarps; carposporangia forming from terminal or all gonimoblast cells, liberating one carpospore each.

Key to the families

1. Plants filamentous to polysiphonous, not foliose 2
1. Plants narrowly to broadly foliose .... Delesseriaceae
2. Plants uniseriate, with or without nodal cortication .... Ceramiaceae
2. Plants polysiphonous, mostly with monosiphonous or polysiphonous determinate branches at tips or over much of the surface 3
3. Growth sympodial; axes clothed with mostly persistent, highly pigmented,
   determinate branches (ramelli); tetrasporangia whorled in stichidia........
   .... Dasyaceae
3. Growth monopodial; apices often bearing hyaline determinate branches
   (trichoblasts); tetrasporangia not whorled in stichidia ...... Rhodomelaceae

Ceramiaceae

Plants erect and/or prostrate, filamentous to reticulate, often delicate, bushy and epiphytic; axes uniaxial, uniseriate in some, partly to wholly corticated and terete to flattened in others; corticating cells smaller than cells of axial filaments, rhizoidal in some; growth from single, often enlarged, apical cell; sporangia sessile or pedicellate on uniseriate plants, borne singly or in clusters in the cortex of others, cruciately or tetrahedrally divided, some species with polysporangia divided into eight to sixty-four spores; spermatangia in small hyaline clusters terminal or lateral on short fertile branchlets in some and in patches covering portions of the cortex in others; two- to four-celled carpogonial branches, borne on supporting cells, with or without sterile cells, and which give rise to one or two auxiliary cells after fertilization; one or two connecting cells often join fertilized carpogonia and auxiliary cells, forming large fusion cells; cells of the gonimoblast develop as carposporangia; cystocarps naked or partially enclosed by filamentous involucres.

Key to the genera

1. Uniseriate axes corticated at the nodes or wholly corticated by small pseudoparenchymatous cells... 2
1. Uniseriate axes ecorticate or loosely corticated basally by descending rhizoids.. 4
2. Cortication by rectangular cells in vertical rows; spines present at nodes
   .... Centroceras
2. Cortication by polygonal to rounded cells not in vertical rows; spines not
   present at nodes... 3
3. Cortication pattern similar in all branches .... Ceramium

3. Cortication complete in main axes, limited to nodes in determinate branches .... Spyridia
4. Plants forming as small nets, without obvious main axes .... Rhododictyon
4. Plants uniseriate, upright to prostrate, main axes obvious 5
5. Diameter of largest axial cells greater than 250 μm 6
5. Diameter of largest axial cells less than 250 μm 7
6. Subapical cells elongate, progressively shortening and slightly tapering to small, caplike apices; spermatangia borne in terminal heads on whorled branchlets at the nodes .... Anotrichium
6. Subapical cells globose to subglobose, abruptly tapering to globose-celled apices; terminal cells on male plants remaining large, covered by spermatangia in caplike sori .... Griffithsia
7. Plants minute, prostrate, and spreading, erect axes less than 5 mm tall ... 8
7. Plants larger, generally erect, greater than 5 mm tall 11
8. Erect axes consistently bearing opposite or whorled branches; gland cells usually present .... Antithamnionella (in part)
8. Erect axes consistently bearing few to many alternate branches, with occasional opposite branches; gland cells lacking 9
9. Upright axes alternately radially branched from each cell, for the most part with simple or dichotomous determinate branches; sporangia cruciately divided.. Callithamniella
9. Upright axes sparingly branched with indeterminate axes; sporangia tetrahedrally divided... 10
10. Cell below procarp bearing a single lateral filament which becomes incorporated into the enclosing pericarp .... Lejolisia
10. Cell below procarp bearing two lateral filaments forming part of loose involucre... Ptilothamnion
11. Branching opposite or alternate................................... 12
11. Branching trichotomous, in part dichotomous .... Calliclavula
12. Branching opposite, distichous to decussate 13
12. Branching alternate, distichous to spiraled.......................... 14
13. Basal cell of each branch distinctly shorter than the cell distal to it; gland cells covering more than one vegetative cell on specialized branchlets .... Antithamnion
13. Basal cell of each branch indistinctly or slightly smaller than the cell distal to it; gland cells covering a portion or all of one vegetative cell, not formed on specialized branchlets .... Antithamnionella (in part)
14. Plants producing only tetrasporangia; carposporophytes lacking involucres 15
14. Plants producing polysporangia and occasionally tetrasporangia; carpo-

Anotrichium Nägeli 1862
Plants erect to prostrate, bushy, ecorticate or with rhizoidal cortication below; axes monosiphonous, sparsely to greatly subdichotomously to laterally
branched, appearing segmented; cells elongate, multinucleate, with numerous
discoid plastids and large central vacuoles; whorls or groups of delicate, hyaline, di-, tri-, to polychotomous whorl branchlets (trichoblasts) produced from
the upper ends of some cells, particularly distally and in conjunction with reproductive structures; tetrahedral sporangia single on oblong pedicels or adaxial on
the upper end of the basal cell of whorled trichoblasts; dioecious, spermatangia
borne in single heads on oblong pedicels or adaxial on the upper end of the basal
cell of whorled trichoblasts; procarps produced apically and becoming laterally
displaced upon continued apical growth, or subapically on three-celled fertile
axes at the nodes, carpogonial branches four celled; involucre filaments one
celled, sometimes two celled, arising from the greatly enlarged hypogynous cell;
carpospores developing from most gonimoblast cells.

Anotrichium tenue (C. Agardh) Nägeli 1862, p. 399.
  Griffithsia tenuis C. Agardh 1828, p. 131.
  Figures 417 and 418
  Plants epiphytic or saxicolous, erect in loose tufts to 7 cm tall, but with obvious prostrate spreading axes, dark red, arising from numerous rhizoids issued
from basal cells; prostrate axes secondarily attached by numerous one-celled
rhizoids, usually issued proximally from the cells, often bearing distal, broad-
stellate to coral-like attachment processes; branches usually arising proximally
from the cells, often not abundant, lateral to subsecund and irregular, fastigiate,
often proliferous from the prostrate axes; cells cylindrical below, 200–300 μm
diameter and 0.5–1.1 mm long, more oblong above, 100–200 μm diameter, each
of the terminal five or six cells of a branch successively shortening to the caplike
apex; trichoblasts of vegetative axes short, only present on the most distal nodes,
longer and more persistent on reproductive branches; tetrasporangia globose,
50–100 μm diameter, borne terminally on elongate one-celled pedicels, whorled
to fifteen per node, produced on as many as the six distal nodes, without involucral branches; dioecious, spermatangial heads borne terminally on one- to
three-celled whorled branches, one to seven per node, spermatangia pyriform to
subglobose, 3 μm diameter; procarps produced apically, becoming laterally dis-

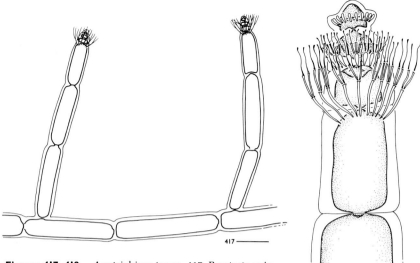

Figures 417, 418. *Anotrichium tenue.* 417. Prostrate axis with uprights, scale 200 μm. 418. Detail of apex, scale 50 μm.

placed, subhypogynous cell lacking trichoblasts; carposporophytes with a large fusion cell and one to three gonimolobes, borne on a prominent one-celled stalk, the majority of cells becoming carposporangia, surrounded by a whorl of six to thirteen large, incurved, one-celled involucral filaments.

Kapraun 1980a; Searles 1987, 1988. As *Griffithsia tenuis*, Blomquist and Humm 1946; Williams 1951; Taylor 1960; Brauner 1975; Schneider 1976; Wiseman and Schneider 1976. As *A. barbatum sensu* Kapraun 1980a.

Known from the shallow subtidal near Beaufort and Wilmington, N.C., more commonly in the drift; from 17 to 47 m offshore in Raleigh Bay, Onslow Bay, and Long Bay, May–December; and from 17 to 21 m on Gray's Reef, July–August.

Distribution: Southern Massachusetts to Virginia, North Carolina, South Carolina, Georgia, southern Florida, Caribbean; elsewhere, widespread in warm temperate to tropical seas.

Although Boudouresque and Coppejans (1982) find Atlantic specimens attributed to this species different from those in the Mediterranean (type locality), we agree with Norris and Aken (1985) in retaining them with plants from South Africa and the Indo-Pacific as *Anotrichium tenue*.

Antithamnion Nägeli 1847

Plants erect to prostrate, filamentous, tufted, ecorticate, branched to three or four orders; axes uniseriate, bearing alternate or opposite branches below, and opposite distichous to decussate determinate branches above; basal cell of branch often only as long as broad, usually lacking determinate laterals; cells uninucleate with small discoid or band-shaped plastids; gland cells present or lacking, usually adaxial, borne on specialized short pinnules; cruciate or tetrahedral tetrasporangia, sessile or pedicellate on lower cells of opposite determinate branches; dioecious, spermatangia clustered on specialized branchlets borne on the adaxial side of proximal cells of branchlets; several procarps borne successively on basal cells of ultimate branchlets near the apex, four-celled carpogonial branches; fertile axes cease to elongate after fertilization, carposporophyte appearing terminal, producing successive gonimolobes; carposporangia developing from most of the gonimoblast cells.

Because of its similarity with species of *Antithamnionella*, the one species of *Antithamnion* known in the flora is included in the key of that genus (p. 362).

Antithamnion cruciatum (C. Agardh) Nägeli 1847, p. 200, table VI, figs 1–6.

  Callithamnion cruciatum C. Agardh 1827, p. 637.

  Figures 419 and 420

  Plants epiphytic, tufted, erect to 5 cm tall and rising from prostrate spreading axes or mostly prostrate with short erect axes; attached by numerous multicellular rhizoids issued from cells of the prostrate axes; rosy red; axes sparsely to much branched, alternate below, and opposite subdistichous to decussate above, the tips usually densely congested with branches; determinate branches ten to twenty cells from tip to base, branched unilaterally on the adaxial surface to alternately, the basal cell obviously shorter than the cell distal to it, quadrate to subquadrate or short rectangular, occasionally bearing a rhizoid; cells of the main indeterminate axes 40–90 μm diameter, 90–300 μm long; the determinate axes 9–28 μm diameter near their origins; gland cells, if present, borne on specialized pinnules covering most if not all of the adaxial surface of two (sometimes three) vegetative cells near the bases of ultimate branches; cruciate tetrasporangia obovoid to ellipsoid, 38–70 μm diameter, 62–100 μm long, sessile or pedicellate, replacing an ultimate vegetative branchlet; spermatangia clustered on ultimate branchlets of four to eight cells; carposporophytes forming two globular gonimolobes to 400 μm diameter, terminal on indeterminate axes.

  Williams 1948a, 1949, 1951; Taylor 1960; Wiseman and Schneider 1976; Kapraun 1977b, 1980a. As *A. cruciatum* var. *radicans*, Schneider 1974, 1976; Wiseman and Schneider 1976.

  Known from throughout the Carolinas from the shallow subtidal to 60 m offshore, June–November.

  Distribution: Newfoundland to Virginia, North Carolina, South Carolina, Bermuda, southern Florida, Caribbean, Brazil, Argentina, Norway to Portugal, Mediterranean, Canary Islands, Black Sea.

  Local plants often lack gland cells, but when present these are found covering two and sometimes three cells on short pinnules. They compare favorably in all respects with specimens from New England and eastern Canada. Culture studies by Whittick and Hooper (1977) have shown that taxonomic status for the prostrate form, var. *radicans* (J. Agardh) Collins and Hervey, is not warranted.

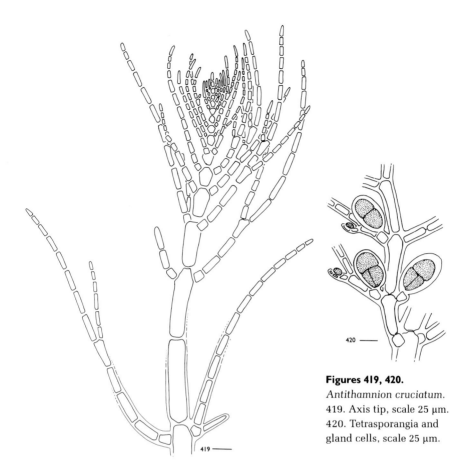

Figures 419, 420.
Antithamnion cruciatum.
419. Axis tip, scale 25 μm.
420. Tetrasporangia and
gland cells, scale 25 μm.

Although some specimens attributed to this variety (Schneider 1974, 1976) lack gland cells, others are figured (Blomquist and Humm 1946) as having gland cells covering only one vegetative cell. Although this is consistent with the description of this subspecific taxon from Bermuda given by Collins and Hervey (1917), an examination of their specimens shows that a majority of the adaxial gland cells cover two vegetative cells on short pinnules, and only a few are restricted to one. Vouchers of the Blomquist and Humm (1946) report have not been located, and based on illustrations, these tristichous specimens are likely *Antithamnionella elegans*.

Antithamnionella Lyle 1922

Plants erect or prostrate and erect, filamentous, ecorticate, branched to two to four orders; axes uniseriate, bearing whorls of one to four branches on each axial cell, basal cell of branch similar in length to cells distal to it; cells uninucleate with small discoid or band-shaped plastids; gland cells present or lacking, usually adaxial; cruciate or tetrahedral sporangia sessile or pedicellate on inner cells of determinate branches; dioecious, spermatangia clustered on specialized branched axes, adaxial, replacing ultimate branches; procarps borne singly on basal cells of immature determinate branches at apices of indeterminate axes, carpogonial branches four celled; fertile axes cease to elongate after fertilization, carposporophyte appearing terminal, producing successive gonimolobes, not protected by surrounding branchlets; carposporangia developing from most gonimoblast cells.

Key to the species (including *Antithamnion*)

1. In older portions, determinate branches greater than ten cells long; gland cells present or lacking 2
1. In older portions, determinate branches ten or fewer cells long; gland cells present 4
2. Basal cells of determinate branches subquadrate, less than half as long as cells distal to them; gland cells often lacking, when present, single, covering two or three cells on short specialized branchlets ....
.... *Antithamnion cruciatum* (see *Antithamnion*, above)
2. Basal cells of determinate branches panduriform to elongate, half as long as, to only slightly smaller than, cells distal to them; gland cells single or often paired, covering one cell on nonspecialized branches 3
3. Determinate branches opposite, flagelliform, two per axial cell, basal cells mostly panduriform; gland cells single or paired, seriate or scattered on determinate branches .... *Antithamnionella flagellata*
3. Determinate branches opposite or whorled, not flagelliform, two or three per axial cell, basal cells elongate; gland cells single, restricted to the second to sixth proximal cells of determinate branches ....
.... *Antithamnionella elegans* (in part)
4. Determinate branches oppositely distichous .... *Antithamnionella atlantica*
4. Determinate branches generally in whorls of three, although oppositely paired in some lower nodes ....
.... *Antithamnionella elegans* (in part)

Antithamnionella atlantica (Oliveira F.) C. W. Schneider 1984, p. 455, fig. 1.
   Antithamnion atlanticum Oliveira F. 1969, p. 37, figs 9–10.
   Figure 421
   Plants epiphytic, prostrate, with erect axes 0.2–1.2 mm tall, pinkish red, attached by numerous multicellular rhizoids issued ventrally from cells of the prostrate axes, replacing determinate branches and opposite to dorsal determinate or indeterminate erect axes, or arising from proximal cells of ventral determinate axes; cells of the main prostrate indeterminate axes 18–31 μm diameter, 46–113 μm long; indeterminate axes distichously branched from each cell, each pair consisting of two determinate branches or, more rarely, one determinate and one indeterminate branch, the tips dense with determinate branches on

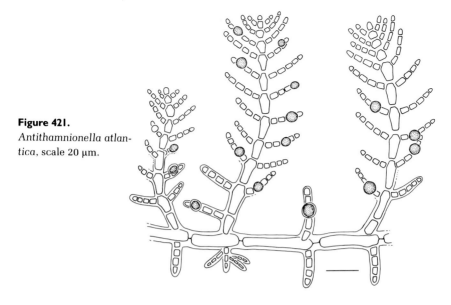

Figure 421.
Antithamnionella atlantica, scale 20 μm.

nonelongated axial cells; cells of the erect indeterminate axes near the base 15–20 μm diameter, determinate branches two to seven cells from tip to base, adaxially curved, rarely to frequently branched one (infrequently to three) times, 8–10 μm diameter, the basal cell indistinguishable in size from the cell distal to it; gland cells completely covering or overlapping the second or third cell from the base of determinate branches, more rarely the fourth cell; tetrahedral sporangia globose to ellipsoid, about 46 μm diameter, sessile, borne on the basal cell of a determinate branch, one per branch; carposporophytes terminal, forming two elongate gonimolobes to 38 μm diameter.

Schneider 1984.

Known on various algae offshore from 20 to 45 m in Onslow Bay, from 50 m in Long Bay, from 30 to 35 m on Snapper Banks, and from 17 to 21 m on Gray's Reef, June–August.

Distribution: North Carolina, South Carolina (new record), Georgia (new record), Brazil.

Antithamnionella elegans (Berthold) Price et John in Price, John et Lawson 1986, p. 16.

Antithamnion elegans Berthold 1882, p. 516.

Figure 422

Plants epiphytic, prostrate, with erect axes 0.5–3(–5) mm tall, rosy red, attached by numerous unicellular and multicellular rhizoids issued usually from basal or occasionally second cell of determinate prostrate axes or replacing determinate axes in an opposite pair or whorl of three; erect indeterminate axes developing from prostrate axial cells and replacing whorled determinate branches, or developing from the basal or second cell of determinate branches; indeterminate axes with all but a few axial cells giving rise to one, two, but usually three whorled determinate axes from their distal ends, if one to two, usually basal and not always in an opposite arrangement, the tips dense with determinate branches on nonelongated axial cells; indeterminate branches infrequently and irregularly replacing determinate branches on erect axes; determinate branches five to ten (rarely to fifteen) cells from tip to base, adaxially curved, simple to alternately branched one to three times, the basal cells slightly smaller to indistinguishable in size from the cells distal to them, branches occasionally ending in long unicellular hairs; cells of the prostrate indeterminate axes 25–36(–50) μm diameter, 70–150(–225) μm long; cells of the erect indeterminate axes near the base 20–36 μm diameter, the determinate axes 8–10(–15) μm diameter near their origins, slightly tapering to their tips; gland cells usually covering one-third to two-thirds of the basal cells of the first lateral of a determinate branch, basal cells of subsequent branches, and/or occasional outer cells of the branches; sporangia tetrahedral, cruciate or irregular, ellipsoid to obovoid, 17–28 μm diameter, 25–40 μm long, sessile, borne singly on the basal cells of determinate branches; spermatangia produced terminally on oppositely branched fertile axes, adaxially near the distal ends of cells of the determinate axes; carposporophytes terminal, forming two (sometimes more) ovoid gonimolobes to 63 μm diameter.

As *A. breviramosa*, Schneider 1984; Searles 1987, 1988. As *Antithamnion cruciatum* var. *radicans sensu* Blomquist and Humm 1946. As *Pterothamnion plumula sensu* Blair and Hall 1981. As *Antithamnionella spirographidis sensu* Searles 1987, 1988.

Known from the shallow subtidal of an offshore buoy and from 20 to 33 m in Onslow Bay, May–August and November; from 50 m in Long Bay, October; from 17 to 21 m on Gray's Reef, and 30 to 35 m on Snapper Banks, July–August.

Distribution: North Carolina, South Carolina, Georgia, Puerto Rico, Brazil; elsewhere, widespread in tropical and subtropical seas.

Searles (1988) provisionally assigned one of his vegetative Gray's Reef specimens of this species to *Antithamnionella spirographidis* (Schiffner) Wollaston (1968), but upon further examination we find it to be closer to *A. elegans* as broadly conceived by Cormaci and Furnari (1988). Because the Georgian specimen has longer determinate axes and a few other differences from specimens more typical of the species, these variations are separated in the preceding key. Cultural studies of the two local morphologies as well as clones from various parts of the world, including type localities of nomenclatural synonyms listed by Cormaci and Furnari (1988), are necessary to elucidate the level of taxonomic significance of branching patterns and other key characters.

Antithamnionella flagellata (Børgesen) Abbott 1979, p. 222, figs 18–20.
   Antithamnion flagellatum Børgesen 1945, p. 5, figs 1–2.
   Figure 423

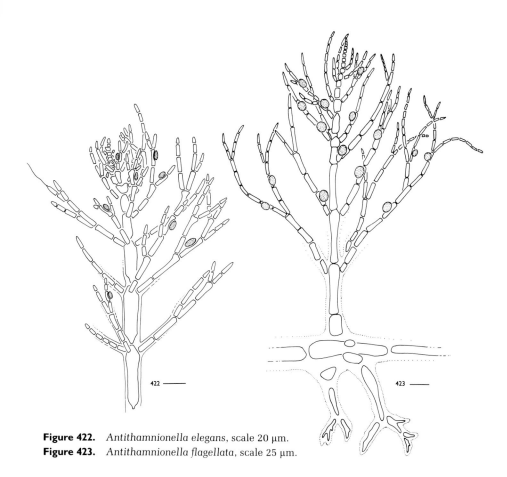

Figure 422. *Antithamnionella elegans*, scale 20 μm.
Figure 423. *Antithamnionella flagellata*, scale 25 μm.

Plants epiphytic, prostrate, with erect axes to 8 mm tall, rosy red, attached by numerous multicellular rhizoids issued ventrally from cells of the prostrate axes, replacing a branch and opposite to dorsal determinate or indeterminate erect axes, or arising from proximal cells of ventral determinate axes, simple or dichotomously branched, deeply penetrating host tissue; rhizoidal cells lightly pigmented, elongate, 10–30 μm diameter; prostrate axes indeterminate, becoming erect at the distal end; erect indeterminate axes with all but a few cells giving rise to an opposite pair of branches, mostly determinate, but occasionally indeterminate, decussate or in a one-third spiral; some erect indeterminate axes with determinate branches infrequently alternate or in whorls of three; cells of prostrate and major erect indeterminate axes long cylindrical, 20–60 μm diameter, 110–130 μm long, producing branches near their distal ends; cells of lesser erect axes and distal portions of all indeterminate axes with narrow panduriform protoplasm; determinate axes ten to twenty cells from tip to base, simple and flagelliform to sparingly alternately or pseudodichotomously branched, 12–20 μm diameter proximally, tapering to 3–4 μm distally, basal cells slightly shorter than cells distal to them, narrow panduriform; gland cells abundant, adaxial to abaxial on determinate axes, single or paired, the second often smaller, seriate to scattered, occasionally found on the proximal cell, covering the distal three-fourths or completely covering each cell which bears them; sporangia cruciate or tetrahedral, obovoid to ellipsoid, 27–37 μm diameter, 37–48 μm long (including thick walls to 7.5 μm), sessile, one to three borne in an adaxial secund series proximal on determinate branches in upper portions of plants; dioecious, spermatangia produced in inconspicuous elongate, loose sori on determinate branches; carposporophytes terminal on indeterminate axes, forming three to five globose to ellipsoid gonimolobes to 150 μm in greatest dimension; carposporangia thirty-five or fewer per gonimolobe, isodiametric, rounded polygonal to ellipsoid, to 40 μm in greatest dimension at maturity.

Schneider 1984.

Known as a common epiphyte of *Halymenia floridana*, and infrequently on other algae, from 29 to 45 m in Onslow Bay, June–August.

Distribution: North Carolina, Virgin Islands, Belize, Mauritius, Hawaii.

Calliclavula C. W. Schneider in Searles et C. W. Schneider 1989

Plants erect, bushy, ecorticate, attached by unicellular rhizoids; axes monosiphonous, growth monopodial, appearing dichotomously to trichotomously branched and becoming fan shaped; cells elongate and club shaped with numerous discoid plastids and large central vacuoles; one or two indeterminate branches cut off of the upper ends of subterminal axial cells, and two to four whorls of lightly pigmented determinate branchlets simultaneously produced from the upper ends of subterminal cells; whorl branchlets with greater pigmentation below, mostly trichotomous at first node and dichotomous where branched above, three cells long, eventually overtopping the apices.

Calliclavula trifurcata C. W. Schneider in Searles et C. W. Schneider 1989, p. 732, figs 1–4.

Figure 424

Plants epiphytic, erect to 2.5 cm tall, rosy red, attached by unicellular rhizoids to 2 mm long and 20–30 μm diameter; indeterminate axes mostly pseudotrichotomously branched, otherwise pseudodichotomous, branches occurring every three to seven segments; cells of indeterminate axes club shaped, 40–120 μm diameter at distal ends and 60–500 μm long below apices, becoming wider and greatly elongate below, 130–240 μm diameter, to 1.3 mm long; two to four whorl branchlets issued from subapical nodes, persistent in upper half of plants; cells of whorl branchlets tapering proximally and greatly lengthening with age, to 450 μm long, cells 5–12.5 μm diameter in basal segments, 2.5 μm diameter in long terminal cells, apices rounded.

Searles and Schneider 1989.

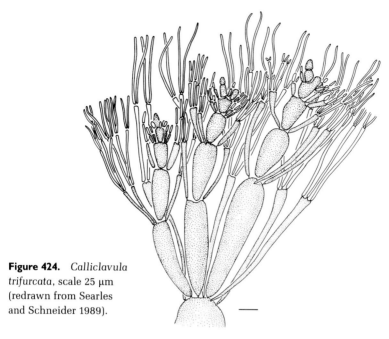

Figure 424. *Calliclavula trifurcata*, scale 25 μm (redrawn from Searles and Schneider 1989).

Known offshore from 29 m in Onslow Bay, July.
Distribution: Endemic to North Carolina as currently known.
Reproductive material of this monotypic genus is unknown.

Callithamniella Feldmann-Mazoyer 1938

Plants filamentous, dorsiventrally organized with erect determinate and indeterminate axes arising irregularly from cells of prostrate indeterminate axes; all axes uniseriate, ecorticate; prostrate axes attached by irregularly issued, mostly simple, ventral, unicellular or multicellular rhizoids; basal cells of determinate axes often giving rise to a rhizoid and later an indeterminate erect axis; other cells of prostrate axes with either a single erect axis without a rhizoid or an erect axis opposite a rhizoid; prostrate axes potentially becoming erect at distal ends; determinate and a few indeterminate axes arising in a regular double spiral from each cell of erect indeterminate axes; determinate axes simple or dichotomously branched, seven to thirty cells long; cells uninucleate; cruciate or irregularly cruciate sporangia pedicellate, borne alternately and irregularly, rarely oppositely, at the distal end of cells of erect indeterminate axes, obscuring the spiral pattern of vegetative axes; spermatangia borne in elongate heads on determinate branches, usually with terminal sterile tips; procarps, carposporophytes unknown.

Key to the species

1. Determinate axes simple or with one to three distal adaxial branchlets, to 700 μm long; tetrasporangia borne basally on extended determinate axes
.... *C. silvae*
1. Determinate axes simple or once dichotomous from the third cell from the

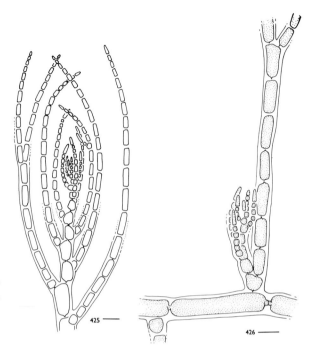

Figures 425, 426.
Callithamniella silvae.
425. Indeterminate axis tip, scale 25 μm. 426. Origin of indeterminate branch from base of determinate branch, scale 25 μm.

base, to 470 μm long; tetrasporangia borne on short indeterminate axes replacing determinate branches.................................... *C. tingitana*

Callithamniella silvae Searles in Searles et C. W. Schneider 1989, p. 733, figs 5–10.

Figures 425 and 426

Plants epizoic, rosy red, prostrate, with erect axes to 5 mm tall, attached by multicellular rhizoidal branches, 18–27 μm diameter, with or without terminal digitate haptera; each cell of the prostrate axes bearing a hapteral branch from which a determinate or an indeterminate erect axis may arise; occasionally, additional indeterminate axes replace determinate branches on erect axes; prostrate cells cylindrical, 20–38 μm diameter, 80–120 μm long; basal cells of determinate branches distinctly shorter than the cells distal to them; cells of erect indeterminate axes 23–32 μm diameter and 85–120 μm long near the base, tapering and shortening to ultimate segments 6 μm diameter and 4 μm long, apices obtuse; erect indeterminate axes issuing alternately one determinate branch or occasionally an indeterminate branch per axial cell in a one-third spiral; determinate branches commonly simple or with one to three adaxial distal branchlets, or with additional determinate or indeterminate laterals produced by basal cells, incurved and overtopping the apices, to twenty cells and 700 μm long, finer than indeterminate axes, 18–20 μm diameter at the base and slightly tapering to the obtuse apices, 7–11 μm diameter; where known, sporangia undivided, ellipsoid, 18–30 μm diameter, 27–45 μm long, pedicellate on one or rarely two cells,

Figures 427, 428. *Callithamniella tingitana.* 427. Prostrate tetrasporangial axis bearing determinate and indeterminate branches, scale 50 μm. 428. Origin of indeterminate branch from base of determinate branch, scale 25 μm.

one to three borne on proximal cells of determinate branches; gametangia unknown.

Searles and Schneider 1989. As *Callithamniella* sp., Searles 1987, 1988.

Known from 17 to 21 m on Gray's Reef, July.

Distribution: Endemic to Georgia as currently known.

Due to slight disparities with generic characteristics and a lack of reproductive material, Searles only tentatively assigned this species to *Callithamniella* (Searles and Schneider 1989).

Callithamniella tingitana (Schousboe ex Bornet) Feldmann-Mazoyer 1938, p. 1119.

Callithamnion tingitanum Schousboe ex Bornet 1892, p. 329.

Figures 427 and 428

Plants saxicolous or epizoic, rosy red, prostrate, with erect axes to 4 mm tall, attached by numerous ventral unbranched unicellular and multicellular rhizoids to 20 μm diameter, one issued from the distal end of scattered cells of the prostrate axes, with or without terminal lobate to discoid haptera; each cell of the prostrate axes bearing a rhizoid and/or determinate and/or indeterminate erect axis; prostrate cells irregularly cylindrical, 24–35 μm diameter, 100–

310 µm long; basal cells of branches generally with panduriform protoplasm, shorter than the cells distal to them; erect indeterminate axes often finer than prostrate axes, 15–25 µm diameter, 75–140 µm long, tapering and shortening to ultimate segments 5 µm diameter and 5–10 µm long, apices obtuse; erect indeterminate axes issuing alternately few indeterminate and mostly determinate branches in a double, one-fourth–one-sixth spiral; determinate axes commonly simple or once dichotomous from the third cell of the branch, rarely twice dichotomous, seven to twenty cells and 270–470 µm long, finer than indeterminate axes, 10–13 µm diameter at the base and occasionally slightly expanding in central portions, slightly tapering to obtuse or acute apices, 5–10 µm diameter; sporangia cruciate or irregularly cruciate, ellipsoid to subglobose, 25–38 µm diameter, 38–55 µm long, single, borne on one-celled pedicels opposite to, adjacent to, or replacing a lateral on indeterminate axes; spermatangia whorled on several successive, persistent axial cells forming elongate heads on determinate branches, usually with short to long sterile tips, fertile branches formed on both prostrate and erect indeterminate axes, mostly near apices, opposite or adjacent to other determinate branches; carpogonia and carposporophytes unknown.

Schneider 1984.

Known offshore from 18 to 60 m in Onslow Bay, June–August, and on Cape Lookout jetty from 4 to 6 m, June–July.

Distribution: North Carolina, Brazil, Portugal, Mediterranean, Ghana.

Previously removed to *Grallatoria* (Abbott 1976), this species has been retained in *Callithamniella* based on tetrasporic and male plants collected in the Carolinas (Schneider 1984; Wynne and Ballantine 1985).

Callithamnion Lyngbye 1819

Plants erect, filamentous, bushy, ecorticate or corticated by rhizoids, attached by basal discs or rhizoids; axes uniseriate, alternately distichous to pseudodichotomous in the main axes and laterals, mostly pseudodichotomous below, lesser branches pseudodichotomous or alternate to secund; cells uninucleate to multinucleate with several small discoid to band-shaped plastids, with or without terminal hairs; tetrahedral or cruciate sporangia usually sessile, single or paired on the adaxial surface of lateral branches; dioecious, spermatangia formed in hyaline clusters embedded in a gelatinous matrix on the adaxial surface of lateral branches, produced from one to three initials, appearing sessile; procarps borne laterally on the main axes in the upper portions of the plant, consisting of two opposite pericentrals, one of which bears a four-celled carpogonial branch; carposporophyte consisting of two or four gonimolobes, involucre lacking; carposporangia developing from most gonimoblast cells.

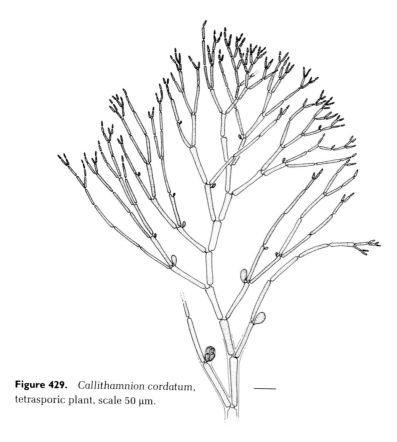

Figure 429. *Callithamnion cordatum*, tetrasporic plant, scale 50 μm.

Key to the species

1. Branching pseudodichotomous near the apices, ultimate cells less than 10 μm diameter . . . . *C. cordatum*
1. Branching alternate near the apices, ultimate cells greater than 10 μm diameter . . . . *C. pseudobyssoides*

Callithamnion cordatum Børgesen 1909, p. 10, figs 5–6.
Figure 429

Plants epiphytic, epizoic, lignicolous, bushy, erect to 6 cm tall, pink to pinkish red, ecorticate, attached by unicellular and multicellular, simple or branched rhizoids issued from the base and nodes of lower segments; main axes straight and alternately branched below, flexuous and pseudodichotomous above, branched to many orders; branches occasionally ending in unicellular hairs, 50–450 μm long; cells 200 μm diameter, short and quadrate at the extreme base, cylindrical but slightly enlarged at the nodes, 50–120(–180) μm diameter and 1–1.6 mm long in the main axes, tapering to 7–10 μm diameter and 12–30 μm long in the ultimate segments; ultimate axes straight or slightly incurved, apices obtuse; sporangia tetrahedral, sessile, single or in groups of two or three, obovate or subglobose, 37–50 μm diameter, 40–63 μm long (including thick cell walls), borne distally and adaxially on cells of the last two orders of branching in the

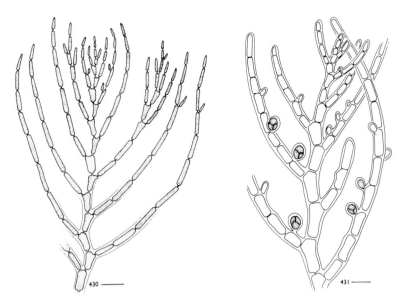

Figures 430, 431. *Callithamnion pseudobyssoides.* 430. Axis tip, scale 50 μm. 431. Detail of tetrasporic axis, scale 50 μm.

upper portions of the plant; spermatangia in single or occasionally confluent, flat sori, originating from a single distal initial in position similar to sporangia, each sorus partially overlapping the cell distal to it; carposporophytes binate, each half cordate to irregularly ovoid, consisting of two gonimolobes borne at the forkings in upper portions of the plant; carposporangia subglobose, rounded polygonal, elongate to irregular, 30–50 μm in greatest dimension at maturity.

Schneider 1980.

Known only from 19 to 32 m offshore in Onslow Bay, June–August.

Distribution: North Carolina, Bermuda, southern Florida, Caribbean, Japan, Pacific Mexico, California.

A detailed morphological study of this taxon in North Carolina can be found in Schneider (1980).

Callithamnion pseudobyssoides P. Crouan et H. Crouan 1852, no. 132.

Figures 430 and 431

Plants epiphytic, epizoic, or saxicolous, globose tufted, erect from 0.5–3 cm tall, pinkish to rosy red, ecorticate to corticated loosely at the base by descending rhizoids from the basal cells of lower lateral branches; main axes straight below, flexuous above, repeatedly alternately distichous throughout lesser branches of main axes, main axes markedly thinner in lower portions; ultimate branches rarely unilateral, always incurved; cells narrowly panduriform to cylindrical, 40–215 μm diameter, 115–440 μm long, in lower portions of the main axes tapering to 11–18 μm diameter and 25–50 μm long; ultimate cells slightly tapering or not tapering at all, apices obtuse; sporangia tetrahedral, rarely cruciate, single, oblique obovate or, less frequently, ellipsoid or subglobose, 27–49 μm diameter, (38–)48–65 μm long (including thin or thick cell walls), borne distally on cells in a proximal adaxial series on mostly ultimate branches throughout the plant; spermatangia in single cushionlike, rarely short, upright sori, originating from one to three distal initials on adaxial surfaces of ultimate branch cells, in series, each sorus partially overlapping the cell distal to it; carposporophytes binate, each half globose or subglobose, to 170 μm diameter, borne at the forkings in upper portions of the plants; carposporangia subglobose, rounded polygonal, elongate to irregular, 25–40 μm in greatest dimension at maturity.

As *C. byssoides*, Williams 1948a, 1949, 1951; Taylor 1957, 1960; Schneider 1974, 1976; Brauner 1975; Wiseman and Schneider 1976; Kapraun 1978b, 1980a; Richardson 1986, 1987. As *C. polyspermum sensu* Harvey 1853; ?Curtis 1867; Hoyt 1920; Wiseman and Schneider 1976. As *C. halliae sensu* Searles 1981, 1988. As *Callithamnion* sp., Searles 1987.

Known from the shallow subtidal and lower intertidal of jetties from Cape Lookout to Cumberland Island, March–July; and offshore from 17 to 60 m in Raleigh Bay, Long Bay, and Onslow Bay, from 17 to 21 m on Gray's Reef, and from 30 to 35 m on Snapper Banks, May–October.

Distribution: Nova Scotia to Virginia, North Carolina, South Carolina, Georgia, Bermuda, southern Florida, Gulf of Mexico, ?Caribbean, ?Brazil, France to Portugal.

Questions have existed for some time as to whether North American specimens identified as *Callithamnion byssoides* Arnott ex Harvey (1833) should be labeled *C. pseudobyssoides* (Rueness and Rueness 1980; Spencer et al. 1981). We note the obvious distinctions between the two (Halos 1965; Dixon and Price 1981) and find the globose to subglobose gonimolobes and method of spermatangia production in local plants similar only to *C.pseudobyssoides*. None of the specimens from the western Atlantic have the young pointed gonimolobes or extended upright spermatangial branches typical of *C. byssoides*. A systematic comparison of eastern and western Atlantic specimens of *C. pseudobyssoides* could show the two areas supporting distinct species, but for the present we break with tradition in calling the North American specimens *C. pseudobyssoides*, finding it impossible to continue to name them *C. byssoides*.

Local specimens of *Callithamnion pseudobyssoides* agree with *C. byssoides sensu* Harvey (1846, 1853) but do not conform with the description of West Indian plants undoubtedly misapplied to the taxon by Børgesen (1917). Interestingly, Børgesen (1930) modified his own concept of the species to include typical *C. byssoides* plants collected in the Canary Islands. Hoyt (1920) and others have reported *C. polyspermum* C. Agardh from our area, but these records are doubtful and were probably based on specimens of *C. pseudobyssoides*. After looking at authentic Florida specimens of *Callithamnion halliae* Collins (1906a), we find that the plants from Georgia described as that species (Searles 1981, 1988) were incorrectly labeled and fall within the range of variability of *C. pseudobyssoides*.

Culture studies indicate that temperature and light intensity are primary factors in controlling the periodicity of inshore populations of *Callithamnion pseudobyssoides* (as *C. byssoides*) in North Carolina (Kapraun 1978b).

Centroceras Kützing 1841

Plants erect and bushy or entangled, or prostrate and matted, terete, filamentous, attached by basal rhizoids and simple multicellular rhizoids issued from nodes of prostrate axes; axes uniseriate, dichotomously and adventitiously branched, corticated completely by regular longitudinal rows of rectangular cells, tiered only at the nodes, originating from the distal end of the larger axial cells; nodes commonly spinulose; cells uninucleate; tetrahedral or cruciate sporangia whorled at nodes, projecting, sometimes on adventitious branches; dioecious, spermatangia formed in terminal clusters on tufted, adventitious branches arising from pericentral cells at the nodes; procarps formed laterally at the nodes, carpogonial branches four celled; cystocarps composed of two gonimolobes and surrounded by short involucral filaments; carposporangia develop from most gonimoblast cells.

Centroceras clavulatum (C. Agardh) Montagne 1846, p. 140.

Ceramium clavulatum C. Agardh in Kunth 1822, p. 2.

Plants epiphytic or saxicolous, prostrate, forming mats with short, stiff, upright axes, or erect, lax, and entangled, to 10–20 cm tall, dark rosy to purplish red, wholly corticated by rectangular cells and bearing whorls of one- or two-celled spines at the nodes, occasionally lacking below, arising from basal and nodal rhizoids, with or without terminally expanded tips; axes dichotomously branched, often with adventitious branches, 50–200 μm diameter, segmented, internodes 300–750 μm long; apices forcipate or circinate, the individual apical cells exposed and short conical; sporangia tetrahedral, less often cruciate, obovate, oblique obovate, subglobose, to globose, 45–50 μm diameter, 50–63 μm long with thick cell walls, whorled at the nodes in the upper portions of the plant, emergent, subtended by short involucral filaments produced at the nodes; gametangia as for the genus.

Blomquist and Pyron 1943; Humm 1952.

Known from the drift near Morehead City, August; and common on the intertidal coquinoid outcroppings at Marineland, year-round.

Distribution: North Carolina, Bermuda, northeastern Florida to Gulf of Mexico, Caribbean, Brazil; elsewhere, widespread in tropical and subtropical seas and north to Santa Cruz, California, on the Pacific Coast of North America.

This species is considered one of the characteristic species at Marineland due to its year-round abundance (Humm 1952). North of Florida, *Centroceras clavulatum* has been once collected in the drift in North Carolina after a summer storm (Blomquist and Pyron 1943). It was attached to a dominant benthic seaweed offshore, *Sargassum filipendula* (Schneider and Searles 1979).

Ceramium Roth 1797, nom. cons.

Plants erect and bushy or entangled, or prostrate and matted, terete, filamentous, attached by basal rhizoids and simple unicellular and multicellular rhizoids issued from the nodes of prostrate axes; axes uniseriate, pseudodichotomously, adventitiously, and less often alternately branched, corticated in all orders of branching at the nodes by small transversely banded, darkly pigmented cells, these spreading to cover the internodes of lower main axes to the entire plant in some species; apices forcipate or circinate to straight; cells uninucleate with several small discoid, fusiform, or elongate plastids; gland cells rarely present in the corticated regions; tetrahedral or cruciate sporangia whorled or single and secund at the nodes to scattered in wholly corticated species, embedded to projecting, with or without protective sterile cells; monoecious or dioecious, spermatangia formed in a hyaline layer over nodal corticating cells; procarps borne laterally on nodal corticating cells, carpogonial branches single or paired on the supporting cell, four celled; carposporophytes subglobular, consisting of one to three gonimolobes surrounded at least partially by few to several involucral filaments; carposporangia developing from most cells of the gonimoblast.

Key to the species

1. Cortication restricted to the nodes, or slightly spreading 2
1. Cortication completely covering all axes . . . . C. rubrum
2. Branching pseudodichotomous, often with adventitious laterals 3
2. Branching alternate . . . . C. floridanum
3. Cortical band consisting of one or two transverse rows of cells 4
3. Cortical band consisting of more than two transverse rows of cells 5
4. Cortical band of a single transverse row of cells . . . . C. leptozonum
4. Cortical band of two transverse rows of cells . . . . C. fastigiatum f. flaccidum
5. Lowermost transverse cortical row consisting of transversely elongate cells
   . . . . C. byssoideum
5. Lowermost transverse cortical row without transversely elongate cells 6
6. Cortical band with a central inner layer of larger globose cells, covered or left
   exposed by an outer layer of cells; sporangia immersed or slightly emergent

.... *C. diaphanum*
6. Cortical band with two layers of cells, generally the lower transverse rows
   intermediate in size, the middle largest, and the upper smallest; sporangia
   projecting ....*C. fastigiatum*

Ceramium byssoideum Harvey 1853, p. 218.

Figure 432

Plants epiphytic or epizoic, prostrate and widespreading, erect, and 0.5–4(–10) mm tall, sparingly yet regularly pseudodichotomously branched, light pinkish red, attached by numerous unicellular and multicellular rhizoids issued from the nodes; apices straight to incurved, not forcipate; cortication limited to the nodes, composed of three to six transverse rows of cells, gland cells occasionally present; the upper rows with subglobose, quadrate, to polyhedral cells, the lower one or two rows composed of obvious transversely elongated cells; axial cells 50–90 μm diameter in lower portions, 100–300 μm long; tetrahedral sporangia globose to subglobose, to 60 μm diameter (including thick cell walls), single or paired and subsecund, or whorled at the nodes, emergent but subtended by short involucral filaments; spermatangia formed in tufts at the nodes; carposporophytes one or two lobed, apparently terminal on short clavate branchlets, subtended by few-celled involucral filaments and lateral axes which extend beyond.

Humm 1952; Taylor 1960; Brauner 1975; Wiseman and Schneider 1976; Kapraun 1980a; Hall and Eiseman 1981. As *C. transversale sensu* Williams 1951.

Known from the shallow subtidal near Beaufort and Wilmington, N.C., North Island, S.C., and Haulover Cove, Fla., June–November; infrequently offshore in Onslow Bay from 18 to 21.5 m, June–August, and from intertidal coquinoid outcroppings at Marineland, year-round.

Distribution: North Carolina, South Carolina, Bermuda, northeastern Florida to Gulf of Mexico, Caribbean, Brazil.

Considerable taxonomic confusion surrounds this species, which is possibly conspecific with, or at least closely related to, *Ceramium transversale* Collins and Hervey (1917) from Bermuda, *C. gracillimum* (Griffiths) Harvey from the Mediterranean (Feldmann-Mazoyer 1941), *C. flaccidum* (Kützing) Ardissone from Australia (Womersley 1978), and *C. dawsonii* Joly from Brazil (Taylor 1960). Wynne (1986) effects these synonymies for the western Atlantic based on the work of Womersley (1978). The nodal cortication of local plants is consistent and distinctly different from nodal cortication illustrated by Womersley (1978) for Australian plants. Therefore we agree with Wiseman (1966), who surveyed Carolina plants, and Taylor (1960) in retaining the Harvey taxon until more definitive studies can be carried out on a worldwide basis.

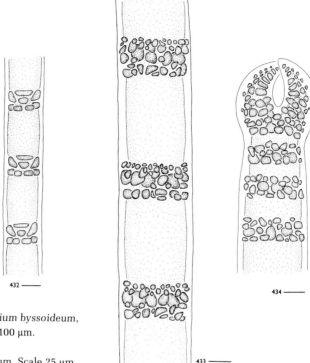

Figure 432. *Ceramium byssoideum*, detail of axis, scale 100 μm.
Figures 433, 434.
Ceramium diaphanum. Scale 25 μm.
433. Detail of axis. 434. Axis tip.

Ceramium diaphanum (Lightfoot) Roth 1806, p. 154.

    Conferva diaphana Lightfoot 1777, p. 996.

    Figures 433 and 434

    Plants epiphytic, epizoic, saxicolous, lignicolous, with prostrate and erect axes, to 20 cm tall, regularly pseudodichotomous with more or less short adventitious branches, branching spreading to fastigiate, dull rosy red to brownish red, attached by numerous rhizoids issued from basal cells and nodes of proximal cells and randomly from nodes of prostrate axes; apices forcipate to erect; cortication only at the nodes, obvious to the unaided eye, composed of an inner central row of larger rounded to globose cells covered or left exposed by an outer layer of smaller angular to rounded cells, altogether consisting of four or more transverse rows; nodes 210–450 μm diameter, broader than long; internodes lightly pigmented, 185–460 μm diameter, 0.5–1.5 mm long below; tetrahedral sporangia globose to short ellipsoid, 50–75 μm diameter, whorled, immersed centrally in nodal cortication or slightly emergent in the upper half of the node; spermatangia formed in tufts on the cortical cells, occasionally circumscribing the node; carposporophytes consisting of one or two subglobular gonimolobes borne on short lateral branches, usually in distal portions of the plants, subtended by one to six short, similarly corticated involucral branches, which often do not overtop the gonimolobes.

?Curtis 1867; Williams 1948a, 1949; Taylor 1960; Brauner 1975; Wiseman and Schneider 1976; Kapraun 1980a, 1980b; Richardson 1986; 1987. As *C. strictum*, Hoyt 1920; Williams 1948a, 1949; Taylor 1957, 1960; Brauner 1975; Wiseman and Schneider 1976.

Known from shallow subtidal, intertidal, and estuarine habitats throughout the Carolinas and Georgia, and rarely from 29 m offshore of Cape Fear, year-round.

Distribution: Newfoundland to Virginia, North Carolina, South Carolina, Georgia, southern Florida, Gulf of Mexico, Caribbean, Brazil, Uruguay, Argentina, Arctic to Portugal, Azores, Mediterranean, Canary Islands, and probably much more widespread (further taxonomic investigations are necessary).

Wiseman (1966, 1978) and Kapraun (1980a) have both found intermediates between *Ceramium diaphanum* and *C. strictum* (Kützing) Harvey in their collections of Carolina plants and have retained all plants of this "complexe *diaphanum-strictum*" (Feldmann-Mazoyer 1941) under the earlier epithet, *C. diaphanum*. Varietal designations used elsewhere (Feldmann-Mazoyer 1941) likewise intergrade in our plants and therefore are not considered useful. It is also possible that *C. diaphanum* intergrades with specimens identified as *C. fastigiatum* in the flora. For the present they are retained as separate entities, but they require further study.

Ceramium fastigiatum (Wulfen ex Roth) Harvey in Hooker 1834, p. 303.
Conferva fastigiata Wulfen ex Roth 1800, p. 224.
Figure 435

Plants epiphytic, epizoic, saxicolous, with prostrate and erect axes, to 10 cm tall, regularly pseudodichotomous with occasional adventitious branches, branching fastigiate, usually wide spaced, purplish, brownish, pinkish, to bright rosy red, attached by simple rhizoids, with or without expanded haptera from the bases and nodes of prostrate axes, often more than one hapteron per node; apices erect, incurved, forcipate or divaricate; cortication limited to the nodes, obvious to the unaided eye, composed of two layers of four to six transverse rows, the lowest generally intermediate in size, the middle largest, and the upper smallest and globose to longitudinally ellipsoid; nodes 60–155 μm diameter, 55–65 μm long, occasionally bearing gland cells; internodes lightly pigmented, 75–150 μm diameter, 0.6–1.4 mm long below; tetrahedral and cruciate sporangia globose to ellipsoid, 33–65 μm diameter, 50–68 μm long, single or paired and seriate, or whorled at the nodes, emergent to strongly projecting, usually subtended by a few short, narrow involucral filaments; carposporophytes consist-

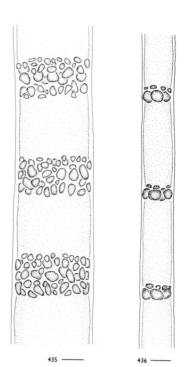

Figure 435. *Ceramium fastigiatum*, detail of axis, scale 25 μm.
Figure 436. *Ceramium fastigiatum* f. *flaccidum*, detail of axis, scale 25 μm.

435 ——— 436 ———

ing of one or two subglobular to ellipsoid gonimolobes, borne on short lateral branches usually in distal portions of the plant, subtended by two to four short, similarly corticated involucral branches, which usually do not overtop the gonimolobes.

C. fastigiatum f. *flaccidum* H. Petersen in Børgesen 1915, p. 16, fig. 231.
Figure 436

Plants 1–7 cm tall; apices straight to incurved, not forcipate; cortication at the nodes consisting of only one or two transverse layers, the lower larger than the upper; nodal and internodal dimensions on the average smaller than the typical form, gland cells lacking; tetrasporangia subglobose to 60 μm diameter (including thick cell walls), single and scattered to clustered and seriate, prominently projecting, subtended by a few short cortical cells; spermatangia covering the nodes; carposporophytes generally with three gonimolobes overtopped by involucres.

Williams 1948a, 1949, 1951; Taylor 1960; Brauner 1975; Schneider 1976; Richardson 1987. As f. *flaccidum*, Williams 1948a; Taylor 1960; Chapman 1971; Kapraun 1980a, 1980b; Searles 1987, 1988. As *C. tenuissimum sensu* Hoyt 1920; Taylor 1960.

Both forms are common in shallow subtidal environments throughout North Carolina, from the jetty on Cumberland Island, and offshore from 14 to 40 m in Onslow Bay and Georgia, year-round. During some years, f. *flaccidum* is common offshore; in other years it is apparently scarce.

Distribution: Newfoundland to Virginia, North Carolina, Georgia, Bermuda, southern Florida, Gulf of Mexico, Caribbean, Norway, British Isles, France, Galapagos Islands.

Specimens labeled *Ceramium tenuissimum* (Lyngbye) J. Agardh by Hoyt (1920) are indistinguishable from *C. fastigiatum*.

Ceramium floridanum J. Agardh 1894, p. 46.

Plants epiphytic, with prostrate and erect axes, to 7.5 cm tall, irregularly alternately branched and complanate below, more regular above, branches spreading, bright rosy red, attached by numerous rhizoids issued from basal cells and from nodes of prostrate axes; branches to several orders, major branches widely spaced, ultimately developing ovate to lanceolate, determinate, closely spaced, flattened, branch systems, 2–3 mm wide to 10 mm long; adventitious branches occasionally present on fertile plants, ultimate axes divergent, cortication closely spaced, apices abruptly conical; cortication only at the nodes, obvious to the unaided eye, composed of one to three rows of larger rounded to globose cells irregularly covered by an outer layer of smaller angular to rounded cells; nodes 57–437 μm diameter, 95–470 μm long, not swollen; internodes lightly pigmented, 200–300 μm diameter, 330–390 μm long, decreasing above to about 150 μm long; tetrahedral sporangia globose, 70–75 μm diameter, whorled, greatly swelling the nodal cortication, slightly emergent in the upper half of the node; carposporophytes borne on short lateral branches in distal portions of the plants, subtended by a few short, similarly corticated involucral branches slightly overtopping the gonimolobes.

Humm 1952; Taylor 1960.

Known only from the intertidal of coquinoid outcroppings at Marineland, year-round.

Distribution: Northeastern Florida to Caribbean.

Ceramium leptozonum Howe 1918, p. 531.

Plants epiphytic, prostrate, and erect to 1–3 cm tall, regularly pseudodichotomous, branching fastigiate, purplish to rosy red, attached by simple rhizoids from the bases and nodes of prostrate axes, often more than one per node; dichotomies acute, apices commonly forcipate, some straight, axes slightly tapering from base to apex; cortication limited to the nodes and composed mostly of a single layer of cells, elongate parallel with the axis, 20–40 μm long, appearing four across the filaments, occasionally with an incomplete row of small cells at the distal ends of nodes; internodes lightly pigmented, 40–72 μm diameter, 1.5–4 times as long as wide; tetrahedral sporangia globose, 50–65 μm diameter, single or two to three per node, secund on the outer side of nodes, occa-

Figure 437. *Ceramium rubrum*, detail of axis, scale 200 μm.

sionally subverticillate, naked and projecting, subtended by two to four small cells, secondary sporangia formed by regeneration, becoming exerted and appearing substipitate.

Chapman 1971 (n.v.).

Known only from 19 to 21 m off Georgia, July–November.

Distribution: Georgia, Bermuda.

Ceramium rubrum (Hudson) C. Agardh 1817, p. 60.

Conferva rubra Hudson 1762, p. XXVII.

Figure 437

Plants epiphytic, epizoic, saxicolous, lignicolous, erect, 5–40 cm tall, regularly pseudodichotomous with shorter adventitious branches, dark purplish to dull rosy red, attached by basal rhizoidal holdfast; apices straight, incurved, commonly forcipate, axes gradually tapering from base to apex; cortication of small angular to rounded cells completely covering the uniseriate filaments, internodal cells often elongated, the plant often appearing banded to the unaided eye; filaments coarse, to 1 mm diameter, segments 2–3 mm long in lower portions; tetrahedral and cruciate sporangia ellipsoid to subglobose, 30–80 μm diameter, 40–80 µm long, in rows over the nodes or scattered throughout, immersed, producing a nodulose appearance of the axes; spermatangia in hyaline patches over the nodes in distal portions; carposporophytes globose to ellipsoid, borne on short lateral branches, subtended by two to six short, similarly corticated involucral branches often overtopping the gonimolobes.

Harvey 1853; ?Curtis 1867; Hoyt 1920; Taylor 1957, 1960; Wiseman and Schneider 1976; Kapraun 1980a. ?As *C. areschougii*, Humm 1952.

Known from the drift and sounds near Morehead City and Charleston, and from several estuarine, brackish-water environments in the Wilmington area during the spring.

Distribution: Arctic to North Carolina, South Carolina, ?northeastern Florida, Argentina, Arctic to Portugal, Azores, Mediterranean, Black Sea, Japan.

Reported as *Ceramium areschougii* Kylin from Marineland (Humm 1952), Taylor (1960) found this report and other more southern reports of *C. rubrum* doubtful. We have not collected specimens from south of the Carolinas.

Great morphological variability has been demonstrated for this species in culture (Garbary et al. 1978).

Compsothamnion Nägeli 1862
Plants erect, filamentous, ecorticate; axes uniseriate, much and alternately branched in a single plane, only one branch produced by each axial cell; lateral axes branched to three orders in a similar fashion, forming a regular distichous habit; tetrahedral sporangia borne terminally on lateral branches; dioecious, spermatangia clustered, sessile and adaxial or terminal on lateral branches; procarps terminal or subterminal on the lateral axes, carpogonial branches two celled; carposporophytes consisting of one or two gonimolobes, involucre lacking; carposporangia developing from most gonimoblast cells.

Compsothamnion thuyoides (Smith) Schmitz in Schmitz et Hauptfleisch 1897, p. 491.
Conferva thuyoides Smith 1810, tab. 2205.
Figures 438 and 439
Plants epiphytic or epizoic, delicate, usually gregarious, to 7.5 cm tall but usually much shorter, pinkish to brownish red, attached by rhizoids issuing from basal cells; branching distinct, alternately distichous; cells of branches in lower portions obviously narrower (50–100 µm diameter at the base) than those of the axial row (160–230 µm diameter), basal cells of lateral branches issuing rhizoids proximally; each branched system lanceolate to ovate in outline; tetrasporangia globose, 37–53 µm diameter (including thick cell walls), 50–60 µm long, terminal on short branches; procarps terminal or subterminal on lateral

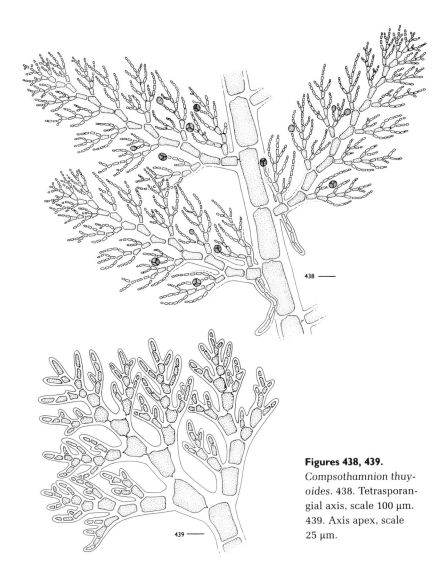

Figures 438, 439.
Compsothamnion thuy-oides. 438. Tetrasporan-gial axis, scale 100 μm.
439. Axis apex, scale 25 μm.

axes, carposporophytes lobed, naked, and solitary or in pairs.

Schneider 1975a, 1976; Wiseman and Schneider 1976; Blair and Hall 1981.

Known offshore from 40 to 60 m in Long Bay and Onslow Bay, June–September.

Distribution: North Carolina, South Carolina, Bermuda, southern Florida, Norway to Portugal, Mediterranean, Canary Islands.

European and Mediterranean plants are larger than those found in the western Atlantic; ours reach only 2 cm. Carolina specimens have immature ovoid sporangia, becoming globose at maturity (Schneider 1975a). Procarpic plants have been collected in our flora; however, cystocarps and spermatangia are unknown.

Griffithsia C. Agardh 1817, nom. cons. prop.

Plants erect to prostrate, bushy, ecorticate or with rhizoidal cortication below; axes monosiphonous, sparsely to greatly subdichotomously or laterally branched, large celled, appearing beaded or segmented; cells elongate to ovoid or subglobose, multinucleate with numerous discoid plastids and large central vacuole; whorls or groups of delicate, hyaline, trichotomous to polychotomous trichoblasts produced from the upper ends of some cells, particularly distally and in conjunction with reproductive structures; tetrahedral or cruciate sporangia whorled at fertile nodes, clustered on polychotomous fascicles, naked or covered by simple involucral branches; dioecious or occasionally monoecious, spermatangia whorled at nodes in polychotomous fascicles, naked or involucrate, or in caplike clusters on terminal cells of vegetative axes; procarps produced subapically on a three-celled fertile axis, single or paired, carpogonial branches four celled; involucral filaments two celled, arising from the non-enlarged hypogynous cell; carpospores developing from most gonimoblast cells and enveloped in a gelatinous matrix.

Griffithsia globulifera Harvey ex Kützing 1862, vol. XII, p. 10, fig. 30.

Figures 440 and 441

Plants saxicolous or epiphytic, erect, forming semiglobular tufts to 6.5 cm tall, bright rosy red, attached by numerous rhizoids issued from basal cells; branching subdichotomous, fastigiate, occasionally proliferous; branches appearing segmented to moniliform, the cells clavate near the base, 0.5–0.9 mm diameter, 0.6–3.2 mm long, pyriform to globose at the apices, to 1.4 mm diameter and 2.1 mm long, female and sporangial plants tapering abruptly to the apex, male plants with the largest cell of each branch located at the terminal end; tetrasporangia globose to obovoid, 40–110 µm diameter, whorled at the node, usually borne three per one-celled branch, enveloped by numerous one-celled, incurved, sausage-shaped involucral branches; dioecious, rarely monoecious, gametangial plants rarely bearing sporangia; spermatangia in caplike sori over the outer one-third to one-half of the terminal cell of a fertile branch, found in lesser quantities at the outer ends of the one or two cells below the apex, ellipsoid to obovoid, 2.5–3.2 µm diameter, 3.8 µm long; procarps produced mostly on uppermost nodes; carposporophytes with a central fusion cell and radiating gonimoblast filaments, the majority of cells becoming carposporangia that are ellipsoid, ovoid, subglobose, to irregular, 15–40 µm diameter in greatest dimension, and surrounded by large, incurved, sausage-shaped involucral cells.

Schneider and Searles 1975; Schneider 1976; Kapraun 1980a; Searles 1987, 1988. As *G. tenuis sensu* Blomquist and Pyron 1943.

Known offshore from 20 to 40 m in Onslow Bay, and occasionally in the drift

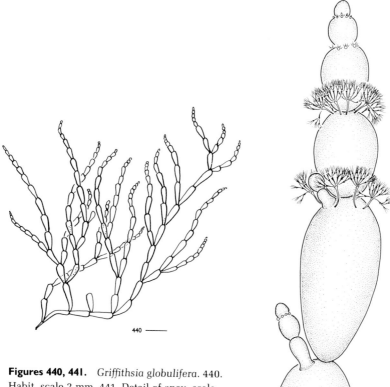

Figures 440, 441. *Griffithsia globulifera*. 440. Habit, scale 2 mm. 441. Detail of apex, scale 100 μm.

in the Beaufort and Wilmington, N.C., areas, June–November; from 17 to 21 m on Gray's Reef and 30 to 35 m on Snapper Banks, June–August.

Distribution: Southeastern Canada, Massachusetts to New York, North Carolina, Georgia, Bermuda, southern Florida, Caribbean, Brazil.

Lejolisia Bornet 1859

Plants filamentous, prostrate with erect axes, spreading to tufted, ecorticate, attached by unicellular rhizoids with terminal lobed discs; axes uniseriate, oppositely to alternately and unilaterally branched from few to several orders; cells uninucleate, containing several discoid plastids; tetrahedral or cruciate sporangia, single or grouped, sessile or pedicellate, terminal or lateral on the adaxial surfaces of branches; monoecious or dioecious, spermatangia in ovoid to cylindrical heads, sessile or pedicellate, terminal or lateral, often near the base, on erect axes; procarps subapical on erect or lateral axes, carpogonial branches four celled; carposporophytes becoming surrounded by distinct gelatinous pericarps; carposporangia developing from the outer gonimoblast cells.

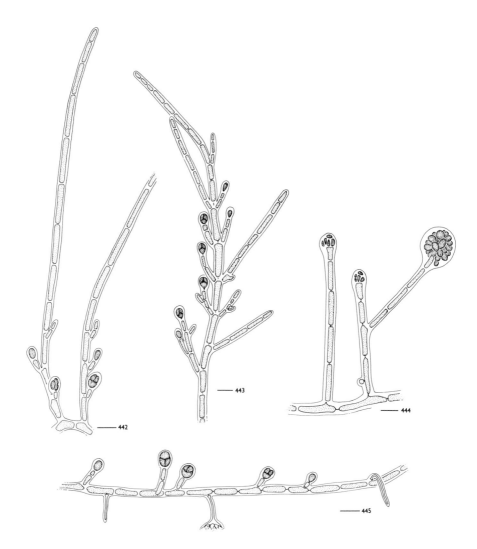

Figures 442–445. *Lejolisia exposita*, scales 50 μm. 442–444. Tetrasporangia axes. 445.
Procarps and carposporophyte (redrawn from Searles and Schneider 1989).

Lejolisia exposita C. W. Schneider et Searles in Searles et C. W. Schneider 1989,
p. 736, figs 18–28.

Figures 442–445

Plants epiphytic or epizoic, rosy red, spreading or tufted, erect to 1–3(–10) mm
tall, ecorticate, attached by one (sometimes two) unicellular digitate rhizoid per
cell issued from the prostrate axes; prostrate axes 18–40 μm diameter; erect
axes arising from few to most of the cells of the prostrate axes; erect axes
simple to sparingly alternately and, rarely, oppositely branched, lateral branches
usually simple, especially in vegetative specimens; cells of erect axes cylindri-

cal, not constricted at the nodes, 13–24 μm diameter, 70–150 μm long below, lengthening and slightly tapering or expanding toward the apices; tetrasporangia subglobose, ovoid, obovoid, to ellipsoid, tetrahedrally or cruciately divided, 30–42 μm diameter, 38–68 μm long, single or paired, terminal on short one- to four-celled branches, those branches often rebranched and sporangiate; tetrasporangial branches single, alternate to secund and irregular, or paired and opposite; monoecious or dioecious, gametangial plants occasionally bearing sporangia; spermatangia in oblong-ovoid to cylindrical heads, 16–36 μm diameter, 45–72 μm long, terminal on the ultimate segments of erect main or short lateral axes; carposporophytes enclosed in gelatinous pericarps, carposporangia pyriform, obovoid, to subglobose, 12.5–22.5 μm diameter, 32.5–42.5 μm long; cystocarps obovoid to subglobose, 86–150 μm diameter, 90–122 μm long, borne on short, simple or little-branched erect axes.

Searles and Schneider 1989. As *Lejolisia* sp., Searles 1987, 1988. As *Spermothamnion investiens* sensu Hoyt 1920; Williams 1948a, 1951; Taylor 1960; Kapraun 1980a.

Known as an epiphyte from the shallow subtidal near Beaufort and Wilmington, N.C.; offshore from 20 to 35 m in Onslow Bay, April–October; and from 17 to 21 m on Gray's Reef, March–July.

Distribution: North Carolina (type locality), Georgia.

Erect filaments of plants collected in the Gray's Reef Sanctuary are narrower (13–20 μm diameter) than those collected in North Carolina (18–23 μm) and have smaller spermatangial sori (16–32 by 45–56 μm vs. 31–36 by 49–72 μm), but these differences do not present a difficulty in recognizing both as a single entity (Searles and Schneider 1989).

Nwynea Searles in Searles et C. W. Schneider 1989

Plants erect, filamentous, ecorticate; axes uniseriate, bearing numerous branched determinate branches; determinate branches whorled, paired, or alternate and radial below, alternate and radial above on indeterminate branches; indeterminate branches issued directly from axial cells; tetrahedral sporangia borne directly on indeterminate axes; sexual reproduction unknown.

Nwynea grandispora Searles in Searles et C. W. Schneider 1989, p. 739, figs 29–33.

Figure 446

Plants delicate, main axes clothed in pigmented, flagellate, determinate axes; indeterminate axes fastigiate, corymbose above, bearing alternate, radial determinate branches, naked below or bearing determinate branches in whorls of two or three or alternate and radial; indeterminate axial cells to 110 μm diameter and

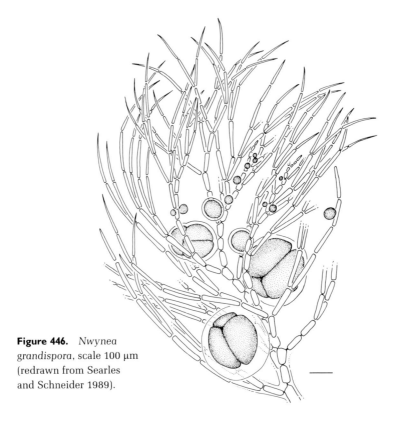

Figure 446. *Nwynea grandispora*, scale 100 μm (redrawn from Searles and Schneider 1989).

550 μm long below, shortening and narrowing to 8 μm below the apices; determinate axes alternately to pseudodichotomously branched, cells elongate cylindrical, 25 μm diameter proximally, not shortening, but narrowing to 8 μm diameter at the bases of the slender, sharply acute tip cells; tetrasporangia globose to 240 μm diameter (not including thick cell walls), sessile and single or paired directly on indeterminate axial cells in upper portions of plants.

Searles and Schneider 1989.

Known only offshore from 40 m Onslow Bay in the Frying Pan Shoals area, June.

Distribution: Endemic to North Carolina as currently known.

This species is based on a single distinctive specimen and is unique among the Ceramiaceae in the flora in producing tetrasporangia directly on axial cells (Searles and Schneider 1989).

Pleonosporium Nägeli 1862, nom. cons.
Plants erect, filamentous, bushy, ecorticate or corticated in basal portions, attached by numerous multicellular branched rhizoids; axes uniseriate, alternately distichous, tristichous, to polystichous, lateral axes similarly branched to several orders or pinnately to unilaterally branched to several orders; polyhedral and/or tetrahedral sporangia, sessile or pedicellate, adaxial and secund or alternately adaxial and abaxial on lateral axes; dioecious, spermatangia in elon-

gate heads, terminal on ultimate branches or sessile to pedicellate in positions similar to sporangia; procarps subapical on upper portions of indeterminate axes or lateral on upper portions of determinate axes, carpogonial branches four celled; carposporophyte consisting of four to eight successive gonimolobes surrounded by involucral filaments; carposporangia developing from most gonimoblast cells.

Norris (1985) reviewed the distinctions used to separate *Mesothamnion* (Børgesen 1917) from *Pleonosporium*, including branching patterns, sporangial types, and development of carpogonial branches. Finding overlap in the first two characters and no distinctions among female branches, he concluded that *Mesothamnion* was congeneric with *Pleonosporium*. We concur with that report; however, the expanded genus *Pleonosporium* now contains both distichous and radial taxa.

Key to the species

1. Plants with alternately radial branching; polysporangia and tetrasporangia present . . . . *P. boergesenii*
1. Plants with alternately distichous branching; polysporangia and occasionally tetrasporangia present . . . . *P. flexuosum*

Pleonosporium boergesenii (Joly) R. Norris 1985, p. 61.

Mesothamnion boergesenii Joly 1957, p. 142, pl. XVI, figs 1a–1c; pl. XVIII, figs 3a–3c.

Figures 447 and 448

Plants epiphytic, saxicolous, bushy, erect to 7 cm tall, brownish to rosy red, ecorticate, attached by multicellular, uniseriate branched rhizoids arising from bases and nodes of lower segments; branching alternate to pseudodichotomous below, alternate in a one-third spiral above; in lower segments the lateral branches markedly smaller than the main axis, above only slightly smaller; cells cylindrical, 210–325 µm diameter, 600–900 µm long in proximal parts of main axes; lateral axes gradually tapering from their origins, 70–120 µm diameter, to the ultimate segments, 10–20 µm diameter, appearing distichous and subdichotomous in distal portions; branchlets incurved, apices obtuse; plants with both tetrasporangia and polysporangia, sessile, single, subglobose, and short ellipsoid, 40–73 µm in greatest dimension (including a thick cell wall), borne distally and adaxially on branch cells; polysporangia divided into eight (or more?) spores, generally larger and less frequent than tetrasporangia; spermatangia in subcylindrical to ellipsoid heads, sessile or pedicellate, on the inner forkings of lateral axes; carposporophytes formed as five or six successively produced gonimolobes surrounded by incurved involucral filaments; carposporan-

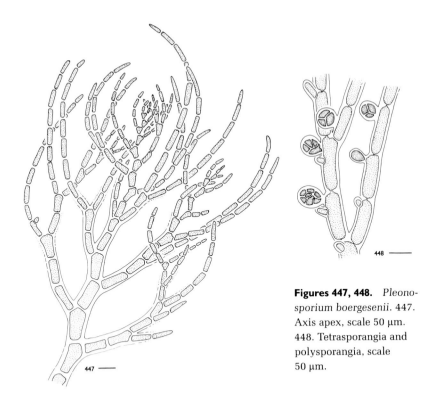

Figures 447, 448. *Pleonosporium boergesenii.* 447. Axis apex, scale 50 μm. 448. Tetrasporangia and polysporangia, scale 50 μm.

gia subglobose to polyhedral, 12–25 μm in greatest dimension at maturity.

Searles 1987, 1988. As *Mesothamnion boergesenii*, Schneider 1975b, 1976.

Known offshore from 21 to 30 m in Onslow Bay, May–June; from just below the low-water mark on Radio Island jetty, Beaufort, N.C., June–July; and from 17 to 21 m on Gray's Reef, August–September.

Distribution: North Carolina, Georgia, Brazil.

Pleonosporium flexuosum (C. Agardh) Bornet ex De Toni 1903, p. 1305.

Ceramium flexuosum C. Agardh 1824, p. 141.

Figures 449–451

Plants epiphytic, bushy, erect to 6 cm tall, brownish red, ecorticate or slightly corticated by rhizoids below, attached by multicellular, uniseriate branched rhizoids, 45–105 μm diameter, arising from the base and nodes of lower segments; branching pseudodichotomous below, alternately distichous above, in lower segments the lateral branches markedly smaller than the main axis, above only slightly smaller and slightly tapering to the ultimate segments; lateral axes short and rigid to longer and corymbose; cells clavate, less often cylindrical, 300–625 μm diameter, 930–1560 μm long below, tapering to 41–62 μm diameter in the ultimate segments; branchlets slightly incurved, apices obtuse; polysporangia sessile, single, ellipsoid to subglobose, 65–110 μm diameter, 85–130 μm long (including thick cell walls to 11 μm), divided into eight to thirty-two spores, borne distally and adaxially, or less commonly abaxially, on branch cells; spermatan-

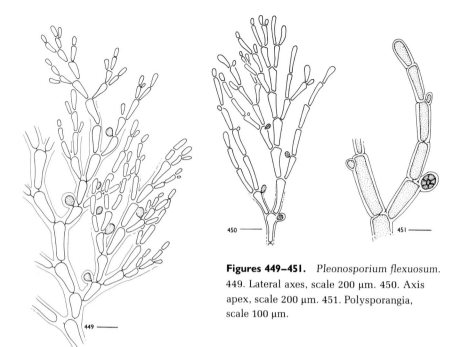

Figures 449–451. *Pleonosporium flexuosum.* 449. Lateral axes, scale 200 µm. 450. Axis apex, scale 200 µm. 451. Polysporangia, scale 100 µm.

gia in ellipsoid to ovoid heads, terminal on the ultimate segments, occasionally becoming laterally displaced by a subtending cell; carposporophytes globose, to 510 µm diameter, borne at branch origins in the upper parts of the plant, forming two gonimolobes and surrounded by incurved involucral filaments; carposporangia globose, subquadrate to greatly elongate, 30–55 µm in greatest dimension at maturity.

Schneider and Searles 1975; Schneider 1976.

Known offshore from 18 to 27 m in Onslow Bay, June–August.

Distribution: North Carolina, France to Portugal, Mediterranean, South Africa.

Species Excluded

Pleonosporium borreri (J. E. Smith) Nägeli

Reported for the Carolinas by Williams (1951); we have been unable to locate vouchers of this species. Because of its great similarity to *Pleonosporium boergesenii* (Schneider 1975b) and its clearly established position in the flora of cooler northerly waters (Taylor 1957; Parke and Dixon 1976), *P. borreri* is removed from the flora until it can be reestablished by new collections.

Ptilothamnion Thuret in Le Jolis 1863

Plants filamentous, prostrate with erect axes, spreading to tufted, ecorticate, attached by unicellular rhizoids with terminal lobed discs; axes uniseriate, simple to pinnately, unilaterally, or irregularly branched; cells multinucleate, containing several discoid plastids; tetrahedral sporangia and/or polysporangia, single or grouped, sessile or pedicellate, and terminal on lateral branches; monoecious or dioecious, spermatangia in ovoid to globose heads, sessile or pedicellate, and terminal or lateral on erect or lateral axes; procarps subapical on erect or lateral axes, carpogonial branches four celled; carposporophytes with one to several lobes, becoming surrounded by a loose involucre developed from basal cells of fertile segment; carposporangia developing from the outer gonimoblast cells.

Ptilothamnion occidentale Searles in Searles et C. W. Schneider 1989, p. 735, figs 11–17.

Figures 452 and 453

Plants epiphytic, rosy red, spreading or tufted, erect to 1 mm tall, ecorticate, attached by unicellular digitate rhizoids issued from the prostrate axes; prostrate axes (22–)25–41(–49) μm diameter; erect axes arising from several of the cells of the prostrate axes; erect axes simple to sparsely branched, sometimes becoming prostrate; cells of erect axes cylindrical, not constricted at the nodes, 18–29 μm diameter, 1–1.5 times as long as broad below, lengthening and slightly or not at all tapering toward the apices; sporangia cruciately or tetrahedrally divided, single or paired on short lateral branches, obovoid, 27–45 μm diameter, 34–54 μm long; spermatangia covering lateral branches, forming ovoid, sessile sori to 31 μm diameter and 45 μm long; procarps formed on subapical cells, involucres of one or two filaments, one to four cells long, forming on the cell below the fertile segment, carposporophytes unknown.

Searles and Schneider 1989. As *Ptilothamnion* sp., Searles 1987, 1988. As *Spermothamnion investiens* var. *cidaricola* sensu Schneider 1976; Wiseman and Schneider 1976.

Known offshore from 40 to 54 m in Long Bay, June; from 17 to 21 m on Gray's Reef, July; and from 30 to 35 m on Snapper Banks, August.

Distribution: South Carolina, Georgia (type locality).

Differences exist between Georgia and Carolina plants (Searles and Schneider 1989), and culture studies are necessary to delineate the boundaries of this species. Asexual and vegetative specimens of this species and *Lejolisia exposita* are impossible to distinguish.

Spermothamnion Areschoug 1847

Species Excluded

Spermothamnion repens (Dillwyn) Rosenvinge

Reported for the Carolinas as *Callithamnion turneri* Mertens (Curtis 1867) and

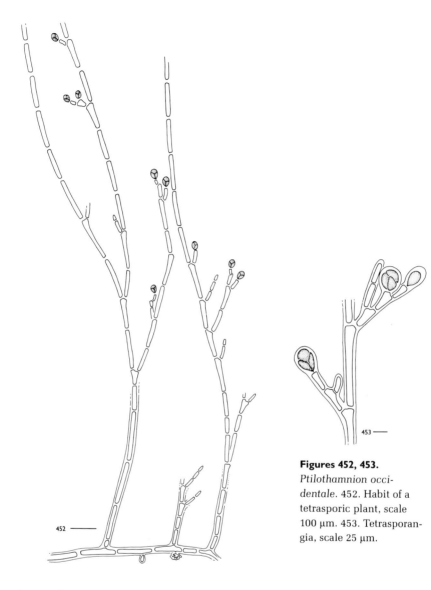

Figures 452, 453.
Ptilothamnion occidentale. 452. Habit of a tetrasporic plant, scale 100 μm. 453. Tetrasporangia, scale 25 μm.

Spermothamnion turneri (Mertens) Areschoug (Williams 1948a, 1951), no specimens have been located to verify this distinct species in the flora. A common member of the cold-water flora of southern Canada and New England, *S. repens* has been reported for Florida and Jamaica, but these subtropical records have been questioned (Taylor 1960). It is probable that the Carolina collections of Curtis and Williams in fact were *Lejolisia exposita*. Filament widths of these species overlap, but *S. repens* is generally much broader. *L. exposita* has occasional opposite branching, which is common in *S. repens*. Until we can locate a voucher specimen, *S. repens* is excluded from the flora.

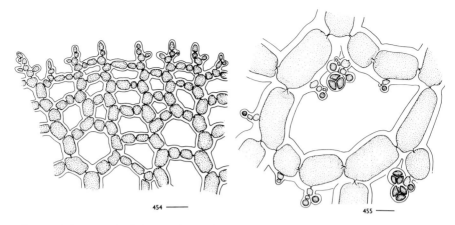

Figures 454, 455. *Rhododictyon bermudense.* 454. Habit of margin, scale 100 μm. 455. Detail of tetrasporic plant, scale 50 μm.

Rhododictyon W. R. Taylor 1961

Plants erect, consisting of planar single-layered nets formed by regularly anastomosing lateral branches of several radiating uniaxial filaments; apical cells distinct and numerous on the distal margin; tetrahedral sporangia borne terminally on specialized fertile branches; gametangia unknown.

Rhododictyon bermudense W. R. Taylor 1961, p. 278, figs 1–4.

Figures 454 and 455

Plants epiphytic, epizoic, saxicolous, to 2.5 cm tall, rosy red, attached by rhizoidal filaments issued from basal portions of the netlike blades, secondarily attached by rhizoidal filaments randomly issued from the mesh or even distal margins; nets irregularly lobed, generally as wide or wider than tall, each mesh typically of four sides, two cells per side, but occasionally stretched into polygons of five to ten sides; inside diameter of meshes 0.3–0.5 mm proximally, much smaller near the growing margin; cells 80–180 μm diameter, of varying lengths to 415 μm proximally, 40–50 μm diameter and of equal lengths, 40–50 μm, distally; tetrasporangia globose, to 50 μm diameter (including a thick wall), clustered, each terminal on specialized short free branches projecting into the mesh in the lower portions of the net.

Schneider 1975b, 1976.

Uncommon offshore from 45 to 54 m in Long Bay and Onslow Bay, June.

Distribution: North Carolina, South Carolina, Bermuda, Barbados, Curaçao.

Spyridia Harvey 1833

Plants erect, bushy, terete to flattened, filamentous, attached by basal rhizoidal discs; axes uniseriate, corticated by transverse series of longitudinally elongated pigmented cells, later subdivided and appearing banded to the unaided eye, cells in lower portions covered and intermixed with descending rhizoids; main axes generally percurrent, branched and rebranched at varying distances with greater and lesser indeterminate axes, and surrounded on all sides by simple, attenuate

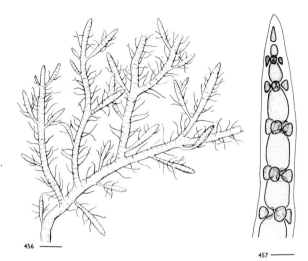

Figures 456, 457.
Spyridia clavata. 456.
Habit, scale 1 mm. 457.
Apex of determinate
branch, scale 20 μm.

456

457

determinate branches; determinate branches deciduous, uniseriate, corticated at the nodes by small pigmented cells with or without spines or hairs, ending in blunt or acute tips with or without recurved spines on the nodes below the tip; tetrahedral sporangia single and seriate or whorled, on the nodes of determinate branches; dioecious, spermatangia formed in hyaline circumscribing, separate, or contiguous patches on nodes of determinate branches; procarps borne laterally or terminally on short indeterminate axes, the axes ceasing to elongate after fertilization, carpogonial branches four celled; carposporophytes consisting of one to three gonimolobes surrounded by slender involucral filaments; carposporangia developing from most gonimoblast cells.

Key to the species

1. Branching distichous, lesser indeterminate branches constricted at base, at least some club shaped .... *S. clavata*
1. Branching radial, lesser indeterminate branches not constricted at the base, tapering to the apices, greater ones with or without occasional hamate tips .... *S. hypnoides*

Spyridia clavata Kützing 1841, p. 744.
   Figures 456 and 457
   Plants saxicolous, pinkish red, erect to 20 cm tall, generally 5–10 cm, terete and approximately 1 mm broad below, flattened above to 1–2 mm; indeterminate and most determinate branches distichous, usually irregular or alternate to secund, occasionally opposite; greater and lesser indeterminate branches intermixed, the lesser ones tapering proximally, expanding distally and appearing club shaped, 1–4 mm long, apices acute to obtuse; cortication continuous to the apices with shorter oval cells over the nodes of axial cells, and elongate cylindrical cells later interspersed among the shorter cells and tiered over the internodes, obscured below; axial cells to 500 μm long; determinate branches deciduous in lower portions of main axes and on clavate branch tips, simple and

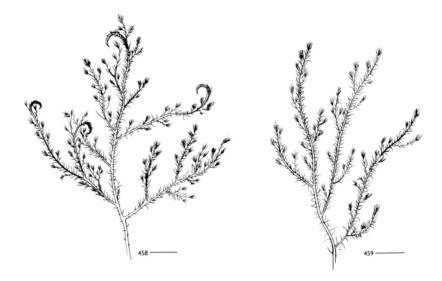

Figures 458–462. *Spyridia hypnoides.* 458, 459. Habits of plants with and without hamate tips, scales 0.5 cm. 460. Detail of indeterminate axis and origin of determinate axis, scale 50 μm. 461, 462. Apex of determinate branches with and without recurved spines, scales 25 μm.

corticated only at the nodes, incurved to straight, 0.6–1.75 mm long, 30–55 μm diameter proximally, tapering to 17.5–25 μm diameter a few segments below the tips; apices with acuminate terminal spines, without recurved spines; sporangia globose to 50 μm diameter, borne adaxially on lower nodes of determinate branches.

Hoyt 1920; Schneider 1976.

Uncommon offshore from 25 to 35 m in Onslow Bay, June–September.

Distribution: North Carolina, southern Florida, Caribbean, Brazil, West Africa, Red Sea.

Spyridia hypnoides (Bory) Papenfuss 1968, p. 281.

Thamnophora hypnoides Bory 1834, p. 175.

Figures 458–462

Plants epiphytic or saxicolous, erect to entangled in other plants, purplish pink, pinkish, or dull red to straw colored, erect to 25 cm tall, generally 5–10 cm, terete to somewhat compressed, 1–2 mm broad; indeterminate and determinate branches alternately radial, occasionally distichous, greater and lesser indeterminate branches intermixed, each gradually tapering from base to acute apex; greater branches occasionally ending in hamate tips, but some plants free of such tips; lesser indeterminate branches adventitious, generally 0.3–0.5 mm broad; cortication continuous to the apices, with alternating transverse bands of unequal numbers of shorter ovoid to rectangular cells over the nodes of axial cells, and attenuate, cylindrical cells over the internodes, obscured below; axial cells to 500 μm long; determinate branches deciduous in lower portions and hamate

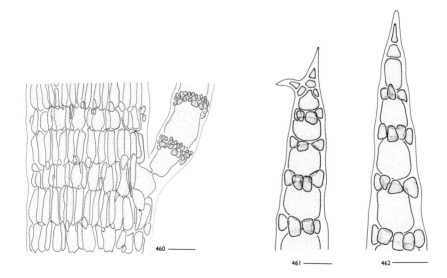

460 ———   461 ———   462 ———

tips, tufted on straight apices, simple and corticated only at the nodes, incurved, 0.3–2 mm long, 30–50 μm diameter proximally, tapering to 20–30 μm diameter a few segments below the tips; apices with terminal acuminate spines, with or without one or more lateral recurved spines on the first and second nodes; sporangia globose or subglobose, 50–85 μm diameter, borne singly or in clusters of two or three on the lower nodes of determinate branches; carposporophytes with two or three lobes, terminal on short, adventitious indeterminate branches surrounded by an incurved involucre.

Schneider 1976; Wiseman and Schneider 1976; Kapraun 1980a; Searles 1987, 1988. As *S. aculeata*, Blomquist and Pyron 1943; Williams 1948a, 1949, 1951; Taylor 1960; var. *disticha*, Blomquist and Humm 1946; var. *hypnoides*, Blomquist and Humm 1946. As *S. filamentosa sensu* Curtis 1867; Hoyt 1920; Blomquist and Pyron 1943; Blomquist and Humm 1946; Schneider 1976; Wiseman and Schneider 1976.

Known from marshes and jetties and offshore from 15 to 60 m in Long Bay and Onslow Bay, May–December; and rarely from 17 to 21 m on Gray's Reef, July–August.

Distribution: North Carolina, South Carolina, Georgia, Bermuda, southern Florida, Gulf of Mexico, Caribbean, Brazil, West Africa, Mediterranean, Red Sea, South Africa, Indian Ocean, Central Pacific islands.

Historically, species in *Spyridia* have been separated on the basis of cortical development, disposition of branches on the main axes, and presence or absence of recurved spines at the tips of determinate axes (Hommersand 1963). Our collections have corroborated the findings of Wiseman (1966), who showed that these characters were highly variable on specimens from the area previously assigned to *S. aculeata* (Schimper) Kützing (see Papenfuss 1968) and *S. filamentosa* (Wulfen) Kützing. Plants have been observed in which some of the "critical" characters vary from branch to branch, and a continuum can be found to the extremes of each character when specimens are compared with one another. Because we cannot distinguish separate entities in the flora, all plants are listed as *S. hypnoides*, despite the fact that not all plants contain hamate tips or recurved spines. Subspecific taxa are likewise difficult to separate. Further work, including examination of type material and culture studies, is necessary to determine if there are reproductive or morphological characters which can be used to segregate these widespread species elsewhere.

Delesseriaceae

Plants membranous, simple or alternately to dichotomously branched, or branches arising from the midrib; blades linear to broadly ovate, often with midribs and/or veins, arising from discoid holdfasts, often secondarily attached by rhizoids; growth from an axial row or rows and subsequent distichous lateral cell rows of varying degrees of branching, in some the blade completed by a few to many intercalary cell divisions, the apical cell distinct, however, only in the earliest stages of development in some; tetrasporangia clustered in laminal or marginal sori, occasionally in special leaflets or proliferations, tetrahedrally divided; spermatangia arranged in laminal or marginal sori; procarps borne either on the midrib or on nonaxial cells, on supporting cells which produce a three- or four-celled carpogonial branch and one or two groups of one- to five-celled sterile branches; the auxiliary cell formed after fertilization fuses with the carpogonium and contiguous cells, usually forming a large fusion cell from which monopodial gonimoblasts are produced terminal to all gonimoblast cells producing carposporangia; carposporophytes surrounded by thin to several-layered pericarps with obvious ostioles.

Key to the genera

1. Plants with midribs 2
1. Plants lacking midribs, with or without veins 7
2. Plants usually large, simple, broad foliaceous blades with an obvious midrib in at least the lower half . . . . Grinnellia
2. Plants variously branched, delicate, linear-lanceolate to ligulate blades with a midrib throughout 3
3. Plants producing branches from midrib 4
3. Branching otherwise 5
4. Lateral veins lacking, apical cells of each order of lateral cell rows reaching the blade margin . . . . Hypoglossum
4. Lateral veins present, apical cells of each order of lateral cell rows not all reaching the blade margin . . . . Apoglossum

5. Plants with branches arising in a subapical position, ultimately appearing dichotomous, growing in brackish water . . . . Caloglossa
5. Plants with marginal branching, not growing in brackish water 6
6. Midribs conspicuous, apical cells of each order of lateral cell rows reaching the blade margin . . . . Branchioglossum
6. Midribs inconspicuous, apical cells of each order of lateral cell rows not all reaching the blade margin . . . . Searlesia
7. Microscopic veins present 8
7. Microscopic veins absent 9
8. Plants large, to 25 cm tall, the veins in the lower portions macroscopic, obvious to the unaided eye . . . . Calonitophyllum
8. Plants smaller, to 5 cm tall, no macroscopic veins present . . . . . Acrosorium
9. Apical cells dividing obliquely, becoming indistinguishable from marginal row initials, blades becoming two layered . . . . Myriogramme
9. Apical cells dividing transversely, remaining obvious at maturity, blades remaining single layered 10
10. Plants erect to 7 cm tall, lower margins with numerous proliferations; marginal rhizoids uncommon or lacking; cystocarps and tetrasporangial sori within or at the base of marginal dentations or proliferations . . . Calloseris
10. Plants spreading or entangled, to 3 cm long, lower margins with or without occasional proliferations; marginal rhizoids common; cystocarps scattered and tetrasporangial sori in median positions on the blades . . . . Haraldia

Acrosorium Zanardini ex Kützing 1869
Plants consisting of entangled epiphytic or erect and prostrate marginally branched membranous blades; branching subdichotomous to alternate and irregular; blades single layered with simple microscopic veins, macroscopic veins lacking; apices without a distinct apical cell, growth by marginal initials and numerous intercalary divisions; tetrahedral sporangia in a single large, round sorus at the tip of each branch; procarps, cystocarps distributed over the entire plant surface, carposporophyte with terminal carposporangia; spermatangia formed in oval sori near the tips of the blades.

Acrosorium venulosum (Zanardini) Kylin 1924, p. 77, fig. 60.
   Nitophyllum venulosum Zanardini 1866, p. 144, pl. 49A.
   Figures 463–465
   Plants epiphytic or saxicolous, consisting of erect or entangled blades 1–5(–8) cm tall, deep rosy red, arising from rhizoidal holdfasts, secondarily attached by marginal rhizoids; blades membranous 1–4(–20) mm wide, highly branched,

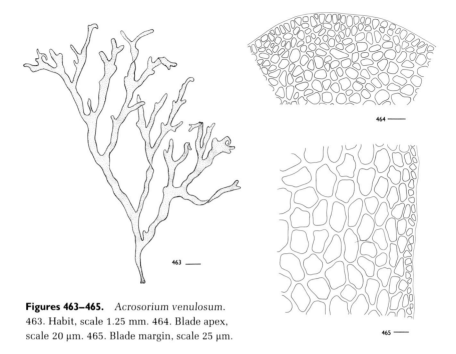

Figures 463–465. *Acrosorium venulosum.*
463. Habit, scale 1.25 mm. 464. Blade apex,
scale 20 μm. 465. Blade margin, scale 25 μm.

occasionally ending in attaching hooks; margins entire to irregularly dentate, the teeth intergrading in form with smaller branchlets; microscopic veins irregularly longitudinal, forking, occasionally anastomosing.

As *A. uncinatum sensu* Schneider 1975b, 1976; Wiseman and Schneider 1976.

Known from 40 to 54 m offshore in Long Bay and Onslow Bay, June.

Distribution: North Carolina, South Carolina, Venezuela, Brazil; elsewhere, widespread in warm temperate to tropical seas.

For a clarification of the nomenclature, refer to Wynne (1989). Only vegetative plants have been collected in the flora.

Apoglossum J. Agardh 1898

Plants consisting of single-layered, membranous, highly branched blades arising from small discoid holdfasts; branches arising from an endogenous budding of central axial cells of the midrib, new blades smaller than the main blade, often rebranched and smaller again; blades with conspicuous midribs, the axial cells obscured by small-celled, corticating, descending rhizoids; lateral microscopic nerves obliquely arising from the midrib; apical cells dividing transversely, not all of the tertiary apical initials reaching the plant margin; intercalary divisions present only in the lateral cell rows of the second and higher orders; tetrahedral sporangia in elongate sori along both sides of the midrib, elevated on both blade surfaces; dioecious, procarps forming on the midribs in apical portions, generally only one per blade forming a cystocarp; spermatangia formed in elongate

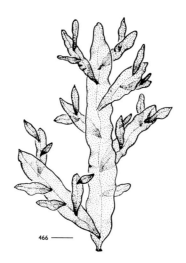

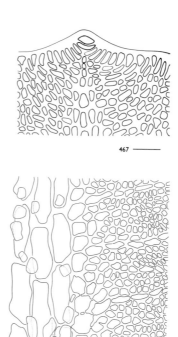

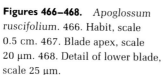

Figures 466–468. *Apoglossum ruscifolium*. 466. Habit, scale 0.5 cm. 467. Blade apex, scale 20 μm. 468. Detail of lower blade, scale 25 μm.

parallel sori midway between the midrib and the margin, raised on both blade surfaces.

Apoglossum ruscifolium (Turner) J. Agardh 1898, p. 190.
  Fucus ruscifolius Turner 1802, p. 127.
  Figures 466–468
  Plants 2–10 cm tall, rosy red; main blades 4–6 mm wide, highly branched from the midrib; blades linear oblong, obtuse, to apiculate, undivided, the margins entire, often undulate; microscopic veins most obvious in older blades; tetrasporangia globose to ovoid, 30–55 μm in longest dimension, in paired elongated sori confluent with the midrib, to 1 mm long; spermatangial sori obliquely set from the midrib, between the midrib and margin, separate but parallel with others over the entire blade; cystocarps forming on the midrib, usually one per blade but more than one on the major axes, situated near the proximal end of the ultimate blades, hemispherical, 500–800 μm diameter, carpospores globose to ovoid, 22.5–30 μm in greatest dimension.
  Schneider 1976; Wiseman and Schneider 1976.
  Infrequent, southeast and southwest of Cape Fear from 30 to 32 m in Onslow Bay, and 40 m in Long Bay, June–August.
  Distribution: North Carolina (new record), South Carolina, Bermuda, southern Florida, Norway to Portugal, Mediterranean, West Africa, South Africa, Indian Ocean.

Only tetrasporangial plants have been collected in the flora. Dimensions of carposporophyte characteristics were taken from specimens collected in Florida south of Cape Canaveral by the Harbor Branch Foundation and by the authors in Bermuda.

Branchioglossum Kylin 1924

Plants consisting of erect and prostrate, simple or marginally branched membranous blades; branching mostly irregular, occasionally appearing opposite, alternate, or secund; blades single layered with prominent multilayered midribs, with or without veins; apical cells dividing transversely; apical initials of all orders of lateral cell rows reaching blade margins, becoming compressed and crowded, intercalary divisions within axial and lateral cell rows not occurring; tetrahedral sporangia in elongate sori, marginal, median, or contiguous with the midrib, elevated on both blade surfaces; dioecious, procarps forming sequentially on the midrib below the apex, only a few per blade developing into cystocarps, cystocarps projecting on either plant surface; spermatangia formed in irregular, transverse, elongate sori, confluent or interrupted, midway between the midrib and the margin of the blade.

Key to the species

1. Erect plants consisting of much to sparsely branched oblanceolate blades with undulate margins and apiculate or obtuse apices; spermatangia in broad confluent sori over the top half of the blade of male plants . . . . *B. minutum*
1. Prostrate plants consisting of much-branched, spreading, linear-lanceolate blades with smooth margins and acute apices; spermatangia in irregularly spaced elongate sori over the entire length of the blade of male plants . . . . *B. prostratum*

Branchioglossum minutum C. W. Schneider in C. W. Schneider et Searles 1975, p. 92, figs 13, 15–16.

Figures 469 and 470

Plants erect, pinkish red, consisting of clustered, oblanceolate blades to 2.2 cm tall, usually less than 1 cm, attached by common discoid or rhizoidal holdfasts; branching sparse to profuse, irregular, alternate to opposite, of one order, occasionally two or three orders, mostly marginal with a rare branch from the midrib, especially in wounded areas; margins undulate, apices apiculate, occasionally obtuse; main branches 1–3 mm broad, corticated at the base, midrib conspicuous, axial cells from 18–40 μm diameter and 80–120 μm long with a row of broad, shorter lateral and transverse pericentral cells surrounding it;

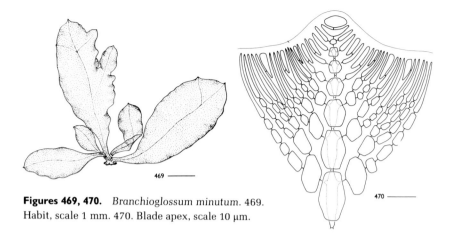

Figures 469, 470. *Branchioglossum minutum.* 469. Habit, scale 1 mm. 470. Blade apex, scale 10 μm.

tetrasporangia 20–40 μm diameter, occasionally germinating in situ, in semi-circular to circular, opposite sori, contiguous with the midrib in the apical portions of the blades; dioecious, spermatangia in irregular confluent sori, midway between the midrib and the margin, covering this region for the entire upper half of the blade; cystocarps forming on the midrib near the apical end of the blade, projecting, hemispherical, one or two per blade, 250–650 μm diameter, carpospores globose to ovoid, 25–32 μm in diameter, occasionally germinating in situ.

Schneider and Searles 1975; Schneider 1976; Kapraun 1980a; Searles 1981, 1987.

Known from rocks near the lowtide mark on the Masonboro Inlet jetty and more commonly on other algae, hydrozoans, bryozoans, and rocks from 17 to 37 m on offshore shipwrecks and outcroppings of Onslow Bay, Gray's Reef, and Snapper Banks, May–October. Also restricted to deep water in Florida, as yet known only from south of Cape Canaveral (Eiseman 1979).

Distribution: North Carolina (type locality), Georgia, southern Florida.

Searles (1981) reported occasional branching from the midrib, an important generic consideration in the Delesseriaceae (see *Hypoglossum*). The implications warrant further attention.

Branchioglossum prostratum C. W. Schneider 1974, p. 1094, figs 1A, 2.
Figures 471 and 472

Plants pinkish red, prostrate, to 2 cm long, consisting of spreading linear-lanceolate blades, occasionally arising from short-stalked rhizoidal holdfasts, but attached more commonly by numerous secondary attachments of multicellular uniseriate rhizoids which arise randomly from the margins; branches alternate, rarely opposite, to secund, to two or three orders, all marginal; the margins smooth, apices acute; main part of the plant 0.8–1.6 mm diameter, the branches to 0.7 mm; midrib inconspicuous, axial cells to 40 μm diameter, 185 μm long, with a row of shorter lateral cells on each side and a row of shorter transverse cells above and below when viewing the flat surface, corticated only at the base; tetrasporangia 30–40 μm diameter, in elongated opposite sori; sori contiguous

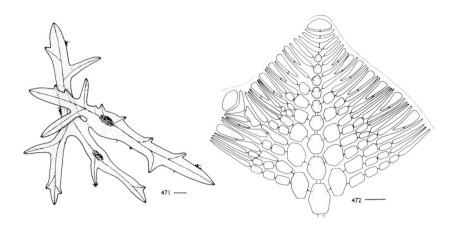

Figures 471, 472. *Branchioglossum prostratum.* 471. Habit, scale 1 mm (from Schneider 1974). 472. Blade apex, scale 10 μm.

with the midrib, causing a broadening of the plant; dioecious, spermatangia in irregularly spaced elongate clusters midway between the midrib and the margin over the entire length of the blade, female plants with procarps formed seriately from the apex along the midrib; cystocarps unknown.

Schneider 1974, 1976; Wiseman and Schneider 1976.

Known only from 40 to 60 m offshore on sponges and larger seaweeds, June–July.

Distribution: North Carolina (type locality), South Carolina, southern Florida, Puerto Rico.

This taxon has been collected at 48 m south of Cape Canaveral during March (Eiseman 1979), and from 61 m off Puerto Rico during January (Ballantine and Wynne 1987). Earlier reports (Schneider 1974; Schneider and Searles 1975) that both area *Branchioglossum* species were monoecious have proven to be false upon reexamination of all fertile material.

Calloseris J. Agardh 1898
Plants consisting of entangled, erect, marginally branched, membranous blades arising from small rhizoidal holdfasts; branching subdichotomous to trichotomous and irregular; blades mostly single layered above, few layered to tubular near the extreme base, midribs and veins lacking; apical cells dividing transversely, the apical row quickly obscured by periclinal divisions; intercalary divisions responsible for mature blade growth; margins serrulate near the apices, grading below to margins with mixed dentations and laciniate proliferations by limited growth of marginal initials; tetrahedral sporangia in corticated sori, elevated on both blade surfaces, in or at the base of the laciniate marginal proliferations, obvious to the unaided eye; cystocarps formed in positions similar to tetrasporangial sori, projecting on one blade surface; spermatangia unknown.

Calloseris halliae J. Agardh 1898, p. 213.
Figures 473 and 474
Plants to 7 cm tall, pinkish to rosy red; blades ligulate, 0.5 cm wide proximally

to 0.1 cm wide distally, 40–50 μm thick; in surface view, blade cells rounded iso-diametric, 50–120 μm in greatest dimension; blades mostly single layered, but with occasional scattered clusters of two to four cells forming dorsal or ventral pericentral cells; blades beset with scattered patches of darker pigmented cells, obvious to the unaided eye; apices commonly angular to acute, with several transversely dividing apical cell rows quickly obscured by periclinal cell divisions; tetrasporangia globose to subglobose, 45–110 μm diameter in oblong to subcircular and somewhat irregular corticated sori to 200 μm thick, 1.3 mm diameter, 2.0 mm long, transverse to the axis of the laciniate marginal prolifera-tions and parallel to the margins of the ligulate blades, those along the margin at the base of the proliferations discrete or confluent, forming sori to 3.0 mm long; cystocarps forming at the base of marginal proliferations, projecting, hemi-spherical, to 1.3 mm diameter, carpospores pyriform, 26–32 μm diameter, 56–70 μm long.

Schneider 1984.

Collected from 30 m offshore in the Frying Pan Shoals region off Cape Fear, July.

Distribution: North Carolina, southern Florida, Tobago.

A recent report of previously misidentified plants from Tobago allowed for the first description of cystocarps in this rarely collected species (Wynne 1990).

Caloglossa J. Agardh 1876
Plants consisting of regularly branched, spreading prostrate membranous blades;

Figures 473, 474. *Calloseris halliae*. 473. Habit of tetrasporic blade, scale 2 mm. 474. Blade apex, scale 25 μm (redrawn from Schneider 1984).

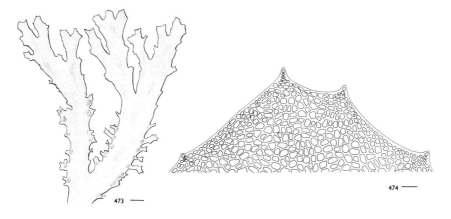

473 —— 474 ——

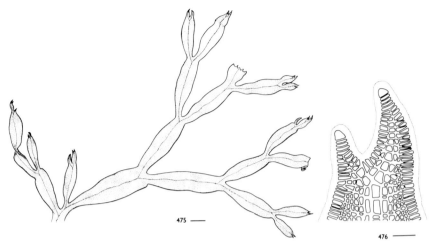

Figures 475, 476. *Caloglossa leprieurii.* 475. Habit, scale 0.5 mm.
476. Blade apex, scale 20 μm.

blades subdichotomously branched, constricted at the nodes and attached by rhizoids issuing from the ventral midrib in these regions; secondary branching proliferous from the midribs; single layered blades interrupted by a prominent midrib of elongated cells, lateral veins lacking; broad apical cells dividing transversely; apical cells of all three orders of branching reaching the plant margin, intercalary divisions lacking; tetrahedral sporangia developed in oblique series on the adaxial surface of lateral secondary branches, the series confluent in the upper portions of the blade; dioecious or monoecious, procarps and cystocarps forming on the midribs, sessile, with thin pericarps; spermatangial sori confluent with the midrib in terminal and subterminal segments, raised on both the plant surfaces.

Caloglossa leprieurii (Montagne) J. Agardh 1876, p. 499.
   Delesseria leprieurii Montagne 1840, p. 196, pl. 5, fig. 1.
   Figures 475 and 476
   Plants spreading up to 5 cm across to somewhat upright, purplish red to brown, the blades linear-lanceolate to linear-attenuate or rarely ovate, generally 1(–2) mm wide, each segment 1–2(–6) mm long, constricted at the nodes; apices acute to somewhat forcipate; tetrasporangia globose, 30–40 μm diameter.
   forma *pygmaea* (Martius) Post 1936, p. 49.
   Plants with narrow blades, 0.6 mm wide or less.
   Taylor 1960; Chapman 1971; Wiseman and Schneider 1976; Kapraun 1980a; Richardson 1987. As *Delesseria leprieurii*, Bailey 1851; Harvey 1853; Melvill 1875.
   On mud, rock, wood, pilings, *Spartina* and other marsh plants in the intertidal zone of brackish water estuarine environments, commonly associated with *Bostrychia radicans* (Montagne) Montagne, from throughout the area during the entire year.

Figure 477. *Calonitophyllum medium*, scale 1 cm.

Distribution: Connecticut to Virginia, North Carolina, South Carolina, Georgia, Bermuda, southern Florida, Gulf of Mexico, Caribbean, Brazil, West Africa.

Both forma *leprieurii* and forma *pygmaea* have been found in North Carolina, South Carolina, and Georgia. Undoubtedly, they are also present in the estuaries throughout Florida, although they are unreported north of Cape Canaveral. Only tetrasporic individuals have been collected in the flora. For a comprehensive study of the structure and reproduction of this species, refer to Papenfuss (1961).

Calonitophyllum Aregood 1975
Plants consisting of branched membranous erect blades arising from a distinct, branched holdfast; blades single layered above, multilayered near the base and in the stipe, traversed by numerous anastomosing microscopic and multilayered macroscopic veins, midribs absent; apical cell dividing transversely, intercalary divisions in the axial rows frequent, margins entire, undulate, and proliferous, dentate to serrulate; tetrahedral sporangia in corticated sori over the entire plant, elevated on both blade surfaces; dioecious, procarps scattered over nonaxial portions of the blade; cystocarps mammillate, ostiolate, projecting from either blade surface; spermatangial sori raised on both plant surfaces.

Calonitophyllum medium (Hoyt) Aregood 1975, p. 348, figs 1–23, 24b.
 Nitophyllum medium Hoyt 1920, p. 494, fig. 35, pls CV, CXIV figs 4–5.
 Figure 477
 Plants to 25 cm tall, bushy, rosy red, arising from distinct holdfasts; holdfasts prominently branched, lightly calcified, perennial, creamy pink in color; blades membranous, 0.4–2.0 cm wide, subdichotomously or alternately branched from the margins, apices acute to obtuse, margins occasionally thickened; tetrasporangia globose to subglobose, 48–61 μm diameter, in scattered, ovate to ellipsoid sori, occasionally forming parallel or radiating lines; spermatangia two to

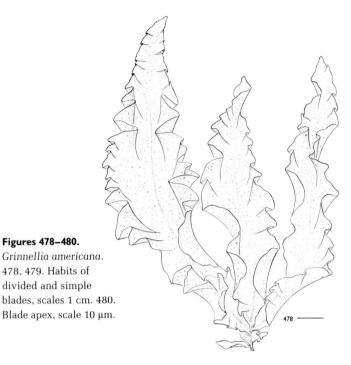

Figures 478–480.
Grinnellia americana.
478, 479. Habits of
divided and simple
blades, scales 1 cm. 480.
Blade apex, scale 10 μm.

478 ———

three per mother cell, ovoid to globose in small, irregular, often confluent sori; carposporangia irregular, terminal in chains of one to three, 15–19.5 μm diameter, cystocarps asymmetrical, pericarps eight to ten layered; gametophytes somewhat dimorphic, male plants lacking macroscopic veins, being smaller, more regularly dichotomous and less proliferous than females and sporophytes.

Aregood 1975; Wiseman and Schneider 1976; Kapraun 1980a. As *Nitophyllum medium*, Hoyt 1920; Blomquist and Pyron 1943; Williams 1951. As *Hymenena media*, Taylor 1960; Schneider 1976.

Known from the shallow subtidal at Beaufort and Masonboro inlets and more commonly from 20–60 m on the offshore rock outcroppings and shipwrecks in Onslow Bay, year round.

Distribution: North Carolina (type locality), South Carolina, Puerto Rico, Venezuela.

If not quickly preserved and removed from sunlight, *Calonitophyllum* turns pinkish orange and becomes translucent. Summer inshore plants are often greenish in basal portions due to massive infestations of *Entocladia viridis* Reinke.

A detailed morphological study of this species can be found in Aregood (1975).

Grinnellia Harvey 1853

Plants producing large, usually simple membranous blades, sometimes irregularly branched, but more often proliferous near the base; blades single layered with prominent multilayered midribs, inconspicuous or lacking above, veins lacking; apical cell dividing transversely, evident in young plants, disappearing in older plants, intercalary divisions responsible for mature blade growth, margins entire, undulate to crenate; tetrahedral sporangia in corticated sori over

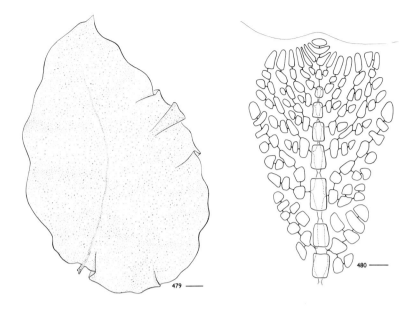

479 ——

480 ——

nonapical portions of the plant, elevated on either plant surface; dioecious, procarps appearing scattered but actually arising from fertile axial cells in specialized islets of cells across the single-layered blade, one per islet; cystocarps produced directly on the blade or upon a proliferated fertile leaf from the islet, thus appearing stalked, hemispherical, prominently ostiolate, projecting on either blade surface, carposporangia seriate and terminal; scattered spermatangial sori raised on one or both plant surfaces.

Grinnellia americana (C. Agardh) Harvey 1853, p. 92, pl. 21B.
    Delesseria americana C. Agardh 1822, p. 173.
    Figures 478 and 480
    Plants consisting of mostly simple blades, 1–50(–110) cm tall, 1–17 cm wide, arising from small discoid holdfasts and short stipes, often gregarious; blades narrowly to broadly lanceolate or ovate-oblong to broadly ovate and cordate, translucent pink to rosy pink, rarely purplish; tetrasporangia globose, 50–80 μm diameter in scattered elongate, irregular sori, to 1 by 2 mm; spermatangia ovoid to subglobose, in small scattered, elongate irregular, separate or confluent sori to 0.5 mm long; cystocarps to 1.5 mm diameter, with terminal chains of immature to mature carposporangia, globose to irregular, 5–15 μm diameter, pericarps thin; gametophytes somewhat dimorphic, male plants small, generally 1–3 cm long.
    Johnson 1900; Hoyt 1920; Williams 1948a, 1949, 1951; Taylor 1960; Chapman 1971; Brauner 1975; Schneider 1976; Wiseman and Schneider 1976; Kapraun 1980a; Searles 1981, 1987. As var. *caribaea*, Williams 1948a; Taylor 1960; Schneider 1976.
    Collected subtidally on Carolina and Georgia jetties, seawalls, and pilings

year-round, reaching maximum development during the fall-spring; known from 15–50 m offshore in Raleigh Bay and Onslow Bay, April-November, from 17–21 m on Gray's Reef, and from 30–35 m on Snapper Banks, March-August.

Distribution: Massachusetts to Virginia, North Carolina, South Carolina, Georgia, southern Florida, Caribbean.

We have included the var. *caribaea* Taylor (1942, p. 127) within the description of the species proper. Examination of North Carolina material alone shows a continuum of variation from the var. *americana* to var. *caribaea*. The varieties are therefore separable only in extreme forms, and should not be maintained.

Summer inshore plants are often greenish in basal portions due to massive infestations of *Entocladia viridis* Reinke.

Haraldia Feldmann 1939

Plants consisting of prostrate to erect, marginally lobed to branched, membranous blades arising from small rhizoidal holdfasts, some briefly stipitate, secondarily attached by rhizoids issued from blade margins or lower surfaces; blades irregularly lobed, producing lanceolate, irregular ovate to clavate ultimate segments; blades single layered, stipes tubular when present, midribs and veins lacking; apical cells dividing transversely, the axial cell row quickly obscured by periclinal divisions; intercalary divisions responsible for mature blade growth; margins serrulate near the apices, grading below to margins with mixed dentations to longer hairlike extensions, often undulate, proliferations by limited growth of marginal initials; tetrahedral sporangia in corticated sori, elevated on both blade surfaces, formed centrally in the laminae and lobes; dioecious, spermatangia formed in scattered to central, small, irregular to subcircular sori in laminae; procarps scattered over the laminae, cystocarps elevating the blade surface.

Haraldia lenormandii (Derbès et Solier) J. Feldmann 1939, p. 5, pl. 1, figs 1–3.
    Aglaophyllum lenormandii Derbès et Solier in Castagne 1851, p. 107.
    Figures 481 and 482

Plants prostrate and entangled, in part becoming erect, to 3.5 cm tall or long, rosy red; blades lobate to laciniate, 1 mm wide proximally, expanding distally to 3 mm wide, lobes overlapping, secondarily attached by rhizoids issued from blade margins, tips, or lower surfaces; lobes lanceolate, irregularly ovate to clavate, margins dentate, laciniate, in part undulate, appearing irregularly pinnate to flabellate; in surface view, blade cells rounded isodiametric to polygonal, 25–68 μm in greatest dimension; blades single layered, apices commonly angular to acute, with transversely dividing apical cells, axial cell rows quickly obscured by periclinal cell divisions; tetrasporangia globose to short ellipsoid,

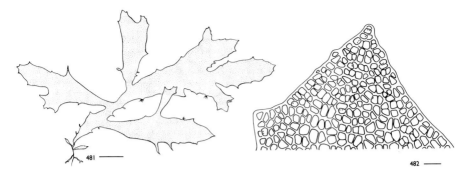

Figures 481, 482. *Haraldia lenormandii.* 481. Habit, scale 1 mm. 482. Blade apex, scale 25 μm.

30–45 μm diameter in subcircular to irregular corticated sori to 0.35 mm diameter and 0.6 mm long, mostly in the middle of blades on dorsal surfaces or restricted to ultimate lobes; dioecious, spermatangia globose, 1 μm diameter, in small subcircular to irregular sori, in similar position to tetrasporangial sori; procarps scattered on the dorsal surfaces.

As *H. prostrata sensu* Searles 1987, 1988.

Collected from 30–40 m offshore in Onslow Bay and the Snapper Banks region, July-August.

Distribution: North Carolina (new record), Georgia, Bermuda, Azores, France, Spain, Mediterranean.

A study of material from the flora (Wynne 1990) has shown these plants best placed under the above name rather than *Haraldia prostrata* Dawson, Neushul, et Wildman. Cystocarpic plants are unknown in the flora.

Hypoglossum Kützing 1843

Plants consisting of erect and/or prostrate, membranous, branched blades, occasionally stipitate, arising from small discoid holdfasts, secondarily attached by marginal rhizoids; branches arising from an endogenous budding of central axial cells of the midrib, single or paired; blades single layered with prominent three-layered midribs above, corticated by rhizoids below and lacking lateral veins; apical cells dividing transversely; apical initials of all orders of lateral cell rows reaching blade margins; intercalary divisions within axial and lateral cell rows lacking; tetrahedral sporangia in elongate sori along both sides of the midrib in the upper half of the blade, elevated on both blade surfaces; dioecious, procarps forming sequentially on the midrib below the apex, only a few per blade developing into cystocarps, cystocarps projecting on either plant surface; spermatangia formed in irregular, transverse sori, interrupted or confluent, along or midway between the midrib and blade.

Key to the species

1. On blades, cells of the second order rows always giving rise to third order initials; tetrasporangia developing in a random fashion in elongate sori along both sides of the midrib, cover cells abundant, corticating the sori . . . .
. . . . H. *hypoglossoides*

1. On blades, cells of the second order rows not always giving rise to third order initials; tetrasporangia developing in an acropetal series from the apex, forming in oblique symmetrical lateral rows consisting of 2 or 3 sporangia on each side of the midrib, cover cells not corticating the sori .... *H. tenuifolium*

Hypoglossum hypoglossoides (Stackhouse) Collins et Hervey 1917, p. 116.
   Fucus hypoglossoides Stackhouse 1795, p. 76, pl. XIII.
   Figures 483–486
   Plants epizoic, epiphytic or saxicolous, light pink to translucent, forming upright tufts to 20 cm tall, though often much smaller, or prostrate, spreading, entwined mats; secondarily attached by marginal or terminal rhizoids; blades linear lanceolate, to 5 cm long, 1–4 mm wide, sparsely to greatly branched from the midrib, generally to three orders; blades undivided, cells of second order rows always giving rise to third order initials, initials compressed and crowded at margins; margins entire to subcrenate and undulate, apices acute to emarginate, rarely obtuse; midrib in younger portions usually having one to two rows of elongate hexagonal to cylindrical cells laterally bordering the cylindrical axial row, three cells thick, later becoming corticated and much thicker; tetrasporangia globose to subglobose, 40–80 μm diameter, forming randomly in elongate, half-ellipsoid, heavily corticated sori along one or usually both sides of the midrib in the upper portions of the blade; dioecious, spermatangia in interrupted to confluent sori covering most or at least the upper portions of the blade, midway between the midrib and margins; cystocarps globose, sessile on the midrib, one to two (or more) per blade.
   Searles 1987, 1988. As *Delesseria hypoglossum*, Bailey 1851; Harvey 1853. As *Hypoglossum tenuifolium* var. *carolinianum*, Williams 1951; Taylor 1960; Schneider 1976; Kapraun 1980a. As *Hypoglossum tenuifolium sensu* Williams 1948a; Taylor 1960 *pro parte*; Chapman 1971; Wiseman and Schneider 1976; Searles 1981; Richardson 1987.
   Known from the shallow subtidal of nonprotected jetties from Cape Lookout to Cumberland Island, and from St. Augustine, Florida, year-round; from 60 m offshore in the Carolinas, May-November; and from 17–21 m on Gray's Reef and from 30–35 m on Snapper Banks, March-September.
   Distribution: North Carolina, South Carolina, Georgia, Bermuda, northeastern Florida to Gulf of Mexico, Caribbean, Brazil, British Isles to Portugal, Azores, Mediterranean, Canary Islands.
   First reported from our area in Charleston as *Delesseria hypoglossum* (Bailey 1851), this taxon has had a confused nomenclatural history (refer to Wynne 1984). The Charleston plants have likewise had a confused taxonomic history, being one of the many records of *Hypoglossum* to be combined by Taylor (1942,

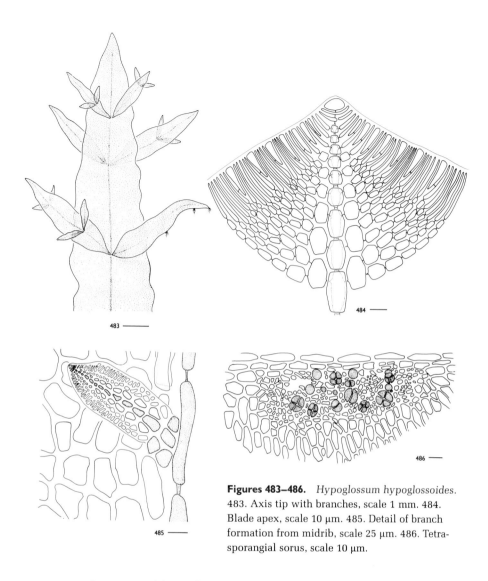

Figures 483–486. *Hypoglossum hypoglossoides.*
483. Axis tip with branches, scale 1 mm. 484.
Blade apex, scale 10 μm. 485. Detail of branch
formation from midrib, scale 25 μm. 486. Tetrasporangial sorus, scale 10 μm.

1960) under *H. tenuifolium*, despite contention by others (Collins and Hervey
1917, Børgesen 1930) that the European taxon was found in the western Atlantic. Wiseman (1966, 1978) examined Harvey's plants from Charleston and recognizing the earlier distinction of the two taxa by Harvey, found all Carolina
plants known at that time to be closer to *H. hypoglossoides* than to *H. tenuifolium*. This separation was further substantiated by the collection of a distinctly
different *Hypoglossum tenuifolium* from deep offshore water in North Carolina
(Schneider 1975a).

When describing *Hypoglossum tenuifolium* var. *carolinianum* from North
Carolina, Williams (1951, p. 156) likened his new plant to the European taxon it
is now placed under, *H. hypoglossoides* (Wynne and Ballantine 1986). We have
specimens of *H. hypoglossoides* in the flora which range from the small, narrow

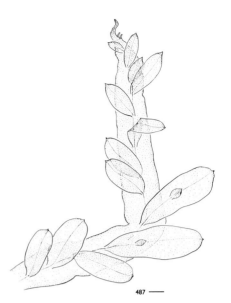

Figures 487–489. *Hypoglossum tenui-folium*. 487. Habit, scale 1 mm. 488. Blade apex, scale 10 μm. 489. Tetrasporangial sorus, scale 100 μm.

487 ——

prostrate 1 cm plants described by Williams to more robust, erect 5 cm plants typical of *H. hypoglossoides*, with similar midrib assemblages and tetrasporangial sori, and find no reason to retain the variety *carolinianum*.

Hypoglossum tenuifolium (Harvey) J. Agardh 1898, p. 186.
  Delesseria tenuifolia Harvey 1853, p. 97, pl. XXII.
  Figures 487–489
  Plants to 10 cm tall, translucent light pink to greenish; blades linear ellipsoid to lanceolate, 3–5 cm long, 2.5–3.5 mm wide, arising from rhizoidal holdfasts, greatly branched from the midrib, to three to four orders; blades undivided, cells of the second order rows not always giving rise to third order initials, initials not all compressed at the margins, margins entire to subcrenate, apices acute, obtuse, or emarginate; midrib in younger portions usually having broadly rectangular to half hexagonal cells laterally bordering the axial row, three cells thick, later becoming corticated and much thicker; tetrasporangia globose to subglobose, 70–120 μm diameter, forming in oblique, symmetrical lateral rows consisting of two to three sporangia on each side of the midrib, developing in an acropetal series from the apex, the newest forming while the oldest have already dispersed; dioecious, spermatangia forming in oblique, interrupted to confluent sori, covering the upper half of the blade, midway between the midrib and margin; cystocarps urn shaped and sessile on the midribs, one per blade.
  Schneider 1975a, 1976.
  Rare, known only from 40–45 m offshore in Onslow Bay, June-August.
  Distribution: North Carolina, southern Florida, Caribbean, Brazil.
  Only tetrasporic and male plants have been collected in our area.
  Harvey (1853) reported three species of *Delesseria* from the southern United States. Only one was listed for the Carolinas, the European *Delesseria hypoglossum* Lamouroux (=*Hypoglossum hypoglossoides* [Stackhouse] Collins and Hervey) from Charleston. Unfortunately, despite Harvey's contention that this

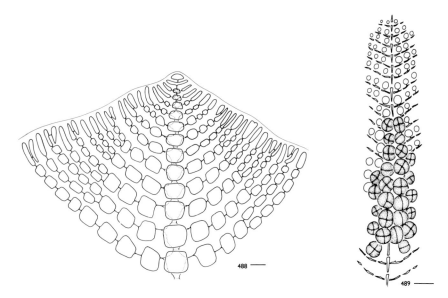

species could clearly be distinguished from his newly described *D. tenuifolia* from Florida, Taylor (1942, 1960) combined all American records of both taxa as *Hypoglossum tenuifolium*, no doubt because of the difficulty in separating vegetative specimens. Because offshore collections of *Hypoglossum* conform to *D. tenuifolia* Harvey (1853) and Børgesen (1910, 1919) and are distinct from plants previously recorded as *H. tenuifolium* for the Carolinas (Schneider 1975a), all records of this taxon from our area prior to 1975 are recognized as *H. hypoglossoides*. Those records of *H. tenuifolium* from tropical and subtropical America not already verified should be held in question until they can be separated into the two species as outlined by Wynne and Ballantine (1986) and here.

We have found the midrib distinctions outlined by Harvey (1853) for the above two species inconsistent and therefore unreliable. In extreme conditions, the two taxa are separable based upon their midribs; however, the distinction as to whether all second order cells produce third order initials (Wynne and Ballantine 1986) and tetrasporangial sori are the best and most definitive characters for identification. In the flora, *Hypoglossum hypoglossoides* is by far the more common of the two species.

Myriogramme Kylin 1924

Plants foliose or palmately to subdichotomously lobed, membranous, both erect and prostrate blades arising from small holdfasts, occasionally secondarily attached by marginal or laminal rhizoids, either in groups or single; vegetative blades single to few layered throughout, mostly multilayered near the base, some with macroscopic veins, microscopic veins always lacking; apical growth by a marginal row of initials, lacking single transversely divided apical cells, margins entire to serrulate, undulate to smooth, occasionally proliferous; tetrahedral sporangia in circular to linear and irregular sori scattered over upper portions of both surfaces of the blades or on marginal proliferations; dioecious, procarps

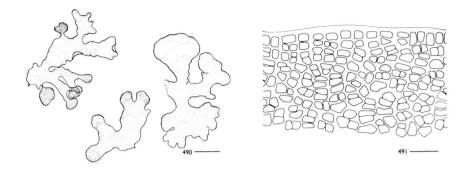

Figures 490–493. *Myriogramme distromatica*. 490. Habit of three plants, scale 0.5 cm. 491. Blade apex, scale 20 µm. 492. Detail of distromatic nature of blade, scale 20 µm. 493. Rhizoid issuing from ventral surface, scale 25 µm.

scattered over the blades, cystocarps ostiolate, projecting from either blade surface, carposporangia produced in chains; spermatangia formed in small, circular to crescent-shaped sori over upper portions of the blade.

Myriogramme distromatica Rodriguez ex Boudouresque 1971, p. 80, figs 1–15.
    Figures 490–493
    Plants prostrate with occasional unattached lobes, spreading to 3.5 cm in length, pink to mauve, drying pinkish to rosy red; blades cuneate and lobate to short ligulate and spatulate, 2–12 mm broad, the margins undulate, smooth to erose, occasionally issuing single or clustered multicellular rhizoids; branching di- to polychotomous and irregular, the apices obtuse; near apices blades single layered, elsewhere double layered; dorsal surfaces with several smaller cells forming a divided pattern over ventral undivided cells; dorsal cells polygonal, 10–37 µm in greatest dimension, ventral cells linear polygonal 20–65 µm in length, giving rise to numerous, scattered, multicellular, long uniseriate rhizoids; reproduction unknown.
    As *Nitophyllum wilkinsoniae sensu* ?Williams 1951, *pro parte*; Schneider 1976; Wiseman and Schneider 1976.
    Uncommon offshore in Onslow Bay and Long Bay from 40–100 m, June-November.
    Distribution: North Carolina (new record), South Carolina (new record), western Mediterranean.
    Eight collections of deep-water plants from the Carolinas compare well with descriptions and illustrations of those from off Corsica, Minorca, and the Mediterranean coast of France (Boudouresque 1971). *Myriogramme distromatica* has a diagnostic pattern of smaller divided dorsal cells overlying undivided ventral cells (figure 492), and numerous ventrally issued multicellular rhizoids (figure 493). The Atlantic plants also display oval or spatulate tips similar to those found in Mediterranean specimens. This is the first report of this species outside of the western Mediterranean where it is uncommon from 3–90 m in depth (Bou-

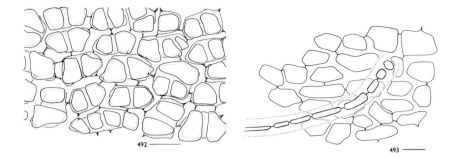

492
493

douresque 1971). Local specimens had previously been misidentified as *Nitophyllum wilkinsoniae* (Schneider 1976, Wiseman and Schneider 1976).

Nitophyllum Greville 1830, nom. cons.

Species Excluded

Nitophyllum punctatum (Stackhouse) Greville

Although specimens of this taxon have not been found to verify early records (Harvey 1853), it is likely that those reports would now be attributed to either *Myriogramme distromatica* or *Calonitophyllum medium* (young plants). The only Curtis (1867) specimen found was *Meristotheca floridana*.

Searlesia C. W. Schneider et Eiseman 1979

Plants consisting of erect and prostrate, marginally branched, single-layered blades; erect blades attached by small discoid holdfasts, prostrate blades secondarily attached by multicellular rhizoids which issue from margins and apices; single-layered blades interrupted by a three-tiered simple or branched midrib and inconspicuous veins; apical cells dividing transversely, intercalary divisions in the axial rows infrequent, margins serrulate; tetrahedral sporangia in corticated sori, elevated on both blade surfaces, midway between the midrib and margin, over the entire plant surface; dioecious, procarps scattered over nonaxial single-layered portions of the plant; cystocarps mammillate, ostiolate, projecting from either blade surface; spermatangial sori raised on both surfaces of the blade, in similar positions to sporangial sori.

Searlesia subtropica (C. W. Schneider) C. W. Schneider et Eiseman 1979, p. 324, figs 1–16.

Membranoptera subtropica C. W. Schneider 1974, p. 1097, figs 1B, 3.

Figures 494 and 495

Plants epiphytic or epizoic, to 4 cm tall, pink to pinkish red; blades ligulate to lobate, 0.3–2.0 cm wide, to 75 µm thick; branching dichotomous to subdichotomous below, alternate, opposite, subdichotomous or irregular above, the apices acute to obtuse; tetrasporangia globose to ellipsoid, 30–75 µm diameter in circular to oval sori, 210–220 µm thick, to 625 µm long; spermatangia pyriform, ovoid to ellipsoid, 4–5 µm diameter, 8–9 µm long in oval to irregular sori, 0.8–1.2 mm

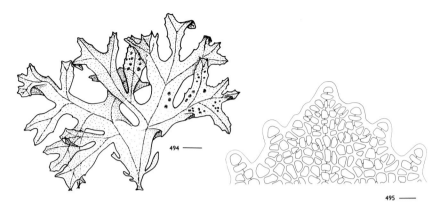

Figures 494, 495. *Searlesia subtropica.* 494. Habit of tetrasporic plant, scale 0.5 cm. 495. Blade apex, scale 20 μm (from Schneider 1974).

in longest dimension, often concentrated in the distal portions of the plant, but found in positions similar to sporangial sori; carposporangia globose to ovoid, to 50 μm diameter, cystocarps symmetrical, to 1 mm diameter.

Schneider and Eiseman 1979. As *Membranoptera subtropica*, Schneider 1974; Schneider 1976; Wiseman and Schneider 1976.

Collected from 23–63 m offshore on rock, coral, sponges, and other algae, May-August.

Distribution: North Carolina (type locality), South Carolina, southern Florida.

Male plants have not been collected north of Cape Canaveral (Schneider and Wiseman 1979).

Dasyaceae

Plants erect, sometimes partly prostrate, epiphytic or saxicolous, bushy or sparsely branched, attached by small to large fleshy, discoid holdfasts; indeterminate axes terete, uniaxial, usually becoming polysiphonous and corticated mostly by rhizoids, but also by division of the pericentral cells; indeterminate axes clothed with monosiphonous, pigmented, determinate pseudolaterals (ramelli); pseudolaterals either free and simple or branched, or united into a network; indeterminate growth sympodial, the apex continually displaced by a lateral axis which continues axial development; branching radial or distichous, occasionally dorsiventrally organized; tetrahedrally divided sporangia whorled on specialized determinate pseudolaterals in long series forming cylindrical to compressed stichidia; spermatangia hyaline, clustered apically for the most part, on specialized determinate pseudolaterals, with or without sterile cell tips, occasionally adventitiously derived; carpogonial branches four celled, borne on a pseudolateral or axial cell, with two groups of sterile cells, the supporting cell giving rise to an auxiliary cell after fertilization; one to two connecting cells joining fertilized carpogonia and auxiliary cells, forming large fusion cells; carposporangia formed in short chains on monopodial gonimoblasts issued from the fusion cell; carposporophytes surrounded by distinct pericarps with obvious ostioles, often beaked.

Key to the genera

1. Branching distichous 2
1. Branching radial .... Dasya
2. Determinate branches alternately produced on successive axial segments, basal segments undivided; tetrasporangia only partially covered by corticating cells of stichidia .... Dasysiphonia
2. Determinate branches alternately produced on every second axial segment, basal segments polysiphonous; tetrasporangia completely covered by corticating cells of stichidia .... Heterosiphonia

Dasya C. Agardh 1824, nom. cons.

Plants erect, sympodially developed and radially organized, appearing virgate to bushy; indeterminate axes terete, polysiphonous, with or without rhizoidal cortication, mostly clothed with pigmented determinate pseudolateral branches (ramelli); pseudolaterals monosiphonous distally, occasionally polysiphonous proximally, simple to dichotomously branched, spirally arranged and produced on every axial segment; polysiphonous axes uniaxial with four to five pericentral cells; four to six tetrahedral sporangia formed in each of several successive whorled series on specialized determinate branches, overall forming organized stichidia; stichidia ovoid, oblong, conical to linear-lanceolate, occasionally with sterile tips, single or paired; sporangial mother cells initially with two to four cover cells which may divide but never completely cover the sporangia at maturity; dioecious, hyaline spermatangia formed in stichidia, conical to linear-lanceolate, often with sterile tips; procarps formed on pericentral cells, one per node, usually near the apices of short lateral determinate branches, carpogonial branches four celled; carposporophytes developing as monopodial gonimoblasts from a large fusion cell, surrounded by large, corticated pericarps with obvious ostioles; carposporangia formed terminally on short gonimoblasts, either single or in short, simple to branched chains.

Key to the species

1. Main axes bearing short spinose indeterminate axes .... *D. spinuligera*
1. Main axes lacking short spinose indeterminate axes 2
2. Indeterminate axes ecorticate above, corticated in the lower segments of main axes.. *D. rigidula*
2. Indeterminate axes corticated throughout........................... 3
3. Plants 2–7 cm tall at maturity; determinate pseudolaterals conspicuously

tufted at the apices of indeterminate axes; basal cells of pseudolaterals generally large, 40–80 μm diameter; tetrasporangia 30–40 μm diameter . . . .
. . . . D. ocellata

3. Plants 20–90 cm tall at maturity; determinate pseudolaterals densely and evenly covering all, or at least the upper portions of indeterminate axes; basal cell of pseudolaterals smaller, 10–20 μm diameter; tetrasporangia 40–80 μm diameter . . . . D. baillouviana

Dasya baillouviana (Gmelin) Montagne 1841, p. 165.
  Fucus baillouviana Gmelin 1768, p. 165.
  Figure 496
  Plants saxicolous, conchicolous, rarely epiphytic or epizoic, erect, 20–90 cm tall at maturity, virgate to bushy, pinkish, brownish, rosy, to deep purplish red, arising from small discoid holdfasts; indeterminate axes corticated, to 2–3(–6) mm diameter, sparsely to freely and alternately branched; determinate branches generally abundant over all indeterminate axes, some lower indeterminate axes sparsely covered to completely denuded; determinate branches usually three times dichotomous, 2–8(–14) mm long; basal cells of determinate branches 10–40 μm diameter, 20–50 μm long; determinate branches with or without an expansion in width to the median cells, 20–60 μm diameter, to 200 μm long, then tapering to the acute to obtuse tips, 5–12 μm diameter; tetrasporangial stichidia borne on one to two celled pedicels on lower segments of determinate branches, lanceolate to acute linear-lanceolate, 80–160 μm diameter and 0.6–1.0(–1.2) mm long at maturity; tetrasporangia globose, 40–80 μm diameter; spermatangial stichidia with same shape and position as tetrasporangial stichidia, occasionally with terminal sterile cells, 60–75 μm diameter, 0.2–0.6 mm long; cystocarps single, rarely grouped two to three, borne near the tips of short determinate branches, urceolate, to 1.1 mm diameter, with an obvious neck 100–200 μm diameter.
  Schneider 1976, Wiseman and Schneider 1976, Kapraun 1980a, Richardson 1981, 1986, 1987, Searles 1987, 1988. As *D. elegans*, Bailey 1848, Harvey 1853, Curtis 1867, Melvill 1875, Johnson 1900. As *D. pedicellata*, Hoyt 1920, Williams 1948a, 1949, Taylor 1942, 1960, Chapman 1971.
  Known from abundant collections in shallow subtidal of jetties and sounds from Cape Lookout through Georgia from winter to spring, disappearing in May-June; offshore in Onslow Bay and Long Bay, from 18–40 m, June-August and from 17–21 m on Gray's Reef and 30–35 m on Snapper Banks, July-August.
  Distribution: Massachusetts to Virginia, North Carolina, South Carolina, Georgia, southern Florida, Gulf of Mexico, Caribbean, Brazil; elsewhere, widespread in warm temperate to tropical seas.

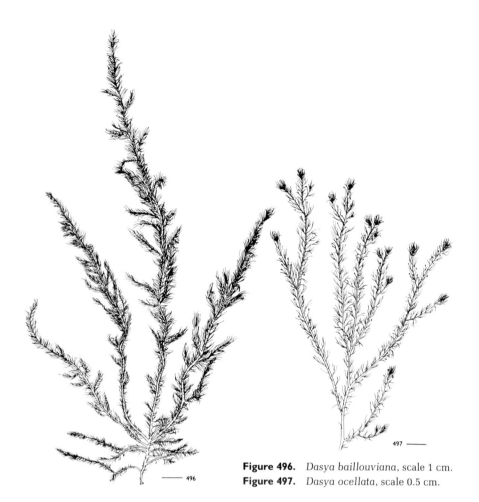

Figure 496. *Dasya baillouviana*, scale 1 cm.
Figure 497. *Dasya ocellata*, scale 0.5 cm.

Richardson (1981) speculated that two races of this taxon may reside in the western Atlantic, one centered in temperate Cape Cod waters, the other in tropical Caribbean waters. An overlap of the two races in North Carolina could explain the phenology of inshore winter-spring and offshore summer populations. Because bottom temperatures offshore are greatly warmed by the Gulf Stream (Schneider 1976), it is unlikely that this deeper environment is a summer refuge for the inshore plants. Richardson (1981) found, using artificial substrates, that *Dasya baillouviana* produced several generations during its active winter-spring period. These inshore populations perennate by maintaining both dormant microscopic germlings and pseudoparenchymatous discs of holdfast material during the summer and fall. Neither of these structures was destroyed by fish or urchin grazing.

Dasya ocellata (Grateloup) Harvey 1833, p. 335.
    Ceramium ocellatum Grateloup 1806, pl. [1], fig. 2.
    Figure 497
    Plants saxicolous, conchicolous, or epiphytic, erect, 2–4(–7) cm tall, bushy, rosy to purplish red, arising from small discoid holdfasts; indeterminate axes

Figures 498–500.
Dasya rigidula. 498.
Habit, scale 2 mm. 499.
Origin of determinate laterals from ecorticate axis,
scale 150 µm. 500. Lightly
corticated axis bearing determinate branches with
tetrasporangial stichidia,
scale 100 µm.

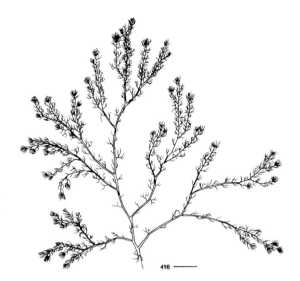

498 ——

corticated, to 0.3 mm diameter, sparsely to densely and alternately to pseudo-dichotomously branched; determinate branches produced in a loose spiral about the indeterminate axes, densely tufted at axial tips, one to three times dichotomous, to 5 mm long below, less than 1 mm long on upper segments; basal cells of determinate branches (30–)40–80 µm diameter, 25–80 µm long, median cells 20–30 µm diameter, 60–140 µm long; determinate branches gradually tapering to the acute to obtuse tips, 10–30 µm diameter; tetrasporangial stichidia borne on one- to three-celled pedicels on lower segments of determinate branches, short, conical, and approximately 200 µm long in early stages, becoming linear-lanceolate, 100–150 µm diameter and 0.95–1.1 mm long at maturity, often with sterile tips of one to four cells; tetrasporangia globose, 30–40 µm diameter; spermatangial stichidia in positions similar to sporangial stichidia, 60–70 µm diameter, to 200 µm long; cystocarps sessile on short determinate branches, ovate-globose.

Schneider 1975b, 1976.

Known from 23–40 m offshore in Onslow Bay, June-August.

Distribution: North Carolina, Bermuda, Caribbean, British Isles to Portugal, Azores, Mediterranean, Canary Islands, West Africa.

Female and cystocarpic plants have not been collected in the flora.

Dasya rigidula (Kützing) Ardissone 1878, p. 140.

Eupogonium rigidulum Kützing 1843, p. 415.

Figures 498–500

Plants epiphytic, saxicolous or conchicolous, erect or in part prostrate, 1–8 cm tall, purplish to rosy red, arising from a small rhizoidal holdfast, secondarily attached by adventitious rhizoids or tips of determinate branches; indeterminate axes ecorticate above, lightly corticated in lower portions of main axes on large plants, 0.3–0.5 mm diameter, freely and alternately to irregu-

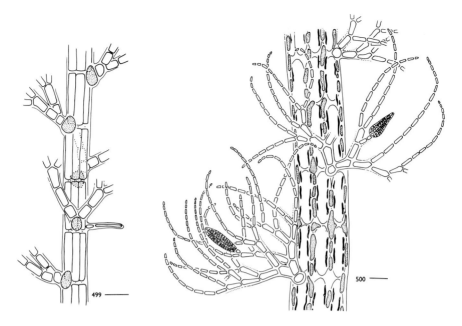

larly and adventitiously branched; lower axial cells 70–180 μm diameter, 350–400 μm long; determinate branches spirally arranged about the indeterminate axes, three to five times widely dichotomous, incurved, 0.4–1.0 mm long, more compacted on newer portions of indeterminate axes, appearing slightly tufted at the apices, becoming in part deciduous on lower portions; basal cells of determinate branches globose to ovoidal, 50–140 μm diameter, other proximal cells 30–100 μm diameter, 70–190 μm long, the branching systems usually tapering to long, hairlike, lightly pigmented tips, 15 μm diameter, occasional determinate axes truncated, with little tapering, above the lower dichotomies; tetrasporangial stichidia single, rarely paired, sessile or on one- to three-celled pedicels, replacing branches at the second through fourth dichotomies of determinate axes, short to long conical to linear-lanceolate and cylindrical subapiculate, tipped with a sterile cell or cells, or occasionally with long filaments, 30–60 μm diameter, 80–220 μm long; tetrasporangia globose, to 20 μm diameter; spermatangial stichidia with same shape and position as tetrasporangial stichidia, to 47 μm diameter, 180 μm long.

Schneider 1975a, 1976.

Known from 25–40 m offshore in Onslow Bay, July-August.

Distribution: North Carolina, Bermuda, southern Florida, Gulf of Mexico, Caribbean, Brazil, Mediterranean.

This taxon is similar to another deep-water species, *Heterosiphonia crispella* var. *laxa* (Børgesen) Wynne, which has distichous, as opposed to radial, branching, and polysiphonous, as opposed to undivided, basal segments of determinate branches. Male and female plants are unknown in the flora. Some Onslow Bay specimens are significantly larger (to 8 cm) than those previously reported from the western Atlantic by Taylor (1960, 2 cm) and Oliveira F. and Ugadim (1974, 3 mm). These bear considerably more cortication on main axes than the smaller specimens which are mostly ecorticate.

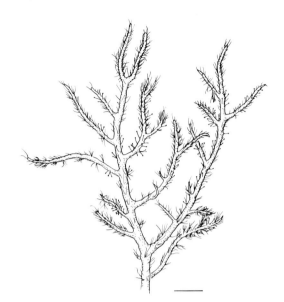

Figure 501.
Dasya spinuligera, scale
2 mm.

Dasya spinuligera Collins et Hervey 1917, p. 130, figs 24–25.

Figure 501

Plants saxicolous or conchicolous, erect or in part prostrate, 2–5 cm tall, deep rosy red, arising from small discoid holdfasts, secondarily attached by rhizoidal pads; indeterminate axes corticated, 0.5–1.2 mm diameter, beset with few, mostly short, spinose indeterminate branches; indeterminate axes densely covered in upper portions by determinate axes, sparingly covered to naked below; determinate branches two to three times dichotomous proximally, upper portions flagelliform, 0.3–1.7 mm long; determinate branches deciduous or partially truncated near their dichotomies, with tips often regenerating; basal cells of determinate branches pyriform to quadrate, 15–25 μm diameter and long, proximal cells 15–22.5 μm diameter, 25–40 μm long, shortening and tapering little or not at all to the obtuse to acute tips, 5–10 μm diameter; tetrasporangial stichidia single or paired, borne on one- to five-celled pedicels, replacing branches at the most proximal dichotomies of determinate branches, conical to linear-lanceolate and cylindrical, 50–150 μm diameter, 140–350(–800) μm long; tetrasporangia globose, 20–40 μm diameter; spermatangial stichidia unknown; cystocarps forming terminally on short spinose, indeterminate branches, urceolate, to 1.35 mm diameter, with an obvious neck 150–160 μm diameter.

As *Dasyopsis spinuligera*, Schneider 1975b, 1976.

Known from 25–40 m offshore in Onslow Bay, May-August.

Distribution: North Carolina, Bermuda.

Plants from the type collection (Bermuda) and Onslow Bay are spirally organized and therefore do not fit the bilateral concept of *Dasyopsis* Zanardini (= *Eupogodon* Kützing 1845, refer to Silva et al. 1987, p. 129), the genus in which they have previously been placed (Howe 1920, Taylor 1960). This taxon quite possibly represents a dwarf variety of *Dasya baillouviana*. Male plants are unknown for this species.

Dasysiphonia Lee et West 1979

Plants erect or in part prostrate, sympodially developed and dorsiventrally organized, appearing plumose; indeterminate axes terete, polysiphonous, with or without rhizoidal cortication, alternately beset with determinate pseudo-lateral branch systems from every axial segment; pseudolaterals pigmented, monosiphonous throughout, alternately to subdichotomously branched once to several times, distichous; polysiphonous axes uniaxial, with five pericentral cells; five tetrahedral sporangia formed in each of several successive whorled series on monosiphonous or polysiphonous pedicels on pseudolaterals, overall forming organized stichidia; stichidia linear-lanceolate, with sterile tips, single, occasionally forked; sporangia partially enclosed by cover cells at maturity; dioecious, hyaline spermatangia formed in stichidia in similar positions to tetra-sporangial stichidia, linear to lanceolate with sterile tips of several cells; pro-carps formed on short polysiphonous branches in the upper half of main axes, carpogonial branches four celled with two sterile groups of cells; carposporo-phytes develop as monopodial gonimoblasts from a fusion cell, surrounded by large, corticated pericarps with obvious ostioles; carposporangia formed termi-nally, then laterally, on gonimoblast filaments.

Key to the species

1. Determinate pseudolaterals nine to thirteen cells long, with one to two orders of branching; cells of pseudolaterals broadest centrally, barrel shaped, length-ening from base to apex at maturity; tetrasporangial stichidia replacing mono-siphonous laterals at the first or second dichotomy . . . . *D. doliiformis*
1. Determinate pseudolaterals fifteen to eighteen cells long, with 4 orders of branching; cells of pseudolaterals tapering from base to apex, with the short-est cells distally; tetrasporangial stichidia replacing monosiphonous laterals at the third dichotomy . . . . *D. concinna*

Dasysiphonia concinna C. W. Schneider 1989, p. 522, figs 1–3, 9–13.
Figure 502

Plants epiphytic or epizoic, pinkish red, prostrate and spreading, with erect axes, to 1.5 cm in length, attached by small discoid holdfasts and secondarily by rhizoids issued from pericentral cells; indeterminate axes 80–260 μm diame-ter, surrounded by five pericentral cells, 130–240 μm long, ecorticate through-out, irregularly producing distichous indeterminate branches at distances of eight to eleven segments; indeterminate axes beset with alternately distichous, filiform determinate pseudolateral branching systems, 0.9–1.3 mm long, pro-duced on every segment, appearing more dense and compacted at the apices; pseudolaterals tapering from base to tip, consisting of fifteen to eleven cells,

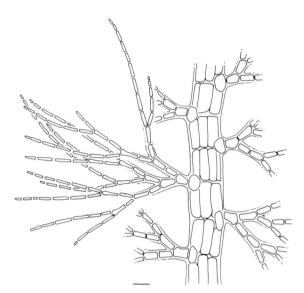

Figure 502.
Dasysiphonia concinna,
scale 100 μm.

four times pseudodichotomously branched, proximal cells above basal rounded cells somewhat rectangular, 40–80 μm diameter and 80–100 μm long, bearing the first dichotomies, cells attenuating and becoming cylindrical above, centrally 25–30 μm diameter and 75–95 μm long, distally 10–12.5 μm diameter and 7.5–48 μm long, apices obtuse to conical; tetrasporangial stichidia on one- to three-celled pedicels, lanceolate, 50–75 μm diameter, 120–200 μm long, mostly one per pseudolateral system, replacing an adaxial monosiphonous branchlet at the third dichotomy; tetrasporangia globose, 22–30 μm diameter.

Schneider 1989. As *Heterosiphonia wurdemannii* var. *laxa sensu* Schneider 1974, 1976, Wiseman and Schneider 1976.

Uncommon, from 33–60 m offshore in Onslow Bay and Long Bay, June-August.

Distribution: Endemic to the Carolinas as currently known.

Dasysiphonia doliiformis C. W. Schneider 1989, p. 522, figs 4–6, 14–15.
Figure 503
Plants epiphytic, rosy red, prostrate and spreading to entangled with erect axes, to 1.5 cm in length, attached by rhizoids issued from basal cells of determinate pseudolaterals and occasional pericentral cells; indeterminate axes 100–220 μm diameter below, surrounded by five pericentral cells, 200–490 μm long in lower portions, ecorticate throughout, irregularly producing distichous indeterminate branches at distances of ten or more segments; indeterminate axes beset throughout with alternately distichous determinate pseudolateral branching systems, 0.7–1.4 mm long, produced on every segment, appearing more dense and compacted at the apices; pseudolaterals with nine to thirteen cells from tip to base, one to two times pseudodichotomously branched, occasionally simple, composed mostly of short to long barrel-shaped, turgid cells, 75–100 μm diameter and 100–170 μm long centrally, slightly narrowing and shortening to the proximal ends and lengthening distally, 20–90 μm diameter

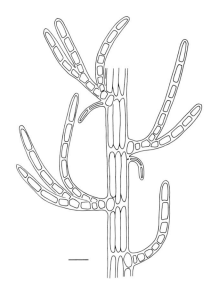

Figure 503.
Dasysiphonia doliiformis,
scale 200 μm.

and 50–230 μm long subapically, apical cells short to long conical on young pseudolaterals, mostly deciduous on older pseudolaterals, thus with apices appearing obtuse; tetrasporangial stichidia on one- to two-celled pedicels, linear-lanceolate, 90–125 μm diameter and 220–310 μm long, replacing an adaxial, or rarely abaxial, monosiphonous branchlet at the first or second dichotomy; tetrasporangia globose, 25–48 μm diameter.

Schneider 1989. As *Heterosiphonia wurdemannii sensu* Schneider and Searles 1973, Schneider 1976.

Uncommon, from 29–50 m offshore in Onslow Bay, July-September.

Distribution: Endemic to North Carolina as currently known.

Heterosiphonia Montagne 1842a, nom. cons.

Plants erect or in part prostrate, sympodially developed and dorsiventrally organized, appearing plumose to bushy; indeterminate axes flattened to terete, polysiphonous, with or without rhizoidal cortication, alternately to subdichotomously and irregularly distichous, alternately and distichously beset with determinate pseudolateral branch systems at intervals of two to nine axial segments; pseudolaterals pigmented, monosiphonous distally, polysiphonous proximally, alternately to subdichotomously branched one to several times, distichous; polysiphonous axes uniaxial with four to twelve pericentral cells; four to six tetrahedral sporangia formed in each of several successive whorled series on specialized determinate branches, overall forming organized stichidia; stichidia oblong apiculate, conical to linear-lanceolate, occasionally with sterile tips, single or paired; sporangial mother cells initially with two cover cells which divide to completely cover the sporangia at maturity; dioecious, hyaline spermatangia formed in stichidia, oblong to linear-lanceolate with sterile tips of one to several cells; procarps formed on cells below dichotomies on pseudolateral branches near the apices of indeterminate axes, carpogonial branches four celled; carposporophytes develop as monopodial gonimoblasts from a fusion cell, surrounded by large, corticated pericarps with obvious ostioles; carposporangia formed terminally, then laterally, on gonimoblast filaments.

Heterosiphonia crispella var. *laxa* (Børgesen) Wynne 1985b, p. 87.

Heterosiphonia wurdemannii (Bailey ex Harvey) Falkenberg var. *laxa* Børgesen 1919, p. 327, figs 327–328.

Figure 504

Plants epiphytic, epizoic or saxicolous, rosy red, prostrate and spreading with erect axes, to 3 cm in length, attached by rhizoidal holdfasts and secondarily by rhizoids issued from pericentral or pseudolateral cells; indeterminate axes 70–170 μm diameter, surrounded by four pericentral cells, 170–220 μm long in lower portions, ecorticate throughout, producing distichous, filiform determinate pseudolateral branching systems, 0.9–1.3 mm long, on every other segment; pseudolaterals tapering from base to tip, twelve to twenty-five cells long, with a polysiphonous proximal segment or segments, monosiphonous above, two to five times alternately, unilaterally to pseudodichotomously branched, proximal cells above polysiphonous segments short to elongate rectangular, 30–60 μm diameter and 40–70 μm long, gradually attenuating above to distal cells 12.5–25 μm diameter and 20–50 μm long, apices obtuse; tetrasporangial stichidia terminal on monosiphonous branchlets, occasionally becoming polysiphonous below stichidia, short conical to long and linear-lanceolate, 110–120 μm diameter and 170–400 μm long; tetrasporangia globose, 25–40 μm diameter, completely covered by small cortical cells at maturity.

Schneider 1989.

Known only from 29–30 m offshore of Cape Fear in the Frying Pan Shoals area of Onslow Bay, July.

Distribution: North Carolina, southern Florida, Gulf of Mexico, Caribbean, Brazil, West Africa, Central Pacific islands, Pacific Mexico.

Only tetrasporophytes are known in the flora. Several authors (Falkenberg 1901; Dawson 1963; Oliveira F. 1969) believe this distinct variety of *Heterosiphonia crispella* (C. Agardh) Wynne (1985b) should be separated at the specific level. Our earlier records of this species in the Carolinas are now attributed to *Dasysiphonia* (Schneider 1989).

Rhodomelaceae

Plants erect or partly to wholly prostrate, epiphytic, epizoic, saxicolous, or parasitic, bushy or sparsely branched, attached by basal holdfasts and/or rhizoids issued from pericentral or cortical cells; indeterminate axes terete to flattened and foliose, development monopodial, uniaxial, polysiphonous, ecorticate or

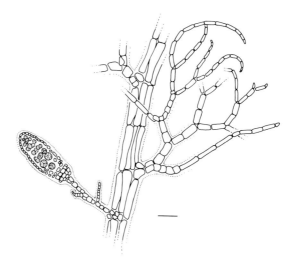

Figure 504. *Heterosiphonia crispella* var. *laxa*, axis bearing determinate branches and tetrasporangial stichidium, scale 100 μm.

corticated by divisions of the pericentral cells or downgrowing rhizoids, radially, distichously or dorsiventrally organized, more or less beset with indeterminate or monosiphonous or polysiphonous, simple to branched, determinate or pigmented axes, the apices or entire plant often bearing monosiphonous, simple or branched, pigmented or nonpigmented filaments, known as trichoblasts; tetrahedrally divided sporangia usually formed from pericentral cells, less commonly from cortical cells or cells of trichoblasts, generally one per segment, often in long series, occasionally resembling stichidia; spermatangia hyaline, clustered mostly on fertile trichoblasts with polysiphonous bases, or occasionally on polysiphonous determinate laterals, with or without sterile cell tips; carpogonial branches four to five celled, borne mostly on suprabasal cells of trichoblasts, which like the basal cells are polysiphonous, with two sterile cells, and which produce an auxiliary cell after fertilization; one to two connecting cells join fertilized carpogonia and auxiliary cells forming fusion cells; carposporangia formed in short chains on sympodial gonimoblasts issued from the fusion cell; carposporophytes surrounded by distinct pericarps with obvious ostioles.

Key to the genera

1. Plants prostrate, dorsiventrally organized 2
1. Plants mostly to wholly erect, radially to bilaterally organized 4
2. Branches produced in distichous pairs, one branched, the other unbranched, the pairs alternately set on prostrate axes .... Dipterosiphonia
2. Branches produced alternately on prostrate axes, not in pairs 3
3. Indeterminate apices inrolled, determinate apices bearing trichoblasts; plants growing in salt water .... Herposiphonia
3. Indeterminate apices not inrolled, determinate and indeterminate apices lacking trichoblasts; plants growing in brackish water .... Bostrychia
4. Indeterminate or determinate axes beset with short to long, spur-, spike-, or spinelike polysiphonous branchlets 5

Acanthophora Lamouroux 1813

Plants erect, widespreading to bushy, radially organized, arising from irregularly lobed to discoid holdfasts; indeterminate axes terete, polysiphonous with five pericentral cells heavily corticated by parenchymatous divisions, beset with spirally arranged, short, spurlike, polysiphonous, determinate branches; determinate branches and occasionally indeterminate axes producing spirally arranged, short, spinelike, polysiphonous branchlets, often appearing tufted; monosiphonous, deciduous trichoblasts remaining only at branch apices and on fertile segments; tetrahedral sporangia developed in swollen spinelike branchlets or determinate branches; dioecious, spermatangia formed in hyaline, flat, circular to irregular sori with sterile cell margins, on trichoblasts of spinelike branchlets, thus appearing stalked; carposporophytes within sessile, ovoid to urceolate pericarps, in the axils of, or at the bases of, spinelike branchlets.

Acanthophora spicifera (Vahl) Børgesen 1910, p. 201, figs 18–19.

Fucus spiciferus Vahl 1802, p. 44.

Plants saxicolous, brownish to purplish red, erect to 20 cm tall, firm, wide-spreading to bushy, arising from large, irregularly lobed holdfasts; axes terete, 0.6–1.5 mm diameter in lower segments, with five pericentral cells; branching irregularly radial, sparse below, often abundant above; branches mostly short and determinate (spurlike), occasionally long and indeterminate, repeating the pattern of main axes; determinate axes obliquely erect, 1–2 mm long, beset with short, sharply acute, simple spinelike, polysiphonous branchlets, to 0.5 mm long, most abundant at apices; indeterminate axes lacking spinelike branches; trichoblasts soon deciduous, dichotomously branched; tetrasporangia ellipsoid, 42–50 µm diameter, 60–80 µm long, in spinelike branchlets; cystocarps single, urceolate, 0.5–1 mm diameter, formed in the axils of short spinelike branchlets on determinate branches.

Hall and Eiseman 1981.

Attached to sea grasses and free-floating in Haulover Cove, Indian River, Fla., year-round, abundant December-January.

Distribution: Bermuda, northeastern Florida to Gulf of Mexico, Caribbean, Brazil; elsewhere, widespread in tropical and subtropical seas.

Bostrychia Montagne 1842b, nom. cons.

Plants prostrate and creeping or erect from spreading stoloniferous axes, dorsiventrally organized, attached by terminal fascicles of filiform, unbranched rhizoids or fascicles issued from ventral tier cells of prostrate axes; indeterminate axes terete to flattened, segmented and polysiphonous, each segment composed of an axial cell and four to ten pericentral cells, each transversely divided one or more times into tier cells, the lowermost retaining pit-connections to the axial cells, with or without cortication, producing branch initials every several segments in an alternately distichous to pseudodichotomous, rarely radial, arrangement; primary axes branched to several orders, polysiphonous except at the tips, apices often incurved; transverse divisions of apical cells slightly oblique, alternating side to side; trichoblasts lacking; tetrahedral sporangia embedded beneath the cortical layer, produced from pericentral cells, two to six per segment, in several whorls at the tips of swollen ultimate branches with tapering ends or stichidia; dioecious, spermatangia formed from superficial cells in cylindrical sori with sterile tips on ultimate branches; procarps formed from pericentral cells in successive series of segments lying at some distance from the apex of ultimate branches, carpogonial branch four celled; carposporophytes within sessile, ovoid to subglobose pericarps, appearing terminal on fertile branches, one (to two) per branch.

Bostrychia radicans (Montagne) Montagne 1842d, p. 661.

Rhodomela radicans Montagne 1840, p. 198, pl. 5, fig. 3.

Figures 505 and 506

Plants terricolous, epiphytic, saxicolous or lignicolous, brownish purple to deep violet, either prostrate forming dense, erect tufts to 4 cm tall, or loosely branched and wide spreading, attached by terminal fascicles of long, unbranched rhizoids and similar haptera issued from ventral tier cells; prostrate and erect indeterminate axes terete to flattened, main axes to 200 μm diameter with six to eight pericentral cells, each divided transversely into two tier cells, ecorticate; apices with somewhat incurved branchlets; indeterminate axes issuing branches from each of several segments in an alternately distichous to pseudodichotomous arrangement, the tips often appearing corymbose, branches spreading on prostrate axes, somewhat fastigiate on erect axes, some rebranched to several orders, ultimate branchlets 1–2 mm and mostly tapering to monosiphonous tips of few to several cells, 30–40 μm diameter; tetrasporangia globose, formed in long, whorled series below the apex to central in ultimate branchlets; cystocarps ovoid, appearing terminal on short, lower branchlets.

Stephenson and Stephenson 1952; Wiseman and Schneider 1976; Kapraun 1980a; Richardson 1987. As f. *moniliforme*, Post 1936; Chapman 1971. As *B. rivularis*, Harvey 1853; Curtis 1867; Melvill 1875; Farlow 1881; Hoyt 1920; Taylor 1960. As *B. scorpioides sensu* Bailey 1851.

Known on mud, pilings, *Spartina* culms, and other substrates in the intertidal zone of brackish water estuarine environments, commonly with *Caloglossa leprieurii*, from throughout the area during the entire year. At the time of his book on the marine algae of the Beaufort, N.C., area, Hoyt (1920) could only speculate that *Bostrychia radicans* should be found in the local estuary. Actually, it is the most common intertidal seaweed in the brackish environment of Adams Creek (Intracoastal Waterway).

Distribution: New Hampshire to Virginia, North Carolina, South Carolina, Georgia, Bermuda, southern Florida, Gulf of Mexico, Caribbean, Brazil; elsewhere, widespread in warm temperate to tropical seas.

Post (1936) united *Bostrychia rivularis* Harvey with *B. radicans* on the basis of similar haptera, number of tier cells, and lack of cortication. As outlined by Taylor (1960), these two species differ most obviously in the overall habit of the plants, either compact or loosely branched and wide-spreading. Local specimens show a habit continuum between these two extremes, corroborating the circumscription of Post (1936).

Although the subspecific taxon for long monosiphonous tips, forma *moniliforme* Post, has been reported for South Carolina (Post 1936, Wiseman 1978) and

Figures 505, 506. *Bostrychia radicans.* 505. Habit, scale 0.3 mm. 506. Detail of apex, scale 50 μm.

Georgia (Chapman 1971), an extensive study of the species in South Carolina by Wiseman (1966) showed an intergradation of all four forms delineated by Post (1936). Therefore, we have elected to list the variation within the species description, not recognizing formal subspecific taxa for the flora.

Gametophytes are uncommon from our area, perhaps due to a lack of a seasonal study of *Bostrychia* from any local estuary.

Bryocladia Schmitz in Schmitz et Falkenberg 1897

Plants erect, wide-spreading to bushy, radially organized, filamentous, arising from prostrate, spreading axes; indeterminate axes terete, alternately laterally branched, polysiphonous with six to sixteen pericentral cells, ecorticate; indeterminate polysiphonous axes densely beset with spirally organized, simple and straight to recurved or pinnately branched indeterminate branches, occasionally bearing hyaline, monosiphonous, branched, deciduous trichoblasts; tetrahedral sporangia, one per segment, acropetally developed in a straight series in the outer portions of fertile branchlets; procarps produced near the tips of short determinate branchlets, carpogonial branches four celled; carposporophytes within pedicellate, ovoid to urceolate pericarps.

Key to the species

1. Axes densely beset with short, straight to recurved, polysiphonous branchlets; producing six to eight pericentral cells .... *B. cuspidata*
1. Axes loosely beset with lax, pinnately branched, polysiphonous branchlets; producing nine to twelve pericentral cells .... *B. thyrsigera*

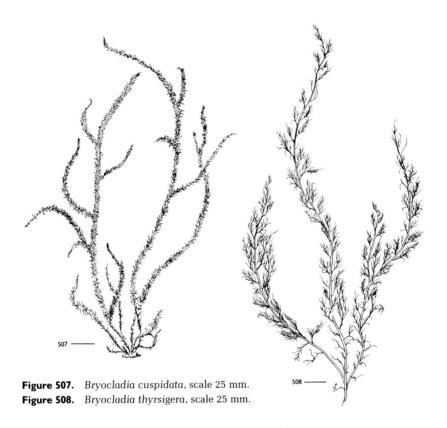

Figure 507. *Bryocladia cuspidata*, scale 25 mm.
Figure 508. *Bryocladia thyrsigera*, scale 25 mm.

Bryocladia cuspidata (J. Agardh) De Toni 1903, p. 968.

   Polysiphonia cuspidata J. Agardh 1852, p. 953.

   Figure 507

   Plants saxicolous, purplish red to black, erect to 8 cm tall, spreading with prostrate axes, bushy, clothed with short polysiphonous axes; erect axes terete, 195–350 µm diameter, with six to eight pericentral cells, ecorticate; major axes alternately branched, segments about twice as broad as long, 80–100 µm long; determinate axes radially and densely coating indeterminate axes, mostly uniform in length, 0.7–1.2 mm long, simple to once branched near the base, straight to sharply recurved, markedly tapering from bases to unicellular apices; trichoblasts, if present, simple or once forked, tapering base to apex, deciduous below the apices of vegetative axes; tetrasporangia globose, to 48 µm diameter, in straight series in upper portions of fertile branchlets; cystocarps ovoid to urceolate, to 320 µm diameter.

   Humm 1952; Taylor 1960; Chapman 1971; Richardson 1987.

   Known from the lower intertidal of Cumberland Island jetty, July-August, and from the intertidal coquinoid outcroppings at Marineland, year-round.

   Distribution: Georgia, northeastern Florida to Gulf of Mexico, Caribbean, Brazil, West Africa.

Bryocladia thyrsigera (J. Agardh) Schmitz in Falkenberg 1901, p. 169.

    Polysiphonia thyrsigera J. Agardh 1847, p. 17.

    Figure 508

    Plants saxicolous, purplish red to black, erect to 10 cm tall, widespreading from prostrate axes; erect axes terete, 165–345 μm diameter, with nine to twelve pericentral cells, ecorticate; major axes lax, alternately branched, segments about twice as broad as long, 70–165 μm long; determinate axes radially arranged around indeterminate axes, 0.6–3.0 mm long, alternately and pinnately branched, markedly tapering from bases to unicellular apices; trichoblasts, if present, simple to once forked, tapering base to apex, deciduous below the apices of vegetative axes; fertile branchlets short, stiff, mostly recurved, and clustered; tetrasporangia globose, 37–46 μm diameter, in straight series in abaxial portions of fertile branchlets.

    Humm 1952; Taylor 1960.

    Known only from the intertidal coquinoid outcroppings at Marineland, year-round.

    Distribution: Northeastern Florida to Gulf of Mexico, Caribbean, Brazil, West Africa.

Bryothamnion Kützing 1843

Plants erect, wide-spreading to bushy, radially to bilaterally organized, cartilaginous, arising from small discoid holdfasts; indeterminate axes terete to triangular or compressed, polysiphonous with six to nine pericentral cells heavily corticated by parenchymatous divisions; lateral axes arising in the axils of short, mostly once forked, hyaline, monosiphonous deciduous trichoblasts, alternately to spirally produced on every second axial segment, mostly developing into short, simple or branched corticated determinate branches interrupted by occasional indeterminate axes repeating the pattern; tetrahedral sporangia, one per segment, acropetally developed in a spiral series in the upper segments of nodulose fertile branches; dioecious, spermatangia formed in ovoid sori on specialized trichoblasts issued from each segment of fertile branches; procarps produced on the second segment of trichoblasts on determinate branches, carpogonial branches four celled; carposporophytes within short pedicellate, ovoid to subglobose pericarps.

Bryothamnion seaforthii (Turner) Kützing 1843, p. 433.

    Fucus seaforthii Turner 1811, pl. 190.

    Figure 509

    Plants saxicolous, brownish red to reddish purple, erect to 20 cm tall, membranous when young, becoming cartilaginous, wide-spreading to bushy; axes

Figure 509. *Bryothamnion seaforthii*, scale 2 mm. ——

compressed, occasionally terete to 2 mm diameter in lower segments, with eight to nine pericentral cells; branching in the axils of trichoblasts, alternately distichous, occasionally radial on every second axial segment, mostly short and determinate, occasionally long and indeterminate, repeating the pattern of main axes; determinate axes obliquely erect, mostly uniform in length, 2–3 mm long, distichous or in rows of three to four beset with short, sharply acute, simple distally to several spined proximally, polysiphonous branchlets; indeterminate tips more or less fastigiate; trichoblasts once forked, four cells long above the fork, tapering base to apex, deciduous several segments below the apices of vegetative axes; tetrasporangia globose, 60–100 μm diameter, in spiral series in clustered fertile branches at axis tips.

Schneider and Searles 1973; Schneider 1976.

Uncommon, collected from 27–35 m offshore in the Frying Pan Shoals region off Cape Fear, August-December.

Distribution: North Carolina, southern Florida, Gulf of Mexico, Caribbean, Brazil, Ceylon.

Gametangia are unknown in the flora. This species does not adhere well to herbarium paper and specimens fade when dried.

In a seasonal study of deep water seaweeds in Onslow Bay (Schneider 1976), *Bryothamnion seaforthii* was the only member of the offshore flora which was restricted to the fall-winter grouping, not being collected in numerous spring and summer dredgings where found later in the year. This taxon is known only from deep water throughout its range.

Chondria C. Agardh 1817, nom. cons.

Plants either erect, often bushy, with one or several axes arising from common discoid holdfasts, or prostrate and spreading, attached by randomly issued bundles of numerous multicellular rhizoids; axes polysiphonous with smaller-celled hyaline cortical layers barely or greatly obscuring generally five pericentral cells and the axial row, outer layer or layers of small cortical cells densely pigmented; axes fleshy to cartilaginous, terete to somewhat flattened, much or sparingly alternately radially or irregularly (rarely, oppositely or verticillately) branched, branches forming on the basal cells of trichoblasts at apices, the lesser branchlets often constricted or tapered at the bases, appearing club or spindle shaped; all axes terminating in single apical cells, exposed on acute tips or sunken in pits of obtuse tips, surrounded by terminal tufts of branched trichoblasts; tetrahedral sporangia embedded beneath the outermost cortical layer, produced from pericentral cells, numerous and regularly to irregularly distributed in the ultimate or penultimate branches; dioecious, spermatangia formed in hyaline, flat, regular to irregular obovoid sori with sterile cell margins, on the lowest lateral of trichoblasts, in clusters at or near the apices; procarps produced on the second cells of truncated, three-celled trichoblasts, carpogonial branches four celled; carposporophytes forming in distal portions of plant, carposporangia arising from large fusion cells, surrounded by pseudoparenchymatous, multilayered (at least below) pericarps; cystocarps short stalked to sessile, urceolate, ovoid to subglobose, displaced to lateral positions on ultimate branchlets.

Although eight species are presently recognized in the flora, taxonomic concepts for this genus are in need of clarification. Several of the below species (*Chondria atropurpurea*, *C. baileyana*, *C. littoralis*, and *C. tenuissima*) are closely related and form an intergrading complex here, with only the most typical specimens being easily identified to species. Until a comprehensive taxonomic study of *Chondria* in the western Atlantic is undertaken, we basically follow the species concepts of Taylor (1960). Intermediate forms with exposed apical cells would best be left as "*C. tenuissima* complex."

Key to the species

1. Plants mostly prostrate, with bundles of numerous rhizoids randomly issued from the axes 2
1. Plants erect, arising from discoid holdfasts 3
2. Apices truncate to obtuse, pericentrals with thickened end walls which can be seen through the cortex . . . . *C. curvilineata*
2. Apices acute, pericentrals without thickened end walls, only seen through the cortex in youngest portions . . . . *C. polyrhiza*
3. Apices obtuse or truncate, apical cell sunken in a terminal pit 4

3. Apices acute to acuminate or cuspidate, apical cell exposed, occasionally masked by numerous trichoblasts 5
4. Branches not constricted at the base, trichoblasts inconspicuous or lacking .... *C. floridana*
4. Branches markedly constricted at the base, trichoblasts conspicuous .... *C. dasyphylla*
5. Apices acuminate or cuspidate 6
5. Apices acute 7
6. Lower main axes 0.2–1.0 mm diameter, ultimate branches less than 350 μm diameter, tetrasporangia 55–105 μm diameter .... *C. baileyana*
6. Lower main axes 0.8–2.0 mm diameter, ultimate branches greater than 350 μm diameter; tetrasporangia 100–150 μm diameter .... *C. littoralis* (in part)
7. Plants dark purplish red, drying blackish .... *C. atropurpurea*
7. Plants dull purplish red to straw colored, not drying blackish 8
8. Plants with branches similar in size to the axes which bear them........... .... *C. tenuissima*
8. Plants with branches markedly more slender than the axes which bear them .... *C. littoralis* (in part)

Chondria atropurpurea Harvey 1853, p. 22, pl. 18E.

Figures 510–512

Plants saxicolous or conchicolous, dark purplish red, often black when dried, erect to 40 cm tall, terete and firm, 0.8–2.0 mm diameter below, subpyramidal in outline, attached by discoid holdfasts; axes somewhat naked of branches below, above beset with numerous intermixed short and long branches to several orders, alternately and irregularly arranged, branches single or fasciculate, linear fusiform 0.2–1(–3) cm long, arising from superficial depressions, narrow at the base, tapering gradually to acute apices; apical cells protruding from short tufts of trichoblasts; tetrasporangia globose, formed in ultimate fusiform, subdentate branches; gametangia unknown; cystocarps subspherical to broadly urceolate, to 1.5 mm diameter, sessile on densely crowded ultimate branches.

Harvey 1853; Hoyt 1920; Taylor 1960; Schneider 1976; Wiseman and Schneider 1976.

Known from the shallow subtidal near Beaufort, N.C., and Charleston, and 15–20 m offshore in Onslow Bay, May-December.

Distribution: North Carolina, South Carolina, Bermuda, southern Florida, Gulf of Mexico, Caribbean, Brazil.

This taxon is probably a form of *Chondria tenuissima* outlined below and is maintained here only for lack of a detailed examination of type material. Wiseman (1966) discussed the taxonomic problems associated with this acute tipped species.

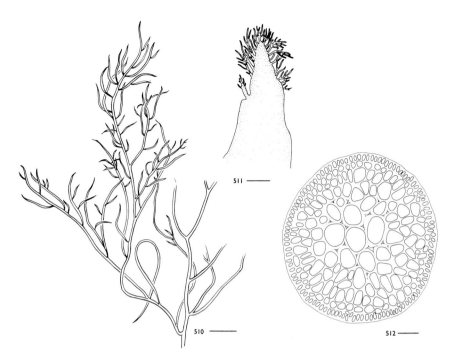

Figures 510–512. *Chondria atropurpurea*. 510. Habit, scale 0.5 cm. 511. Branch apex, scale 100 μm. 512. Cross section, scale 100 μm.

Chondria baileyana (Montagne) Harvey 1853, p. 20, pl. 18A.

    Laurencia baileyana Montagne 1849, p. 63.

    Figures 513 and 514

    Plants epiphytic, conchicolous or saxicolous, gregarious, dull purplish red to straw colored, erect to 25 cm tall, though usually much less, terete and soft in texture, 0.2–1.0 mm diameter below, narrow pyramidal in outline, attached by discoid holdfasts; main axes subsimple or with more or less long branches, beset with numerous shorter branches 80–200 μm diameter, to 5 mm long, markedly tapered at the base, elongate and club shaped with abrupt narrow acuminate tips projecting from obtuse apices, apical cells partially obscured by short to long tufts of trichoblasts; tetrasporangia globose, 55–105 μm diameter, covering the upper three quarters of ultimate branches or in a circumscribing band well below the apices; spermatangial sori broadly obovate 250–400 μm diameter, 240–250 μm long, displaced to lateral positions on the ultimate segments; cystocarps obovate to broadly urceolate to 1 mm diameter, formed laterally on truncated lateral branches which are acute and appear spurlike, subtending the cystocarp at maturity.

    Williams 1948a; Brauner 1975; Taylor 1960. As *C. tenuissima* var. *baileyana*, Hoyt 1920.

    Known from the shallow subtidal of sounds and jetties near Beaufort, N.C., and the intertidal at Fort Fisher, on rock, shell, and *Zostera*, May-June; often found drifting.

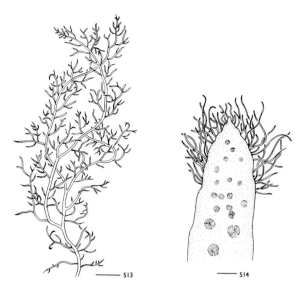

Figures 513, 514.
Chondria baileyana.
513. Habit, scale 0.5 cm.
514. Apex of tetrasporic
branch, scale 100 μm.

Distribution: Maritimes to Virginia, North Carolina, ?southern Florida, ?Gulf of Mexico, ?Caribbean, Mediterranean.

Typical forms with club-shaped branches are easily identified, although most specimens intergrade with *Chondria tenuissima*, under which this taxon has been placed as a variety in the past, and with *C. littoralis*. Records of *C. baileyana* from North Carolina and more southerly locations have been questioned (Taylor 1960, Kapraun 1980a), however, we have observed narrow plants which favorably compare with plants from New England. We retain this taxon for the present, recognizing that a complete study of the taxa with acute apices from tropical and subtropical western Atlantic waters is needed.

Chondria curvilineata Collins et Hervey 1917, p. 120, pl. II figs 10–11.
Figures 515 and 516
Plants epiphytic, dull rosy red to yellowish, prostrate and spreading with erect axes 1–3 cm tall, terete and lax, 200–350(–500) μm diameter, attached by random bundles of numerous short, simple, multicellular rhizoids; axes infrequently to

Figures 515, 516.
Chondria curvilineata.
515. Habit, scale 1 mm.
516. Branch apex, scale
100 μm.

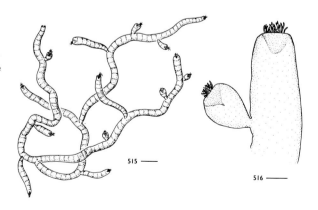

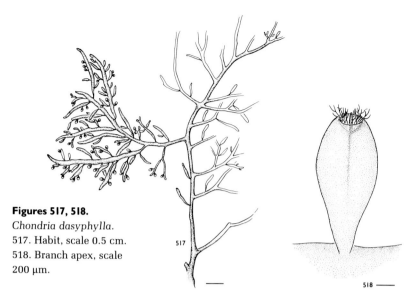

Figures 517, 518.
Chondria dasyphylla.
517. Habit, scale 0.5 cm.
518. Branch apex, scale
200 μm.

frequently and irregularly branched, rarely to more than two orders, ultimate branches often short, constricted or markedly tapered to the base; apices truncate to obtuse, apical cells in terminal pits, or slightly projecting from the pits, surrounded by short tufts of trichoblasts which can be partially seen on most or all apices; distal convex ends of pericentral cells thickened, showing through the cortex, showing the axes as somewhat banded, not obvious to the unaided eye; outer cortical cells 10–30 μm diameter, 40–100 μm long; tetrasporangia globose, formed in the distal portions of ultimate branches; gametangia unknown; cystocarps sessile on the branches.

Schneider 1975a, 1976.

Rare, known only from 29–40 m offshore in Onslow Bay, July-August.

Distribution: North Carolina, Bermuda, Caribbean.

Chondria dasyphylla (Woodward) C. Agardh 1817, p. xviii.

Fucus dasyphyllus Woodward 1794, p. 239, pl. 23, figs 1–3.

Figures 517 and 518

Plants saxicolous or epiphytic, gregarious, bushy, pale straw colored, light brownish or dark purplish to rosy red, erect to 30 cm tall, terete and firm but often brittle, 1.0–2.5 mm diameter below, attached by small discoid holdfasts; main axes of several long similar branches, each broadly pyramidal in outline owing to numerous alternate to opposite longer to shorter secondary branches, these beset with short club- or top-shaped ultimate branches, single or fasciculate, 200–600 μm diameter, 2–10 mm long, markedly constricted at the base, the older ones becoming uneven or irregularly swollen in contour; apical cells sunken in terminal depressions of the obtuse tips occasionally slightly protruding and surrounded by conspicuous tufts of trichoblasts; tetrasporangia globose, 40–165 μm diameter, concentrated near the distal ends to covering the upper three quarters of ultimate branches; spermatangial sori broadly ovate to cordate, 500–600 μm diameter, clustered at the tips on the basal cells of trichoblasts;

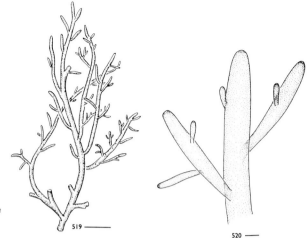

Figures 519, 520.
Chondria floridana.
519. Habit, scale 1 cm.
520. Branch apex, scale
0.5 mm.

519 ———

520 ———

cystocarps subglobose to broadly ovoid, to 1.0 mm diameter, sessile near the distal ends of ultimate branches, the tips often appearing as short spurs.

Curtis 1867; Hoyt 1920; Williams 1948a, 1949, 1951; Taylor 1957, 1960; Schneider 1976; Kapraun 1980a; Hall and Eiseman 1981. As *Laurencia dasyphylla*, Bailey 1851. As *Chondria sedifolia sensu* Hoyt 1920; Williams 1948a, 1951; Taylor 1957, 1960; Schneider 1976, pro parte. As *C. floridana sensu* Williams 1951, pro parte.

Known year-round in the shallow subtidal near Ocracoke, Beaufort, and Wilmington, N.C. (in winter found only as decumbent filaments on rock); from intertidal coquinoid outcroppings at Marineland and on seagrasses at Haulover Cove, Indian River, Fla., year-round; and common offshore from 14–60 m in Onslow Bay, found year-round as upright plants, although often fragmented in winter.

Distribution: New Hampshire to Virginia, North Carolina, Bermuda, northeastern Florida to Caribbean, Brazil, Uruguay; elsewhere, widespread in warm temperate to tropical seas.

Chondria floridana (Collins) Howe in W. R. Taylor 1928, p. 170, pl. 34, fig. 3.
Chondria dasyphylla f. *floridana* Collins 1906a, p. 111.
Figures 519 and 520

Plants conchicolous or saxicolous, pinkish to yellowish red, erect to 20 cm tall, terete and lax, 1.5–3.0 mm diameter below, gradually tapering from greater to lesser branches, subpyramidal in outline, attached by small discoid holdfasts; main axes branched to four to five orders, regularly alternate, the ultimate branches straight to curved, cylindrical to somewhat clavate, but not constricted at their bases, 0.3–0.5 mm diameter, 5–10 mm long; apices obtuse to truncate, apical cells sunken in terminal depressions, trichoblasts remaining inconspicuous, soon deciduous; tetrasporangia globose, forming a circumscribing band about 1–2 mm below the apices, with few scattered below.

?Williams 1951; Duke University in herb.

Verified only from specimens of various collectors and our own, to date exclusively from shrimp trawls in 20 m off Beaufort Inlet, N.C., all from July.

Distribution: North Carolina, southern Florida, Gulf of Mexico, Caribbean, Brazil.

This species has been taken only from a soft shell- and sandy-bottomed area not surveyed by Schneider (1976). It is possible that *Chondria floridana* is present in the flora for a more extended period than shown above, but all known collections have been made during the limited Carolina shrimp-trawling season. All specimens of the Williams (1951) report thus far have been assigned to *C. dasyphylla*—therefore, the report of *C. floridana* from off New River Inlet remains in doubt.

Chondria littoralis Harvey 1853, p. 22.

Figures 521 and 522

Plants saxicolous, dull rosy and brownish red to straw colored, erect to 35 cm tall, terete and firm in texture, 0.8–2.0 mm diameter below, subpyramidal to virgate in habit, attached by discoid holdfasts; axes sparingly branched below, somewhat paniculately branched above, beset with more or less elongate

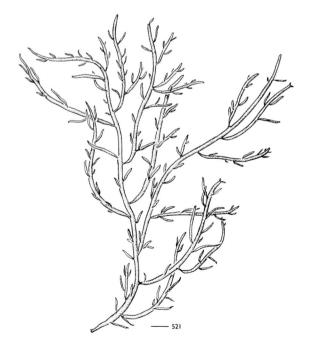

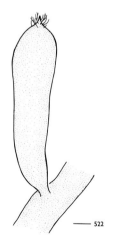

Figures 521, 522.
Chondria littoralis. 521. Habit, scale 0.5 cm. 522. Branch apex, scale 0.5 mm.

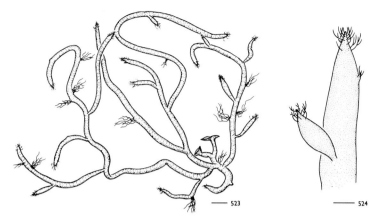

Figures 523, 524. *Chondria polyrhiza.* 523. Habit, scale 1 mm. 524. Branch apex, scale 0.3 mm.

spindle-shaped branches, about 500 μm diameter, 1–15(–25) mm long, occasionally rebranched; apices acute, apical cells either exposed or obtuse with acuminate tips, sometimes obscured by prominent tufts of trichoblasts; tetrasporangia globose, 100–120(–150) μm diameter, concentrated in distal portions of ultimate branches and main axes; spermatangial sori broadly obovate, about 450 μm in greatest dimension, displaced to a lateral position on the ultimate branches; cystocarps single, broadly ovate to urceolate, to 850 μm diameter, formed laterally on truncated lateral branches.

Hoyt 1920; Williams 1948a, 1951; Taylor 1960; Schneider 1976, *pro parte*; Kapraun 1980a. As *C. sedifolia sensu* Schneider 1976, *pro parte*.

Known from the shallow subtidal near Beaufort and Wilmington, N.C., and in sounds, year-round; common offshore from 14–32 m in Onslow Bay, April-September.

Distribution: North Carolina, Georgia, Bermuda, southern Florida, Gulf of Mexico, Caribbean, Brazil.

This species is retained in the flora with doubt as it is similar to *Chondria tenuissima.* Only the most typical coarse specimens are easily distinguished, these often bearing cuspidate apices.

Chondria polyrhiza Collins et Hervey 1917, p. 121, pl. II, fig. 12.

Figures 523 and 524

Plants epiphytic or saxicolous, pale rosy red, prostrate and spreading, with erect axes 5–8 mm tall, terete and somewhat firm, to 525 μm diameter, attached by random bundles of numerous short, simple, unicellular and multicellular rhizoids; axes infrequently to frequently alternately to irregularly branched, somewhat flexuous, these beset with one to two orders of shorter, more slender branches with more or less constricted bases; apices acute, the apical cells exposed or partially covered by short tufts of trichoblasts; pericentral cells not seen through cortical layers in older portions, occasionally in younger portions, the axes not appearing banded; outer cortical cells 5–15 μm diameter, 25–60 μm

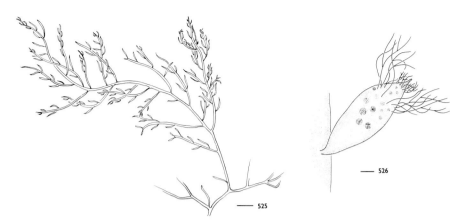

Figures 525, 526. *Chondria tenuissima*. 525. Habit, scale 0.5 cm. 526. Tetrasporic branch, scale 100 μm.

long; tetrasporangia globose, formed in the distal portions of ultimate branches, these swollen and subdentate in outline; gametangia unknown.

Kapraun 1980a; Searles 1987, 1988.

Known from 17–21 m on Gray's Reef and 30–35 m on Snapper Banks, June-August, and as an epiphyte on pelagic *Sargassum* spp. and from the Masonboro Inlet jetty near Wilmington (rare), summer.

Distribution: North Carolina, Georgia, Bermuda, southern Florida, Gulf of Mexico, Caribbean, Brazil.

Chondria tenuissima (Goodenough et Woodward) C. Agardh 1822, p. 352.

Fucus tenuissimus Goodenough et Woodward 1797, p. 215, pl. 19.

Figures 525 and 526

Plants saxicolous or conchicolous, dull purplish red to straw colored, erect to 25 cm tall, terete and firm in the main branches, soft in the ultimate ones, 0.5–2.5 mm diameter below, subpyramidal in outline, attached by discoid hold-fasts; main axes somewhat simple, or with a few similar main branches, beset with numerous long, widely divergent secondary branches 5–8 cm long; ulti-mate branches fusiform to linear-lanceolate, 250–500 μm diameter, 2–7(–15) mm long, narrow at the base, tapering gradually to acute apices, apical cells not ob-scured by short tufts of trichoblasts; tetrasporangia subglobose, 40–60(–150) μm diameter, concentrated near the distal ends of ultimate branches; cystocarps ovoid to broadly urceolate, to 1 mm diameter, subsessile or lateral on truncated lateral branches or distal portions of main axes.

Hoyt 1920; Williams 1948a, 1951; Humm 1952; Taylor 1960; Schneider 1976; Wiseman and Schneider 1976; Kapraun 1980a. As *C. littoralis sensu* Schneider 1976, *pro parte*.

Known from the shallow subtidal near Beaufort Inlet, N.C., and Oregon Inlet, Port Royal Sound, and from 14–40 m offshore in Onslow Bay, April-September; also known from intertidal coquina at Marineland, year-round.

Distribution: Massachusetts to Virginia, North Carolina, South Carolina, Ber-

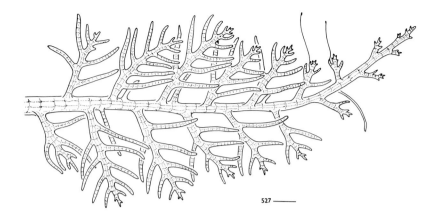

527 ———

muda, Georgia, northeastern Florida to Gulf of Mexico, Caribbean, British Isles to Portugal, Azores, Mediterranean, Red Sea, West Africa, Indian Ocean.

Only the most typical specimens can be distinguished from *Chondria atropurpurea*, *C. baileyana*, and *C. littoralis*, and all four species are probably members of a single species complex under this name.

Species Excluded

Chondria sedifolia Harvey

Plants identified as *Chondria sedifolia* from North Carolina and more southerly locations have been questioned (Taylor 1960, Kapraun 1980a). We have examined specimens first reported for this taxon in the Carolinas (Hoyt 1920) and find them to intergrade with the variable *C. dasyphylla*. Comparison of type and authentic material of these two taxa is necessary; however, we can find no grounds to retain *C. sedifolia* in the flora at present.

Dipterosiphonia Schmitz et Falkenberg 1897

Plants prostrate and spreading, dorsiventrally organized, attached by unicellular rhizoids with discoid or digitate tips, issued from ventral pericentral cells; main indeterminate axes flattened to terete, polysiphonous, composed of four to six pericentral cells, producing a branch initial from each segment, mostly developing into a regular pattern of primary determinate branches with occasional irregularly produced indeterminate branches; determinate branches produced in distichous pairs, one branched, the other unbranched, becoming polysiphonous; first-formed determinate branches dorsally and proximally displaced, the second-formed, ventrally displaced and more distal; branched determinate axes following a distichous pattern of secondary branches similar to that of primary branches on indeterminate axes or with distichous secondary branches alternating on every other segment; trichoblasts uncommon on axial tips; tetrahedral sporangia produced from pericentral cells, one per segment, greatly swelling the fertile branched or unbranched determinate branches; dioecious, spermatangia formed in cylindrical sori at the base of reduced trichoblasts on the tips of determinate axes; procarps produced on the second segment of reduced trichoblasts

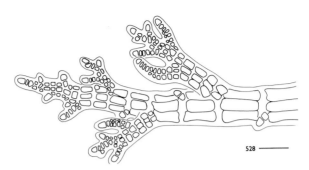

Figures 527, 528.
Dipterosiphonia re-versa. 527. Habit, scale 100 μm. 528. Detail of branch apex, scale 50 μm (redrawn from Schneider 1975b).

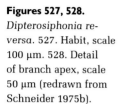

528 ———

near the tips of determinate axes, carpogonial branches four celled; carposporophytes within ovoid pericarps, one to two per determinate branch.

Dipterosiphonia reversa Schneider 1975b, p. 392, figs 3–7.

Figures 527 and 528

Plants epiphytic, brownish red, spreading to 2 cm, following the contour of the substrate, attached by one-celled rhizoids to 1 mm long, issued from the ventral pericentrals; main indeterminate axes terete, 80–105 μm diameter, dorsiventrally organized with five pericentral cells, ecorticate, producing pairs of alternately distichous branches; first-formed branch of each pair dorsally and proximally displaced, developing into a polysiphonous, determinate or indeterminate branch, producing secondary branches alternately distichous on every other segment, 50–75 μm diameter below, 20–40 μm near the apices to greater than 1 mm in length, occasionally branching to the third order; second-formed branch initial ventrally and distally displaced, often not developing or growing to a few cells in length and aborting, rarely developing into a simple, usually unbranched, polysiphonous lateral; sporangia and gametangia unknown.

Schneider 1975b, 1976; Searles 1987, 1988.

Rare; known from 28–40 m offshore in Onslow Bay, August, and from 17–21 m on Gray's Reef, July-August.

Distribution: North Carolina (type locality), Georgia.

This species is unique among members of the genus *Dipterosiphonia* in having the branched and unbranched members of alternate pairs in reversed sequence from all other taxa previously described (Schneider 1975b).

Herposiphonia Nägeli 1846

Plants prostrate and spreading, often with main axis tips and long uprights growing free from the substrate, dorsiventrally organized, attached by mostly unicellular rhizoids with digitate tips, issued from the distal end of ventral pericentral cells, usually one per cell; prostrate indeterminate axes terete to compressed, polysiphonous, composed of six to eighteen pericentral cells, ecorticate, dorsally producing erect indeterminate branches alternately on each fourth seg-

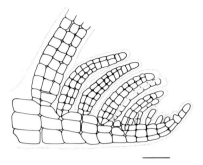

Figure 529. *Herposiphonia delicatula*, prostrate indeterminate axis tip, scale 20 μm.

ment, these often not greatly developed, the apices inrolled toward the distal end of prostrate axes, trichoblasts lacking; erect to subdistichous determinate branches or their rudiments, usually produced alternately on each segment, the first produced on the side opposite the previously formed indeterminate branch, usually arched towards and occasionally overtopping the distal end of prostrate axes, simple or branched, bearing trichoblasts; tetrahedral sporangia, one per segment, usually produced in straight series in the middle of the determinate branches on the side away from prostrate axis tip; dioecious, spermatangia formed in conical to linear-conical sori on or replacing trichoblasts in various positions on determinate branches; procarps produced on the second segment of reduced trichoblasts in various positions on determinate branches, carpogonial branches four celled; carposporophytes within ovoid to globular and urceolate pericarps, sessile or short stalked.

Key to the species

1. Prostrate indeterminate axes 30–45 μm diameter, the tips slightly upturned, mostly not overtopped by young determinate branches; spermatangial sori on a secund series of trichoblasts on the adaxial surface of determinate branches
. . . . *H. delicatula*
1. Prostrate indeterminate axes greater than 45 μm diameter, the tips strongly upturned, overtopped by young determinate branches; spermatangial sori on trichoblasts spiraled on the upper half of determinate branches . . . *H. tenella*

Herposiphonia delicatula Hollenberg 1968b, p. 540, figs 1A–B, 2H, 3.
Figure 529
Plants epiphytic, dark rosy to brownish red, bearing uprights from the spreading, entangled axes, attached by rhizoids with digitate and frequently multicellular tips; prostrate indeterminate axes 30–45 μm diameter with seven to eight pericentral cells, the segments one to two times as long as broad, bearing three upright determinate branches between indeterminate branches, bare nodes uncommon but as many as four in succession, the tips only slightly upturned and not overtopped by young determinate branches; erect indeterminate branches mostly remaining short; determinate branches simple, slightly arched towards the axial tips, 30–40 μm diameter with six to eight pericentral cells and twelve to fifty-three segments long, each one to one and a half (to two) times as long

as broad, apically bearing short tufts of three to four spirally arranged and one to three times dichotomously branched deciduous trichoblasts, occasionally not present on vegetative specimens, leaving obvious spiraled scar cells; tetrasporangia globose, to 60 μm in diameter, formed in a single, straight series of five to thirty-two, giving the fertile branch a nodulose appearance, in various portions of determinate branches; spermatangial stichidia lanceolate, with one to four sterile tip cells, to 25 μm diameter and 80 μm long, formed singly or in an adaxial secund series below the tips of determinate branches; cystocarps ovoid, 0.2 mm diameter, one to two forming in median positions of determinate branches, with obvious ostioles.

Hollenberg 1968b; Morrill 1976.

On shallow subtidal *Codium* at Radio Island jetty, July-August.

Distribution: North Carolina, Central Pacific islands.

Although Cribb (1983) has questioned whether *Herposiphonia delicatula* is distinct from *H. tenella*, Morrill (1976) corroborated and added to the specific sexual distinctions in the original description (Hollenberg 1968b). Vegetative specimens of the two taxa are difficult to separate. It is probable that if more of the collections of *Herposiphonia* from the area contained fertile gametophytes, the geographic range of *H. delicatula* would be extended.

Herposiphonia tenella (C. Agardh) Nägeli 1846, p. 238, pl. 8.

Hutchinsia tenella C. Agardh 1828, p. 105.

Figures 530–533

Plants epiphytic, saxicolous, epizoic, lignicolous or conchicolous, dark rosy to brownish and purplish red, bearing uprights from the spreading, entangled axes, attached by unicellular or few-celled rhizoids, often with digitate tips; prostrate indeterminate axes 50–90(–100) μm diameter with seven to ten pericentral cells, classically bearing three determinate branches between shorter, indeterminate branches or primordia, but bare nodes and irregular branching patterns occasionally common, determinate branch tips strongly upturned and curled back towards the axes, with young determinate branches curling towards them and mostly overtopping; erect indeterminate branches usually remaining short and erect, approximately 0.2 mm long, 20–25 μm diameter at the base, some lengthening and spreading laterally over the substrate; determinate branches simple, slightly constricted at the base and somewhat narrower than prostrate axes, gradually arching towards the axial tips, 1–2(–8) mm tall, 30–70 μm diameter, eleven to forty-five segments long, each one to two (to five) times as long as broad, bearing spirally arranged, dichotomously branched trichoblasts, occasionally not present, but leaving obvious spiraled scar cells; tetrasporangia globose, 35–60 μm in diameter, formed in a single, mostly unbroken

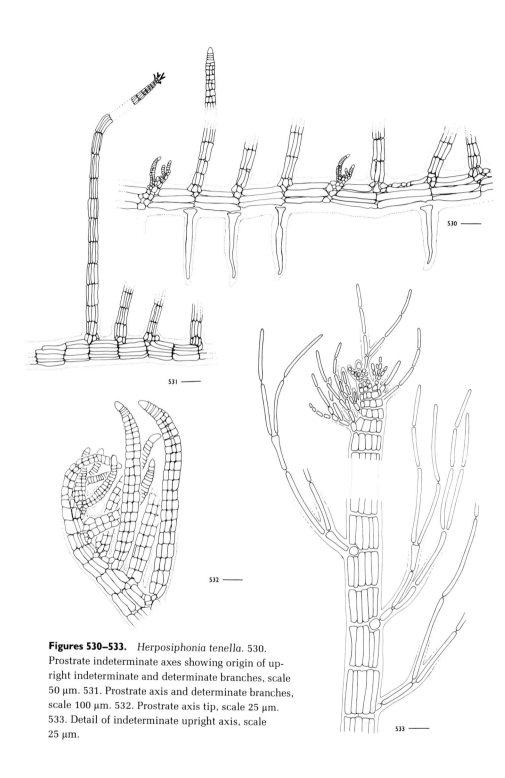

Figures 530–533. *Herposiphonia tenella*. 530.
Prostrate indeterminate axes showing origin of upright indeterminate and determinate branches, scale
50 μm. 531. Prostrate axis and determinate branches,
scale 100 μm. 532. Prostrate axis tip, scale 25 μm.
533. Detail of indeterminate upright axis, scale
25 μm.

straight series of ten to thirty, giving the fertile branch a nodulose appearance, in various portions of determinate branches; spermatangial stichidia cylindrical with one to two sterile tip cells and four to five internal segments, formed on unbranched, spirally arranged trichoblasts, usually one on each segment of the upper half of determinate branches; cystocarps ovoid when immature, becoming globular, 0.3–0.5 mm diameter, single on short polysiphonous stalks with obvious necks and ostioles.

Hoyt 1920; Williams 1948a, 1949; Taylor 1960; Brauner 1975; Morrill 1976; Kapraun 1980a. ?As *H. secunda sensu* Blomquist and Pyron 1943. As *H. tenella* forma *secunda sensu* Wiseman and Schneider 1976.

Common on intertidal and shallow subtidal plants, especially *Padina* and *Dictyota*, and animals, especially *Styela*, *Bugula*, oysters, and barnacles, as well as rock and pilings throughout the Carolinas, year-round, reproducing July-September and losing trichoblasts during the coldest months.

Distribution: North Carolina, South Carolina, Bermuda, northeastern Florida to Gulf of Mexico, Caribbean, Brazil; elsewhere, widespread in tropical and subtropical seas.

Many workers have considered this a variety of *Herposiphonia secunda* (C. Agardh) Falkenberg (Wynne 1985a), but we here follow others (Børgesen 1915; Morrill 1976; Cordeiro-Marino 1978) in recognizing key sexual distinctions between them, specifically spermatangial structure and position and cystocarp position. All of the vegetative specimens located in the flora conform to Morrill's (1976) strict concept of *H. tenella* with spiraled, not strictly tufted and terminal, trichoblasts. Taylor (1960) followed the classical line in distinguishing these similar species: *H. secunda* with few to several bare nodes between uprights, *H. tenella* lacking bare nodes. *H. tenella* specimens from our area with some bare nodes are not uncommon, nor are they elsewhere (Morrill 1976). Thus, position of male and female gametangia as well as trichoblasts on determinate axes, rather than presence or absence of bare nodes, is the determining distinction between *H. tenella* and *H. secunda*. The only report of a collection of *H. secunda* in the flora was from the drift in North Carolina (Blomquist and Pyron 1943), and a voucher could not be located. Other reports merely accepted this drift report for summary listings in North and South Carolina.

Laurencia Lamouroux 1813, nom. cons.

Plants erect, often bushy, with one or several axes arising from common discoid holdfasts, often several plants closely clustered; axes polysiphonous, with large, hyaline, often isodiametric, inner cortical cells obscuring the pericentral cells below the apices, outer layer or layers of smaller, rounded cortical cells densely pigmented; axes fleshy to cartilaginous, terete to flattened, radially or pinnately

branched; branches forming on the basal cells of trichoblasts at apices, sparse
to dense ultimate branchlets often short, clavate, constricted little or not at all
at the bases, but expanding in diameter to the apices; fertile branchlets similar
or occasionally reduced and clustered; all axes terminating in single apical cells
sunken in shallow depressions (pits) and surrounded by tufts of evanescent,
rudimentary, branched trichoblasts; tetrahedral sporangia embedded beneath
the outermost cortical layer, produced from parenchymatous derivatives of pericentral cells; sporangia numerous and irregularly scattered below the branchlet
tips, often giving the surface a rough texture; dioecious, spermatangia hyaline,
formed on paniculate trichoblasts, sori cylindrical, ovoid to dolioform, densely
clustered in terminal bowl-shaped to cryptlike apical pits; procarps produced
on the second cells of truncated, three-celled trichoblasts within terminal pits,
carpogonial branches four celled; carposporophytes radiating from basal fusion
cells surrounded by pseudoparenchymatous, multilayered pericarps; cystocarps
sessile, ovoid to globose, displaced to a lateral position on ultimate branchlets.

Key to the species

1. Plants terete or slightly compressed 2
1. Plants distinctly flattened . . . . *L. pinnatifida*
2. Plants alternately radially to irregularly branched throughout; densely beset
   with short, tuberculate branchlets . . . . *L. poiteaui*
2. Plants dichotomously to pseudodichotomously branched (at least below),
   subcorymbose above; short, cylindrical branchlets clustered near the apices
   . . . . *L. corallopsis*

Laurencia corallopsis (Montagne) Howe 1918, p. 519.
   Sphaerococcus corallopsis Montagne 1842b, p. 49.
   Figures 534 and 535
   Plants saxicolous or conchicolous, brownish to dark rosy red, often encrusted
with calcareous red algae or bryozoans, to 16 cm tall, terete but often compressed in upper portions, firm in texture, attached by either small or overlapping, branched cartilaginous holdfasts; when young, 2–4 cm plants consisting
of compact dichotomously to pseudodichotomously branched clusters without
short clavate branchlets, later developing into longer, more distantly branched
main axes, dichotomously to pseudodichotomously branched below, more irregular, opposite to alternate above becoming subcorymbose, beset with few
to numerous short, clavate, often forked branchlets, usually clustered and becoming densely bunched, 1–4 mm long; axes and branches uniform, 1–2 mm
diameter, the obtuse tips usually swollen and forked, bases slightly constricted;
outermost cortical layer composed of compact, various-sized angular isodiamet-

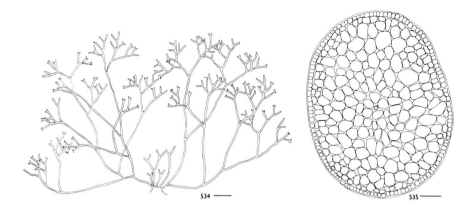

Figures 534, 535. *Laurencia corallopsis.* 534. Habit, scale 1 cm. 535. Cross section, scale 200 μm.

ric to rectangular cells (as viewed from the surface), to 50 μm diameter in greatest dimension; tetrasporangia obovoid to subglobose, 60–100 μm long, concentrated near the distal ends of ultimate branchlets; cystocarps subglobose to broadly ovoid with obvious necks, sessile near the distal end of ultimate branchlets.

Schneider and Searles 1973; Schneider 1976; Kapraun 1980a.

Common offshore from 20–60 m in Onslow Bay and from near-shore ledges off Wilmington, April-December.

Distribution: North Carolina, Bermuda, southern Florida, Caribbean, Philippines.

Older specimens of this species do not adhere well to herbarium paper.

Laurencia pinnatifida (Gmelin) Lamouroux 1813, p. 130.

Fucus pinnatifidus Gmelin 1768, p. 156.

Figure 536

Plants saxicolous, purplish to rosy red, to 15 cm tall, strongly compressed and firm to cartilaginous in texture, attached by small discoid, cartilaginous holdfasts, each issuing one to several upright axes; axes oppositely to alternately pinnate, branching to four orders; ultimate branchlets short and lanceolate to clavate, branchlets 2–3 mm long; axes and branches broadest in the middle, 1–3 mm diameter, tapering slightly to the bases and apices, the tips obtuse; outermost cortical layer composed of irregularly rounded to elongate cells as viewed from the surface, to 45 μm diameter in greatest dimension; tetrasporangia obovoid to subglobose, 25–35 μm long, concentrated in subapical bands in ultimate branchlets; spermatangia in one to four cryptlike apical pits in clavate ultimate branchlets; cystocarps broadly ovoid with obvious necks, to 1.2 mm diameter, sessile near the distal end of ultimate branchlets.

Schneider and Searles 1975; Schneider 1976.

Known only from 24–45 m in Onslow Bay, June-August.

Distribution: North Carolina, southern Florida, ?Caribbean, Brazil, Norway to Portugal, Azores, Mediterranean, ?West Africa.

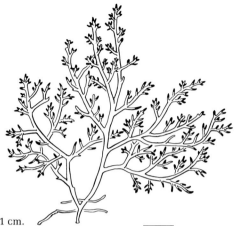

Figure 536. *Laurencia pinnatifida*, scale 1 cm.

This species may have been collected from the area as early as 1906 (Hoyt 1920), but the battered drift specimen could not be positively identified at the time, and vouchers cannot be located. Since its discovery in North Carolina (Schneider and Searles 1975), only tetrasporic specimens have been collected. Our plants are not nearly as broad as the European specimens, nor are they as densely branched (Saito 1982), but the overall habit and anatomy are comparable.

Laurencia poiteaui (Lamouroux) Howe 1905, p. 583.
　　Fucus poiteaui Lamouroux 1805, p. 63, pl. XXXI, figs 2, 3 ["*poitei*"].
　　Figure 537
Plants saxicolous or conchicolous, pale straw colored, pinkish green, or purplish to dark rosy red, often encrusted with calcareous red algae or bryozoans, to 15(–20) cm tall, terete and cartilaginous in texture, attached by small discoid holdfasts; main axes much and alternately to irregularly branched, 0.5–1.5 mm diameter, some branch systems virgate in outline, most irregular; axes generally diminishing in size from one order to the next, tips obtuse to slightly swollen, upper branches occasionally slightly flattened, beset, except at naked tips with few to numerous short, 0.5–2.0 mm, cylindrical branchlets, 0.3–1.0 mm diameter, occasionally tuberculate; outermost cortical layer composed of densely packed, irregularly rounded to elongate cells (as viewed from the surface), to 20 µm diameter in greatest dimension; tetrasporangia obovoid to subglobose, 90–130 µm long, concentrated near the broadened distal ends of ultimate branchlets; cystocarps subglobose to broadly ovoid with obvious necks, sessile near the distal end of ultimate branchlets.

　　Blomquist and Pyron 1943; Taylor 1960; Schneider 1976; Schneider and Searles 1979; Kapraun 1980a. As *L. tuberculosa* and *L. tuberculosa* var. *gemmifera sensu* Hoyt 1920. ?As *L. scoparia sensu* Williams 1951.

　　Common in certain areas offshore from 15–40 µm in Onslow Bay, year-round, and less commonly from shallow subtidal on most jetties and free-living in bays and estuaries during summer.

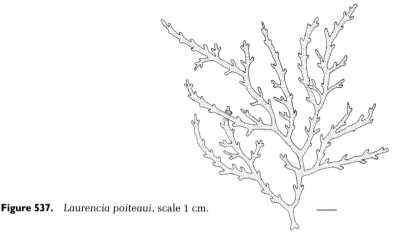

Figure 537. *Laurencia poiteaui*, scale 1 cm. ——

Distribution: North Carolina, Bermuda, southern Florida, Gulf of Mexico, Caribbean, Brazil.

This taxon has previously been listed as *Laurencia poitei* (refer to Silva et al. [1987] for the orthographic clarification). All of the specimens attributed to *L. tuberculosa* and var. *gemmifera* by Hoyt (1920) have been found to conform to more recently collected specimens of *L. poiteaui* (Schneider 1976), none having the mammillate cortical cell projections found in *L. gemmifera* (Taylor 1960). Monographic treatment of the western Atlantic species of *Laurencia* is sorely needed.

In the offshore reef and fishing area first sampled by Hoyt and Radcliffe (Hoyt 1920, map 3; Schneider 1976, fig. 1), *Laurencia poiteaui* is one of the dominant seaweeds, reaching its peak biomass in August (Schneider and Searles 1979). This 30 m site was found to be the most productive area for benthic algae offshore in the Carolinas (Schneider and Searles 1979).

Questionable Record

Laurencia filiformis (C. Agardh) Montagne 1845, p. 125.
    Chondria filiformis C. Agardh 1822, p. 358.
    As *L. scoparia*, Williams 1951.

No voucher specimens are known with the characteristics given for *Laurencia filiformis* (Rodríguez de Ríos and Saito 1985; as *L. scoparia*, Taylor 1960), and numerous offshore collections have not included this taxon reported off New River Inlet (Williams 1951). Until new collections are found to reestablish *L. filiformis* in the Carolinas, the above report is considered questionable.

Lophocladia Schmitz 1893

Species Excluded

Lophocladia trichoclados (Mertens in C. Agardh) Schmitz
This taxon was reported from an offshore reef in Onslow Bay by Williams (1951); however, no vouchers have been located, nor has it been recollected. It is possible that the plants collected were *Wrightiella tumanowiczii*, a related plant

known from offshore depths similar to those of Williams' report (see following account). *Wrightiella* bears few to many spur branchlets, unlike *Lophocladia*; plants with very few such branchlets could easily be confused with *Lophocladia*. Until a new collection of this taxon from the Carolinas is confirmed, it is excluded from the flora.

Micropeuce J. Agardh 1899

Plants erect, bushy, radially organized, arising from modest to massive discoid holdfasts, often bearing one or more uprights; indeterminate axes terete, polysiphonous, with five pericentral cells, heavily corticated below the tips by descending rhizoids, producing at the tips a spiral sequence of branch initials, one on each transversely divided apical cell prior to pericentral cell formation, each initial developing into a persistent, monosiphonous, alternately to pseudodichotomously branched, pigmented, determinate trichoblast; secondary indeterminate axes produced from basal cells of some trichoblasts, becoming long branches similarly beset with trichoblasts and indeterminate branches; tetrahedral sporangia, one per segment, acropetally developed in a spiral series in the upper portions of indeterminate axes, giving the axes twisted, nodulose appearances; dioecious, spermatangia formed in cylindrical sori replacing the entire trichoblast or a branchlet of the trichoblast, borne on short monosiphonous bases, with or without sterile tips; procarps produced on the second segment of reduced trichoblasts near the tips of indeterminate axes, carpogonial branches four celled; carposporophytes within ovoid to urceolate pericarps, borne singly, rarely paired.

Micropeuce mucronata (Harvey) Kylin 1956, p. 511.
  Dasya mucronata Harvey 1853, p. 63.
  Figure 538
  Plants saxicolous, brownish to rosy red, to 20 cm tall, attached by large cartilaginous holdfasts, each bearing one or more uprights; major axes terete, to 1.5 mm diameter in lower portions, indeterminate branching irregular, a branch usually produced at some distance from the previous one, the axes of at least the upper two thirds of the plant beset with persistent, pigmented, determinate trichoblasts, giving the plant as a whole a bushy appearance, the tips often appearing ocellate; trichoblasts issued spirally, one per axial cell, alternately to pseudodichotomously branched to four to six orders, divaricate, 0.6–1.0 mm long, basal cells clavate, 30–70 μm diameter, 50–130 μm long, median cells clavate to dolioform and cylindrical, 35–50 μm diameter, 90–120 μm long, tapering to mucronate tips, some trichoblast apices ending in flagelliform, rhizoidal, lightly pigmented filaments, more common at the tips of indeterminate axes;

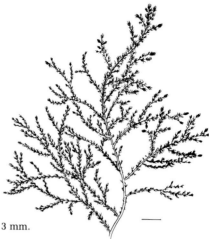

Figure 538. *Micropeuce mucronata*, scale 3 mm.

tetrasporangia globose, 90–150 μm diameter, in long spiral series at the tips of nodulose fertile indeterminate axes; spermatangial stichidia conical to linear-conical, terminal, subterminal or intercalary on trichoblasts, 60–90 μm diameter, 130–150 μm long, many per trichoblast system near the tips of indeterminate axes; cystocarps urceolate, ovate or subspherical, 0.3–0.7 mm diameter, 0.3–1.0 mm long, with obvious necks and ostioles.

Schneider 1976; Kapraun 1980a. As *Brongniartella mucronata*, Hoyt 1920; Taylor 1960.

Common offshore in Onslow Bay from 14–35 m, year-round, reaching peak abundance from July-September; also known from the inshore ledges of the Wilmington area.

Distribution: North Carolina, Bermuda, southern Florida, Gulf of Mexico, Caribbean, Brazil, Pacific Mexico.

Most often confused with *Dasya*, *Micropeuce* can be distinguished by its persistent short trichoblasts with mucronate tips and monopodial growth at the apex. In Brazil, this plant was first described as *Dasya sertularioides* (Howe and Taylor 1931), and later placed in a new genus, *Heterodasya* (Joly and Oliveira F. 1966) before being synonymized with *Micropeuce mucronata* (Oliveira F. 1977).

Polysiphonia Greville, 1823, nom. cons.

Plants erect and sparse to bushy, or prostrate and spreading, radially organized, arising from discoid to fibrous holdfasts or from prostrate axes anchored by unicellular or multicellular rhizoids, one to many per segment; indeterminate axes terete, polysiphonous with four to twenty-four pericentral cells, these straight or spiraled around axial rows, ecorticate to heavily corticated, mostly by division of pericentral cells, producing at the tips a spiral sequence of branch initials on every other, or greater distanced, axial cell prior to pericentral cell formation, each initial developing into an often deciduous, unbranched to alternately or pseudodichotomously branched, hyaline, determinate trichoblast or secondary indeterminate polysiphonous axis; in some, secondary indeterminate axes produced from the base of occasional trichoblasts; deciduous trichoblasts often leaving persistent scar cells; polysiphonous axes branched from one to several orders; tetrahedral sporangia, one per segment, acropetally developed in straight or spiral series in upper portions of indeterminate axes; dioecious, spermatangia formed in short- to long-conical or cylindrical sori on special hyaline branchlets exogenously developed from scar cells, or directly from trichoblasts, borne on short monosiphonous bases, with or without sterile tips; procarps produced on the second segment of reduced trichoblasts near the tips of indeterminate axes, carpogonial branches four celled; carposporophytes within spherical, ovoid to urceolate pericarps, with or without necks, with obvious ostioles, borne singly.

Polysiphonia is represented in the flora by sixteen species, and some are in need of further taxonomic clarification. In a major taxonomic study of *Polysiphonia* in North Carolina, Kapraun (1977a) collected nine species, and later work revealed four more (Kapraun 1980a). Richardson (1987) has reported seven species from coastal Georgia waters. Those reports have been supplemented by additional species records (Brauner 1975, Schneider 1976, Kapraun and Searles 1990). To date, several of the *Polysiphonia* species from North Carolina have been studied in culture (Kapraun 1977c, 1978a, 1978c, 1979). A difficult genus to pin a species tag on (Taylor 1960), vegetative characteristics are the most reliable for distinguishing species (Hollenberg 1942). For species with some overlap in characteristics, several options for separating species are given in the key below. Without enough of the characteristics used in this key, some specimens with four pericentral cells are best determined as simply *Polysiphonia* sp.

Key to the species

1. Plants with four pericentral cells per axial segment 2
1. Plants with more than four pericentral cells per axial segment 14
2. Plants erect, arising from discoid holdfasts, never becoming prostrate and spreading 3
2. Plants prostrate and spreading or initially erect and secondarily attached and spreading 5
3. Branches arising in the axils of trichoblasts 4
3. Branches replacing trichoblasts . . . . *P. gorgoniae*
4. Plants corticated at the base . . . . *P. harveyi*
4. Plants ecorticate throughout . . . . *P. binneyi*
5. Prostrate axes with rhizoids as long or short extensions of ventral pericentral cells, pit-connections lacking 6
5. Prostrate axes with rhizoids cut off, pit-connected to ventral pericentral cells 10

6. Branches replacing trichoblasts in the developmental sequence at apices
.... 7
6. Branches forming in the axils of trichoblasts at apices ...... *P. havanensis*
7. Plants restricted to brackish water environments .... *P. subtilissima*
7. Plants not restricted to brackish water environments 8
8. Plants small, erect axes to 2 cm tall; tetrasporangia 40–60 μm diameter;
   spermatangial branches replacing trichoblasts, not lateral on them 9
8. Plants large, erect axes to 10 cm tall; tetrasporangia 65–90 μm diameter;
   spermatangial branches lateral on lower segments of trichoblasts ....
   .... *P. urceolata*
9. Plants with scar cells, remnant bases of deciduous trichoblasts; trichoblasts
   simple to twice branched; cystocarps urceolate .... *P. atlantica*
9. Plants lacking scar cells; when present, trichoblasts branched to four orders;
   cystocarps ovoid .... *P. scopulorum* v. *villum*
10. Branches forming in the axils of trichoblasts at apices................ 11
10. Branches not forming in the axils of trichoblasts 12
11. Plants 1–2 cm tall, branching alternate; cells surrounding cystocarp ostioles
    similar-sized to other pericarp cells .... *P. flaccidissima*
11. Plants 3–12 cm tall (larger if planktonic), branching pseudodichotomous;
    cells surrounding cystocarp ostioles larger than other pericarp cells ......
    .... *P. breviarticulata*
12. Branches replacing trichoblasts in the developmental sequence at apices, en-
    dogenously developed 13
12. Branches exogenously forming only from scar cells of trichoblasts ....
    .... *P. pseudovillum*
13. Plants small, to 2 cm tall; branching mostly pseudodichotomous; cystocarps
    200–300 μm diameter, the ostioles ringed by cells larger or the same size as
    those in the middle of the pericarp .... *P. sphaerocarpa*
13. Plants taller than 2 cm; branches distinctly alternate; cystocarps to 200 μm
    diameter, the ostioles ringed by cells shorter than those in the middle of the
    pericarp .... *P. ferulacea*
14. Plants with five to six pericentral cells per axial segment ..... *P. denudata*
14. Plants with seven or more pericentral cells per axial segment.......... 15
15. Plants arising from prostrate, spreading filaments, with seven to twelve peri-
    central cells; tetrasporangia less than 60 μm diameter 16
15. Plants erect from discoidal holdfasts, with twelve to sixteen pericentral
    cells; tetrasporangia 60–100 μm diameter .... *P. nigrescens*
16. Plants with seven to eight pericentral cells, branches forming in the axils of
    trichoblasts at apices, trichoblasts without conspicuously broadened supra-
    basal cells; rhizoids issued from the proximal ends of pericentral cells; tetra-

sporangia 30–35 µm diameter.............................. *P. tepida*
16. Plants with eight to twelve pericentral cells, branches replacing trichoblasts in the developmental sequence at apices, trichoblasts with conspicuously broadened suprabasal branch cells above small rounded basal cells; rhizoids issued from the distal ends of pericentral cells; tetrasporangia 40–55 µm diameter .... *P. howei*

Polysiphonia atlantica Kapraun et J. Norris 1982, p. 226, figs 107a–c.
Figures 539–542

Plants saxicolous, lignicolous, or epiphytic, brownish and purplish red to blackish, prostrate indeterminate axes spreading and forming mats, attached by numerous unicellular rhizoids, giving rise to erect indeterminate axes 1–2 cm tall; prostrate axes 60–100 µm diameter, with obvious apical cells and ecorticate segments two to three diameters long with four pericentral cells, the tips curved upwards, axes showing conspicuous scar cells from deciduous determinate trichoblasts and giving rise to erect axes from every fourth to sixth segment; rhizoids mostly issued centrally from pericentral cells and formed in open connection with them, with or without digitate tips; erect axes slightly narrower than prostrate axes, pseudodichotomously branched below, sparsely branched above, branches replacing trichoblasts in developmental sequence at apices, trichoblasts usually lacking, occasionally persistent, simple to twice branched; adventitious branching rare; tetrasporangia globose to ellipsoid, 40–60 µm diameter, in long, straight, usually continuous series in upper half of erect axes, causing the segments to swell when mature; spermatangial sori on one- to few-celled stalks, cylindrical, to 40 µm diameter and 180 µm long, forming from entire trichoblast primordia on tips of erect axes, without sterile tip cells; cystocarps short stalked to sessile, formed centrally on erect axes, urceolate, to 200 µm diameter and 250 µm long, with obvious necks and ostioles.

Searles 1987, 1988. As *P. macrocarpa sensu* Williams 1948a, 1949; Humm 1952; Taylor 1960; ?Schneider 1976; Kapraun 1977a, 1980a; Richardson 1986, 1987. As *P. gorgoniae sensu* Richardson 1987.

Known from lower intertidal rocks and pilings throughout the Carolinas and Georgia and Marineland, Florida, and from the air-water interface of floating buoys, reaching peak size and abundance June-September, but present year-round as decumbent mats; common from 17–21 m on rock at Gray's Reef and 30–35 m on Snapper Banks, June-August.

Distribution: North Carolina, Georgia, Bermuda, northeastern Florida to Gulf of Mexico, Caribbean, Brazil, British Isles to Portugal, Mediterranean, Adriatic, West Africa, Indian Ocean.

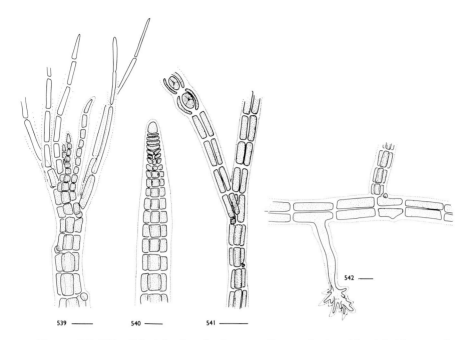

Figures 539–542. *Polysiphonia atlantica.* 539. Erect axis tip with trichoblasts, scale 20 μm. 540. Erect axis tip without trichoblasts, scale 15 μm. 541. Axis with tetrasporangia, scale 100 μm. 542. Prostrate axis with rhizoid, scale 50 μm.

Kapraun and Norris (1982) abandoned usage of the name *Polysiphonia macrocarpa* Harvey in Mackay (1836) for plants classically applied to that taxon as it is a later homonym of *P. macrocarpa* (C. Agardh) Sprengel (1827), a plant in need of further investigation (Womersley 1979). Thus, the new name, *P. atlantica*, was chosen for this widespread taxon.

Plants reported as *Polysiphonia macrocarpa* for offshore in North Carolina (17–30 m, July-September; Schneider 1976) differ greatly from inshore populations of *P. atlantica* as outlined above, but do not closely fit any other species currently reported for the area. Although these specimens have rhizoids in open connection with ventral pericentral cells and branches which replace trichoblasts at apices, the plants are much larger than *P. atlantica* and *P. scopulorum* var. *villum*. Unlike these species and the larger *P. urceolata*, the offshore specimens have compact, spiraled tetrasporangia in numerous short adventitious branches. Clearly, these deep-water plants need further attention to discover whether they belong to a taxon not currently reported for the area or rather represent ecotypic variation in one of the species listed in this section.

Polysiphonia binneyi Harvey 1853, p. 37.

Plants epiphytic, dark brownish to rosy red, 0.5–3 cm tall, attached by small discoid holdfasts, each issuing one or more erect indeterminate stiff axes, ecorticate, lower segments issuing rhizoids for secondary attachment, in open connection with pericentral cells; erect indeterminate axes pseudodichotomous and alternate below, alternate above, becoming bushy, axes 68–380 μm diameter, segments one to two and a half diameters long, with four pericentral cells; branches forming in the axils of trichoblasts at apices, often numerous short, branched, adventitious branches issued directly from scar cells below; trichoblasts persistent at the apices; tetrasporangia globose, 56–80 μm diameter, in short, compact spiral series below apices causing the segments to swell when mature; spermatangial sori borne laterally on lower segments of trichoblasts, cylindrical, 25–37(–50) μm diameter, 100–150(–200) μm long, mostly with two to three sterile tip cells; cystocarps short stalked, formed in upper branches, urceolate to subglobose, 190–420 μm diameter, with broad ostioles, without necks.

Hall and Eiseman 1981.

Known as an epiphyte of sea grasses in Haulover Cove, Indian River, Florida, year-round.

Distribution: Bermuda, northeastern Florida to Gulf of Mexico, Caribbean, Brazil.

In the flora, *Polysiphonia binneyi* is most similar to *P. harveyi* and *P. havanensis*. Several specimens need to be examined, including immature as well as mature fertile ones, to insure proper identification.

Polysiphonia breviarticulata (C. Agardh) Zanardini 1840, p. 203.

Hutchinsia breviarticulata C. Agardh 1824, p. 153.

Figures 543 and 544

Plants epiphytic or planktonic, brownish to purplish red; attached plants with prostrate indeterminate axes spreading and forming mats, affixed by numerous unicellular rhizoids, giving rise to erect indeterminate axes 3–12 cm tall; rhizoids issued proximally from lower pericentral cells, occasionally more than one per cell, pit-connected to them, with or without digitate tips; main axes with four pericentral cells, segments to 500 μm diameter and 250–500 μm long below, older ones sometimes lightly corticated by rhizoids, above 80–140 μm diameter with segments one to one and a half diameters long and ecorticate; erect axes tapering towards the apices, pseudodichotomously branched, fastigiate; branches arising in the axils of trichoblasts at the apices, below arising exogenously from conspicuous scar cells from deciduous determinate trichoblasts, constricted at the base, distinctly narrower than the axes producing them, mostly abundant and narrow angled to the axes; trichoblasts, when present, dichotomously branched to several orders; tetrasporangia globose, 50–75 μm diameter, in spiral series in upper portions of plants causing the segments to swell slightly when mature; spermatangial sori on one-celled stalks, cylindrical, 25–40 μm diameter and 80–120 μm long, forming in place of lower branches

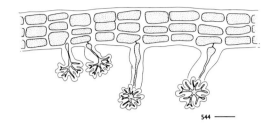

Figures 543, 544. *Polysiphonia breviarticulata.* 543. Axis with lateral branch, scale 50 μm. 544. Prostrate axis with rhizoids, scale 100 μm.

543 ———

on trichoblasts near branch apices, without sterile tip cells; cystocarps short stalked, spherical to ovoid, 340–450 μm diameter, with broad ostioles surrounded by larger sized cells than the rest of the pericarp, without necks.

Kapraun and Searles 1990.

Known as an epiphyte of *Zostera* and macroalgae in the sounds near Wrightsville Beach, December-June, and perennating as free-living plants near the sea floor, at times reaching bloom proportions and cast ashore in the drift on Carolina beaches.

Distribution: North Carolina, South Carolina, West Indies, Mediterranean, Adriatic, Canary Islands.

A planktonic bloom of this species was observed during the spring and summer of 1988, with large quantities washing ashore and into the sounds between Cape Lookout and Myrtle Beach. Kapraun and Searles (1990) noted that the bloom coincided with unseasonably cool surface temperatures in the area which remained below normal into June. Attached plants in southeastern North Carolina have a biomass and reproductive peak in March-April.

Free-living plants have polysiphonous axes branched to several orders and mostly lack trichoblasts and reproductive structures. Unlike attached plants, these planktonic forms have axes only to 210 μm diameter but have much longer segments, five (to eight) diameters long. The identification of these specimens was considered tentative by Kapraun and Searles (1990) as the type specimen from the Adriatic Sea was unavailable.

Polysiphonia denudata (Dillwyn) Greville ex Harvey 1833, p. 332.

Conferva denudata Dillwyn 1809, p. 85.

Figures 545 and 546

Plants epiphytic, saxicolous, lignicolous, or conchicolous, brownish to dark purplish red, mostly to 15, occasionally to 25 cm tall, initially attached by small

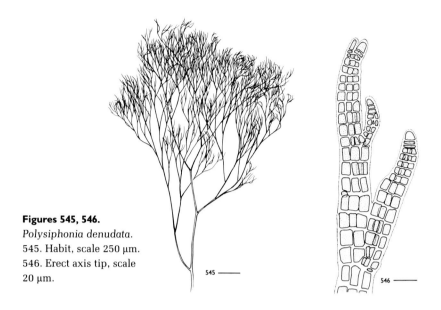

Figures 545, 546.
Polysiphonia denudata.
545. Habit, scale 250 µm.
546. Erect axis tip, scale
20 µm.

545 ———

546 ———

discoid holdfasts, later becoming secondarily attached by numerous unicellular rhizoids, forming large, loose tufts; indeterminate axes sparsely branched
and 100–750 µm diameter below, much and alternately to pseudodichotomously
branched above, ecorticate or slightly corticated at the bases, segments one (to
three) diameter long with five to six pericentral cells, axes showing scar cells
from deciduous determinate trichoblasts; secondary axes widely angled, narrower and more lax than lower portions of primary axes, 100–170 µm diameter,
tapering even more to the ultimate branches, 33–45 µm diameter; rhizoids issued
from the proximal ends of pericentral cells, pit-connected to them; branches
forming in the axils of trichoblasts at apices, trichoblasts usually lacking, occasionally persistent on rapidly developing axes; tetrasporangia subglobose to
wide ellipsoid, 50–85 µm diameter, in long, straight series in upper half of erect
axes causing the segments to swell slightly when mature; spermatangial sori
borne laterally on lower segments of trichoblasts in acropetal series at axis tips,
long and conical to cylindrical and fusiform, 30–80 µm diameter, 100–300 µm
long, with or without one to three sterile tip cells; cystocarps short stalked,
formed in upper branches, globose to broadly ovoid, 200–450 µm diameter, with
narrow to broad ostioles, without necks.

Hoyt 1920; Williams 1948a, 1949, 1951; Taylor 1960; Chapman 1971; Brauner
1975; Schneider 1976; Wiseman and Schneider 1976; Kapraun 1977a, 1978a,
1980a, 1980b; Richardson 1987; Searles 1987, 1988. As *P. variegata*, Harvey
1853; Curtis 1867; Melvill 1875.

Known year-round from the shallow subtidal throughout the Carolinas and
Georgia, epiphytic on larger algae or directly attached to jetties, seawalls, pilings,
and floating buoys, on the open coast or in protected bays and sounds, reaching
peak size and abundance November-March; common offshore from 15–48 m in
Onslow Bay, May-November, and rare from 17–21 m on Gray's Reef and 30–35 m
on Snapper Banks, July-August.

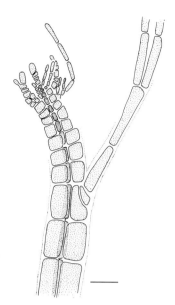

Figure 547. *Polysiphonia ferulacea*, erect axis tip, scale 20 µm.

Distribution: Maine to Virginia, North Carolina, South Carolina, Georgia, Bermuda, southern Florida, Gulf of Mexico, Caribbean, Brazil, British Isles to Portugal, Mediterranean, Adriatic, Black Sea, West Africa, ?Central Pacific islands.

Polysiphonia ferulacea Suhr ex J. Agardh 1863, p. 980.

Figure 547

Plants epiphytic, saxicolous or lignicolous, light brownish to purplish red, to 6(–15) cm tall, initially attached by fibrous discoid holdfasts, later becoming secondarily attached by numerous unicellular rhizoids from adventitious branches; prostrate indeterminate axes lacking trichoblasts, giving rise to erect axes from every fourth to sixth segment; erect indeterminate axes conspicuously alternately branched, becoming bushy, axes 80–100(–300) µm diameter below, ecorticate, segments one diameter long with four pericentral cells, lower axes showing scar cells from ultimately deciduous determinate trichoblasts; secondary axes narrowly angled, slightly thinner than primary axes, often basally constricted, the ultimate branches tapering to acute apices; rhizoids issued from the distal ends of pericentral cells, pit-connected to them; branches replacing trichoblasts in developmental sequence at apices, adventitious branches issued from lower axes directly from scar cells, trichoblasts persisting, conspicuous, and highly branched; tetrasporangia globose, 50–60 µm diameter, in short, spiral series in upper half of erect axes, causing the segments to swell when mature; spermatangial sori borne laterally on lower segments of trichoblasts in acropetal series at axis tips, short and cylindrical, to 60 µm diameter and to 150 µm long, with one to two thick-walled, sterile tip cells; cystocarps sessile to short stalked, formed in upper branches, globose to broadly ovoid, to 200 µm diameter, with broad ostioles surrounded by cells shorter than those below in the pericarp, without necks.

Kapraun 1977a, 1977c, 1978a, 1978c, 1980a, 1980b.

Known only, yet commonly, from the lower intertidal and shallow subtidal of jetty and inlet rocks in the Wrightsville Beach area, spring-fall, reaching peak abundance April-May.

Distribution: North Carolina, Bermuda, southern Florida, Gulf of Mexico, Caribbean, Brazil, France to Portugal, West Africa; elsewhere, widespread in tropical seas.

A plant extensively studied in culture in the Carolinas (Kapraun 1978a, 1978c), this species has at least two genetically isolated populations here, one of which has vegetative propagules similar in form and position to spermatangial sori (Kapraun 1977c).

Polysiphonia flaccidissima Hollenberg 1942, p. 783, fig. 8.

Figures 548 and 549

Plants epiphytic or saxicolous, rosy red, minute, prostrate indeterminate axes spreading, attached by numerous unicellular rhizoids, giving rise to erect in-determinate axes 1(–2) cm tall; prostrate axes 70–300 μm diameter, ecorticate segments one to two diameters long with four pericentral cells, the tips slightly upturned, axes showing conspicuous scar cells from deciduous determinate trichoblasts and giving rise to erect axes; rhizoids mostly issued from the proxi-mal ends of pericentral cells, pit-connected to them, with or without digitate tips, occasionally more than one per cell; erect axes 70–150 μm diameter, alter-nately branched throughout to several orders, branches forming in the axils of trichoblasts at apices, adventitious branches lacking; trichoblasts long, con-spicuous and branched to three orders, soon deciduous; tetrasporangia globose, 40–70 μm diameter, in long and occasionally interrupted spiral series in upper portions of erect axes causing the segments to swell when mature; spermatan-gial sori borne laterally on lower segments of trichoblasts, cylindrical, 30–50 μm diameter, 100–180 μm long, without sterile tip cells; cystocarps short stalked, globose, to 500 μm diameter.

Brauner 1975; Schneider 1976; Kapraun 1980a.

Known as an epiphyte on *Zostera* in the Beaufort, N.C., area and on *Sargassum* at Wrightsville Beach year-round, reaching peak abundance August-November; known offshore as an epiphyte of macroscopic algae from 18–50 m in Onslow Bay, May-August.

Distribution: North Carolina, Belize, Central Pacific islands, Peru, Pacific Mexico, California.

Originally described as having sterile tip cells on spermatangial sori (Hollen-berg 1968a); however, Atlantic and Caribbean specimens do not contain them (Brauner 1975; Kapraun and Norris 1982). This diminutive species is possibly

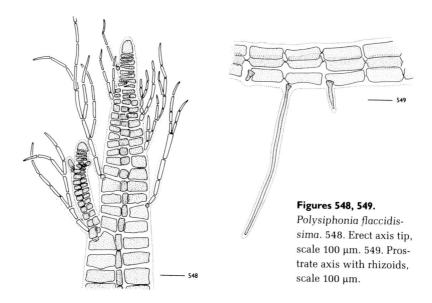

Figures 548, 549.
Polysiphonia flaccidis-sima. 548. Erect axis tip, scale 100 μm. 549. Prostrate axis with rhizoids, scale 100 μm.

more broadly distributed in the western Atlantic Ocean, potentially confused with Børgesen's (1918) concept of *Polysiphonia havanensis* Montagne. Womersley (1979) further suggested that *P. flaccidissima* may be synonymous with the widespread *P. sertularioides* (Grateloup) J. Agardh. Further studies are necessary to clarify these problems.

Polysiphonia gorgoniae Harvey 1853, p. 39.

Plants epizoic, epiphytic, saxicolous, tufted, pale yellowish brown to brownish red, 0.5–2(–3) cm tall, attached by small discoid holdfasts, each issuing one or two erect indeterminate axes; erect indeterminate axes pseudodichotomously branched every fifth to sixth segment to several orders, irregular in upper portions, becoming flabellate in outline, axes tapering from bases to apices, to 105 μm diameter below, to 75 μm in median sections, ecorticate, segments seven tenths to one diameter long below, one and a half to two diameters long above, and very short in the ultimate branches, with four pericentral cells; trichoblasts few, found only at the apices, soon deciduous, leaving obvious scar cells; tetrasporangia globose, about 75 μm diameter, in spiral series causing the segments to swell only slightly when mature; cystocarps sessile or short stalked, formed in upper branches, globose to compressed globose, 250–400(–500) μm diameter, with broad ostioles, without necks.

Chapman 1971; not Richardson 1987 (see *P. atlantica*).

Known only as an epiphyte of larger algae and on rock from the intertidal of Cumberland Island jetty, July-August.

Distribution: Georgia, southern Florida, Caribbean, Brazil.

This species is one of only three entirely erect *Polysiphonias* in the flora and is found only on one jetty, a site where *P. harveyi* and *P. binneyi* have not been collected to date.

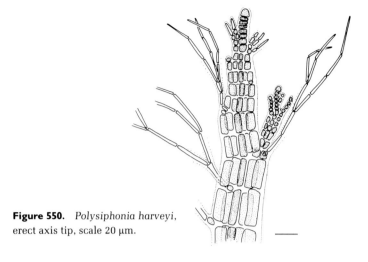

Figure 550. *Polysiphonia harveyi*, erect axis tip, scale 20 μm.

Polysiphonia harveyi Bailey 1848, p. 38.

Figure 550

Plants epiphytic, saxicolous, conchicolous or lignicolous, dark brownish to purplish red and black, 3–11 cm tall, attached by small discoid holdfasts, each issuing one or more erect indeterminate stiff axes; erect indeterminate axes conspicuously alternately to irregularly branched to several orders, becoming bushy and broadly pyramidal in outline, axes 225–675 μm diameter below, only young plants ecorticate, becoming corticated by smaller cells between pericentral cells and descending rhizoids in basal regions, segments three tenths to one diameter long below, seven tenths to two diameters long above, with four pericentral cells; secondary axes only slightly thinner than primary axes, but branching thereafter gradually reducing in size, the ultimate branches tapering to acute apices, often appearing fusiform; branches forming in the axils of trichoblasts at apices, often numerous short, branched adventitious branches issued directly from inconspicuous scar cells below; trichoblasts usually found only at the apices, soon deciduous; tetrasporangia globose, 60–100 μm diameter, in short, compact spiral series below apices causing the segments to swell greatly when mature; spermatangial sori borne laterally on lower segments of trichoblasts in acropetal series at axis tips, short cylindrical to conical, 40–60 μm diameter, 100–180 μm long, with or without 1 sterile tip cell; cystocarps short stalked, formed in upper branches, globose to broadly ovoid, 250–400(–500) μm diameter, with broad ostioles, without necks.

var. *olneyi* (Harvey) Taylor 1937, p. 363.

Polysiphonia olneyi Harvey 1853, p. 40, pl. XVIIb.

Plants light brownish red, lax, and openly branched to 15 cm tall; tetrasporangia less congested due to longer segments in upper portions; spermatangial sori with one to two sterile tip cells; cystocarps urceolate with broadly flared ostioles.

Hoyt 1920; Taylor 1957; Williams 1949; Brauner 1975; Schneider 1976; Wiseman and Schneider 1976; Kapraun 1977a, 1978b, 1980a, 1980b; Richardson 1986, 1987.

Known from the intertidal and shallow subtidal of jetties and breakwaters throughout North Carolina, both on rocks and larger algae, from buoys in open ocean and on *Zostera* and other substrates in protected sounds, year-round, reaching peak abundance December-March; found as massive (to 20 cm diameter) free-living, rolling balls in shallow water of a protected harbor at Ocracoke Inlet, summer; known from South Carolina in dredgings from Port Royal Sound, January; and found inshore in Georgia, November-January; from 18–19 m offshore in Onslow Bay as a rare epiphyte, June.

Distribution: Newfoundland to Virginia, North Carolina, South Carolina, Georgia.

In a study comparing the typical form of this species (as var. *arietina* Harvey) with var. *olneyi* from southern North Carolina habitats, Kapraun (1978b) found the two entities to have different basal chromosome numbers, n = 32 for the common *Polysiphonia harveyi* and n = 28 for var. *olneyi*.

Polysiphonia havanensis Montagne 1837, p. 352 *sensu* Børgesen 1918, p. 266, figs 259–61.

Figures 551 and 552

Plants epiphytic or saxicolous, yellowish to brownish and purplish red, prostrate indeterminate axes spreading, attached by numerous unicellular rhizoids,

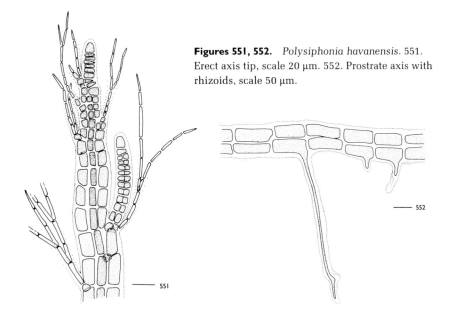

Figures 551, 552. *Polysiphonia havanensis*. 551. Erect axis tip, scale 20 μm. 552. Prostrate axis with rhizoids, scale 50 μm.

giving rise to erect indeterminate axes to 8 cm tall; indeterminate axes ecorticate, with four pericentral cells, showing conspicuous scar cells from deciduous determinate trichoblasts; rhizoids formed on prostrate axes remaining in open connection with pericentral cells, with or without digitate tips; prostrate axes 80–200 µm diameter, erect axes 50–100 µm diameter below, segments one half to two (to three) diameters long, alternately to irregularly branched to few orders, branches forming in the axils of trichoblasts at apices, mostly remaining short and slightly narrowed at the bases, numerous adventitious branches forming from scar cells throughout; trichoblasts conspicuous and highly branched, deciduous below apices; tetrasporangia globose, 50–70 µm diameter, scattered singly or in short series of two to three in upper portions of erect axes; spermatangial sori borne laterally on lower segments of trichoblasts, cylindrical to long and conical, 30–60 µm diameter, 80–300 µm long, without sterile tip cells; cystocarps short stalked, globose to broadly ovoid and urceolate, to 250–300 µm diameter, with narrow ostioles.

Hoyt 1920; Williams 1948a, 1949; Taylor 1960; Kapraun 1977a, 1980a; Richardson 1986, 1987.

Known from the lower intertidal of Beaufort, N.C., and Wilmington area jetties and from the coastal waters of Georgia, including Cumberland Island jetty, binding sand, year-round, but reaching peak size and abundance April-August.

Distribution: North Carolina, Georgia, Bermuda, southern Florida, Gulf of Mexico, Caribbean, Brazil.

This species is possibly synonymous with *Polysiphonia sertularioides* (Grateloup) J. Agardh (see p. 467) from the Mediterranean Sea (see Kapraun 1977a, Womersley 1979), although we remain unconvinced. Further clarification is necessary to elucidate these similar species on a worldwide basis; however, both *P. havanensis sensu* Børgesen and *P. flaccidissima* are found and easily distinguished in the flora (Kapraun 1980a, Kapraun and Norris 1982).

Polysiphonia howei Hollenberg in W. R. Taylor 1945, p. 302, fig. 3.

Plants saxicolous, epiphytic, or lignicolous, rosy red, minute to moderate size, prostrate indeterminate axes spreading, attached by numerous unicellular rhizoids, giving rise to erect indeterminate axes 0.5–1(–5.5) cm tall; prostrate axes 100–170 µm diameter, ecorticate segments seven tenths to one diameter long with eight to twelve pericentral cells, the tips slightly upturned, axes showing conspicuous scar cells from deciduous determinate trichoblasts and giving rise to erect axes; rhizoids mostly issued from the distal ends of pericentral cells, pit-connected to them, with or without digitate tips, occasionally more than one per cell; erect axes 70–150 µm diameter, at first strongly arched towards the prostrate axis tips, simple or sparingly branched, branches replacing tricho-

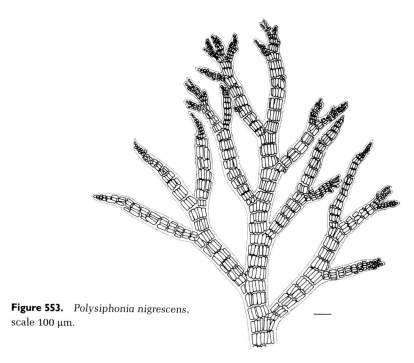

Figure 553. *Polysiphonia nigrescens,*
scale 100 μm.

blasts in developmental sequence at apices, adventitious branches from occasional scar cells in older portions of plants; trichoblasts short but branched to three orders, with conspicuously enlarged suprabasal branch cells above small, rounded basal cells, soon deciduous; tetrasporangia globose, 40–55 μm diameter, in long spiral series in upper portions of erect axes causing the segments to swell when mature; spermatangial sori borne laterally on lower segments of trichoblasts, 22–25 μm diameter, 100–170 μm long; cystocarps globose to subglobose, 140–190 μm diameter.

Williams 1948a, 1949; Hollenberg 1958; Taylor 1960; Kapraun 1980a.

Known only from lower intertidal rock on Cape Lookout jetty and as an epiphyte of *Sargassum filipendula* in the shallow subtidal at Wrightsville Beach, year-round, mats reaching greatest size spring-summer.

Distribution: North Carolina, Bermuda, southern Florida, Caribbean, Brazil, Central Pacific islands, Pacific Central America.

Polysiphonia nigrescens (Hudson) Greville in Hooker 1833, p. 322.
   Conferva nigrescens Hudson 1778, p. 602.
   Figure 553
   Plants saxicolous, lignicolous or epiphytic, dark brown to purple and black, 4–30 cm tall, attached by large, spreading, discoid holdfasts, each issuing one to several stiff, erect indeterminate leading axes; erect indeterminate axes conspicuously alternately branched to several orders above, pinnately to spirally arranged, becoming bushy or broadly triangular in outline, the tips often fasciculate and corymbose, upper axes occasionally beset with short, adventitious, spinelike branchlets, sparingly branched below, axes 500–650(–850) μm diameter below, becoming corticated in basal regions, segments to ten diameters long,

usually one half to two diameters long, with (eight–)twelve–sixteen(–twenty) pericentral cells; secondary axes only slightly thinner than primary axes, but branching thereafter gradually reducing in size, the ultimate branches tapering to acute apices; branches replacing trichoblasts in the developmental sequence at apices; trichoblasts usually found only at the apices, soon deciduous; tetrasporangia globose, 60–100 µm diameter, in long, occasionally interrupted spiral series below apices causing the segments to swell when mature, often formed in paired branchlets, appearing forked; spermatangial sori borne laterally on lower segments of trichoblasts in acropetal series at axis tips, long and conical; cystocarps short stalked, formed in upper branches, subglobose to broadly ovoid, 275–500 µm diameter, with narrow ostioles, without necks.

Hoyt 1920; Williams 1948a, 1949; Taylor 1957, 1960; Brauner 1975; Wiseman and Schneider 1976; Kapraun 1977a, 1980a.

Frequent from lower intertidal rock of jetties in the Beaufort-Cape Lookout area in winter-spring, uncommon elsewhere to the south in the Carolinas; known on floating docks in Ocracoke harbor, June.

Distribution: Newfoundland to Virginia, North Carolina, ?South Carolina, ?Bermuda, USSR to Portugal, Alaska.

This taxon apparently reaches its southern limit of distribution in the western Atlantic in the Carolinas, but the exact location is unknown. *Polysiphonia nigrescens* was verified from Wilmington, North Carolina (Kapraun 1977a), but its report as small plants from Pawleys Island, South Carolina (Hoyt 1920) has been questioned (Wiseman 1966) and a voucher could not be located. Further collections from South Carolina are needed to firmly establish this taxon in the state's flora. Clearly recognized in the winter and spring floras on Cape Lookout jetty and several other breakwaters near Beaufort, N.C., this area is the southern limit of continuous abundance from Newfoundland for the species.

Highly variable throughout its geographic range, several subspecific taxa have been proposed (see Harvey 1853, Taylor 1957). However, continuous variation between certain forms exists in North Carolina, and we do not choose to recognize varieties. Small, distinctly pinnate specimens have a macroscopic similarity to *Pterosiphonia pennata*, which regularly bears a bilateral axis from every other segment and lacks trichoblasts (see p. 479).

Polysiphonia pseudovillum Hollenberg 1968a, p. 73, fig. 3C.

Plants epiphytic, conchicolous or saxicolous, minute, prostrate indeterminate axes spreading, attached by numerous unicellular rhizoids, giving rise to erect indeterminate axes 1–2.7 mm tall; prostrate axes 60 µm diameter, ecorticate segments one to one and a half diameters long with four pericentral cells, giving rise to erect axes from every fourth to eighth segment, showing obvious

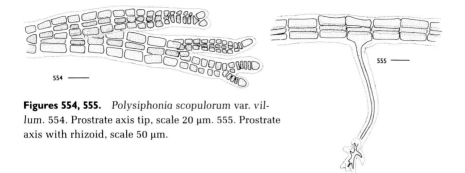

Figures 554, 555. *Polysiphonia scopulorum* var. *villum.* 554. Prostrate axis tip, scale 20 μm. 555. Prostrate axis with rhizoid, scale 50 μm.

scar cells from ultimately deciduous determinate trichoblasts; rhizoids mostly issued from proximal ends of pericentral cells, pit-connected to them, with or without digitate tips; erect axes 40–60 μm diameter, slightly constricted at the nodes, simple or sparingly branched, branches forming exogenously from scar cells in a pseudodichotomous manner; trichoblasts usually conspicuous and highly branched, tardily deciduous; tetrasporangia subglobose, 40–50 μm diameter, in short spiral series in upper portions of erect axes causing the segments to slightly swell when mature; spermatangial sori cylindrical, borne laterally on lower segments of trichoblasts at tips of erect axes; cystocarps ovoid to slightly urceolate, to 150 μm diameter.

Kapraun 1980a.

Rare, known only as an epiphyte of pelagic *Sargassum* washed ashore at Wrightsville Beach, summer.

Distribution: North Carolina, Johnston Island (Central Pacific).

Polysiphonia scopulorum var. *villum* (J. Agardh) Hollenberg 1968a, p. 81, fig. 7a.

Polysiphonia villum J. Agardh 1863, p. 941.

Figures 554 and 555

Plants saxicolous or epiphytic, brownish red, minute, prostrate indeterminate axes spreading, attached by numerous unicellular rhizoids, giving rise to erect indeterminate axes 0.3–1 cm tall; prostrate axes 60–100 μm diameter, ecorticate segments one to one and a half diameters long with four pericentral cells, giving rise to erect axes from every second to sixth segment, scar cells lacking; rhizoids mostly issued from central portions of pericentral cells, in open connection with them, with or without digitate tips; erect axes 40–80 μm diameter, simple or sparingly branched, branches replacing trichoblasts in the developmental sequence at apices, adventitious branches forming at some distance from the apices, not common; trichoblasts usually inconspicuous or lacking, soon deciduous; tetrasporangia globose to wide ellipsoid, 50–60 μm diameter, in long, and occasionally interrupted straight series in upper portions of erect axes causing the segments to swell when mature; spermatangial sori cylindrical, forming from entire trichoblast primordia on tips of erect axes, without sterile tip cells; cystocarps ovoid, 150–190 μm diameter.

Hollenberg 1968a; Brauner 1975.

Known as an epiphyte of *Zostera* in the shallow subtidal of the Beaufort, N.C., area, September-October, and on *Spartina* culms on Pivers Island, July. Only tetrasporic plants (July) are known in the flora.

Distribution: North Carolina, Bermuda, southern Florida, Belize, Puerto Rico, Brazil, British Columbia to Pacific Panama.

Polysiphonia sphaerocarpa Børgesen 1918, p. 271, figs 267–71.

Plants epiphytic or saxicolous, light brownish to rosy red, 0.5–2 cm tall, initially attached by discoid holdfasts, later becoming decumbent, secondarily attached by numerous unicellular rhizoids from lower pericentral cells; erect indeterminate axes pseudodichotomously to alternately branched, occasionally with rhizoids issued above attaching to other erect filaments, axes 60–200 μm diameter below, ecorticate, segments one half to one and a half diameters long with four pericentral cells, showing scar cells from ultimately deciduous determinate trichoblasts; secondary axes narrowly angled, slightly thinner than primary axes, often basally constricted, the ultimate branches tapering to acute apices; rhizoids issued from the distal ends of pericentral cells, pit-connected to them, occasionally more than one per cell, with or without digitate tips; branches replacing trichoblasts in developmental sequence at apices, trichoblasts found only at apices, highly branched to several orders; tetrasporangia globose, 50–60 μm diameter, in short, spiral series in upper half of erect axes causing the segments to swell slightly when mature, often formed in paired branchlets, appearing forked; spermatangial sori borne laterally on lower segments of trichoblasts in acropetal series at axis tips, cylindrical to long and conical, occasionally forked, 50–60 μm diameter, 150–180 μm long, with or without sterile tip cells; cystocarps short stalked, formed in upper branches, globose, 200–300 μm diameter, with broad ostioles surrounded by larger or similar sized cells than the rest of the pericarp, without necks.

Williams 1951; Brauner 1975; Wiseman and Schneider 1976; Kapraun 1977a, 1980a; Richardson 1986, 1987.

Known as an epiphyte of *Zostera* in the Beaufort, N.C., area, March-July; on *Sargassum* stipes from 5–8 m off New River Inlet, summer; on *Gracilaria*, *Sargassum*, and other subtidal macroscopic algae in the Wilmington area, May-September; as an epiphyte from North Island, Winyah Bay, S.C.; from coastal Georgia habitats, March; and probably more widespread in the area, being overlooked due to its small size.

Distribution: North Carolina, South Carolina, Georgia, Caribbean; elsewhere, widespread in tropical seas.

Difficult to distinguish from immature specimens of *Polysiphonia ferulacea*, *P. sphaerocarpa* is smaller, with mostly pseudodichotomous branching, and has

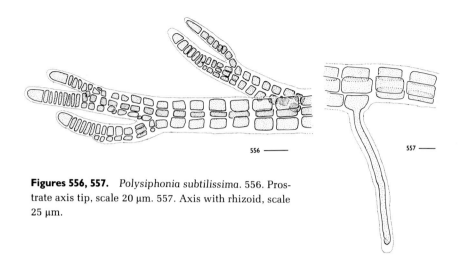

Figures 556, 557. *Polysiphonia subtilissima.* 556. Prostrate axis tip, scale 20 μm. 557. Axis with rhizoid, scale 25 μm.

larger cystocarps, with distinct ostiolar cells and occasional forked spermatangial sori (Kapraun and Norris 1982).

Polysiphonia subtilissima Montagne 1840, p. 199.
   Figures 556 and 557
   Plants epiphytic, saxicolous, lignicolous, or terricolous, olive green to purplish red and blackish, prostrate indeterminate axes spreading and forming mats, attached by numerous unicellular rhizoids, giving rise to erect indeterminate axes 2–4 cm tall; prostrate axes 70–120 μm diameter, with large, obvious apical cells and ecorticate segments one to four diameters long with four pericentral cells, the tips slightly upturned, axes giving rise to erect axes from every seventh to tenth or even greater distanced segment; rhizoids mostly issued from the proximal end of pericentral cells, formed in open connection with them, with or without digitate tips; erect axes narrower than prostrate axes, 30–60 μm diameter, alternately to pseudodichotomously branched, branches replacing trichoblasts in developmental sequence at apices, trichoblasts simple to forked, usually lacking; adventitious branching rare; tetrasporangia globose, 65–75 μm diameter, in short to long series in median to upper portions of erect axes causing the segments to swell when mature; spermatangial sori known only in culture, forming from entire trichoblast primordia at tips of erect axes, long-cylindrical, with or mostly without sterile tip cells.
   Taylor 1960; Chapman 1971; Kapraun 1980a, 1980b; Richardson 1987.
   A brackish-water species known in the Carolinas only from the Cape Fear River estuary, epiphytic on *Spartina* culms, November, and from tidal creeks throughout Georgia; probably throughout the area where salinities are reduced, found in association with *Caloglossa leprieurii* and *Bostrychia radicans*.
   Distribution: Maritimes, Massachusetts to Virginia, North Carolina, Georgia, Bermuda, southern Florida, Gulf of Mexico, Caribbean, Brazil.
   The only gametangial structures for this species known to date are from Brazil

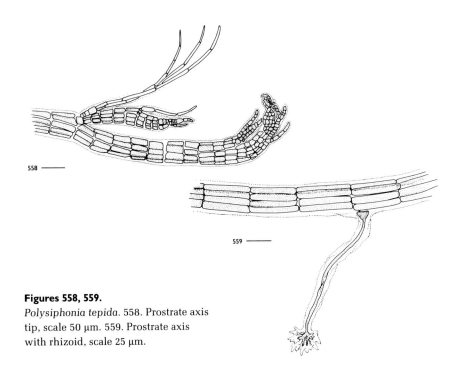

Figures 558, 559.
Polysiphonia tepida. 558. Prostrate axis
tip, scale 50 μm. 559. Prostrate axis
with rhizoid, scale 25 μm.

(Oliveira F. 1969) and from spermatangia produced in culture from North Carolina isolates (Kapraun 1980b). The field specimens from Brazil have spermatangia subtended by branching trichoblasts, while the cultured Carolina plants have spermatangial branches replacing trichoblasts. This disparity suggests that further taxonomic comparisons are necessary for plants referred to *Polysiphonia subtilissima* throughout its purported range.

Polysiphonia tepida Hollenberg 1958, p. 65, fig. 1.
Figures 558 and 559
Plants saxicolous or terricolous, dark brownish red, prostrate indeterminate axes spreading, attached by numerous unicellular rhizoids, giving rise to erect indeterminate axes 1–2(–10) cm tall; prostrate axes 80–140 μm diameter, ecorticate segments one to two diameters long with seven to eight pericentral cells, shorter segments above, giving rise to erect axes from every sixth to eighth or greater distanced segment, scar cells obvious; rhizoids issued from proximal portions of pericentral cells, pit-connected to them, with or without digitate tips; erect axes 80–130 μm diameter, branching mostly alternately distichous, branches forming in the axils of trichoblasts at apices; trichoblasts obvious at apices, ultimately deciduous; tetrasporangia globose to wide ellipsoid, 30–35 μm diameter, scattered or in long spiral series in upper portions of erect axes causing the segments to swell when mature; spermatangial sori borne laterally on lower segments of trichoblasts, cylindrical to long and conical, occasionally forked, 60–80 μm diameter, to 250 μm long, without sterile tip cells.
Hollenberg 1958; Taylor 1960; Schneider 1976; Wiseman and Schneider 1976;

Kapraun 1977a, 1980a. As *P. taylori* (nomen nudum), Williams 1948a, 1949.

Known from lower intertidal and shallow subtidal rock of jetties and mud of inlets from Cape Lookout to Charleston, March-September, reaching maximum size and abundance April-May and persisting as 1–2 cm tall mats through the summer; also known from a single collection of small, dense, sterile mats from 21 m offshore in Onslow Bay, July.

Distribution: North Carolina (type locality), South Carolina, southern Florida, Gulf of Mexico, Puerto Rico, Brazil, Hawaiian Islands.

The type locality is Shackleford jetty near Beaufort, N.C.

Polysiphonia urceolata (Lightfoot ex Dillwyn) Greville 1824, p. 309.

Conferva urceolata Lightfoot ex Dillwyn 1809, no. 156, pl. G.

Figures 560 and 561

Plants terricolous, saxicolous, or epiphytic, light brownish to purplish and rosy red, prostrate indeterminate axes spreading and forming mats, attached by numerous unicellular rhizoids, giving rise to erect indeterminate axes to 10 cm tall; prostrate axes 80–180 µm diameter, with ecorticate segments two to three diameters long with four pericentral cells, giving rise to erect axes, showing conspicuous scar cells from deciduous determinate trichoblasts; apices of prostrate

Figures 560, 561.
Polysiphonia urceolata.
560. Erect axis tip, scale
20 µm. 561. Prostrate axis
with rhizoid, scale 50 µm.

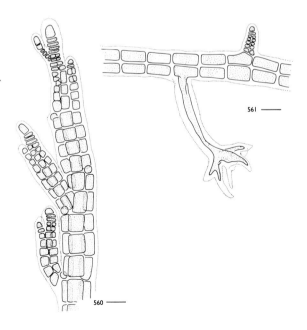

561 ——

560 ——

axes not usually upturned, rhizoids mostly issued centrally from pericentral cells, formed in open connection with them, with or without digitate tips; erect axes slightly narrower than prostrate axes, 50–60 µm diameter, much and alternately branched, branches mostly short, replacing trichoblasts in developmental sequence at apices, trichoblasts usually lacking, occasionally persistent; adventitious branching common, branches issued from scar cells; tetrasporangia globose to wide ellipsoid, 65–90 µm diameter, in long, often interrupted, straight series in upper portions of erect axes causing the segments to swell when mature; spermatangial sori borne laterally on lower segments of trichoblasts, cylindrical to long and conical, 20–40(–60) µm diameter, 75–250 µm long, with or without sterile tip cells; cystocarps short stalked, subglobose to urceolate, 300–480 µm long, with or without obvious necks and ostioles.

Williams 1948a; Taylor 1957; Kapraun 1977a, 1978a, 1979, 1980a, 1980b.

Locally abundant on sand in the lower intertidal at Lockwood's Folly Inlet, reaching peak size and abundance, January-March; less commonly, from area jetties near Wilmington and Beaufort, N.C., year-round, perennating in summer as decumbent mats.

Distribution: Labrador to Maryland, North Carolina, Argentina, USSR to northern Spain, Mediterranean.

In a comparative culture study of *Polysiphonia urceolata* from inshore North Carolina, New England, and Norway, Kapraun (1979) found distinct populational differences in vegetative and reproductive characteristics, most strikingly the shape of cystocarps and origin of spermatangial branches. Although the Carolina population had spermatangial sori subtended by branching trichoblasts and ovoid cystocarps without necks, those isolated from Norway had spermatangial branches replacing trichoblasts and urceolate cystocarps with necks. As shown throughout the discussion of this genus above, these are considered important taxonomic criteria. New Hampshire and Massachusetts isolates further confused the taxonomic concept of this European based species. Kapraun concluded from this study that the geographical isolates of *P. urceolata* were either latitudinally distributed ecotypes or genetically isolated sibling species. The above description reflects the North Carolina populations of this "species."

Pterosiphonia Falkenberg in Schmitz and Falkenberg 1897
Plants erect from rhizomatous, creeping, prostrate axes, bilaterally organized, attached by rhizoids issued from pericentral cells of prostrate axes; indeterminate axis terete to flattened, polysiphonous, composed of four to twenty pericentral cells, with or without cortication, producing branch initials every few to several segments in an alternately distichous arrangement; primary branches rebranched one to several orders, polysiphonous throughout; ultimate branch-

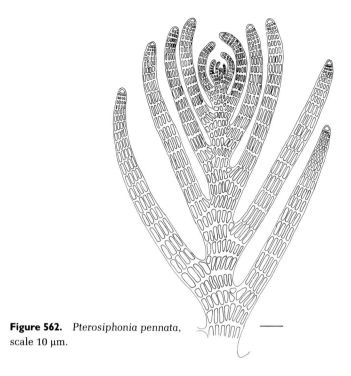

Figure 562. *Pterosiphonia pennata,* scale 10 µm.

lets determinate, generally either short and simple or forked; trichoblasts absent on vegetative axes; tetrahedral sporangia produced from pericentral cells, one per segment, in straight series in the upper portions of branch systems; dioecious, spermatangia formed in conical to cylindrical sori, with or without sterile tips, on reduced trichoblasts or short, determinate branchlets with monosiphonous bases at the apices of branch systems; procarps produced on the second segment of specially produced trichoblasts, commonly near apices of penultimate branches, carpogonial branches four celled; carposporophytes within short pedicellate, ovoid to globular pericarps.

Pterosiphonia pennata (C. Agardh) Falkenberg 1901, p. 263, pl. 2 figs 1–2.

 Hutchinsia pennata C. Agardh 1824, p. 146.

 Figure 562

 Plants saxicolous, conchicolous or epiphytic, brownish to purplish red, drying purplish black and becoming brittle, to 2.5 cm tall, attached by one-celled rhizoids, with simple or digitate tips, issued from ventral pericentrals of prostrate axes; prostrate and erect indeterminate axes terete to somewhat flattened, 125–200(–250) µm diameter, with eight to ten (to twelve) pericentral cells, corticated only rarely on prostrate axes; erect indeterminate axes bearing determinate branches on every other segment from above the base to the tip, in an alternately distichous arrangement; determinate branches simple, rarely forked, of similar length, to 1.5 mm along the lower two thirds of the indeterminate axes, 80–125 µm diameter at the base tapering to the acute to obtuse tips, obliquely erect to incurved when mature, strongly curled and overtopping apical cells of indeterminate branches; tetrasporangia globose, formed in long series in distal portions; spermatangial stichidia incurved, linear and conical, borne alternately on every second segment of short, adaxial branches approximately midway on determinate axes; cystocarps sessile.

Williams 1948a, 1949; Taylor 1960; Wiseman and Schneider 1976; Kapraun 1980a. As *Bostrychia rivularis sensu* Blomquist and Humm 1946.

Known on rock and often covered by sediment; also as a common epiphyte from the lower intertidal and shallow subtidal of Cape Lookout and Radio Island jetties near Beaufort, N.C., and North jetty, Charleston, year-round.

Distribution: North Carolina, South Carolina, Venezuela, Brazil, Argentina, British Isles to Portugal, Mediterranean, Adriatic Sea, Canary Islands, Japan, California, Pacific Mexico, Peru.

This species is codominant with *Rhodymenia pseudopalmata* in summer on certain portions of Cape Lookout jetty, from mean low tide to the spring low tide mark (Williams 1949). Smaller plants and reduced populations of the species are found in all seasons at that location.

Wrightiella Schmitz 1893

Plants erect, sparse to bushy, radially organized, arising from rhizoidal scutate or discoid holdfasts; indeterminate axes terete and polysiphonous with four to five pericentral cells, producing branch initials from each segment, loosely to heavily corticated by descending rhizoids; uniaxial row producing pigmented, monosiphonous determinate branches in a spiral sequence before division into pericentrals and axial row at the apex, these below often deciduous; indeterminate axes more or less developing short, simple or forked, spinelike branchlets endogenously produced in a spiral sequence within mature segments from the axial row, polysiphonous at the base, rapidly tapering to single apical cells and covered, at least initially, by monosiphonous determinate branches; spinelike branchlets occasionally developing into long indeterminate axes bearing spinelike branchlets; monosiphonous determinate branches alternately branched in a spiral sequence, the basal cells of each ultimate branch markedly smaller than the cells producing and borne on them; tetrahedral sporangia borne in a single, spiral line in nodulose, corticated stichidia on determinate branches which remain monosiphonous below the fertile portion; spermatangia unknown; procarps near the tips of reduced monosiphonous determinate branches, carpogonial branches four celled; carposporophytes within broad ovoid, globose to urceolate pericarps on short, polysiphonous stalks.

Wrightiella tumanowiczii (Gatty ex Harvey) Schmitz 1893, p. 222.
    Dasya tumanowiczii Gatty ex Harvey 1853, p. 64.

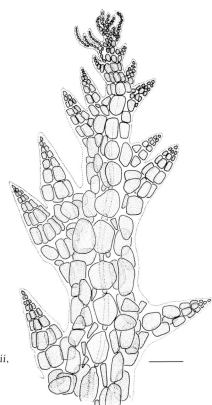

Figure 563. *Wrightiella tumanowiczii,* detail of apex, scale 100 μm.

Figure 563

Plants saxicolous or epizoic, rosy red, 7–75 cm tall, attached by small discoid holdfasts; several long indeterminate axes issued from the lower portions of the plant, no single axis becoming dominant, 0.5–1.3 mm diameter with four (to five) pericentral cells, all long axes above becoming sparingly alternately to irregularly divided with few to several, lax to rigid branches, 0.2–3.0 cm in length; major axes beset with few to several short and broad to long and narrow spinelike branchlets, 30–350 μm diameter at the base and 185–625 μm long, corticated by rhizoids from the basal cells of lateral branches, few above, ensheathing the axes below; monosiphonous determinate branches dichotomously branched proximally, flagelliform above, to 2.0 mm in length, 40–50 μm diameter above the smaller basal cell, tapering to 2.5–10 μm diameter distally, becoming deciduous not much below the branch apices; tetrasporangia globose, 30–60 μm diameter, in long spiral series on uppermost portions of monosiphonous laterals, the stichidia becoming corticated by pericentral divisions, appearing twisted and nodulose in form; cystocarps ovoid to urceolate on short polysiphonous stalks.

Schneider 1975a, 1976.

Uncommon offshore in Onslow Bay from 35–40 m, August.

Distribution: North Carolina, Bermuda, southern Florida, Gulf of Mexico, Caribbean, Brazil.

As pointed out earlier (Schneider 1975a), Carolina specimens of *Wrightiella* are difficult to place specifically in either the above taxon or in *W. blodgettii* (Harvey) Schmitz; the two are more than likely conspecific. Because most of the rarely collected area plants are more similar to *W. tumanowiczii*, we have retained them with some doubt under that epithet until type specimens are compared. Carolina plants range from having many short, broad, conical, spinelike branchlets to having wide-spaced, long, narrow, lax, spinelike branchlets. Unfortunately, we have so few plants we cannot draw greater conclusions from these data, and we do not wish to place them under more than one name, given the fact that the plants are collected with varying morphologies from the same dredge trawls.

ARTIFICIAL KEYS TO THE GENERA

The following three keys to the genera included in this flora are intended to supplement the keys that are presented in the body of the systematic treatment. The keys to classes, orders, and families are natural keys that separate taxa along lines thought to reflect evolutionary relationships. The criteria used are not always easily determined. In the three following keys to genera, there has been an attempt to use characters that are commonly present in specimens and, as an aid to routine identification of specimens, as easy to determine as possible.

The first key includes those seaweeds that are green in color—either grass green or slightly olive. It includes both the Chlorophyta and Chrysophyta. The second key includes those seaweeds that are brown—the Phaeophyta. The last key is for the Rhodophyta, a division whose members may be red, but in shallow water are sometimes purple, blue green, yellowish, and other hues, but not grass green or brown.

ARTIFICIAL KEY TO THE GENERA OF THE CHLOROPHYTA AND CHRYSOPHYTA

1. Plants cellular; divided into uninucleate or multinucleate vegetative units by cross-walls 2
1. Plants acellular; vegetative parts not divided by cross-walls 20
2. Cells separated by empty, colorless tubes .... Blastophysa (in part)
2. Cells not separated by empty, colorless tubes 3
3. Plants composed of branched or unbranched, uniseriate filaments, the filaments not united into blades 4
3. Plants not filamentous, or if filamentous the filaments uniting to form blades, tubes, or discs 10
4. Filaments unbranched 5
4. Filaments branched 7
5. Filament diameter less than 30 μm................................. 6
5. Filament diameter greater than 30 μm .... Chaetomorpha
6. Plastid an incomplete band, cells uninucleate .... Ulothrix
6. Plastid reticulate, cells multinucleate .... Rhizoclonium
7. Cells with aseptate hairs .... Phaeophila
7. Cells lacking aseptate hairs 8
8. Plants creeping, epi-endophytic, cell diameter less than 10 μm .... Entocladia
8. Plants erect or matted, but not creeping, cell diameters large, mostly greater than 10 μm 9
9. Cell diameters less than 1 mm .... Cladophora
9. Cell diameters greater than 1 mm .... Valonia

10. Plants discoidal 11
10. Plants not discoidal 13
11. Pyrenoids lacking, marginal cells forked . . . . Ulvella
11. Pyrenoids present, marginal cells not forked 12
12. Cells with aseptate hairs . . . . Pringshiemiella
12. Cells without aseptate hairs . . . . Pseudendoclonium
13. Plants tubular 14
13. Plants not tubular 15
14. Plastids stellate, with one pyrenoid; base of plant discoidal, lacking rhizoids;
     cell diameter 10 μm or less . . . . Blidingia
14. Plastids not stellate, with one to many pyrenoids; base of plant rhizoidal;
     cells often greater than 10 μm diameter . . . . Enteromorpha (in part)
15. Blade one cell thick 17
15. Blade two cells thick 16
16. Plant tubular at the base of blade . . . . Enteromorpha (in part)
16. Plants not tubular at base of blade . . . . Ulva
17. Blade reticulate 18
17. Blade not reticulate 19
18. Blade stalked . . . . Struvea
18. Blade not stalked . . . . Microdictyon
19. Cells of blade of similar sizes, veins not present . . . . Ulvaria
19. Cells of blade of different sizes, large cells forming veins . . . . Anadyomene
20. Plants boring in shell . . . . Gomontia
20. Plants not boring in shell 21
21. Plants endophytic . . . . Blastophysa (in part)
21. Plants not endophytic 22
22. Plants filamentous, the filaments not united to form larger, distinctive struc-
     tures 23
22. Plants not filamentous, or if filamentous, the filaments united to form larger,
     distinctive structures 26
23. Plants forming short branchlets (pinnules) in more or less regular series
     . . . . Bryopsis
23. Plants branching irregularly or dichotomously, but not in regular lateral
     series 24
24. Filaments constricted above di- or trichotomies . . . . Boodleopsis
24. Filaments lacking constrictions in any regular pattern 25
25. Plants staining dark blue with I$_2$KI, primarily oceanic . . . . Derbesia
25. Plants not staining dark blue with I$_2$KI, primarily estuarine . . . . Vaucheria
26. Structure filamentous, the filaments united to form distinctive shapes . . . 27
26. Structure not filamentous; plants consist of a rhizome, rhizoids, and erect

blades or branches... Caulerpa
27. Plants stalked, the stalks clearly differentiated from the top of the plant
.... 28
27. Plants not stalked, or the stalk similar to the top of the plant ..... Codium
28. Plants calcified .... Udotea
28. Plants not calcified .... Avrainvillea

ARTIFICIAL KEY TO THE GENERA OF
THE PHAEOPHYTA

1. Plants primarily creeping, crustose, or endophytic.................... 2
1. Plants primarily erect 10
2. Growing on rock 3
2. Growing on other algae or on vascular plants 4
3. Plants parenchymatous, creeping blades loosely attached to the substrate
.... Lobophora (in part)
3. Plants pseudoparenchymatous, closely adherent to the substrate ....
.... Pseudolithoderma
4. Basal filaments endophytic....................................... 5
4. Basal filaments epiphytic 7
5. Endophytic filaments occasionally biseriate; emergent reproductive filaments form multicellular propagules .... Onslowia
5. Endophytic filaments uniseriate throughout; emergent reproductive filaments not forming multicellular propagules 6
6. Bases of hairs with trichothallic meristems .... Streblonema
6. Bases of hairs without a trichothallic meristem.............. Herponema
7. Growth of filaments radial 8
7. Growth of filaments irregular, matted .... Myriactula
8. Basal filaments forming a single layer .... Myrionema
8. Basal filaments divided parallel to the substrate to form two or more layers
.... 9
9. Plastids platelike, several per cell......................... Phaeostroma
9. Plastids discoid or in short chains .... Hecatonema
10. Erect part of plant filamentous, uniseriate throughout................ 11
10. Erect part of plant multiseriate, parenchymatous, or tagmatic.......... 15
11. Unilocular and plurilocular reproductive structures intercalary ....
.... Bachelotia
11. Unilocular sporangia terminal; plurilocular sporangia terminated by hairs
.... 12

12. Plurilocular structures in clusters . . . . .Botrytella
12. Plurilocular structures not in clusters 13
13. Growth trichothallic; rhizoids produced perpendicular to filament axis . . . .
. . . . Acinetospora
13. Growth diffuse; no rhizoidal branches produced perpendicular to filament
axis 14
14. Plastids elongate, band shaped . . . . Ectocarpus
14. Plastids discoid . . . . Hincksia
15. Plants compressed or bladelike throughout 16
15. Plants totally or in part terete or inflated and bladderlike 26
16. Plants with midribs throughout length of blades. 17
16. Plants without midribs or with midribs confined to lower parts of blades
. . . . 18
17. Plants with gas-filled floats. . . . . Fucus (in part)
17. Plants lacking gas-filled floats . . . . Dictyopteris
18. Growth by marginal meristems 19
18. Growth diffuse, trichothallic, or from apical cell or cluster of apical cells
. . . . 21
19. Margins of blades curled . . . . Padina
19. Margins of blades flat. 20
20. Plants erect, bushy; lower parts of blades with thickening ribs . . . . Zonaria
20. Plants creeping in part, in part erect; thickening ribs absent . . . .
. . . . Lobophora (in part)
21. Branches hollow . . . . Rosenvingea
21. Branches solid 22
22. Blades with marginal hairs. . . . . Punctaria (in part)
22. Blades without marginal hairs. 23
23. Growth diffuse. 24
23. Growth apical 25
24. Blades two to four cells (less than 100 μm) thick, margins often crisped,
tapered base symmetrical . . . . Punctaria (in part)
24. Blades to seven cells (100–200 μm) thick margins smooth, straight, tapered
base asymmetrical . . . . .Petalonia
25. Growth from a single apical cell . . . . Dictyota
25. Growth from a cluster of apical cells . . . . Spatoglossum
26. Plant axis hollow or having hollow floats 27
26. Plants solid throughout 36
27. Plants spherical or subspherical 28
27. Plants elongate cylinders or more complex and branched. 29

28. Plants pseudoparenchymatous; plastids numerous in each cell ... Leathesia
28. Plants parenchymatous; plastids single in each cell .... Colpomenia
29. Plants forming well defined, gas-filled floats 30
29. Plants lacking individual floats, all or most of the plant hollow 32
30. Gas-filled floats in pairs .... Fucus (in part)
30. Gas-filled floats single or clustered, not in pairs 31
31. Gas-filled floats single .... Ascophyllum
31. Gas-filled floats in clusters .... Sargassum
32. Plants unbranched 33
32. Plants branched 34
33. Plurilocular structures sessile, in large confluent patches; unilocular sporan-
    gia absent .... Scytosiphon
33. Plurilocular sporangia and unilocular sporangia stalked, in small scattered
    sori .... Asperococcus
34. Branching mostly opposite .... Striaria
34. Branching irregular to subdichotomous 35
35. Plurilocular structures sessile, converted from surface cells, unilocular
    sporangia absent .... Rosenvingea
35. Plurilocular structures sessile, converted from surface cells, unilocular
    sporangia absent .... Rosenvingia
36. Surface of plant axes soft, composed of radially oriented filaments which are
    not laterally united 37
36. Surface of plant axes firm, distinct; plants parenchymatous or pseudoparen-
    chymatous.. 38
37. Growth uniaxial, from an apical cell; axes to 0.3 mm diameter ....
    .... Nemacystus
37. Growth multiaxial, without a single apical cell, axes greater than 0.75 mm
    diameter .... Cladosiphon
38. Growth from apical cell.................................. Sphacelaria
38. Growth not from apical cell 39
39. Branches ending in tufts of 50 of more filaments .... Sporochnus
39. Branches ending with few or no hairs 40
40. Plants with large axial filament surrounded with corticating filaments.....
    .... Arthrocladia
40. Plants without axial filament..................................... 41
41. Medulla in cross section with four large cells .... Hummia
41. Distinct medulla lacking; uniseriate filament becoming pluriseriate at matu-
    rity, if inner cells formed then similar in size to outer cells ....
    .... Punctaria (in part)

ARTIFICIAL KEY TO THE GENERA
OF RHODOPHYTA

1. Plants impregnated with calcium carbonate throughout or producing calcium carbonate hypobasally 2
1. Plants not calcareous 17
2. Plants with upright portion articulated (jointed) 3
2. Plants crustose or in part branched, but never articulated 7
3. Intergenicula rigid 4
3. Intergenicula lax to flexible .... Galaxaura
4. Genicula of a single tier of elongated cells, intergenicula with medulla of uniform-sized tiers of cells, conceptacles axial 5
4. Genicula of more than one tier of cells, intergenicula with medulla of variable-sized tiers of cells, conceptacles lateral .... Amphiroa
5. Branching primarily dichotomous, occasionally irregular .... Jania
5. Branching primarily pinnate, occasionally in part dichotomous or irregular 6
6. Plants coarse, main axes mostly greater than 500 µm diameter, conceptacles formed terminally on pinnae, lacking any subtending branches ... Corallina
6. Plants delicate, main axes less than 500 µm diameter, tetrasporic and cystocarpic conceptacles axial, later becoming subtended by new branches issued from the fertile intergeniculum .... Haliptilon
7. Plants calcified throughout 8
7. Plants calcified only hypobasally .... Peyssonnelia
8. Plants unattached and branched .... Lithothamnion
8. Plants crustose without branched uprights 9
9. Crusts epiphytic and thin, perithallus absent or generally less than five cells thick 10
9. Crusts saxicolous and thick, perithallus greater than five cells thick at maturity 13
10. Tetrasporangial conceptacles, each with several pores .... Melobesia
10. Tetrasporangial conceptacles each with a single pore 11
11. Crusts with hypothallial cells vertically elongated, palisadelike; secondary pit-connections present between adjacent cells .... Titanoderma
11. Crusts with horizontally elongated, spreading hypothallial cells, secondary pit-connections absent between adjacent cells, cell fusions present 12
12. Crusts originating from central four-celled germinating elements, trichocytes terminal on hypothallial filaments .... Fosliella
12. Crusts originating from central eight-celled germinating elements, trichocytes absent or intercalary in hypothallial filaments .... Pneophyllum

13. Tetrasporangial conceptacles, each with several pores 14
13. Tetrasporangial conceptacles, each with a single pore 16
14. Hypothallus coaxial .... Mesophyllum
14. Hypothallus noncoaxial 15
15. Conceptacle roofs to three cells thick; gonimoblasts marginally located in conceptacles .... Leptophytum
15. Conceptacle roofs four or more cells thick; gonimoblasts centrally located in conceptacles .... Phymatolithon
16. Trichocytes vertical in perithallus, secondary pit-connections absent between adjacent cells, cell fusions present .... Neogoniolithon
16. Trichocytes lacking, secondary pit-connections present between adjacent cells, cell fusions absent .... Lithophyllum
17. Plants discoid to crustose, of one or more layers, compact to spreading ... 18
17. Plants erect to prostrate, not discoid or crustose 19
18. Crusts of radiating filaments with free ends at the outer margins .... Erythrocladia (in part)
18. Crusts with marginal growth by united border cells .... Sahlingia
19. Plants filamentous and uniseriate or pseudofilamentous 20
19. Plants not filamentous or filamentous, with cortication or pericentral cells (polysiphonous) 41
20. Cells loosely embedded in a gelatinous matrix, not pit-connected to other cells 21
20. Cells regularly arranged and pit-connected 26
21. Plants erect, axes unbranched or branched 22
21. Plants endophytic, endozoic or prostrate, and spreading .... Erythrocladia (in part)
22. Plants unbranched 23
22. Plants sparingly to repeatedly branched 25
23. Plants uniseriate; cells widely spaced and bluish green to grayish green .... Chroodactylon (in part)
23. Plants initially uniseriate, possibly becoming multiseriate above; cells closely spaced, rosy red and purplish red to black or brown 24
24. Plants epiphytic .... Erythrotrichia (in part)
24. Plants saxicolous .... Bangia
25. Plants sparingly branched, cells bluish green to grayish green .... Chroodactylon (in part)
25. Plants repeatedly branched, cells pink to rosy red and purplish red .... Stylonema
26. Plants forming as small nets, without percurrent axes ...... Rhododictyon
26. Plants upright to prostrate, axes percurrent 27

75. Blades perforated with many holes .... Kallymenia
75. Blades not perforated 76
76. Blades with midribs at least in the lower half .... Cryptonemia
76. Blades lacking midribs 77
77. Plants dorsiventrally organized, initially peltate, later attached by secondary haptera issued mostly at the margins .... Halichrysis
77. Plants not dorsiventrally organized, basally attached 78
78. Axes generally less than 3 mm broad 79
78. Axes generally greater than 3 mm broad 85
79. Main axes dichotomously to trichotomously divided 80
79. Main axes subsimple to pinnately branched 82
80. Axes with few to several pinnate branches, medulla filamentous ....
.... Grateloupia (in part)
80. Axes lacking pinnate branches, medulla pseudoparenchymatous 81
81. Plants gelatinous, medulla consisting of a single layer of quadrate cells
.... Gloioderma (in part)
81. Plants firm, medulla consisting of few to several layers ....
.... Rhodymenia (in part)
82. Axes subsimple with few pinnate branches; tetrasporangia formed in hyaline subapical sori .... Gelidium (in part)
82. Axes with regular pinnate branches from the margins, alternate to suboppo- site; tetrasporangia not formed in hyaline subapical sori 83
83. Branching alternate, lateral branches falcate and secund ...... Plocamium
83. Branching opposite to subopposite and palmate 84
84. Plants gelatinous .... Gloioderma (in part)
84. Plants firm .... Lomentaria (in part)
85. Plants dichotomously branched 86
85. Plants simple or subsimple, irregularly to alternately lobed or branched
.... 93
86. Plants 4 cm or less tall, dichotomous branches markedly constricted at the nodes, medulla consisting of a single layer of cells .... Leptofauchea
86. Plants greater than 4 cm tall, dichotomous branches not constricted at the nodes, medulla not consisting of a single layer of cells 87
87. Medulla filamentous 88
87. Medulla pseudoparenchymatous 90
88. Axes smooth and of similar width throughout, repeatedly dichotomously branched to several orders, apices acute .... Sarcodiotheca
88. Axes crispate to undulate and wider at apices than bases, dichotomously branched to few orders, apices broadly rounded to obtuse 89
89. Ultimate branches less than 1 cm broad, margins often inrolled, papillae present .... Mastocarpus

89. Ultimate branches greater than 1 cm broad, margins undulate, papillae lacking .... Cirrulicarpus
90. Margins smooth 91
90. Margins undulate and crenate .... Petroglossum
91. Cystocarps marginal, tetrasporangia formed in distinct apical or subapical sori ....Rhodymenia (in part)
91. Cystocarps laminal, tetrasporangia scattered 92
92. Blades soft and lubricous, medulla with small and large cells intermixed .... Agardhinula (in part)
92. Blades firm and leathery, medulla with uniform-sized cells ....
.... Gracilaria (in part)
93. Axes less than 2 mm thick, lubricous to firm....................... 94
93. Axes greater than 5 mm thick, gelatinous .... Predaea (in part)
94. Medulla filamentous to hollow 95
94. Medulla pseudoparenchymatous 102
95. Blades hollow at maturity..................... Chrysymenia (in part)
95. Blades with a central mass of entangled filaments or with cortex connected with traversing filaments 96
96. Axes ligulate and abundantly branched, with numerous short marginal spines or spurlike branchlets .... Meristiella
96. Axes narrowly to broadly foliose, entire to split or lobed, spines and spur-like branchlets lacking 97
97. Blades borne on short to long stalks.................Halymenia (in part)
97. Blades sessile or briefly stipitate................................. 98
98. Blades 5 cm or less in width 99
98. Blades more than 5 cm in width 101
99. Blades asymmetric, lacking marginal proliferations ....Grateloupia (in part)
99. Blades symmetric, with or without marginal proliferations 100
100. Blades bearing linear-lanceolate marginal proliferations ....
.... Grateloupia (in part)
100. Blades simple or bearing pinnately arranged marginal proliferations .....
.... Halymenia (in part)
101. Blades often papillate; tetrasporangia zonately divided, cystocarps borne on papillae .... Meristotheca
101. Blades lacking papillae; tetrasporangia cruciately divided, cystocarps immersed in blades .... Halymenia (in part)
102. Blades soft and lubricous, medulla with small and large cells intermixed .... Agardhinula (in part)
102. Blades firm and leathery, medulla with uniform-sized cells..............
.... Gracilaria (in part)

103. Plants hollow throughout or in part hollow 104
103. Plants solid throughout 108
104. Plants consisting of hollow, septate or continuous, branched cylindrical axes 105
104. Plants consisting of, or bearing in part, obovoid to obpyriform, mucilage-filled bladders 107
105. Cylindrical axes hollow except at constrictions of branch bases 106
105. Cylindrical or flattened axes divided by regularly spaced, single-layered septa . . . . Champia
106. Occasional inner cortical cells bearing gland cells into central cavities . . . . Chrysymenia (in part)
106. Gland cells lacking . . . . Lomentaria (in part)
107. Plants consisting of solitary or paired large bladders or solid axes bearing lateral and terminal bladders . . . . Botryocladia
107. Plants consisting of hollow axes bearing ever decreasing sized orders of hollow branches . . . . Chrysymenia (in part)
108. Plants gelatinous, cortex loose 109
108. Plants not gelatinous, texture firmer, cortex more compact 113
109. Plants uniaxial (check apical portions) 110
109. Plants multiaxial 111
110. Main axes 1 mm or less in diameter, corticated by rhizoidal downgrowths from basal cells of lateral indeterminate branches . . . . Naccaria
110. Main axes greater than 1 mm in diameter, corticated by dense lateral determinate branches . . . . Dudresnaya
111. Main axes less than 2 mm in diameter, alternately to subdichotomously branched 112
111. Main axes 4 mm or more in diameter, multilobate and bullate . . . . Predaea (in part)
112. Branching alternately radial, axes uniformly pigmented throughout . . . . Gloioderma (in part)
112. Branching subdichotomous, apices more densely pigmented than axes below . . . . Helminthocladia
113. Plants turgid, releasing mucilage when squeezed 114
113. Plants not turgid with mucilage 116
114. Main axes 3 mm or less wide . . . . Scinaia
114. Main axes greater than 3 mm wide 115
115. Branching in a single plane, axes deeply pigmented, medullary ganglia bearing dark-staining glandlike cells . . . . Sebdenia
115. Branching not all in the same plane, axes lightly pigmented, medullary ganglia lacking glandlike cells . . . . Halymenia (in part)

Glossary

abaxial. On the side away from the axis. See also *adaxial.*

accessory. Additional; applied in the Rhodophyta to filaments that form in addition to regular vegetative branches.

acellular. Multinucleate, without division of the vegetative cytoplasm by septa.

acropetal. Produced in succession from the base upward so that the youngest are nearest the apex.

aculeate. Bearing prickles or spines.

acuminate. Tapering gradually to a point.

acute. Sharp, ending abruptly in a point.

adaxial. On the side toward the axis. See also *abaxial.*

adnate. Attached or fused, one structure to another.

adventitious. Developing in an unusual position, out of the usual sequence; said of branches.

akinete. A thick-walled resting spore formed from a vegetative cell.

alternate. Formed singly at regular intervals along an axis.

ampulla, pl. *ampullae.* A special cluster of filaments that bear carpogonial branches or auxiliary cells (Cryptonemiaceae).

anastomosing. Joining of adjacent parts.

androphore. An antheridium-bearing branch (*Vaucheria*).

anisogamous. Having motile gametes of different sizes.

annual. Living for only a single year, during which the life history is completed.

annular. Ringlike.

annulate. Ringed; made up of ringed segments.

antheridium, pl. *antheridia.* A male reproductive structure producing flagellate gametes.

anticlinal. Perpendicular to the surface.

apiculate. Terminating in a short, sharp point.

aplanospore. A nonmotile spore with a wall separate from the parent cell.

apomeiotic cell division. Mitotic division in a cell (meiosporangium or gametangium) that typically divides by meiosis.

arcuate. Curved, bent in a bow.

articulated. Jointed, as in the segmented and flexible calcified Rhodophyta.

aseptate. Without crosswalls. See also *coenocytic.*

asexual reproduction. Reproduction without syngamy.

assimilatory filaments. Pigmented, photosynthetic filaments; contrasted with filaments that are structural and/or nonpigmented.

attenuate. Gradually tapering toward the base or apex.

autospore. A nonmotile spore which is a miniature of the parent cell.

autotrophic. Capable of using light or inorganic compounds as an energy source.

auxiliary cell. In the Rhodophyta, either a cell that produces a carposporophyte after receiving a diploid nucleus from the fertilized carpogonium, or a cell to which a connecting filament attaches. In the latter case, the carposporophyte may be produced from the connecting filament.

axial. Extending longitudinally through the center of a branch or stem.

axial filament. A series of cells extending longitudinally down the center of a branch or blade.

axil. The angle between the axis and a lateral branch.

axillary. Situated in the axil.

basipetal. Produced in succession from the apex downward so that the youngest are nearest the base.

benthic. Growing or living on the sea floor; attached.

biflagellate. Having two flagella.

bifurcate. Forked; dichotomous.

biloculate. With two chambers.

binate. Consisting of two parts; growing in pairs.

biseriate. In two rows.

bisporangium, pl. *bisporangia.* A sporangium producing two spores.

bullate. Having an outward bulging or puckering of the surface.

caducous. Nonpersistent; transitory.

caespitose. Growing in short tufts or mats.

calcareous (calcified). Covered with or impregnated with calcium carbonate (lime).

carpogonium, pl. *carpogonia.* The female gametangium in the Rhodophyta, including the egg and trichogyne.

carposporangium, pl. *carposporangia.* The reproductive cell of the carposporophyte, producing diploid carpospores by mitosis.

carpospore. A diploid spore produced by the carposporophyte.

carposporophyte. The diploid, partially parasitic generation in the life history of the Rhodophyta, initiated by mitotic division of the zygotic nucleus and attached to the gametophyte; some or all of the cells become carposporangia and release diploid carpospores.

carpotetraspore. One of four spores formed by meiosis in a divided carposporangium.

cartilaginous. Firm, tough, elastic.

centric spindle. Organization of the cell at nuclear division in which there are centrioles at the poles of the array of microtubules involved in chromosome movement.

cervicorn. Resembling the horns of a deer.

chrysolaminarin. Beta-linked glucan storage product (Chrysophyta).

circinate. Coiled inward or downward in a flat spiral.

clavate. Club shaped, gradually thickening toward the apex from a slender base.

coaxial filaments. Parallel filaments formed by the simultaneous division of apical cells to produce transverse bands of cells.

coenocytic. Multinucleate as a result of nuclear division without cell division, or as a result of cell fusion.

columella. Tuft of persistent cells in the center of a tetrasporangial or bisporangial conceptacle (Corallinales).

conceptacle. A cavity in the surface tissue of a plant containing the reproductive structures.

conchicolous. Growing on or in shell.

conchocelis stage. Diploid, filamentous, sporophyte phase (Bangiales).

conchospore. Spore of a conchocelis stage that germinates to produce a gametophyte.

concrescent. Joining together of originally separate parts.

connecting filament (ooblast). Filaments that grow from fertilized carpogonia to the auxiliary cells, either directly or indirectly.

conspecific. Belonging to the same species.

convolute. Rolled up longitudinally.

coralline. Calcified and thus resembling coral; a member of the order Corallinales (Rhodophyta).

cordate. Heart shaped, with the point upward or outward.

cornute. Horned.

cortex. Tissue external to the medulla, or external to the axial cell or pericentral cells, where there is no medulla.

corticate. Having a cortex.

corymbose. Branched, with the longer branches toward the base so that tips of all branches are at the same level to make a flat-topped cluster.

costa. A rib in a blade.

cover cell. A special cell, cut off of a pericentral cell and covering tetrasporangia in the Ceramiales; a surface cell in the Corallinales.

crenate. Edged, with rounded, forward-pointing teeth; scalloped.

crenulate. Finely crenate.

crispate. Uneven and ruffled, looking like crumpled paper.

crozier. A hook-shaped branch tip.

cruciate. Divided by two successive planes perpendicular to each other (tetrasporangia in Rhodophyta).

crustose. Flattened, adhering closely to the substrate.

cryptostome, pl. cryptostomata. A sterile cavity opening to the surface and containing a tuft of hairs (Fucales).

cuneate. Wedge shaped.

cuspidate. Having a rigid point.

cystocarp. The carposporophyte, including, if present, surrounding pericarp tissue (Rhodophyta).

deciduous. Falling off, not persistent.

decumbent. Prostrate, with the tip curving upward.

decussate. Branching with laterals in opposite pairs, each pair rotated 90° to the pairs above and below.

deltoid. Triangular.

dendroid. Treelike, with an erect main stem; freely branched.

dentate. Toothed.

denticulate. With very small teeth.

determinate. Having limited growth potential.

dichotomous. Forked, bifurcate.

digitate. Diverging from a single point, as the fingers of a hand.

dimorphic. Having two forms.

dioecious. Having male and female gametangia on separate plants.

diplontic. Having a life history in which mitosis is restricted to diploid nuclei and in which meiosis produces gametes.

diplohaplontic. Having a life history in which the diploid zygote and haploid spores divide by mitosis to produce sporophytes and gametophytes, respectively; meiosis produces spores.

discoid. Rounded and flat.

dissected. Cut deeply into narrow segments.

distal. Away from the point of attachment or origin. See also proximal.

distichous. Arranged in two rows on opposite sides of an axis.

divaricate. Spreading widely.

dolioform. Barrel shaped.

dorsal. Located on the upper surface.

dorsiventral. Having distinct top and bottom surfaces.

dredge. A device for scraping or scooping material from the sea floor when dragged behind a vessel.

ecad. A plant form which is assumed to be adapted to the habitat.

ecophene. Characteristic morphology that is not fixed genetically, but depends on environmental conditions.

ecorticate. Lacking a cortex.

ectocarpoid. Resembling members of the genus *Ectocarpus*.

ellipsoid. Elliptic in side view or longitudinal section, circular in cross-section.

emarginate. Having a slight notch at the tip.

endemic. Native, restricted to a specific region.

endogenous. Arising from within a plant, as from an axial cell after formation of pericentral cells (Ceramiales).

endophytic. Living within the tissue of another plant.

endozoic. Living within an animal.

entire. Having a margin without lobes, indentations, teeth, or spines.

epi-endophytic. Living within and outside the tissue of another plant.

epidermis. The outermost layer of cells of a plant.

epiphyte. A plant growing on another plant, but not usually parasitic.

epiphytic. Growing on a plant.

epithallus (epithallium). Uppermost cells of crustose coralline algae. In very simple crusts, these are formed on the hypothallus; in more complex crusts, a meristem of intercalary cells forms the epithallus above and a perithallus below.

epizoic. Growing on an animal.

erose. Irregularly notched, toothed, or indented.

exogenous. Originating externally from the outermost layer of cells.

exserted. Protruding.

falcate. Sickle shaped.

farinaceous. Having a mealy surface.

fascicle. A cluster or tuft of branches, all arising at about the same place on an axis.

fasciculate. In bundles or clusters consisting of similar members.

fastigiate. Having many branches that are erect and appressed or parallel, giving a narrow elongate habit.

filamentous. Composed of threadlike structures, or filaments.

filiform. Threadlike.

flabellate. Fan shaped.

flaccid. Limp.

flagelliform. Whiplashlike.

flexuous. Changing direction in a zigzag pattern.

foliose (foliaceous). Broad, flat, leaflike.

forcipate. Forked and incurved, pincerlike.

frond. An erect blade or branch of an alga.

fucosin (fucosan). Tanninlike compounds deposited in small refractive vesicles in the cells of the Phaeophyta.

fusiform. Spindle shaped, elongate, and tapering toward each end.

fusion cell. A multinucleate cell formed by the union of several cells; characteristic of vegetative tissue in some Corallinales and carposporophytes in many Rhodophyta.

gametangium, pl. gametangia. A cell or multicellular structure that produces gametes.

gametophyte. A gamete-producing generation in a life history, generally haploid.

geniculum, pl. genicula. An uncalcified joint in the axis of articulated Rhodophyta.

girdling. Encircling.

glabrous. Smooth, hairless.

gland cells. Distinctive refractive cells presumed to either release extracellular compounds or possibly serving as storage cells.

glaucous. Grayish or bluish green.

gonimoblast. Filaments forming the diploid carposporophyte generation in the Rhodophyta.

gonimolobe. Branch of a carposporophyte in which the carposporangia mature synchronously.

gregarious. Growing together in a group.

hamate. Bent at the tip; hooked.

haplontic. Having a life history in which only haploid nuclei divide by mitosis; the zygote divides by meiosis.

hapteron, pl. *haptera.* A rootlike, multicellular part of a holdfast.

heterokaryotic. Having nuclei of more than one genotype.

heteromorphic. Having a life history in which the gametophyte and sporophyte are dissimilar.

heterothallic. Requiring two different clones for sexual reproduction.

heterotrichous. Formed of two different systems of filaments, one prostrate and one erect.

holdfast. Basal attachment structure; may be unicellular or multicellular.

holotype. A single specimen or element, chosen by the author as the nomenclatural type of a new taxon; the single specimen or element used by the author.

hyaline. Transparent, colorless, nonpigmented.

hypobasal. Below the base or hypothallus.

hypogynous cell. The cell bearing the carpogonium.

hyposaline. Less saline than ocean water.

hypothallus (hypothallium). Basal layer(s) of cells of crustose Rhodophyta.

incised. Cut sharply and deeply at the margins.

incrassate. Thickened.

indeterminate. Having unlimited growth potential, with the ability to resemble the primary axis in structure and function.

intercalary. Lying somewhere along the length of a branch or filament, but not at the apex or base.

intergeniculum, pl. *intergenicula.* Calcified segments of articulated Rhodophyta.

internode. Part of a branch or filament between the lateral branch points or nodes.

interzonal spindle. Microtubules formed at mitosis and extending between the poles of the cell.

involucre. A group of sterile cells or filaments subtending a carposporophyte and usually overtopping it.

involute. Having blade edges rolled inward.

isodiametric. Having equal sides.

isogamous. Having gametes of similar size and form which may or may not differ genetically and physiologically.

isolectotype. A duplicate of a lectotype.

isomorphic. Having a life history in which the gametophyte and sporophyte are similar.

isotype. A duplicate of a holotype.

karyogamy. Fusion of gamete nuclei.

labyrinthine. Intricate; mazelike.

laciniate (lacerate). With the edge divided or cut into narrow strips; fringed.

lamellate. Made up of thin layers.

lamina, pl. *laminae*. The flat, thin part of a leaf or blade.

laminal. Pertaining to the flat blade, as contrasted with marginal.

laminarin. Beta-linked glucan storage compound (Phaeophyta).

lanceolate. Lance shaped; flattened, narrow, two or three times as long as broad, widest in the middle or toward the base and tapering toward each end.

lateral. Coming from or situated on the side of an axis.

lectotype. A specimen or other element, chosen from the original material to serve as a nomenclatural type either when no holotype was designated at the time of publication or as long as the designated holotype is missing.

leucoplast. A colorless plastid frequently containing starch.

lignicolous. Growing on wood.

ligulate. Tongue or strap shaped.

linear. Long and narrow with parallel sides.

littoral. Of or existing on the shore; intertidal.

locule. A compartment or space within a structure.

lubricous. Slippery.

mammillate. Nipple shaped.

mannitol. A sugar alcohol storage product of the Phaeophyta.

marginal. Situated on or pertaining to the edge.

medulla. A central core of tissue surrounded by the cortex.

mega-. A prefix indicating "large size."

megacell. An enlarged cell formed from a trichocyte (Corallinales).

meiosporangium. A reproductive structure in which spores (meiospores) are produced by meiosis (sometimes followed by mitosis).

meristem, adj. *meristematic*. The region of a multicellular plant in which cell division is localized.

midrib. The thickened, central, longitudinal costa of a blade.

mitosporangium. A reproductive structure in which spores (mitospores) are produced by mitosis.

moniliform. Resembling a string of spherical beads.

monocarpogonial. Having a single carpogonium per auxiliary cell or supporting cell.

monoecious. Having male and female gametangia on the same plant.

monopodial. Having a single dominant apical cell or apical meristem and a dominant axis from which laterals arise.

monosiphonous. Composed of a single row of cells; uniseriate.

monosporangium, pl. *monosporangia*. A sporangium that forms a single spore (monospore).

monotypic. Having only one included taxon.

mucronate. Terminated by a short, sharp point, a mucro.

multiaxial. Having an axis composed of many elongate filaments with no evidence of a primary filament.

multilateral. Having many sides, or derived from many sides.

multinucleate. Having several to many nuclei.

multiporate. With a separate pore above each sporangium in a tetrasporangial conceptacle (Corallinales).

multiseriate. With several rows of cells in one or more planes.

nemathecium, pl. *nemathecia*. A raised tissue containing or bearing reproductive cells.

nerves. Fine veins in the blades.

node. The point along an axis where branches originate.

nodose. Having knotlike swellings.

nodulose. Having small local thickenings.

ob-. A prefix meaning reversed, upside down.

oblanceolate. Lanceolate, but broader toward the apex.

oblate. Globose, with the diameter at the equator greater than that between the poles.

oblique. At an angle to the axis, but not at right angles.

oblong. Elliptical, blunt at each end, having nearly parallel sides, and two to four times longer than broad.

obovate. Having the shape of a longitudinal section of a bird's egg, with the narrow end basal.

obovoid. Egg shaped, attached at the narrow end.

obtuse. Blunt or rounded on the end.

ocellate. Resembling an eye; having a circular, distinctive structure.

ooblast. See *connecting filament.*

oogamous. Reproducing by sperm or spermatia and egg.

oogonium, pl. *oogonia.* Egg-forming female gametangium.

opposite. Inserted at the same level, but separated by an axis.

orbicular. Flat, with a circular or nearly circular outline.

ostiole. An opening or pore in the tissue around a conceptacle or cystocarp, through which spores or gametes are released.

ovate. Having the shape of a longitudinal section of an egg, the broad end basal.

ovoid. Egg shaped, attached at the broad end.

palmelloid. Nonmotile and forming a gelatinous mass.

palmate. Having several lobes or segments spreading from the same point. See also *digitate.*

panduriform. Fiddle shaped.

paniculate. Loosely, repeatedly, and irregularly branched.

papilla, pl. *papillae.* A small, rounded, nipplelike process on a surface or margin.

paraphysis, pl. *paraphyses.* A sterile hair associated with reproductive structures.

parasporangium. A reproductive structure producing many spores, but not homologous to a tetrasporangium (Rhodophyta).

parenchymatous. Composed of thin-walled, isodiametric cells produced by cell divisions in three dimensions. See also *pseudoparenchymatous.*

parietal. Near the cell wall.

parthenogenesis. Reproduction by gametes without syngamy.

pectinate. Comblike; having a unilateral series of closely set branches.

pedicel. A small stalk, sometimes a single cell, that supports a reproductive structure.

pedicellate. Borne on or having a pedicel.

pelagic. Living in the open sea.

peltate. More or less flattened, circular, and attached to a stalk in the center of the lower side.

penicillate. Brushlike; having a terminal cluster or tuft of filaments.

percurrent. Running through the entire plant or axis.

perennial. A plant living for three or more years, usually becoming reproductive in the second and subsequent years.

perforate. Pierced by holes.

pericarp. Sterile tissue united around and enclosing a carposporophyte.

pericentral cell. A cell cut off from the axial cell by a longitudinal division and initially as long as the axial cell.

periclinal. Parallel to the surface.

perithallium (perithallus). Erect filaments borne on the hypothallus and forming the bulk of the tissue in thick crusts (Rhodophyta).

phenology. Study of the seasonal, periodic changes in organisms.

phragmoplast. Array of microtubules parallel to the spindle at telophase.

phycobilins. Water-soluble biliprotein photosynthetic pigments of Rhodophyta.

pinna, pl. *pinnae.* The ultimate branchlets in a pinnately branched plant. In *Bryopsis* they are tubular, coenocytic branches; in the Rhodophyta they are multicellular.

pinnate. Arranged in two rows on opposite sides of an axis, either paired or alternate; featherlike.

pinnule. A secondary pinna; one of the ultimate divisions of a bipinnate (twice-pinnate) axis.

pit-connection. A connection between daughter cells in Rhodophyta that is usually plugged by material (a pit plug) other than the usual cell wall compounds.

placenta. A complex fusion cell in the Rhodophyta, linking a carposporophyte and its gametophyte.

planar. Flat.

plasmogamy. Fusion of protoplasts without karyogamy.

plastid. A cell organelle, often containing photosynthetic pigments.

pleiomorphic. Having more than one form in the same generation of a life history.

plicate. Folded lengthwise, like a fan.

plumose. Feathery.

plurilocular. Having many compartments within a sporangium or gametangium.

polycarpogonial. Having more than one carpogonium on a supporting cell or associated with an auxiliary cell (Rhodophyta).

polychotomous. Having the axis divided into several equal branches.

polyhedral. Having three or more planes intersecting at a common vertex.

polyphyletic. Containing members with separate evolutionary histories, the member taxa less closely related to one another than they are to other taxa.

polysiphonous. Having a tier of pericentral cells forming a ring around the axial cell, each cell being secondarily connected to the cells above and below it in the axis.

polysporangium, pl. *polysporangia.* A sporangium containing more than four spores.

polyspore. A spore formed in a polysporangium.

polystromatic. Bladelike, having many cell layers.

primordium, pl. *primordia.* The first recognizable beginning of a structure.

procarp. The association of specific carpogonia with specific auxiliary cells (Rhodophyta).

proliferous. Producing adventitious outgrowths similar to, but generally smaller than, the structure bearing them.

propagation. Increase in number of individuals by vegetative reproduction.

propagulum, pl. *propagula (propagules).* A modified, deciduous, vegetative branch that functions as a means of vegetative reproduction.

prostrate. Procumbent; growing along the substrate.

protonema. Early, filamentous, prostrate stage in germination.

proximal. Nearer the point of origin or attachment.

pseudodichotomous. Appearing dichotomous, but formed when a lateral branch partially displaces the apex, becoming equal in appearance and position to the apex.

pseudofilament. A thread of loosely arranged cells, frequently held together by mucilage.

pseudolateral. The displaced apex, generally of determinate growth, that in a sympodially branched axis is replaced by a true lateral to continue the main axis.

pseudoparenchymatous. Appearing parenchymatous, but composed of parallel, closely packed, uniseriate filaments.

pulvinate. Cushion shaped.

pyrenoid. A proteinaceous structure within a plastid, associated with food reserves.

pyriform. Pear shaped, broader at the bottom than at the top.

quadrate (quadrangular). Having four sides and four angles; rectangular or square.

quadriflagellate. Having four flagella.

raceme. A reproductive structure with sporangia or gametangia borne along a main axis in acropetal sequence.

racemose. In the form of a raceme.

rachis. The main axis.

radial branching. Formation of branches singly and along different radii around the axis.

ramelli. Small branches or branchlets.

receptacle. The terminal branchlet or part of a branch in which conceptacles are embedded (Fucales).

reflexed. Bent backward.

reniform. Kidney shaped.

reticulate. Forming a network, a reticulum.

retuse. Having a broad or rounded, notched apex.

rhizoid. A unicellular or single-stranded filamentous attachment or corticating structure.

rhizoplast. Striated fibers associated with flagellar basal bodies; also called *system II fibers.*

rosette. A roselike cluster of cells or blades.

saccate. Saclike.

saxicolous. Growing on rock.

SCUBA. Self-Contained Underwater Breathing Apparatus; used for free-swimming underwater breathing, usually using compressed air.

scutate. Shield shaped.

secund. Unilateral; arranged along one side of an axis.

segregative cell division. Simultaneous cleavage of a multinucleate cell with individual nuclei forming new cells within the old cell.

septate. Having cross walls, divided.

seriate. Arranged in a row.

serpentine. Sinuous; winding.

serrate. Toothed on the margin, the teeth pointing toward the apex.

serrulate. Minutely serrate.

sessile. Lacking a stalk.

seta, pl. *setae.* A thin, stiff, hairlike structure.

simple. Unbranched; undivided.

sinus. A notch between two lobes of a blade.

snout. An acellular tube of wall material formed around the ostiole of a coralline algal conceptacle.

siphonous. Having a tubular, acellular structure (Chlorophyta).

sorus, pl. *sori.* A cluster of reproductive structures that are raised only slightly or not at all above the surface of the plant.

spatulate. Spoon shaped, with the rounded end distal.

spermatangium, pl. *spermatangia*. A gametangium containing a single male spermatium.

spermatium, pl. *spermatia*. A nonflagellated male gamete.

spindle shaped. Elongate, cylindrical, and tapered toward each end.

spinelike. Long, tapering.

spinulose. Bearing fine spines.

spongiose. Like a sponge; elastic and porous.

sporangium, pl. *sporangia*. A reproductive structure producing spores by mitosis (mitospores) or by meiosis (meiospores). Sporangia are often named according to the number of spores formed (monosporangia, bisporangia, tetrasporangia, polysporangia).

sporophyte. Diploid generation; the stage in the life history in which spores are usually produced by meiosis.

spurlike. Short, tapering.

stellate. Star shaped.

stephanokont. Having many flagella borne in a ring around the distal end of a cell.

stichidium, pl. *stichidia*. A specialized reproductive branch, generally somewhat swollen, which produces spermatia or tetraspores in some Rhodophyta.

stipe. A stem or stalk.

stipitate. Having a stem or stalk.

stolon. A horizontal axis growing from the base of the parent plant, capable of producing new plants or shoots.

sub-. A prefix indicating "less than, almost, or approaching."

subsidiary cell. A vegetative cell borne on a supporting cell and becoming part of a fusion cell.

supporting cell. A cell that bears a carpogonial branch.

supralittoral. Above the high tide line, but affected by the sea.

surficial. Occurring on the surface.

sympodial. With the apex displaced by a subtending lateral branch, the apex forming a pseudolateral and the lateral continuing the direction of growth of the main axis.

synonymy. A listing of the scientific names applied to a particular taxon.

tagmatic. Filamentous organization with filaments united to form a larger structure that may be clearly filamentous to pseudoparenchymatous.

taxon, pl. *taxa*. Unit of biological classification: for example, "species," "genus," "family," etc.

tela arachnoidea. Network of stretched cells on the inside of the pericarp, surrounding the carposporophyte (Rhodymeniales).

terete. Cylindrical.

terricolous. Living on or in soil, sand, peat, or mud.

tetrahedral. Divided simultaneously (tetrasporangia in Rhodophyta), the four spores touching at the center and only three walls being visible from any direction.

tetrasporangium. A sporangium containing four spores, usually the products of meiosis.

tetraspore. One of the four spores formed in a tetrasporangium.

thallus, pl. *thalli*. The body of a macroscopic alga.

tier. One of a series of transverse cell rows placed one above the other.

tortuous. Bending irregularly.

torulose. Cylindrical with slight constrictions, very nearly the same as moniliform.

trabecula, pl. *trabeculae*. A strand of cell-wall material crossing a space within a plant or a cell.

transverse. At right angles to the long axis.

trapezoid. Four sided, with two sides parallel.

trawl. A device used in fishing; a conical net dragged behind a vessel.

trichoblast. A colorless, branched or unbranched, hairlike filament in the Rhodophyta.

trichocyte (trichocyst). An uncalcified, hair-forming cell in the coralline algal hypothallus or perithallus.

trichogyne. Receptive extension of the carpogonium to which spermatia attach.

trichothallic. Growing by means of an intercalary meristem in a uniseriate filament.

trifurcate. Having three prongs or branches diverging from the same point.

truncate. Ending abruptly, as if cut straight across.

tuberculate. Having irregular warty outgrowths.

turbinate. Shaped like a top (a cone with a rounded base) and attached at the point.

umbilicate. Having a central depression resembling a navel.

undulate. Having a wavelike form.

uniaxial. Having a single, central, longitudinal filament from which the remaining tissue or branches are derived.

unicarpogonial. Having one carpogonium per supporting cell or auxiliary cell.

unicellular. Consisting of a single cell.

unilateral. Formed on one side of an axis.

unilocular. With a single compartment, undivided by walls.

uniporate. With a single opening per conceptacle (Corallinales).

uniseriate. Formed in a single row or chain.

urceolate. Urn shaped.

utricle. Enlarged, cortical, photosynthetic end of a sympodially branched coenocytic filament (*Codium*).

vacuous. Empty.

vasiform. Resembling a duct.

vegetative. Not reproductive.

ventral. Pertaining to the lower surface of a dorsiventral plant.

vermiform. Wormlike.

verticillate. Whorled.

vesicle. A thin-walled, hollow, swollen branch.

villose. Covered with long, soft hairs.

virgate. Long, slender, stiff, and much branched.

voucher. A representative specimen from a collection deposited for future reference.

whorled. With several branches at the same level around the axis.

zonate. Divided by more or less parallel walls to form a short row of cells; in the zonate tetrasporangia of the Rhodophyta, four spores are formed in a row.

zooidangium, pl. *zooidangia.* A cell within which are formed flagellated cells that may be either zoospores or gametes.

zoosporangium, pl. *zoosporangia.* A cell within which zoospores are formed.

zoospore. A motile, flagellated spore.

zygospore. A thick-walled resting spore that develops from a zygote.

zygote. A cell formed by the union of two gametes.

Collection Sites Mentioned in the Text

	Latitude	Longitude
North Carolina		
Adams Creek	34°57′N	76°40′W
Atlantic Beach	34°42′N	76°44′W
Beaufort	34°43′N	76°40′W
Beaufort Inlet	34°41′N	76°40′W
Cape Lookout jetty	34°37′N	76°35′W
Fort Fisher	33°59′N	77°55′W
Fort Macon	34°42′N	76°41′W
Frying Pan Shoals	33°40′N	77°55′W
Lockwoods Folly Inlet	33°55′N	78°14′W
Masonboro Inlet	34°11′N	77°49′W
Mile Hammock Rock	34°31′N	77°19′W
Morehead City	34°43′N	76°43′W
Newport River Estuary	34°43′N	76°41′W
New River Inlet	34°32′N	77°21′W
Ocracoke Inlet	35°03′N	76°01′W
Oregon Inlet	35°47′N	75°31′W
Pivers Island	34°43′N	76°40′W
Radio Island	34°42′N	76°41′W
Salvo	35°31′N	75°30′W
Shackleford Bank jetty	34°41′N	76°39′W
Surf City	34°26′N	77°32′W
Topsail Island	34°26′N	77°32′W
Wilmington	34°14′N	77°57′W
Wrightsville Beach	34°12′N	77°47′W
Wrightsville Beach jetty	34°11′N	77°49′W
South Carolina		
Bears Bluff (Wadmalaw Island)	32°38′N	80°15′W
Charleston	32°47′N	79°56′W
Ft. Johnson	32°45′N	79°56′W
Georgetown	33°13′N	79°11′W
Myrtle Beach	33°42′N	78°53′W
North Island	33°13′N	79°11′W
Pawleys Island	33°24′N	79°08′W
Port Royal Sound	32°15′N	80°40′W
Sullivans Island	32°46′N	79°51′W
Winyah Bay	33°17′N	79°15′W
Georgia		
Cumberland Island	30°51′N	81°26′W
Gray's Reef	31°24′N	80°52′W
Snapper Banks	31°35′N	80°23′W
Sapelo Island	31°27′N	81°15′W
Florida		
Haulover Cove	28°41′N	80°46′W
Marineland	29°39′N	81°12′W
St. Augustine	29°54′N	81°19′W

Table 1. Seaweed species endemic to the southeastern United States.

Species known only from the region
between Cape Hatteras
and Cape Canaveral

Cladophora pseudobainesii
Derbesia turbinata
Vaucheria acrandra
Vaucheria adela
Hincksia onslowensis
Streblonema invisible
Porphyra carolinensis
Porphyra rosengurttii
Audouinella affinis
Audouinella hoytii
Helminthocladia andersonii
Peyssonnelia atlantica
Dudresnaya georgiana

Trematocarpus papenfussii
Craspedocarpus humilis
Petroglossum undulatum
Champia parvula var. prostrata
Calliclavula trifurcata
Callithamniella silvae
Lejolisia exposita
Nwynea grandispora
Ptilothamnion occidentale
Dasysiphonia concinna
Dasysiphonia doliiformis
Dipterosiphonia reversa

Species also present only in other parts of Florida (F)
or in Bermuda (B)

Codium carolinianum (F)
Onslowia endophytica (F)
Padina profunda (F)
Naccaria corymbifera (F & B)
Agardhinula browneae (F)
Cirrulicarpus carolinensis (F)

Gloioderma blomquistii (F)
Gloioderma rubrisporum (F)
Branchioglossum minutum (F)
Searlesia subtropica (F)
Dasya spinuligera (B)
Ceramium leptozonum (B)

Table 2. Local seaweed species with disjunct distributions in the regions listed.

Species	Region of Disjunct Distribution
Vaucheria longicaulis	Brazil, Indian Ocean, Pacific Ocean
Vaucheria erythrospora	Europe, Japan
Herponema solitarium	Europe
Myriactula stellulata	Europe
Dictyopteris hoytii	Venezuela, South Florida
Audouinella botryocarpa	Europe, Africa, Australia, New Zealand
Lithophyllum subtenellum	Europe, Mediterranean
Phymatolithon tenuissimum	Canary Islands, West Africa
Peyssonnelia stoechas	Mediterranean
Rhodymenia divaricata	Gulf of California, Galapagos, South Florida
Pleonosporium boergesenii	Brazil
Pleonosporium flexuosum	Europe, Mediterranean, South Africa
Myriogramme distromatica	Mediterranean
Herposiphonia delicatula	Central Pacific
Polysiphonia pseudovillum	Central Pacific

Literature Cited

Abbott, I. A. 1976. On the red algal genera *Grallatoria* Howe and *Callithamniella* Feldmann-Mazoyer (Ceramiales). *Brit. Phycol. J.* 11:143–49.

———. 1979. Some tropical species related to *Antithamnion* (Rhodophyta, Ceramiaceae). *Phycologia* 18:213–27.

Abbott, I. A., and G. J. Hollenberg. 1976. *Marine Algae of California*. xii + 827 pp. Stanford University Press, Stanford, Calif.

Adanson, M. 1763. *Familles des plantes*. Pt. 2. [18] + 640 + [5] pp. Paris.

Adey, W. H. 1966. The genera *Lithothamnion*, *Leptophytum* (nov. gen.) and *Phymatolithon* in the Gulf of Maine. *Hydrobiologia* 28:321–70.

———. 1970. A revision of the Foslie crustose coralline herbarium. *Kongel. Norske Vidensk. Selsk. Skr. (Copenhagen)* 1970(1):1–47.

Adey, W. H., and H. W. Johansen. 1972. Morphology and taxonomy of the Corallinaceae with special reference to *Clathromorphum*, *Mesophyllum* and *Neopolyporolithon* gen. nov. (Rhodophyta, Cryptonemiales). *Phycologia* 11:159–80.

Agardh, C. A. 1817. *Synopsis algarum Scandinaviae*. Pp. I–XL, 1–135. Lund.

———. 1820. *Species algarum*. 1(1):[i–iv], 1–168. Lund. (Also issued at Greifswald and dated 1821.)

———. 1822. *Species algarum*. 1(2):[i–viii], 169–531. Lund. (Also issued at Greifswald and dated 1823.)

———. 1824. *Systema algarum*. Pp. i–xxxviii, 1–312. Lund.

———. 1827. Aufzählung einiger in den östreichischen Ländern gefundenen neuen Gattungen und Arten von Algen, nebst ihrer Diagnostik und beigefügten Bemerkungen. *Flora* 10:625–46.

———. 1828. *Species algarum*. 2(1):i–lxxvi, 1–189.

Agardh, J. G. 1841. In historiam algarum symbolae. *Linnaea* 15:1–50, 443–57.

———. 1842. *Algae maris mediterranei et adriatici*. Pp. i–x, 1–164. Paris.

———. 1847. Nya alger från Mexico. *Öfvers. Förh. Kongl. Svenska Vetensk.-Akad.* 4:5–17.

———. 1848. *Species genera et ordines algarum*. 1:i–viii, 1–363. Lund.

———. 1849. Algologiska bidrag. *Öfvers. Förh. Kongl. Svenska Vetensk.-Akad.* 6:79–89.

———. 1851. *Species genera et ordines algarum*. 2(1):i–xii, 1–351. Lund.

———. 1852. *Species genera et ordines algarum*. 2(2:1):337(bis)–51(bis), 352–504. Lund.

———. 1863. *Species genera et ordines algarum*. 2(3:2):787–1291 (1139–58 omitted). Lund.

———. 1872. Bidrag till Florideernes systematik. *Acta Univ. Lund.* 8(6):1–60.

———. 1876. *Species genera et ordines algarum*. Vol. 3(1), *Epicrisis systematis floridearum*. Pp. I–VII, 1–724.

———. 1882. Till algernes systematik, Nya bidrag. (Andra afdelningen). *Acta Univ. Lund.* 17(4):1–136, 3 pls.

———. 1883. Till algernes systematik, Nya bidrag. (Tredje afdelningen.) *Acta Univ. Lund.* 19(2):1–177, 4 pls.

———. 1885. Till algernes systematik, Nya bidrag. (Fjerde afdelningen.) *Acta Univ. Lund.* 21(8):1–117, 1 pl.

———. 1892. Analecta algologica. *Acta Univ. Lund., Andra Afd.* 28(6):1–182, 3 pls.

———. 1894. Analecta algologica. Continuatio I. *Acta Univ. Lund., Andra Afd.* 29(9):1–144, 2 pls.

———. 1898. *Species genera et ordines algarum*. 3(3):[i–vi], 1–239. Lund.

———. 1899. Analecta algologica. Continuatio V. *Acta Univ. Lund., Andra Afd.* 35(4):1–160, 3 pls.

Aleem, A. A., and E. Shulz, 1952. Über Zonierung von Algengemeinschaften. (Ökologische Untersuchungen im Nord-Ostsee-Kanal, I.) *Kieler Meeres.* 9:70–76.

Alexander, N. J. 1970. *The taxonomy of Enteromorpha at Beaufort, North Carolina.* A.M. thesis, vi + 1–86 pp. Duke University, Durham, N.C.

Amsler, C. D. 1983. *Culture and field studies of* Acinetospora crinita, Giffordia mitchelliae, *and G.* irregularis *(Phaeophyceae, Ectocarpaceae) at Wrightsville Beach, North Carolina, U.S.A.* M.S. thesis, v + 149 pp. University of North Carolina, Wilmington.

――――. 1984a. Culture and field studies of *Acinetospora crinata* (Carmichael) Sauvageau (Ectocarpaceae, Phaeophyceae) in North Carolina (USA). *Phycologia* 23:377–82.

――――. 1984b. Culture and field observations of the controlling effects of temperature on the seasonality of three *Giffordia* species (Ectocarpaceae, Phaeophyceae). *J. Phycol.* 20 (suppl.):4 (abstract).

――――. 1985. Field and laboratory studies of *Giffordia mitchelliae* (Phaeophyceae) in North Carolina. *Bot. Mar.* 28:295–301.

Amsler, C. D., and D. F. Kapraun. 1985. *Giffordia onslowensis* sp. nov. (Phaeophyceae) from the North Carolina continental shelf and the relationship between *Giffordia* and *Acinetospora. J. Phycol.* 21:94–99.

Amsler, C. D., and R. B. Searles. 1980. Vertical distribution of seaweed spores in a water column offshore of North Carolina. *J. Phycol.* 16:617–19.

Ardissone, F. 1878. Studi sulle alghe italiche della famiglia della Rhodomelacee. *Atti Soc. Crittog. Ital.* 1:41–159.

――――. 1883. Phycologia mediterranea. *Mem. Soc. Crittog. Ital.* 1:1–516.

Ardré, F., and P. Gayral. 1961. Quelques *Grateloupia* de l'Atlantique et du Pacifique. *Rev. Algol.*, N.S. 6:38–48, 3 pls.

Aregood, C. C. 1975. A study of the red alga, *Calonitophyllum medium* (Hoyt) comb. nov. [=*Hymenena media* (Hoyt) Taylor]. *Brit. Phycol. J.* 10:347–62.

Areschoug, J. E. 1842. Algarum minus rite cognitarum pugillus primus. *Linnaea* 16:225–36.

――――. 1843. Algarum (phycearum) minus rite cognitarum pugillus secundus. *Linnaea* 17:257–69, 1 pl.

――――. 1847. Phycearum, quae in maribus Scandinaviae crescunt, enumeratio. Sectio prior Fucaceas continens. *Nova Acta Regiae Soc. Sci. Upsal.* 13:223–382, 9 pls.

――――. 1850. Phycearum, quae in maribus Scandinaviae crescunt. Sectio posterior Ulvaceas continens. *Nova Acta Regiae Soc. Sci. Upsal.* 14:385–454, 3 pls.

Aziz, K. M. S. 1965. *Acrochaetium and Kylinia in the Southwestern North Atlantic Ocean.* Ph.D. thesis, ix + 235 pp. Duke University, Durham, N.C.

――――. 1967. The life cycle of *Acrochaetium dasyae* Collins. *Brit. Phycol. Bull.* 3:408.

Aziz, K. M. S., and H. J. Humm. 1962. Additions to the algal flora of Beaufort, N.C., and vicinity. *J. Elisha Mitchell Sci. Soc.* 78:55–63.

Bailey, J. W. 1848. Continuation of the list of localities of algae in the United States. *Am. J. Sci. Arts* 6:37–42.

――――. 1851. Microscopical observations made in South Carolina, Georgia, and Florida. *Smithsonian Contr. Knowl.* 2(Art. 8) 48 pp., 3 pls.

Balakrishnan, M. S. 1960. Reproduction in some Indian red algae and their taxonomy. Pp. 85–98. In P. Kachroo, ed., *Proceedings of the Symposium on Algology.* Indian Council of Agricultural Research, New Delhi.

――――. 1961. Studies on Indian Cryptonemiales. III. *Halymenia* C. Ag. *J. Madras Univ.*, 31:183–217, 1 pl.

Ballantine, D. L. 1985. *Botryocladia wynnei* sp. nov. and *B. spinulifera* Taylor and Abbott

(Rhodymeniales, Rhodophyta) from Puerto Rico. *Phycologia* 24:199–204.

Ballantine, D. L., and M. J. Wynne. 1986. Notes on the marine algae of Puerto Rico. I. Additions to the flora. *Bot. Mar.* 29:131–35.

———. 1987. Notes on the marine algae of Puerto Rico. III. *Branchioglossum pseudoprostratum* new species and *B. prostratum* Schneider (Rhodophyta, Delesseriaceae). *Bull. Mar. Sci. Gulf Caribbean* 40:240–45.

Basson, P. W. 1979. Marine algae of the Arabian Gulf coast of Saudi Arabia (second half) *Bot. Mar.* 22:65–82.

Batters, E. A. L. 1902. A catalogue of the British marine algae. *J. Bot.* 40(Suppl.):[2] + 1–107.

Berthold, G. 1882. Über die Vertheilung der Algen im Golf von Neapel nebst einem Verzeichnis der bisher daselbst beobachteten Arten. *Mitt. Zool. Sta. Neapel* 3:394–536.

Bird, C. J., and E. C. de Oliveira F. 1986. *Gracilaria tenuifrons* sp. nov. (Gigartinales, Rhodophyta), a species from the tropical western Atlantic with superficial spermatangia. *Phycologia* 25:313–20.

Bird, N., J. McLachlan and D. Grund. 1977. Studies on *Gracilaria* sp. from the Maritime Provinces. *Canad. J. Bot.* 55:1282–90.

Bivona-Bernardi, A. de. 1822. *Scinaia* algarum marinarum novum genus. Pp. [1–3], 1 pl. Palermo.

Blair, S. M. 1983. Taxonomic treatment of the *Chaetomorpha* and *Rhizoclonium* species (Cladophorales: Chlorophyta) in New England. *Rhodora* 85:175–211.

Blair, S. M., and M. O. Hall. 1981. Ten new records of deep water marine algae from Georgia and South Carolina. *Northeast Gulf Sci.* 4:127–29.

Blair, S. M., A. C. Mathieson, and D. P. Cheney. 1982. Morphological and electrophoretic investigations of selected species of *Chaetomorpha* (Chlorophyta; Cladophorales). *Phycologia* 21:164–72.

Bliding, C. 1963. A critical survey of European taxa in Ulvales. I. *Capsosiphon, Percursaria, Blidingia, Enteromorpha*. *Opera Bot.* 8(3):1–160.

———. 1968. A critical survey of European taxa in Ulvales. II. *Ulva, Ulvaria, Monostroma, Kornmannia*. *Bot. Not.* 121:535–629.

Blomquist, H. L. 1954. A new species of *Myriotrichia* Harvey from the coast of North Carolina. *J. Elisha Mitchell Sci. Soc.* 70:37–41.

———. 1955. *Acinetospora* Born. new to North America. *J. Elisha Mitchell Sci. Soc.* 71:46–49.

———. 1958a. *Myriotrichia scutata* Blomquist conspecific with *Ectocarpus subcorymbosus* Holden in Collins. *J. Elisha Mitchell Sci. Soc.* 74:24.

———. 1958b. The taxonomy and chromatophores of *Pylaiella antillarum* (Grunow) DeToni *J. Elisha Mitchell Sci. Soc.* 74:25–30.

Blomquist, H. L., and H. J. Humm. 1946. Some marine algae new to Beaufort, North Carolina. *J. Elisha Mitchell Sci. Soc.* 62:1–8.

Blomquist, H. L., and J. H. Pyron. 1943. Drifting "seaweed" at Beaufort, N.C. *Amer. J. Bot.* 30:28–32.

Blum, J. L. 1972. Vaucheriaceae. *North American Flora*. Ser. 2, pt. 8. Pp. 1–63. New York Botanic Gardens.

Blum, J. L., and J. T. Conover. 1953. New or noteworthy Vaucheriae from New England salt marshes. *Biol. Bull. Mar. Biol. Lab. Woods Hole* 105:395–401.

Børgesen, F. 1909. Some new or little known West Indian Florideae I. *Bot. Tidsskr.* 29:1–19.

———. 1910. Some new or little known West Indian Florideae II. *Bot. Tidsskr.* 30:177–207.

———. 1913. The marine algae of the Danish West Indies. Part 1: Chlorophyceae. *Dansk Bot. Ark.* 1(4)):1–158 + [2], 1 map.

———. 1914. The marine algae of the Danish West Indies. Part 2: Phaeophyceae, *Dansk Bot. Ark.* 2(2):1–66 + [2].

———. 1915. The marine algae of the Danish West Indies. Part 3: Rhodophyceae (1), *Dansk Bot. Ark.* 3:1–80.

———. 1916. The marine algae of the Danish West Indies. Part 3: Rhodophyceae (2), *Dansk Bot. Ark.* 3:81–144.

———. 1917. The marine algae of the Danish West Indies. Part 3: Rhodophyceae (3), *Dansk Bot. Ark.* 3:145–240.

———. 1918. The marine algae of the Danish West Indies. Part 3: Rhodophyceae (4), *Dansk Bot. Ark.* 3:241–304.

———. 1919. The marine algae of the Danish West Indies. Part 3: Rhodophyceae (5), *Dansk Bot. Ark.* 3:305–68.

———. 1920. The marine algae of the Danish West Indies. Part 3: Rhodophyceae (6), *Dansk Bot. Ark.* 3:369–504.

———. 1930. Marine algae from the Canary Islands, especially from Teneriffe and Gran Canaria. III. Rhodophyceae; Part 3. Ceramiales. *Biol. Meddel. Kongel. Danske Vidensk. Selsk.* 9(1):1–159.

———. 1932. Revision of Forsskål's algae. *Dansk Bot. Arkiv.* 8(2): 14 pp., l pl., 4 figs.

———. 1941. Some marine algae from Mauritius. II. Phaeophyceae. *Biol. Meddel. Kongel. Danske Vidensk. Selsk.* 24(3):1–81.

———. 1945. Some marine algae from Mauritius. III. Rhodophyceae. Part 4. Ceramiales. *Biol. Meddel. Kongel. Danske Vidensk. Selsk.* 19(10):1–68.

———. 1950. A new species of the genus *Predaea. Dansk Bot. Ark.* 14(4):1–8.

Bold, H. C., and M. J. Wynne. 1985. *Introduction to the Algae*, second edition. xvi + 720 pp. Prentice-Hall, Englewood Cliffs, N.J.

Bornet, E. 1859. Description d'un nouveau genre de Floridées des côtes de France. *Ann. Sci. Nat. Bot.*, Sér. 4, 11:80–92, pls 1,2.

———. 1891. Note sur quelques *Ectocarpus. Bull. Soc. Bot. France* 38:353–72, pls 6–8.

———. 1892. Les algues de P. K. A. Schousboe. *Mém. Soc. Sci. Nat. Cherbourg* 28:165–376, 3 pls.

Bornet, E., and C. Flahault. 1888. Note sur deux nouveaux genres d'algues perforantes. *J. Bot. (Morot)* 2:161–65.

———. 1889. Sur quelques plantes vivant dans le test calcaire des mollusques. *Bull. Soc. Bot. France* 36:CXLVII–CLXXVI, pls 6–12.

Bory de Saint-Vincent, J. B. 1822. Botrytelle. *Dict. Class. Hist. Nat.* 2:425–26.

———. 1823. Céramiaires. *Dict. Class. Hist. Nat.* 3:339–41.

———. 1834. Hydrophytes, Hydrophytae. In C. Bélanger, ed., *Voyage aux Indes-Orientales... 1825, 1826, 1827, 1828 et 1829; Botanique, IIe Partie. Cryptogamie [2^{me}livr.].* Pp. 159–78, pls 15,16.

Boudouresque, C.-F. 1971. Sur le *Nitophyllum distromaticum* Rodriguez mscr. (*Myriogramme distromatica* [Rodriguez] comb. nov.). *Soc. Phycol. de France, Bull.* 16:76–81.

Boudouresque, C.-F., and E. Coppejans. 1982. Végétation marine de L'Ile de Port-Cros (Parc National). XXIII. Sur deux espèces de *Griffithsia* (Ceramiaceae, Rhodophyta). *Bull. Soc. Roy. Bot. Belg.* 115:43–52.

Boudouresque, C.-F., and M. Denizot. 1975. Révision du genre *Peyssonnelia* (Rhodophyta) en Méditerranée. *Bull. Mus. Hist. Nat. (Marseille)* 35:7–92.

Brauner, J. F. 1975. The seasonality of epiphytic algae on *Zostera marina* at Beaufort, North Carolina. *Nova Hedwigia* 26:125–33.

Bumpus, D. F. 1973. A description of the circulation on the continental shelf of the east coast of the United States. *Prog. Oceanogr.* 6:111–57.

Calderón-Sánez, E., and R. Schnetter. 1989. The life histories of *Boodleopsis vaucherioidea* sp. nov. and *B. pusilla* (Caulerpales) and their phylogenetic implications. *Phycologia* 28:476–90.

Cardinal, A. 1964. Étude sur les Ectocarpacées de la Manche. *Beih. Nova Hedwigia* 15:1–86.

Castagne, L. 1851. *Supplément au catalogue des plantes qui croissent naturellement aux environs de Marseille.* 125 pp., pls 8–11. Aix.

Causey, N. B., J. P. Prytherch, J. McCaskill, H. J. Humm, and F. A. Wolf. 1946. Influence of environmental factors upon the growth of *Gracilaria confervoides*. *Duke Univ. Mar. Sta. Bull.* 3:19–24.

Cerame-Vivas, M. J., and I. E. Gray. 1966. The distributional pattern of benthic invertebrates of the continental shelf off North Carolina. *Ecology* 47:260–70.

Chamberlain, Y. M. 1983. Studies in the Corallinaceae with special reference to *Fosliella* and *Pneophyllum* in the British Isles. *Bull. Brit. Mus. (Nat. Hist.) Bot.* 11:291–463.

——. 1985. The typification of *Melobesia membranacea* (Esper) Lamouroux (Rhodophyta, Corallinaceae). *Taxon* 34:673–77.

Chapman, R. L. 1971. The macroscopic marine algae of Sapelo Island and other sites on the Georgia coast. *Bull. Georgia Acad. Sci.* 29:77–89.

——. 1973. An addition to the macroscopic marine algal flora of Georgia. The genus *Cladophora. Bull. Georgia Acad. Sci.* 31:147–50.

Chauvin, J. F. 1842. *Recherches sur l'organisation, la fructification et la classification de plusieurs genres d'algues.* 132 pp. Caen.

Cheney, D. P. 1988. The genus *Eucheuma* J. Agardh in Florida and the Caribbean. In I. A. Abbott, ed., *Taxonomy of Economic Seaweeds with reference to some Pacific and Caribbean Species,* 2:209–19. California Sea Grant Program, La Jolla, Calif.

Chihara, M. 1961. Life cycle of the Bonnemaisoniaceous algae in Japan (1). *Sci. Rep. Tokyo Kyoiku Daigaku, Sec. B.* 10:121–53.

Christensen, T. 1956. Studies on the genus *Vaucheria.* III. Remarks on some species from brackish water. *Bot. Not.* 109:275–80.

Clayton, M. N. 1974. Studies on the development, life history, and taxonomy of the Ectocarpales (Phaeophyta) in Southern Australia. *Austral. J. Bot.* 22:743–813.

Clemente y Rubio, S. de R. 1807. *Ensayo sobre las variedades de la vid comun que vegetan en Andalucía.* Pp. i–xviii, 1–324.

Codomier, L. 1974. Recherches sur la structure et le développement des *Halymenia* C. Ag. (Rhodophycées, Cryptonemiales) des côtes de France et de la Méditerranée. *Vie & Milieu, sér. A., Biol. Mar.* 24:1–42.

Colijn, F., and C. van den Hoek. 1971. The life-history of *Sphacelaria furcigera* Kütz. (Phaeophyceae). II. The influence of daylength and temperature on sexual and vegetative reproduction. *Nova Hedwigia* 21:899–922.

Coll, J., and J. Cox. 1977. The genus *Porphyra* C. Ag. (Rhodophyta, Bangiales) in the American North Atlantic. I. New species from North Carolina. *Bot. Mar.* 20:155–59.

Collins, F. S. 1896. Notes on New England marine algae. *Bull. Torrey Bot. Club* 23:458–62.

——. 1901. *Scinaia furcellata* var. *complanata.* In F. S. Collins, I. Holden, and W. A. Setchell, *Phycotheca Boreali-Americana* (Exsiccatae). Vol. 17, no. 836.

———. 1906a. New species in the Phycotheca. *Rhodora* 8:104–13.

———. 1906b. *Acrochaetium* and *Chantransia* in North America. *Rhodora* 8:189–96.

———. 1908. Two new species of *Acrochaetium*. *Rhodora* 10:133–35.

———. 1909. The green algae of North America. *Tufts Coll. Stud., Sci. Ser.* 2:79–480, pls 1–18.

———. 1918. The green algae of North America. Second supplement. *Tufts Coll. Stud., Sci. Ser.* 4(7):1–106, pls 1–3.

Collins, F. S., and A. B. Hervey. 1917. The algae of Bermuda. *Proc. Amer. Acad. Arts* 53:1–195.

Collins, F. S., and M. A. Howe. 1916. Notes on some species of *Halymenia*. *Bull. Torrey Bot. Club.* 43:169–82.

Cordeiro-Marino, M. 1978. Rhodofíceas bentônicas marinhas do Estado de Santa Catarina. *Rickia* 7:1–243.

Cormaci, M., and G. Furnari. 1988. *Antithamnionella elegans* (Berthold) Cormaci et Furnari (Ceramiaceae, Rhodophyta) and related species, with the description of two new varieties. *Phycologia* 27:340–46.

Cotton, A. D. 1907. New or little known marine algae from the east. *Kew Bull.* 7:260–64.

Coutinho, R., and R. Zingmark. 1987. Diurnal photosynthetic responses to light by macroalgae. *J. Phycol.* 23:336–43.

Cribb, A. B. 1983. *Marine Algae of the Southern Great Barrier Reef. I: Rhodophyta*. Pp. 1–173 + [2], pls 1–71. Australian Coral Reef Society, Brisbane.

Crouan, P. L., and H. M. Crouan. 1835. Observations microscopiques sur le genre *Mesogloia*, Agardh. *Ann. Sci. Nat., Bot.*, Sér. 2, 3:98–99, pl. 2.

Crouan, P. L., and H. M. Crouan. 1852. *Algues marines du Finistère*. 404 specimens with printed labels. Brest (Exsiccatae).

Crouan, P. L., and H. M. Crouan. 1859. Notice sur quelques espèces et genres nouveaux d'algues marines de la rade de Brest. *Ann. Sci. Nat., Bot.*, Sér. 4, 12:288–92.

Crouan, P. L., and H. M. Crouan. 1867. *Florule du Finistère*. Pp. [i]–x, [1]–262, pls 1–31 [32]. Paris, Brest.

Curtis, M. A. 1867. Algae. Pp. 155–56. In *Geological and natural history survey of North Carolina. III: Botany; containing a catalogue of the indigenous and naturalized plants of the state*. Raleigh, N.C.

Dangeard, P. 1939. Le genre *Vaucheria*, spécialement dans la région du sud-ouest de la France. *Botaniste* 29:183–264, pls 11–15.

———. 1958. La reproduction et le développement de l'*Enteromorpha marginata* Ag. et le rattachement de cette espèce au genre *Blidingia*. *Compt. Rend. Hébd. Séances Acad. Sci.* 246:347–51, figs A–K.

Dawes, C. J. 1974. *Marine Algae of the West Coast of Florida*. Pp. i–xviii, 1–201. University of Miami Press, Coral Gables, Florida.

Dawson, E. Y. 1941. A review of the genus *Rhodymenia* with descriptions of new species. *Allan Hancock Pacific Exped.* 3:123–81.

———. 1949. Studies of northeast Pacific Gracilariaceae. *Allan Hancock Found. Publ. Occas. Pap.* 7:1–105.

———. 1953. On the occurrence of *Gracilariopsis* in the Atlantic and Caribbean. *Bull. Torrey Bot. Club* 80:314–16.

———. 1963. Marine red algae of Pacific Mexico, Part 8. Ceramiales: Dasyaceae, Rhodomelaceae. *Nova Hedwigia* 6:401–81, pls 126–71.

Dawson, E. Y., C. Acleto, and N. Foldvik. 1964. The seaweeds of Peru. *Beih. Nova Hedwigia* 13:1–111, 81 pls.

Decaisne, J. 1841. Plantes de l'Arabie Heureuse. *Arch. Mus. Hist. Nat.* 2:89–199, pls 5–7.

——. 1842. Mémoire sur les Corallines ou Polypiers calcifères. *Ann. Sci. Nat., Bot.*, Sér. 2, 18:96–128.

De Candolle, A. P. 1801. Extrait d'un rapport sur les Conferves, fait à la Société Philomatique. *Bull. Sci. Soc. Philom.* 3:17–21.

Delile, A. R. 1813. Flore d'Égypte, explication des planches. In France, Commission d'Égypte, *Description de l'Égypte*,... *Histoire Naturelle*, 2:145–320 (atlas, 62 pls, 1826). Paris.

DeLoach, W. S., O. C. Wilton, H. J. Humm, and F. A. Wolf. 1946a. Preparation of an agar-like gel from *Hypnea musciformis*. *Duke Univ. Mar. Sta. Bull.* 3:31–39.

DeLoach, W. S., O. C. Wilton, J. McCaskill, H. J. Humm, and F. A. Wolf. 1946b. *Gracilaria confervoides* as a source of agar. *Duke Univ. Mar. Sta. Bull.* 3:25–30.

Denizot, M. 1968. Les algues Floridées encroutantes (à l'exclusion des Corallinacées). Pp. 1–310. Published privately by the author, Paris.

Derbès, F., and A. J. J. Solier. 1850. Sur les organes reproducteurs des algues. *Ann. Sci. Nat., Bot.*, Sér. 3, 14:261–82, pls 32–37.

——. 1851. Algues, Pp. 93–121. In L. Castagne, ed., *Supplément au catalogue des plantes qui croissent naturellement aux environs de Marseille*. Aix.

——. 1856. Mémoire sur quelques points de la physiologie des algues. *Compt. Rend. Hébd. Séances Acad. Sci.* Suppt 1:1–120, 23 pls.

Desvaux, N. A. 1809. Neues Journal für die Botanik, herausgegeben, vom Professor Schrader; avec des observations sur le genre *Fluggia*, Rich. (*Slateria*, Desv.). *J. Bot. (Desvaux)* 1:240–46.

DeToni, G. B. 1889. *Sylloge algarum*. Vol. I. *Chlorophyceae*. Pp. 1–12, i–cxxxix, 1–592 + [2], 593–1315. Published by the author, Padua.

——. 1897. Current Literature. Notes for students. [Remarks on J. G. Agardh's Analecta Algologica.] *Bot. Gaz. (Crawfordsville)* 23:63–64.

——. 1903. *Sylloge algarum*. Vol. IV. *Florideae* (3). Pp. 775–1525, 1 pl. Padua.

——. 1905. *Sylloge algarum*. Vol. IV. *Florideae* (4). Pp. 1523–1973. Padua.

——. 1936. *Noterelle di nomenclatura algologica VII, Primo elenco di Floridee omonime*. [8] pp. Published by the author, Brescia.

Dillwyn, L. W. 1802–1809. *British Confervae*. Pp. [i–iv], 1–87 + [5, index], pls 1–109, A–G. London.

Dixon, P. S. 1961. On the classification of the Florideae with particular reference to the position of the Gelidiaceae. *Bot. Mar.* 3:1–16.

——. 1976. Appendix I. p. 590. In M. Parke and P. S. Dixon, eds., Check-list of British marine algae, third revision. *J. Mar. Biol. Assoc. U.K.* 56:527–94.

Dixon, P. S., and L. M. Irvine. 1977. *Seaweeds of the British Isles*. Vol. I. *Rhodophyta*. Part 1, *Introduction, Nemaliales, Gigartinales*. Pp. i–xi, 1–252. British Museum (Natural History). London.

Dixon, P. S., and J. H. Price. 1981. The genus *Callithamnion* (Rhodophyta: Ceramiaceae) in the British Isles. *Bull. Brit. Mus. (Nat. Hist.) Bot.* 9:99–141.

Dixon, P. S., and G. Russell. 1964. Miscellaneous notes on algal taxonomy and nomenclature. I. *Bot. Not.* 117:279–84.

Doty, M. S. 1947. The marine algae of Oregon. I. Chlorophyta and Phaeophyta. *Farlowia* 3:1–65, pls 1–10.

Drew, K. M. 1928. A revision of the genera *Chantransia, Rhodochorton*, and *Acrochaetium*. *Univ. Calif. Publ. Bot.* 14:139–224, pls 37–48.

——. 1956. *Conferva ceramicola* Lyngbye. *Bot. Tidsskr.* 53:67–74.

Duby, J. E. 1830. Aug. *Pyrami de Candolle botanicon gallicum*, second edition. 2:[i–vi], 545–1068, i–lviii. Paris.

Duke, C. S., B. E. Lapointe, and J. Ramus. 1986. Effects of light on growth, RuBPCase activity and chemical composition of *Ulva* species (Chlorophyta). *J. Phycol.* 22:362–70.

Earle, L. C., and H. J. Humm. 1964. Intertidal zonation of algae in Beaufort Harbor. *J. Elisha Mitchell Sci. Soc.* 80:78–82.

Earle, S. A. 1969. The Phaeophyta of the eastern Gulf of Mexico. *Phycologia* 7:71–254.

Edelstein, T., L. C.-M. Chen, and J. McLachlan. 1971. On the life histories of some brown algae from eastern Canada. *Canad. J. Bot.* 49:1247–51.

———. 1974. The reproductive structures of *Gigartina stellata* (Stackh.) Batt. (Gigartinales, Rhodophyceae) in nature and culture. *Phycologia* 13:99–107.

———. 1978. Studies on *Gracilaria*: reproductive structures of *Gracilaria* (Gigartinales, Rhodophyta). *J. Phycol.* 14:92–100.

Eiseman, N. J. 1976. Benthic plants of the east Florida continental shelf. *Ann. Rep. Harbor Branch Fdn* 1976. Pp. A1–A13.

———. 1979. Marine algae of the east Florida continental shelf. I: Some new records of Rhodophyta including *Scinaia incrassata* sp. nov. (Nemaliales, Chaetangiaceae). *Phycologia* 18:355–61.

Eiseman, N. J., and R. L. Moe. 1981. *Maripelta atlantica* sp. nov. (Rhodophyta, Rhodymeniales) a new deep-water alga from Florida. *J. Phycol.* 17:299–308.

Eiseman, N. J., and J. N. Norris. 1981. *Dudresnaya patula* sp. nov., an unusual deep-water red alga from Florida. *J. Phycol.* 17:186–91.

Ellis, J., and D. Solander. 1786. *The Natural History of Many Curious and Uncommon Zoophytes, Collected from Various Parts of the Globe*. Pp. i–xii, 1–208, 63 pls. London.

Endlicher, S. L. 1836–1840. *Genera plantarum secundum ordines naturales diposita*. Pp. 1–1483. Vienna.

Esper, E. J. C. 1806. *Fortsetzungen der Pflanzenthiere*. Vol. 2, part 10 (pp. 25–48). Nürnberg.

Eubank, L. 1946. Hawaiian representatives of the genus *Caulerpa*. *Univ. Calif. Publ. Bot.* 18:409–31.

Falkenberg, P. 1879. Die Meeres-Algen des Golfes von Neapel. Nach Beobachtungen in der Zool. Station während der Jahre 1877–78 zusammengestellt. *Mitt. Zool. Sta. Neapel.* 1:218–77.

———. 1901. Die Rhodomelaceen des Golfes von Neapel und der angrenzenden Meeres-Abschnitte. *Fauna Flora Golfes Neapel, Monographie* 26:I–XVI, 1–754, 24 pls.

Farlow, W. G. 1876. List of marine algae of the United States. *Rep. U.S. Comm. Fish and Fisheries for 1873–4 and 1874–5*. Appendix E, Art. 32. Pp. 691–718.

———. 1881. The marine algae of New England. In *Report of the U.S. Fish Commission for 1879*. Appendix A-1. Pp. 1–210, 15 pls.

Feldmann, J. 1939. *Haraldia*, nouveau genre de Delesseriacées. *Bot. Not.* 1939:1–5.

Feldmann, J., and G. Hamel. 1934. Observations sur quelques Gélidiacées. *Rev. Gén. Bot.* 46:528–49.

Feldmann-Mazoyer, G. 1938. Sur un nouveau genre de Céramiacées de la Méditerranée. *Compt. Rend. Hébd. Séances Acad. Sci.* 207:1119–21.

———. 1941. *Recherches sur les Céramiacées de la Méditerranée occidentale*. Pp. 1–510, 4 pls. Alger.

Fiore, J. 1969. *Life History Studies of Phaeophyta from the Atlantic Coast of the United States*. Ph.D. thesis. Pp. i–xiii, 1–242. Duke University. Durham, N.C.

———. 1975. A new generic name for *Farlowia onusta* (Phaeophyta). *Taxon* 24:497–98.

———. 1977. Life history and taxonomy of *Stictyosiphon subsimplex* Holden (Phaeophyta, Dictyosiphonales) and *Farlowia onusta* (Kützing) Kornmann in Kuckuck (Phaeophyta, Ectocarpales). *Phycologia* 16:301–11.

Fletcher, R. L. 1987. *Seaweeds of the British Isles*. Vol. 3, *Fucophyceae (Phaeophyceae)*, Part 1, p. [10] + 1–35. British Museum (Natural History), London.

Forsskål, P. 1775. *Flora aegyptico-arabica*. Pp. 1–32, I–CXXVI, [1] + 1–219, frontispiece map. Copenhagen.

Foslie, M. 1898. Some new or critical lithothamnia. *Kongel. Norske Vidensk. Selsk. Skr. (Copenhagen)* 1898(6):1–19.

———. 1900a. New or critical calcareous algae. *Kongel. Norske Vidensk. Selsk. Skr. (Copenhagen)* 1899(5):1–34.

———. 1900b. Revised systematical survey of the Melobesieae. *Kongel. Norske Vidensk. Selsk. Skr. (Copenhagen)* 1900(5):1–22.

———. 1901. New Melobesieae. *Kongel. Norske Vidensk. Selsk. Skr. (Copenhagen)* 1900(6):1–24.

———. 1904. I. Lithothamnioneae, Melobesieae, Mastophoreae. Pp. 10–77, pls 1–13. In A. Weber van Bosse and M. Foslie, The Corallinaceae of the Siboga Expedition. *Siboga Exped.* 61:1–110.

———. 1906. Algologiske Notiser II. *Kongel. Norske Vidensk. Selsk. Skr. (Copenhagen)* 1906(2):1–28.

———. 1908. Nye kalkalger. *Kongel. Norske Vidensk. Selsk. Skr. (Copenhagen)* 1908(12): 1–9.

Foslie, M., and M. A. Howe. 1906. New American coralline algae. *Bull. New York Bot. Gard.* 4:128–136, pls 80–93.

Fredericq, S., and M. H. Hommersand. 1989a. Proposal of the Gigartinales ord. nov. (Rhodophyta) based on an analysis of the reproductive development of *Gracilaria verrucosa*. *J. Phycol.* 25:213–27.

———. 1989b. Comparative morphology and taxonomic status of *Gracilariopsis* (Gracilariales, Rhodophyta). *J. Phycol.* 25:228–41.

Freshwater, D. W., and D. F. Kapraun. 1986. Field, culture and cytological studies of *Porphyra carolinensis* Coll et Cox (Bangiales, Rhodophyta) from North Carolina. *Jap. J. Phycol.* 34: 251–62.

Gabrielson, P. W. 1983. Vegetative and reproductive morphology of *Eucheuma isiforme* (Solieriaceae, Rhodophyta). *J. Phycol.* 19:45–52.

———. 1985. *Agardhiella* versus *Neoagardhiella* (Solieriaceae, Rhodophyta): another look at the lectotypification of *Gigartina tenera*. *Taxon* 34:275–80.

Gabrielson, P. W., and D. P. Cheney. 1987. Morphology and taxonomy of *Meristiella* gen. nov. (Solieriaceae, Rhodophyta). *J. Phycol.* 23:481–93.

Gabrielson, P. W., and M. H. Hommersand. 1982a. The Atlantic species of *Soliera* (Gigartinales, Rhodophyta): their morphology, distribution and affinities. *J. Phycol.* 18:31–45.

———. 1982b. The morphology of *Agardhiella subulata* representing the Agardhielleae, a new tribe in the Solieriaceae (Gigartinales, Rhodophyta). *J. Phycol.* 18:46–58.

Gaillon, F. B. 1828. Thalassiophytes. *Dict. Sci. Nat.*. 53:350–406.

Ganesan, E. K. 1976. On *Kallymenia westii* sp. nov. (Rhodophyta, Cryptonemiales) from the Caribbean Sea. *Bol. Inst. Oceanogr. (Cumaná)* 15:169–75.

Ganesan, E. K., and A. J. Lemus. 1975. Presence of *Predaea* DeToni (Rhodophyta, Gigartinales) in Venezuela. *Bol. Inst. Oceanogr. (Cumaná)* 14:157–64.

Garbary, D. J. 1979. Numerical taxonomy and generic circumscription in the Acrochaetiaceae (Rhodophyta). *Bot. Mar.* 22:477–92.

Garbary D. J., D. Grund, and J. McLachlan. 1978. The taxonomic status of *Ceramium rubrum* (Huds.) C. Ag. (Ceramiales, Rhodophyceae) based on culture experiments. *Phycologia* 17:85–94.

Garbary, D. J., G. I. Hansen, and R. F. Scagel. 1980. A revised classification of the Bangiophyceae (Rhodophyta). *Nova Hedwigia* 33:145–66.

Garbary, D. J., G. I. Hansen, and R. F. Scagel. 1981 ["1980"]. The marine algae of British Columbia and northern Washington; Division Rhodophyta (red algae), Class Bangiophyceae. *Syesis* 13:137–95.

Garbary, D., and H. W. Johansen. 1982. Scanning electron microscopy of *Corallina* and *Haliptilon* (Corallinaceae, Rhodophyta): Surface features and their taxonomic implications. *J. Phycol.* 18:211–19.

Gayral, P. 1964. Sur le démembrement de l'actuel genre *Monostroma* Thuret (Chlorophycées, Ulotrichales s. l.). *Compt. Rend. Hebd. Séances Acad. Sci.* 258:2149–52.

———. 1965. *Monostroma* Thuret, *Ulvaria* Rupr. emend Gayral, *Ulvopsis* Gayral (Chlorophycées, Ulotrichales): Structure, réproduction, cycles, position systématique. *Rev. gen. Bot.* 72:627–38.

Geesink, R. 1973. Experimental investigations on marine and freshwater *Bangia* (Rhodophyta) from the Netherlands. *J. Exp. Mar. Biol. Ecol.* 11:239–47.

Gepp, A., and E. S. Gepp. 1911. The Codiaceae of the Siboga Expedition, including a monograph of Flabellarieae and Udoteae. *Monogr. Siboga-Exped.* 62:1–150, 22 pls.

Gerloff, J. 1959. *Bachelotia* (Bornet) Kuckuck ex Hamel oder *Bachelotia* (Bornet) Fox? *Nova Hedwigia* 1:37–39.

Gibb, S. P. 1957. The free-floating forms of *Ascophyllum nodosum* (L.) LeJol. *J. Ecol.* 45:49–83.

Gmelin, S. G. 1768. *Historia fucorum.* Pp. [i–xiii], 1–239, [1–6], pls 1–33. St. Petersburg (Leningrad).

Godfrey, R. K., and J. W. Wooten. 1979. *Aquatic Wetland Plants of the Southeastern United States. Monocotyledons.* Pp. i–xii, 1–712. The University of Georgia Press, Athens.

———. 1981. *Aquatic Wetland Plants of the Southeastern United States. Dicotyledons.* Pp. i–x, 1–933. The University of Georgia Press, Athens.

Golden, L., and D. J. Garbary. 1984. Studies on *Monostroma* (Monostromataceae, Chlorophyta) in British Columbia with emphasis on spore release. *Jap. J. Phycol.* 32:319–32.

Goodenough, S., and T. J. Woodward. 1797. Observations on the British Fuci, with particular descriptions of each species. *Trans. Linn. Soc. London* 3:84–235, pls 16–19.

Goor, A. C. J. van. 1923. Die holländischen Meeresalgen (Rhodophyceae, Phaeophyceae und Chlorophyceae) imbesonders der Umgebung von Helder, des Wattenmeeres un der Zuidersee. *Verh. Kon. Ned. Akad. Wetensch. Afd. Natuurk., Tweede Sect.* 23(2):1–232.

Grateloup, J. P. A. S. 1806. Descriptiones aliquorum Ceramiorum novorum, cum iconum explicationibus. [Appendix, with one unnumbered page and one unnumbered plate, to:] *Observations sur la Constitution de l'Été de 1806.* Montpellier.

Gray, I. E., and M. J. Cerame-Vivas. 1963. The circulation of surface waters in Raleigh Bay, North Carolina. *Limnol. & Oceanogr.* 8:330–37.

Gray, J. E. 1864. *Handbook of the British water-weeds or algae: The Diatomaceae by W. Carruthers.* Pp. i–iv, 1–123. London.

———. 1866. On *Anadyomene* and *Microdictyon*, with the description of three new allied genera, discovered by Menzies in the Gulf of Mexico. *J. Bot.* 4:41–51, 65–72, pl. 44.

Gray, S. F. 1821. *A Natural Arrangement of British Plants.* 1:i–xxviii, 1–824, 21 pls. London.

Gregory, B. D. 1934. On the life-history of *Gymnogongrus griffithsiae* Mart. and *Ahnfeltia plicata* Fries. *J. Linn. Soc., Bot.* 49:531–51.

Greville, R. K. 1822–28. *Scottish Cryptogamic Flora.* 6 vols. Vol. 1, 60 pls, (1822–23); Vol. 2, 60 pls, (1823–24); Vol. 3, 60 pls, (1824–25); Vol. 4, 60 pls, (1825–26); Vol. 5, 60 pls, (1826–27); Vol. 6, 60 pls, synopsis [1]–82. (1827–28). Edinburgh.

―――. 1830. *Algae britannicae.* Pp. i–lxxxviii, 1–218, 19 pls. Edinburgh.

Grunow, A. 1867. Algae. In E. Fenzl, ed., *Reise der Österreichischen Fregatte Novara um die Erde, in... 1857, 1858, 1859... Botanischer Theil.* 1:1–104, 12 pls. Wein.

―――. 1916. Additamenta ad cognitionem Sargassorum. *Verh. Zool.-Bot. Ges. Wein.* 66:136–85.

Guiry, M. D., W. R. Kee, and D. J. Garbary. 1987. Morphology, temperature and photoperiodic responses in *Audouinella botryocarpa* (Harvey) Woelkerling (Acrochaetiaceae, Rhodophyta) from Ireland. *Giorn. Bot. Ital.* 121:229–46.

Guiry, M. D., and J. A. West. 1983. Life history and hybridization studies on *Gigartina stellata* and *Petrocelis cruenta* (Rhodophyta) in the North Atlantic. *J. Phycol.* 19:474–94.

Guiry, M. D., J. A. West, D.-H. Kim, and M. Masuda. 1984. Reinstatement of the genus *Mastocarpus* Kützing (Rhodophyta). *Taxon* 33:53–63.

Hall, M. O., and N. J. Eiseman. 1981. The seagrass epiphytes of the Indian River, Florida. I. Species list with descriptions and seasonal occurrences. *Bot. Mar.* 24:139–46.

Halos, M. T. 1965. Sur trois Callithamniées des environs de Roscoff. *Cah. Biol. Mar.* 6:117–34.

Hamel, G. 1929. Contribution à la flore algologique des Antilles. *Ann. Cryptog. Exot.* 2:53–58.

―――. 1935. *Phéophycées de France.* Fasc. 2. Pp. 81–176. Paris.

―――. 1939. Sur la classification des Ectocarpales. *Bot. Not.* 1939:65–70.

Hanisak, M. D., and S. M. Blair. 1988. The deep-water macroalgal community of the east Florida continental shelf (USA). *Helgoländer. Wiss. Meeresuntersuch.* 42:133–63.

Hansen, G. I. 1977a. *Cirrulicarpus carolinensis*, a new species in the Kallymeniaceae (Rhodophyta). *Occas. Pap. Farlow Herb. Cryptogam. Bot. Harvard. Univ.* 12:1–22.

―――. 1977b. A comparison of the species of *Cirrulicarpus* (Kallymeniaceae, Rhodophyta). *Occas. Pap. Farlow Herb. Cryptogam. Bot. Harvard. Univ.* 12:23–34.

Hansgirg, A. 1885. Ein Beitrag zur Kenntnis von der Verbreitung der Chromatophoren und Zellkerne bei den Schizophyceen (Phycochromaceen). *Ber. Deutsch. Bot. Ges.* 3:14–22.

Harlin, M. M., W. J. Woelkerling and D. I. Walker. 1985. Effects of a hypersalinity gradient on epiphytic Corallinaceae (Rhodophyta) in Shark Bay, Western Australia. *Phycologia* 24:389–402.

Harvey, W. H. 1833. Confervoideae. Pp. 259–61, 322–85. In W. J. Hooker, ed., *British Flora*, vol. 2., part 1.

―――. 1834. Algological illustrations. 1.–Remarks on some British algae, and descriptions of new species recently added to our flora. *J. Bot.* (Hooker) 1:296–305, pls 138, 139.

―――. 1836. Algae. In J. T. Mackay, ed., *Flora Hibernica.* 2:154–254. Dublin.

―――. 1841. *A manual of the British algae.* Pp. i–lvii, 1–229. London.

————. 1846–51. *Phycologia britannica*. Vols I, II, and III [1846 = pp. i–xv, i–viii, pls 1–120; 1847 = pls 121–44; 1848 = pls 145–216; 1849 = pls 217–306; 1850 = pls 307–54; 1851 = pls 355–60]. London.

————. 1852. Nereis boreali-americana. Part I. Melanospermeae. *Smithsonian Contr. Knowl.* 3(4): 1–150, pls 1–12. (Also printed separately in 1851 for the Smithsonian Institution.)

————. 1853. Nereis boreali-americana. Part II. Rhodospermeae. *Smithsonian Contr. Knowl.* 5(5):1–258, pls 13–36.

————. 1855. Some account of the marine botany of Western Australia. *Trans. Roy. Irish Acad.* 22(Sci.):525–66.

————. 1857. Short descriptions of some new British algae, with two plates. *Nat. Hist. Rev.* 4(Proc.):201–4, pls 12,13.

————. 1858. Nereis boreali-americana. Part III. Chlorospermeae. *Smithsonian Contr. Knowl.* 10(2):1–140, pls. 37–51.

Hauck, F. 1876. Verzeichniss der im Golfe von Triest gesammelten Meeresalgen. *Österr. Bot. Z.* 26:24–26, 54–57, 91–93.

————. 1883–85. Die Meeresalgen Deutschlands und Österreichs. Pp. ii–xxiv, 1–575. In L. Rabenhorst, ed., *Kryptogamen-Flora von Deutschland, Österreich und der Schweiz*, second edition, 2:ii–xxiv, 1–575. Leipzig.

Hawkes, M. W. 1978. Sexual reproduction in *Porphyra gardneri* (Smith et Hollenberg) Hawkes (Bangiales, Rhodophyta). *Phycologia* 17:329–53.

Hay, M. E. 1986. Associational plant defenses and the maintenance of species diversity: turning competitors into accomplices. *Amer. Naturalist* 128:617–41.

Hay, M. E., J. E. Duffy, C. A. Pfister, and W. Fenical. 1987. Chemical defense against different marine herbivores: Are amphipods insect equivalents? *Ecology* 68:1567–80.

Hay, M. E., P. E. Renaud, and W. Fenical. 1988. Large mobile versus small sedentary herbivores and their resistance to seaweed chemical defenses. *Oecologia* (Berlin) 75:264:252.

Hay, M. E., and J. P. Sutherland. 1988. The ecology of rubble structures of the South Atlantic Bight: A community profile. *U.S. Fish Wildl. Serv. Biol. Rep.* 85(7.20). 67 pp.

Heerebout, G. R. 1968. Studies on the Erythropeltidaceae (Rhodophyceae-Bangiophycidae). *Blumea* 16:139–57.

Henry, E. C. 1987a. The life history of *Onslowia endophytica* (Sphacelariales, Phaeophyceae) in culture. *Phycologia* 26:175–81.

————. 1987b. Morphology and life histories of *Onslowia bahamensis* and *Verosphacela ebrachia* gen et. sp. nov., with a reassessment of the Choristocarpaceae (Sphacelariales, Phaeophyceae). *Phycologia* 26:182–91.

Heydrich, F. 1892. Beiträge zur Kenntniss der Algenflora von Kaiser-Wilhelmsland (Deutsch Neu Guinea). *Ber. Deutsch. Bot. Ges.* 10:458–85, pls 24–26.

————. 1897. Melobesieae. *Ber. Deutsch. Bot. Ges.* 15:403–20, pl. 18.

Hoek, C. van den. 1963. *Revision of the European species of* Cladophora. Pp. 1–248, 55 pls. Leiden.

————. 1975. Phytogeographic provinces along the coasts of the northern Atlantic Ocean. *Phycologia* 14:317–30.

————. 1982. A taxonomic revision of the American species of *Cladophora* (Chlorophyceae) in the North Atlantic Ocean and their geographic distribution. *Verh. Kon. Ned. Akad. Wetensch., Afd. Natuurk., Tweede Sect.* 78:1–236.

Hoek, C. van den and A. M. Cortel-Breeman. 1970. Life history studies on Rhodophyceae. III: *Scinaia complanata* (Collins) Cotton. *Acta Bot. Neerl.* 19:457–67.

Hoek, C. van den and A. Flinterman. 1968. The life-history of *Sphacelaria furcigera* Kütz. (Phaeophyceae). *Blumea* 16:193–242.

Hoek, C. van den and R. B. Searles. 1988. *Cladophora pseudobainesii* nov. sp. (Chlorophyta): an addition to the N.W. Atlantic species of *Cladophora*. *Bot. Mar.* 31:521–24.

Hoek, C. van den and H. B. S. Womersley. 1984. Genus *Cladophora* Kuetzing 1843: 262, nom. cons. In H. B. S. Womersley, ed., *The Marine Benthic Flora of Southern Australia.* Part I. Handbook of the Flora and Fauna of South Australia, pp. 185–213. Woolman, Government Printer, South Australia.

Höhnel, F. von. 1920. Mykologische Fragmente. *Ann. Mycol.* 18:71–97.

Hollenberg, G. J. 1942. An account of the species of *Polysiphonia* on the Pacific coast of North America. I. *Oligosiphonia. Amer. J. Bot.* 29:772–85.

———. 1943. New marine algae from Southern California. II. *Amer. J. Bot.* 30:571–79.

———. 1945. *Polysiphonia*. Pp. 298–303. In W. R. Taylor, ed., Pacific marine algae of the Allan Hancock Expeditions to the Galapagos Islands. *Allan Hancock Pacific Exped.* 12:i–iv, 1–528, pls 1–100.

———. 1958. Phycological notes. II. *Bull. Torrey Bot. Club* 85:63–69.

———. 1968a. An account of the species of *Polysiphonia* of the central and western tropical Pacific Ocean. I. *Oligosiphonia. Pacific Sci.* 22:56–98.

———. 1968b. An account of the species of the red alga *Herposiphonia* occurring in the central and western tropical Pacific Ocean. *Pacific Sci.* 22:536–59.

Hommersand, M. H. 1963. The morphology and classification of some Ceramiaceae and Rhodomelaceae. *Univ. Calif. Publ. Bot.* 35:165–366, pls 1–6.

Hooker, J. D. 1875 ["1874"]. Contributions to the botany of the expedition of H.M.S. 'Challenger.' *J. Linn. Soc., Bot.* 14:311–90.

Hooker, W. J. 1833. Class XXIV, Cryptogamia. Part 1. Mosses, Hepaticae, lichens, Characeae and algae. In J. E. Smith, ed., *The English Flora,* 5(1):i–x, 1–432. London.

Hoppaugh, K. W. 1930. A taxonomic study of species of the genus *Vaucheria* collected in California. *Amer. J. Bot.* 17:329–47, pls 24–27.

Hörnig, I., and R. Schnetter. 1988. Notes on *Dictyota dichotoma, D. menstrualis, D. indica* and *D. pulchella* spec. nov. (Phaeophyta). *Phyton (Horn)* 28:277–91.

Howe, M. A. 1904. Notes on Bahaman algae. *Bull. Torrey Bot. Club* 31:93–100, pl. 6.

———. 1905. Phycological studies–II. New Chlorophyceae, new Rhodophyceae, and miscellaneous notes. *Bull. Torrey Bot. Club* 32:563–86, pls 23–29.

———. 1911. Phycological studies. V. Some marine algae of Lower California, Mexico. *Bull. Torrey Bot. Club* 38:489–514, pls 27–34.

———. 1914. The marine algae of Peru. *Mem. Torrey Bot. Club* 15:1–185, 66 pls.

———. 1918. Class 3, Algae. In N. L. Britton, ed., *Flora of Bermuda,* 489–540. New York.

———. 1920. Algae. In N. L. Britton and C. F. Millspaugh, *The Bahama Flora,* 553–618. New York.

Howe, M. A., and W. D. Hoyt. 1916. Notes on some marine algae from the vicinity of Beaufort, North Carolina. *Mem. New York Bot. Gard.* 6:105–23, pls 11–15.

Howe, M. A., and W. R. Taylor. 1931. Notes on new or little known marine algae from Brazil. *Brittonia* 1:7–33, pls 1,2.

Hoyt, W. D. 1907. Periodicity in the production of the sexual cells of *Dictyota dichotoma*. *Bot. Gaz. (Crawfordsville)* 43:383–97.

———. 1910. Alternation of generations and sexuality in *Dictyota dichotoma. Bot. Gaz.* 49:55–57.

———. 1920. The marine algae of Beaufort, N.C., and adjacent regions. *U.S. Bull. Bur. Fish.* 36:367–556, pls 84–119.

———. 1927. The periodic fruiting of *Dictyota* and its relation to the environment. *Amer. J. Bot.* 14:592–619.

Hudson, W. 1762. *Flora Anglica.* Pp. i–viii, [1–7], 1–506, [1–22]. London.

———. 1778. *Flora Anglica.* Editio altera. Pp. i–xxxviii, 1–690. London.

Humm, H. J. 1942. Seaweeds at Beaufort, N.C., as a source of agar. *Science* 96:230–31.

———. 1944. Agar resources of the south Atlantic and east Gulf coasts. *Science* 100:209–12.

———. 1951. The seaweed resources of North Carolina. In H. F. Taylor, ed., *Survey of the Marine Fisheries of North Carolina*, 231–50. University of North Carolina Press, Chapel Hill.

———. 1952. Notes on the marine algae of Florida. I. The intertidal rocks at Marineland. *Florida St. Univ. Studies* 7:17–23.

———. 1969. Distribution of marine algae along the Atlantic coast of North America. *Phycologia* 7:43–53.

———. 1979. *The Marine Algae of Virginia.* Pp. 1–263. Univ. Press of Virginia: Charlottesville.

Humm, H. J., and M. J. Cerame-Vivas. 1964. *Struvea pulcherrima* in North Carolina. *J. Elisha Mitchell Sci. Soc.* 80:23–24.

Humm, H. J., and S. R. Wicks. 1980. *Introduction and Guide to Marine Bluegreen Algae.* Pp. i–x, [2] + 1–194. Wiley and Sons, New York.

Humm, H. J., and F. A. Wolf. 1946. Introduction to agar and its uses. *Duke Univ. Mar. Sta. Bull.* 3:2–18.

Huvé, P., and H. Huvé. 1977. Notes de nomenclature algae. I. Le genre *Halichrysis* (J. Agardh 1851 emend. J. Agardh 1876) Schousboe mscr. in Bornet 1892 (Rhodyméniales, Rhodyméniacées). *Bull. Soc. Phycol. France* 22:99–107.

Irvine, L. M. 1983. *Seaweeds of the British Isles.* Vol. 1. Rhodophyta; Part 2A: Cryptonemiales (sensu stricto), Palmariales, Rhodymeniales. Pp. i–xii, 1–115. British Museum (Natural History): London.

Itono, H. 1977. Studies on the Ceramiaceous algae (Rhodophyta) from southern parts of Japan. *Bibloth. Phycol.* 35:1–499.

Jacquin, N. J. 1786. *Collectanea ad botanicam, chemiam, et historiam naturalem spectantia.* 1:1–386, 22 pls. Wein.

Jessen, C. F. W. 1848. *Prasiolae generis algarum monographie.* 20 pp., 2 pls. Kiliae.

Johansen, H. W. 1981. *Coralline Algae, A First Synthesis.* Pp. [i–x], 1–239. CRC Press, Boca Raton, Fla.

John, D. M., J. H. Price, C. Maggs, and G. W. Lawson. 1979. Seaweeds of the western coast of tropical Africa and adjacent islands: a critical assessment. III. Rhodophyta (Bangiophyceae). *Bull. Brit. Mus. ⟨Nat. Hist.⟩, Bot.* 7:69–82.

Johnson, D. S. 1900. Notes on the flora of the banks and sounds at Beaufort, N.C. *Bot. Gaz. (Crawfordsville)* 30:405–9.

Joly, A. B. 1957. Contribuição ao conhecimento da flora ficológica marinha da Báia de Santos e arredores. *Bol. Fac. Filos. Univ. São Paulo, Bot.* 14:1–196, 19 pls, 1 map.

———. 1965. Flora marinha do litoral norte do Estado de São Paulo e regiões circunvizinhas. *Bol. Fac. Filos. Univ. São Paulo, Bot.* 21:1–393, 59 pls, 3 maps.

Joly, A. B., and E. C. de Oliveira F. 1966. *Spyridiocolax* and *Heterodasya*, two new genera of the Rhodophyceae. *Sellowia* 18:115–25.

Jones, P. L., and W. J. Woelkerling. 1984. An analysis of trichocyte and spore germination attributes as taxonomic characters in the *Pneophyllum-Fosliella* complex (Corallinaceae, Rhodophyta). *Phycologia* 23:183–94.

Jürgens, G. H. B. 1822. *Algae aquaticae... Dynastium Jeveranam Frisiam.* Decad 13. Jever.

Kapraun, D. F. 1977a. The genus *Polysiphonia* in North Carolina, USA. *Bot. Mar.* 20:313–31.

———. 1977b. Studies on the growth and reproduction of *Antithamnion cruciatum* (Rhodophyta, Ceramiales) in North Carolina. *Norw. J. Bot.* 24:269–74.

———. 1977c. Asexual propagules in the life history of *Polysiphonia ferulacea* (Rhodophyta, Ceramiales). *Phycologia* 16:417–26.

———. 1978a. Field and cultural studies on selected North Carolina *Polysiphonia* species. *Bot. Mar.* 21:143–53.

———. 1978b. Field and culture studies on growth and reproduction of *Callithamnion byssoides* (Rhodophyta, Ceramiales) in North Carolina. *J. Phycol.* 14:21–24.

———. 1978c. A cytological study of varietal forms in two species of *Polysiphonia* (Rhodophyta, Ceramiales) in North Carolina. *Phycologia* 17:152–56.

———. 1979. Comparative studies of *Polysiphonia urceolata* from three North Atlantic sites. *Phycologia* 18:319–24.

———. 1980a. *An Illustrated Guide to the Benthic Marine Algae of Coastal North Carolina. I. Rhodophyta.* Pp. i–viii, 1–206. University of North Carolina Press: Chapel Hill, North Carolina.

———. 1980b. Floristic affinities of North Carolina inshore benthic marine algae. *Phycologia* 19:245–52.

———. 1984. An illustrated guide to the benthic marine algae of coastal North Carolina. II. Chlorophyta and Phaeophyta. *Bibloth. Phycol.* 58:1–173.

Kapraun, D. F., and P. W. Boone. 1987. Karyological studies of three species of Scytosiphonaceae (Phaeophyta) from coastal North Carolina. *J. Phycol.* 23:318–22.

Kapraun, D. F., and W. D. Freshwater. 1987. Karyological studies of five species in the genus *Porphyra* (Bangiales, Rhodophyta) from the North Atlantic and Mediterranean. *Phycologia* 26:82–87.

Kapraun, D. F., and G. M. Gargiulo. 1987. Karyological studies of four *Cladophora* (Cladophorales, Chlorophyta) species from coastal North Carolina. *Giorno Bot. Ital.* 121:1–26.

Kapraun, D. F., and D. G. Luster. 1980. Field and culture studies of *Porphyra rosengurttii* Coll et Cox (Rhodophyta, Bangiales) from North Carolina. *Bot. Mar.* 23:449–57.

Kapraun, D. F., and D. J. Martin. 1987. Karyological studies of three species of *Codium* (Codiales, Chlorophyta) from coastal North Carolina. *Phycologia* 26:228–34.

Kapraun, D. F., and J. N. Norris. 1982. The red alga *Polysiphonia* (Rhodomelaceae) from Carrie Bow Cay and vicinity, Belize. *Smithsonian Contrib. Mar. Sci.* 12:225–38.

Kapraun, D. F., and R. B. Searles. 1990. Planktonic bloom of an introduced species of *Polysiphonia* (Ceramiales, Rhodophyta) along the coast of North Carolina, USA. *Hydrobiologia* (in press).

Kapraun, D. F., and F. W. Zechman. 1982. Seasonality and vertical zonation of benthic marine algae on a North Carolina coastal jetty. *Bull. Mar. Sci. Gulf Caribbean* 32:702–14.

Kerr, G. A. 1976. Indian River coastal zone study inventory. *Ann. Rep. Harbor Branch Consortium Indian River Coastal Zone Study* 2:1–105.

Kim, C. S., and H. J. Humm. 1965. The red alga, *Gracilaria foliifera,* with special reference to the cell wall polysaccharides. *Bull. Mar. Sci. Gulf Caribbean* 15:1036–50.

Kjellman, F. R. 1877. Om Spetsbergens marina, klorofyllförande Thallophyter. II. *Bih. Kongel. Svenska Vetensk.-Akad. Handl.* 4(6):1–61, 5 pls.

Kjellman, F. R., and N. Svedelius. 1911. *Lithodermataceae.* In A. Engler and K. Prantl, eds.,

Die Natürlichen Planzenfamilien, Nachträge zum 1. Teil 2. Abt, 173–76. Leipzig.

Kornmann, P. 1953. Der Formenkreis von *Acinetospora crinita* (Carm.) nov. comb. *Helgoländer Wiss. Meeresuntersuch.* 4:205–24.

———. 1962. Zur Entwicklung von *Monostroma grevillei* und zur systematische Stellung von *Gomontia polyrhiza*. *Vortr. Gesamtgeb. Bot.* N.F. 1:37–39.

———. 1963. Die Lebenzyklus einer marinen *Ulothrix*-Art. *Helgoländer Wiss. Meeresuntersuch.* 8:357–60.

———. 1989. *Sahlingia* nov. gen. based on *Erythrocladia subintegra* (Erthropeltidales, Rhodophyta). *Brit. Phycol. J.* 24: 223–28.

Kornmann, P., and P.-H. Sahling. 1977. Meeresalgen von Helgoland, benthische Grün-, Braun- und Rotalgen. *Ibid.* 29:1–289.

———. 1985. Erythropeltidaceen (Bangiophyceae, Rhodophyta) von Helgoland. *Helgoländer Wiss. Meeresuntersuch.* 39:213–36.

Kraft, G. T. 1984. The red algal genus *Predaea* (Nemastomaceae, Gigartinales) in Australia. *Phycologia* 23:3–20.

Kraft, G. T., and I. A. Abbott. 1971. *Predaea weldii*, a new species of Rhodophyta from Hawaii, with an evaluation of the genus. *J. Phycol.* 7:194–202.

Kraft, G. T., and P. A. Robbins. 1985. Is the Order Cryptonemiales (Rhodophyta) defensible? *Phycologia* 24:67–77.

Kraft, G. T., and M. J. Wynne. 1979. An earlier name for the Atlantic American red alga *Neoagardhiella baileyi* (Solieriaceae, Gigartinales). *Phycologia* 18:325–29.

Kuckuck, P. 1929. Fragmente einer Monographie der Phaeosporeen. *Wiss. Meeresuntersuch., Abt. Helgoland*, N.F. 17(4):1–93.

———. 1961. Ectocarpaceen-Studien–VII. *Giffordia. Helgoländer Wiss. Meeresuntersuch.* 8:119–52. ("herausgegeben von P. Kornmann").

Kunth, C. S. 1822. *Synopsis plantarum*. Part 6, I. *Botanique*. Pp. i–iv, 1–491. Paris.

Kuntze, O. 1898. *Revisio generum plantarum*. Part 3[3]. Pp. i–iv, 1–576. Leipzig.

Kützing, F. T. 1833. Algologische Mittheilungen. *Flora* 16:513–21.

———. 1841. Über *Ceramium* Ag. *Linnaea* 15:727–46.

———. 1843. *Phycologia generalis*. Pp. I–XXXII, 1–458, 80 pls. Leipzig.

———. 1845. *Phycologia germanica*. Pp. I–X, 1–340. Nordhausen.

———. 1847. Diagnosen und Bemerkungen zu neuen order kritischen Algen. *Bot. Zeitung* 5:1–5, 22–25, 33–38, 52–55, 164–67, 177–80, 193–98, 219–23.

———. 1849. *Species algarum*. Pp. i–vi, 1–922. Leipzig.

———. 1858. *Tabulae phycologicae*. 8:i–ii, 1–48, 100 pls. Nordhausen.

———. 1859. *Tabulae phycologicae*. 9:i–viii, 1–42, 100 pls. Nordhausen.

———. 1862. *Tabulae phycologicae*. 12:i–iv, 1–30, 100 pls. Nordhausen.

———. 1863. Diagnosen und Bemerkungen zu drei und siebenzig neuen Algenspecies. Pp. [1]–19. Nordhausen. [Reprinted, *Hedwigia*, 2:86–95 (1863).]

———. 1869. *Tabulae phycologicae*. 19:i–iv, 1–36, 100 pls. Nordhausen.

Kylin, H. 1906. Zur Kenntnis einiger schwedischen *Chantransia*-Arten. In *Botaniska Studier tillägnade F. R. Kjellman den 4 November 1906*, 113–26. Uppsala.

———. 1924. Studien über die Delesseriaceen. *Acta Univ. Lund.* N.F. Avd. 2, 20(6):1–111.

———. 1930. Über die Entwicklungsgeschichte der Florideen. *Acta Univ. Lund.* N.F. Avd. 2, 26(6):1–104.

———. 1931. Die Florideenordnung Rhodymeniales. *Acta Univ. Lund.* N.F. Avd. 2, 27(11):1–48, pls 1–20.

———. 1932. Die Florideenordnung Gigartinales. *Acta Univ. Lund.* N.F. Avd. 2, 28(8):1–88, 28 pls.

———. 1940. Die Phaeophyceenordnung Chordariales. *Acta Univ. Lund. N.F. Avd. 2*, 36(9):1–67, 8 pls.

———. 1944. Die Rhodophyceen der schwedischen Westküste. *Acta Univ. Lund. N.F. Avd. 2*, 40(2):1–104, 32 pls.

———. 1947. Über die Fortpflanzungsverhältnisse in der Ordnung Ulvales. *Förh. Kungl. Fysiogr. Sällsk.* 17:174–82.

———. 1956. *Die Gattungen der Rhodophyceen.* Pp. i–xv, 1–673. C. W. K. Gleerup Förlag, Lund.

Lägerheim, G. 1885. *Codiolum polyrhizum* n. sp. *Öfvers. Förh. Kongl. Svenska Vetensk.-Akad.* 42(8):21–31, pl. 28.

Lamouroux, J. V. F. 1805. *Dissertations sur plusiers espèces de Fucus.* Pp. i–xxiv, 1–83, 36 pls. Agen.

———. 1809a. Exposition des caractères du genre *Dictyota* et tableau des espèces qu'il renferme. *J. Bot. (Desvaux)* 2:38–44.

———. 1809b. Mémoire sur trois nouveaux genres de la famille des algues marines. *J. Bot. (Desvaux)* 2:129–35.

———. 1809c. Mémoires sur les Caulerpes, nouveau genre de la famille des algues marines. *J. Bot. (Desvaux)* 2:136–46.

———. 1809d. Observations sur la physiologie des algues marines, et description de cinq nouveaux genres de cette famille. *Nouv. Bull. Sci. Soc. Philom. Paris* 1:330–33, pl. 6.

———. 1812. Sur la classification des Polypiers coralligènes non entièrement pierreux. *Nouv. Bull. Sci. Soc. Philom. Paris* 3:181–88.

———. 1813. Essai sur les genres de la famille des Thallassiophytes non articulées. *Ann. Mus. Natl. Hist. Nat.* 20:21–47, 115–39, 267–93, pls 7–13.

———. 1816. *Histoire des polypiers coralligènes flexibles.* Pp. i–lxxxiv, chart, 1–559, 19 pls. Caen.

Lawson, G. W., and D. M. John. 1982. The marine algae and coastal environment of tropical West Africa. *Beih. Nova Hedwigia* 70:1–455.

Lee, I. K., and J. A. West. 1979. *Dasysiphonia chejuensis* gen. et sp. nov. (Rhodophyta, Dasyaceae) from Korea. *Syst. Bot.* 4:115–29.

Lee, Y. P., and I. K. Lee. 1988. Contribution to the generic classification of the Rhodochortonaceae (Rhodophyta, Nemaliales). *Bot. Mar.* 31:119–31.

Le Jolis, A. 1863. Liste des algues marines de Cherbourg. *Mém. Soc. Sci. Nat. Cherbourg* 10:5–168, 6 pls.

Lemoine, P. 1917. Corallinaceae, Melobesieae. In F. Børgesen, Marine Algae of the Danish West Indies. Rhodophyceae (pp. 147–82). *Dansk Bot. Ark.* 3:145–204.

———. 1928. Un nouveau genre de Mélobésiées: *Mesophyllum. Bull. Soc. Bot. France* 75:251–54.

Lemus, A. J., and E. K. Ganesan. 1977. Morphological and cultural studies in two species of *Predaea* G. De Toni (Rhodophyta, Gymnophloeaceae) from the Caribbean Sea. *Bol. Inst. Oceanogr. (Cumaná)* 16:63–77.

Levring, T. 1937. Zur Kenntnis der Algenflora der norwegischen Westküste. *Acta Univ. Lund. N.F. Avd. 2*, 33(8):1–147.

Lewin, R. A., and J. A. Robertson. 1971. Influence of salinity on the form of *Asterocytis* in pure culture. *J. Phycol.* 7:236–38.

Lightfoot, J. 1777. *Flora Scotica.* Pp. i–xli, 1–1151, 35 pls. London.

Lindley, J. 1846. *The Vegetable Kingdom.* First edition, Pp. i–lxviii, 1–908. London.

Link, J. H. F. 1820. Epistola... de algis aquaticis, in genera disponendis. In C. G. Nees, ed., *Horae physicae berolinenses*, pp. 1–8, pl. 1. Bonn.

————. 1833. *Handbuch zur Erkennung der nutzbarsten und am häufigsten vorkommenden Gewächse*. Part 3, pp. i–xviii, 1–536. Berlin.

Linnaeus, C. 1753. *Species plantarum*. 2:561–1200 + [1–31]. Stockholm.

————. 1755. *Flora suecica*. Second edition. Pp. [I–IV], I–XXXII,1–464 + [1–30], 1 pl. Stockholm.

————. 1758. *Systema naturae*. Tenth ed. 1:[i–iv], 1–823. Stockholm.

————. 1761. *Fauna Suecica*. Second ed. Pp. [i–xlvi], 1–578, 2 pls. Stockholm.

————. 1763. *Species plantarum*. Second ed. 2:785–1684. Stockholm.

————. 1767. *Systema naturae*. Twelfth ed. 1(2):533–1327 + [1–36]. Stockholm.

Litaker, W., C. S. Duke, B. E. Kenney, and J. Ramus. 1987. Short-term environmental variability and phytoplankton abundances in a shallow tidal estuary. I. Winter and summer. *Mar. Biol.* 96:115–21.

Loiseaux, S. 1967a. Morphologie et cytologie des Myrionématacées. Critères taxonomiques. *Rev. Gén. Bot.* 74:329–47.

————. 1967b. Recherches sur les cycles de développement des Myrionématacées (Phéophycées). I–II. Hecatonématées et Myrionématées. *Rev. Gén. Bot.* 74:529–76.

————. 1969. Sur une espèce de *Myriotrichia* obtenue en culture à partir de zoides d'*Hecatonema maculans* Sauv. *Phycologia* 8:11–15.

Lund, S. 1959. The marine algae of east Greenland. I. Taxonomical part. *Meddel. Grønland* 156(1):1–247, 42 figs.

Lyle, D. 1922. *Antithamnionella*, a new genus of algae. *J. Bot.* 60:346–50.

Lyngbye, H. C. 1819. *Tentamen hydrophytologiae danicae*. Pp. I–XXXII, 1–248, 70 pls. Copenhagen.

McLachlan, J. 1979. *Gracilaria tikvahiae* sp. nov. (Rhodophyta, Gigartinales, Gracilariaceae) from the northwestern Atlantic. *Phycologia* 18:19–23.

McLachlan, J., J. P. van der Meer, and N. L. Bird. 1977. Chromosome numbers of *Gracilaria foliifera* and *Gracilaria* sp. (Rhodophyta) and attempted hybridizations. *J. Mar. Biol. Assoc., U.K.* 57:1137–41.

Marchewianka, M. 1924. Z flory glónow polskiego baltyku. *Spraw. Komis. Fizjogr.* 58/59:33–45.

Martius, C. F. P. von. 1828. *Icones plantarum cryptogamicarum*. Fasc. 1. Pp. 1–30, pls. 1–14. München.

Masaki, T. 1968. Studies on the Melobesioideae of Japan. *Mem. Fac. Fish. Hokkaido Univ.* 16:1–80.

Melvill, J. C. 1875. Notes on the marine algae of South Carolina. *J. Bot.* 12:258–65.

Meneghini, G. 1840a. Lettera del Prof. Giuseppe Meneghini al Dott. Iacob Corinaldi a Pisa. Pisa. Folded sheet without pagination, Bibliothèque Thuret-Bornet, Muséum National d'Histoire Naturelle, Paris.

————. 1840b. Botanische Notizen. *Flora* 23:510–12.

————. 1841. Memoria sui rapporti di organizzazione fra le alghe propriamente dette o ficee e le alghe terrestri o licheni. *Atti Riunione Sci. Ital.* 3:417–31.

Micara, F. A. E. 1946. Suitability of extractive from *Hypnea musciformis* for the cultivation of microorganisms. *Duke Univ. Mar. Sta. Bull.* 3:40–42.

Millar, A. J. K. 1990. *Marine red algae of Coffs Harbour region, northern New South Wales*. *Austr. Syst. Bot.* 3:293–593.

Montagne, C. 1837. Centurie de plantes cellulaires exotiques nouvelles. *Ann. Sci. Nat. Bot., Sér 2*, 8:345–70.

————. 1839–42. Plantes cellulaires. In P. Barker-Webb and S. Berthelot, ed., *Histoire naturelle des Iles Canaries*, 3(2):i–xv, 1–208, 9 pls. Paris.

——. 1840. Seconde centurie de plantes cellulaires exotiques nouvelles. Décades I et II. *Ann. Sci. Nat. Bot.*, sér. 2, 13:193–207, pls. 5–6, figs 1, 3.

——. 1842a. Troisième centurie de plantes cellulaires exotiques nouvelles. *Ann. Sci. Nat., Bot.*, sér. 2, 18:241–82, pl. 7.

——. 1842b. Botanique: Plantes cellulaires. In R. de la Sagra, ed., *Histoire physique, politique, et naturelle de l'île de Cuba*, 11: 1–104, pls 1–5. Paris.

——. 1842c. *Prodromus generum specierumque phycearum novarum.* Pp. 1–16. Paris.

——. 1842d. *Bostrychia. Dictionnaire Universel d'Histoire Naturelle* [Orbigny]. 2:660–61.

——. 1845. Plantes cellulaires. In J. B. and H. Jacquinot, eds., *Voyage au Pôle Sud et dans l'Océanie sur les corvettes l'Astrolabe et la Zelée... pendant les années 1837–1838–1839–1840, sous le commandement de M. J. Dumont-D'Urville: Botanique*, I:i–xiv, 1–349 pp, 20 pls. Paris.

——. 1846. Phyceae. In M. C. Durieu de Maisonneuve, ed., *Exploration scientifique de l'Algérie... Botanique, Partie 1. Cryptogamie*, pp. 1–197, pls 1–16. Paris.

——. 1849. Sixième centurie de plantes cellulaires nouvelles, tant indigènes qu'exotique. *Ann. Sci. Nat. Bot. Sér.* 3, 11:33–66.

——. 1856. *Sylloge generum specierumque cyrptogamarum.* Pp. i–xxiv, 1–498. Paris.

Mook, D. H. 1980. Seasonal variation in species composition of recently settled fouling communities along an environmental gradient in the Indian River Lagoon, Florida. *East Coast Mar. Sci.* 11:573–81.

Morrill, J. F. 1976. *Comparative morphology and taxonomy of some dorsiventral and parasitic Rhodomelaceae.* Ph.D. thesis, i–ix, 1–224 pp., pls. 1–90. University of North Carolina, Chapel Hill.

Müller, O. F. 1778. *Icones plantarum... Flora danicae.* 5(13):1–8, pls. 721–80. Copenhagen.

Murray, G., and L. A. Boodle. 1888. A structural and systematic account of the genus Struvea. *Ann. Bot. (London)* 2:265–82, pl. 16.

——. 1889. A systematic and structural account of the genus *Avrainvillea* Decaisne. I. *Systematic. J. Bot.* 27:67–72.

Nägeli, C. 1846. Herposiphonia. *Z. Wiss. Bot.* 1(3/4):238–56, pl.8.

——. 1847. *Die neuern Algensysteme.* Pp. [i], 1–275, pls. 1–10. Zürich.

——. 1855. Part 1. In *Pflanzenphysiologische Untersuchungen.* Pp. [i–vi], 1–120. Zürich.

——. 1858. Part 2. In *Pflanzenphysiologische Untersuchungen.* Pp. i–x, 1–623, pls. 11–26.

——. 1862 ["1861"]. Beiträge zur Morphologie und Systematik der Ceramiaceae. *Sitzungsber. Bayer. Akad. Wiss. München* 1861(2):297–415.

Nakamura, Y. 1972. A proposal on the classification of the Phaeophyta. In I. A. Abbott and M. Kurogi, eds., *Contributions to the systematics of the marine algae of the North Pacific*, 147–56. Japanese Society of Phycology, Kobe.

Nardo, G. D. 1834. De novo genere algarum cui nomen est *Hildenbrandia prototypus. Isis* [Oken] 6:675–76.

Neal, W. J., W. C. Blakeney, Jr., O. H. Pilkey, Jr., and O. H. Pilkey, Sr. 1984. *Living with the South Carolina Shore.* Pp. 1–205. Durham, N.C.

Nees, C. G. 1820. *Horae physicae berolinenses.* [i–xii] + 1–123 + [1–4 (index)] pp., 27 pls. Bonn.

Nelson, W. A. 1982. A critical review of the Ralfsiales, Ralfsiaceae and the taxonomic position of *Analipus japonicus* (Harv.) Wynne (Phaeophyta). *Brit. Phycol. J.* 17:311–20.

Nielsen, R. 1972. A study of the shell-boring marine algae around the Danish island Læsø. *Bot. Tidsskr.* 67:245–69.

————. 1979. Culture studies on the type species of *Acrochaete*, *Bulbocoleon*, and *Entocladia* (Chaetophoraceae, Chlorophyceae). *Bot. Not.* 132:441–49.

Nordstedt, O. 1879. Algologiska småsaker. 2. *Vaucheria*-studier, 1879. *Bot. Not.* 1879:177–90, 2 pls.

Norris, J. N., and K. E. Bucher. 1976. New records of marine algae from the 1974 R/V *Dolphin* cruise to the Gulf of California. *Smithsonian Contr. Bot.* 34:i–iv, 1–22.

————. 1982. Marine algae and seagrasses from Carrie Bow Cay, Belize. In K. Rutzler and I. G. Macintyre, eds., The Atlantic barrier reef ecosystem at Carrie Bow Cay, Belize. I. Structure and communities, pp. 167–223. *Smithsonian Contr. Mar. Sci.* 12:i–xiv + 1–539.

Norris, J. N., and H. W. Johansen. 1981. Articulated coralline algae of the Gulf of California, Mexico. I. *Smithsonian Contr. Mar. Sci.* 9:1–28.

Norris, R. E. 1985. Studies on *Pleonosporium* and *Mesothamnion* (Ceramiaceae, Rhodophyta) with a description of a new species from Natal. *Brit. Phycol. J.* 20:59–68.

Norris, R. E., and M. E. Aken. 1985. Marine benthic algae new to South Africa. *S. African Tydsk. Plank.* 51:55–65.

Nummedal, D., G. F. Oertel, D. K. Hubbard, and A. C. Hine. 1977. Tidal inlet variability —Cape Hatteras to Cape Canaveral. Coastal Sediments '77, Proc. 5th Symp., WPCO Div. of ASCE, Charleston, S.C., Nov. 2–4, 1977, pp. 543–62.

Ogburn, M. V. 1984. *Feeding ecology and the role of algae in the diet of the sheephead Archosargus probatocephalus (Pisces: Sparidae) on two North Carolina jettys.* M.S. thesis, pp. 1–68, University of North Carolina, Wilmington.

Ohba, H., and S. Enomoto. 1987. Culture studies on *Caulerpa* (Caulerpales, Chlorophyta). II. Morphological variation of *C. ramosa* var. *laetevirens* under various culture conditions. *Jap. J. Phycol.* 35:178–88.

Ohmi, H. 1958. The species of *Gracilaria* and *Gracilariopsis* from Japan and adjacent waters. *Mem. Fac. Fish. Hokkaido Univ.* 6:1–66, 10 pls.

O'Kelly, C. J., and G. L. Floyd. 1984. Correlations among patterns of sporangial structure and development, life histories, and ultrastructural features in the Ulvophyceae. In D. E. G. Irvine and D. M. John, eds., *Systematics of the Green Algae*, pp. 121–56. Systematics Assoc. Special Vol. No. 27, Academic Press, London.

O'Kelly, C. J., and C. Yarish. 1980. Observations on marine Chaetophoraceae (Chlorophyta). I: Sporangial ontogeny in the type species of *Entocladia* and *Phaeophila*. *J. Phycol.* 16:549–58.

————. 1981. On the circumscription of the genus *Entocladia* Reinke. *Phycologia* 20:32–45.

Oliveira F., E. C. de. 1969. Algas marinhas do sul do estado do Espírito Santo (Brasil). I. Ceramiales. *Bol. Fac. Filos. Univ. São Paulo, Bot.* 26:1–277 + 1 map + indices.

————. 1977. *Algas marinhas bentônicas do Brasil.* Pp. [1–4], 1–407. Ph.D. thesis, Universidade de São Paulo, Instituto de Biociências: São Paulo.

Oliveira F., E. C. de and Y. Ugadim. 1974. New references of benthic marine algae to the Brazilian flora. *Bol. Bot. Univ. São Paulo* 2:71–91.

Olsen-Stojkovich, J. 1985. A systematic study of the genus *Avrainvillea* Decaisne (Chlorophyta, Udoteaceae). *Nova Hedwigia* 41:1–68.

Ott, D. W., and M. H. Hommersand. 1974. Vaucheriae of North Carolina. I. Marine and brackish water species. *J. Phycol.* 10:373–85.

Pankow, H. 1971. *Algenflora der Ostsee. I: Benthos. (Blau-, Grün-, Braun-, und Rotalgen).* Pp. 1–419. Gustav Fischer, Jena.

Papenfuss, G. F. 1950. Review of the genera of algae described by Stackhouse. *Hydrobiologia* 2:181–208.

———. 1961. The structure and reproduction of *Caloglossa leprieurii*. *Phycologia* 1:1–31.

———. 1968. Notes on South African algae. V. *J. S. Afr. Bot.* 34:267–87.

Parke, M., and E. M. Burrows. 1976. Chlorophyceae. In M. Parke and P. S. Dixon, eds., Check-list of British marine algae, third revision, pp. 566–70. *J. Mar. Biol. Assoc., U.K.* 56:527–94.

Parke, M., and P. S. Dixon. 1964. A revised check-list of British marine algae. *J. Mar. Biol.* 44:499–542.

———. 1968. Check-list of British marine algae, second revision. *J. Mar. Biol.* 48:783–832.

———. 1976. Check-list of British marine algae, third revision. *J. Mar. Biol.* 56:527–94.

Parkinson, P. G. 1980. *Halymenia. Phycologiae Historiae Analecta Autodiactica Fasciculus Primus.* 20 pp. Pettifogging Press, Auckland.

Parr, A. E. 1933. A geographic ecological analysis of the seasonal changes in temperature conditions in shallow water along the Atlantic coast of the U.S. *Bull. Bingham Oceanogr. Collect.* 4(3):11–42.

Peckol, P. 1982. Seasonal occurrence and reproduction of some marine algae of the continental shelf, North Carolina. *Bot. Mar.* 25:185–90.

———. 1983. Seasonal physiological responses of two brown seaweed species from a North Carolina continental shelf habitat. *J. Exp. Mar. Biol. Ecol.* 72:147–55.

Peckol, P., and J. Ramus. 1985. Physiological differentiation of North Carolina nearshore and offshore populations of *Sargassum filipendula* C. Ag. *Bot. Mar.* 28:319–25.

Peckol, P., and R. B. Searles. 1983. Effects of seasonality and disturbance on population development in a Carolina continental shelf community. *Bull. Mar. Sci. Gulf Caribbean* 33:67–86.

———. 1984. Temporal and spatial patterns of growth and survival of invertebrate and algal populations of a North Carolina continental shelf community. *Estuarine Coastal Shelf Sci.* 18:133–43.

Pedersen, P. M. 1984. Studies on primitive brown algae. *Opera Bot.* 74:1–76.

Perrot, Y. 1972. Les *Ulothrix* marins de Roscoff et le problème de leur cycle de reproduction. *Mém. Soc. Bot. France* 1972:67–74.

Philippi, R. A. Beweis, dass die Nulliporen Pflanzen sind. *Arch. Naturgesch.* 3(1):387–93, pl. IX: figs 2–6.

Piccone, A. 1884. Contribuzioni all'algologia eritrea. *Nuovo Giorn. Bot. Ital.* 16:281–332, pls 7–9.

Pilger, R. 1911. Die Meeresalgen von Kamerun. Nach der Sammlung von C. Ledermann. Pp. 294–323. In A. Engler, Beiträge zur Flora von Afrika. XXXIX. *Bot. Jahrb. Syst.* 46:293–464.

Pilkey, O. H., Jr., W. J. Neal, O. H. Pilkey, Sr., and S. R. Riggs. 1980. *From Currituck to Calabash: Living with North Carolina's Barrier Islands.* Pp. 1–245. Durham, N.C.

Post, E. 1936. Systematische und pflanzengeographische Notizen zur *Bostrychia-Caloglossa*-Assoziation. *Rev. Algol.* 9:1–84.

Price, J. H., and D. M. John. 1979. The marine benthos of Antigua (Lesser Antilles). II. An annotated list of algal species. *Bot. Mar.* 17:322–31.

Price, J. H., D. M. John, and G. W. Lawson. 1978. Seaweeds of the western coast of tropical Africa and adjacent islands: A critical assessment. II. Phaeophyta. *Bull. Brit. Mus. ⟨Nat. Hist.⟩, Bot.* 6:87–182.

———. 1986. Seaweeds of the western coast of tropical Africa and adjacent islands: A critical assessment. IV. Rhodophyta (Florideae). 1. Genera A–F. *Bull. Brit. Mus. (Nat. Hist.), Bot.* 115:1–122.

Printz, H. 1926. Die Algensvegetation des Trondhjemsfjordes. *Norske Vidensk.-Acad. Mat.-Naturvidenskr. Kl., Avh.* 1926(5):1–274, pls 1–10, l chart.

Prud'homme van Reine, W. F. 1982. *A taxonomic revision of the European Sphacelariaceae (Sphacelariales, Phaeophyceae).* Leiden Botanical Series Vol. 6. Leiden. i–ix, 1–293, 6 pls.

Ramus, J. 1983. A physiological test of the theory of complementary chromatic adaptation. II. Brown, green and red algae. *J. Phycol.* 19:173–78.

Ramus, J., and J. P. van der Meer. 1983. A physiological test of the theory of complementary chromatic adaptation. I. Color mutants of a red seaweed. *J. Phycol.* 19:86–91.

Ramus, J., and G. Rosenberg. 1980. Diurnal photosynthetic performance of seaweeds measured under natural conditions. *Mar. Biol.* 56:21–28.

Ramus, J., and M. Venable. 1987. Temporal ammonium patchiness and growth rate in *Codium* and *Ulva* (Ulvophyceae). *J. Phycol.* 23:518–23.

Reading, R. P. and C. W. Schneider. 1986. On the male conceptacles of two terete species of *Gracilaria* (Rhodophyta, Gigartinales) from North Carolina. *J. Phycol.* 22:395–98.

Reinbold, T. 1893a. Die Phaeophyceen (Brauntange) der Kieler Föhrde. *Schriften Naturwiss. Vereins Schlewig-Holstein* 10:21–59.

———. 1893b. Revision von Jürgens' algae aquaticae. I. Die Algen des Meeres-und des Brackwassers. *Nuova Notarisia* 4:192–206.

Reinke, J. 1879. Zwei parasitische Algen. *Bot. Zeitung (Berlin)* 37:473–78, pl. 6.

———. 1888. Einige neue braune und grüne Algen der Kieler Bucht. *Ber. Deutsch. Bot. Ges.* 6:240–41.

———. 1889. *Atlas deutscher Meeresalgen.* 1:[i–iv], 1–34, pls. 1–25. Berlin.

Reinsch, P. F. 1875. *Contributiones ad algologiam et fungologiam.* Pp. [i–xii], 1–103 + [1], 131 pls. Leipzig.

Rhyne, C. F. 1973. Field and experimental studies on the systematics and ecology of *Ulva curvata* and *Ulva rotundata. Sea Grant Publ. UNC-SG-73-09:* pp. i–ii, 1–124.

Richardson, J. P. 1978. *Effects of Environmental Factors on the Life Histories and Seasonality of some Inshore Benthic Marine Algae in North Carolina.* Ph.D. thesis. Pp. 1–141. University of North Carolina, Chapel Hill.

———. 1979. Overwintering of *Dictyota dichotoma* (Phaeophyceae) near its northern distribution limit on the east coast of North America. *J. Phycol.* 15:22–26.

———. 1981. Persistence and development of *Dasya baillouviana* (Gmelin) Montagne (Rhodophyceae, Dasyaceae) in North Carolina. *Phycologia* 20:385–91.

———. 1982. Life history of *Bryopsis plumosa* (Hudson) Agardh (Chlorophyceae) in North Carolina, USA. *Bot. Mar.* 25:177–83.

———. 1986. Additions to the marine macroalgal flora of coastal Georgia. *Georgia J. Sci.* 44:131–35.

———. 1987. Floristic and seasonal characteristics of inshore Georgia macroalgae. *Bull. Mar. Sci. Gulf Caribbean* 40:210–19.

Richardson, N. 1970. Studies on the photobiology of *Bangia fuscopurpurea. J. Phycol.* 6:215–19.

Richardson, N., and P. S. Dixon. 1968. Life history of *Bangia fuscopurpurea* (Dillw.) Lyngb. in culture. *Nature* 218:496–97.

Rodríguez de Ríos, N., and Y. Saito. 1985. *Laurencia scoparia* J. Agardh, nuevo sinonimo de *Laurencia filiformis* (C. Agardh) Montagne (Rhodophyta, Ceramiales). *Ernstia* 32:19–28.

Rosanoff, S. 1866. Recherches anatomiques sur les Mélobésiées (*Hapalidium, Melobesia, Lithophyllum* et *Lithothamnion*). *Mém. Soc. Sci. Nat. Cherbourg.* 12:5–112, pls 1–7.

Rosenberg, G., and J. Ramus. 1982a. Ecological growth strategies in the seaweeds *Gracilaria foliifera* (Rhodophyceae) and *Ulva* sp. (Chlorophyceae): Photosynthesis and antenna composition. *Mar. Ecol. Prog. Ser.* 8:233–41.

———. 1982b. Ecological growth strategies in the seaweeds *Gracilaria foliifera* (Rhodophyceae) and *Ulva* sp. (Chlorophyceae): Soluble nitrogen and reserve carbohydrates. *Mar. Biol.* 66:251–59.

———. 1984. Uptake of inorganic nitrogen and seaweed surface area:volume ratios. *Aquatic Bot.* 19:65–72.

Rosenvinge, L. K. 1893. Grønlands Havalger. *Meddel. Grønland.* 3:795–981, 2 pls.

———. 1909. The marine algae of Denmark: Contributions to their natural history. Part I: Introduction; Rhodophyceae I (Bangiales and Nemalionales). *Kongel. Danske Vidensk. Selsk. Skr.*, ser. 7, Naturv. Math. Afd., 7:1–151, pls. I, II.

Roth, A. W. 1797. *Catalecta botanica.* Fasc. 1, pp. i–viii, 1–244, 8 pls. Leipzig.

———. 1800. *Catalecta botanica.* Fasc. 2, pp. [i–x], 1–258, [1–12], 9 pls. Leipzig.

———. 1806. *Catalecta botanica.* Fasc. 3. Pp. [i–viii], 1–350, [1–9], 12 pls. Leipzig.

Rueness, J., and M. Rueness. 1980. Culture and field observations on *Callithamnion bipinnatum* and *C. byssoides* (Rhodophyta, Ceramiales) from Norway. *Sarsia* 65:29–34.

Russell, G. 1966. The genus *Ectocarpus* in Britain. I. The attached forms. *J. Mar. Biol. Assoc. U.K.* 44:601–12.

———. 1967. The ecology of some free-living Ectocarpaceae. *Helgoländer Wiss. Meeresuntersuch.* 15:155–62.

Russell, G., and R. L. Fletcher. 1975. A numerical taxonomic study of the British Phaeophyta. *J. Mar. Biol. Assoc. U.K.* 55:763–83.

Saint-Hilaire, A. 1833. *Voyage dans le district des Diamans et sur le littoral du Brésil.* 2:1–457. Paris.

Saito, Y. 1982. Morphology and infrageneric position of three British species of *Laurencia* (Ceramiales, Rhodophyta). *Phycologia* 21:299–306.

Santelices, B. 1976. Taxonomic and nomenclatural notes on some Gelidiales (Rhodophyta). *Phycologia* 15:165–73.

Sauvageau, C. 1892. Sur quelques algues pheosporées parasites. *J. Bot. (Morot)* 6:1–10, 36–43, 55–59, 76–80, 90–96, 97–106, 124–31.

———. 1897. Sur quelques Myrionémacées. *Ann. Sci. Nat., Bot.*, Sér. 8, 5:161–288.

———. 1927. Sur les problèmes du *Giraudya.* *Bull. Sta. Biol. Arcachon* 24:3–74, 18 figs.

———. 1933. Sur quelques algues phéosporées de Guéthary (Basses-Pyréenées). *Bull. Sta. Biol. Arcachon* 30:1–28.

Scagel, R. F. 1966. Marine algae of British Columbia and northern Washington. I. Chlorophyceae (green algae). *Bull. Natl. Mus. Canada* 207:i–vii, 1–257.

Scagel, R. F., D. J. Garbary, L. Golden and M. W. Hawkes. 1986. *A Synopsis of the Benthic Marine Algae of British Columbia, Northern Washington and Southeast Alaska.* Phycological contribution No. 1., University of British Columbia. vi + 827 pp. Vancouver.

Schmitz, F. 1879. Squamarieen. Pp. 163–65. In Falkenberg, der Zool. Station während der Jahre 1877–78 zusammengestellt. *Mitt. Zool. Sta. Neapel.* 1:218–77.

———. 1889. Systematische Übersicht der bisher bekannten Gattungen der Florideen. *Flora* 72:435–56, pl. 21.

———. 1893. Die Gattung *Lophothalia* J. Ag. *Ber. Deutsch. Bot. Ges.* 11:212–32.

———. 1896. Bangiaceae. In A. Engler and K. Prantl, eds., *Die natürlichen Pflanzenfamilien*, 1(2):305–16. Leipzig.

Schmitz, F., and P. Falkenberg. 1897. Rhodomelaceae. In A. Engler and K. Prantl, eds., *Die natürlichen Pflanzenfamilien*, 1(2):421–80. Leipzig.

Schmitz, F., and H. Hauptfleisch. 1896. Rhodophyllidaceae. In A. Engler and K. Prantl, eds., *Die natürlichen Pflanzenfamilien*, 1(2):366–82. Leipzig.

———. 1897. Ceramiaceae. In A. Engler and K. Prantl, eds., *Die natürlichen Pflanzenfamilien*, 1(2):481–504. Leipzig.

Schneider, C. W. 1974. North Carolina marine algae. III. A community of Ceramiales (Rhodophyta) on a glass sponge from 60 meters. *Bull. Mar. Sci. Gulf Caribbean* 24:1093–1101.

———. 1975a. North Carolina marine algae. V. Additions to the flora of Onslow Bay, including the reassignment of *Fauchea peltata* Taylor to *Weberella* Schmitz. *Brit. Phycol. J.* 10:129–38.

———. 1975b. North Carolina marine algae. VI. Some Ceramiales (Rhodophyta), including a new species of *Dipterosiphonia*. *J. Phycol.* 11:391–96.

———. 1975c. Taxonomic notes on *Gracilaria mammillaris* (Mont.) Howe and *Gracilaria veleroae* Dawson (Rhodophyta, Gigartinales). *Taxon* 24:87–90.

———. 1975d. *Spatial and Temporal Distributions of Benthic Marine Algae on the Continental Shelf of the Carolinas.* Ph.D. thesis, pp. i–xiv, 1–196. Duke University, Durham, N.C.

———. 1976. Spatial and temporal distributions of benthic marine algae on the continental shelf of the Carolinas. *Bull. Mar. Sci. Gulf Caribbean* 26:133–51.

———. 1980. North Carolina marine algae. VIII. The reproductive morphology of *Callithamnion cordatum* Børgesen (Rhodophyta, Ceramiaceae). *Rhodora* 82:321–30.

———. 1983. The red algal genus *Audouinella* Bory (Nemaliales: Acrochaetiaceae) from North Carolina. *Smithson. Contr. Mar. Sci.* 22:i–iv, 1–25.

———. 1984. Studies on *Antithamnionella*, *Callithamniella* and *Calloseris* (Rhodophyta, Ceramiales) from North Carolina, USA. *Phycologia* 23:455–64.

———. 1988. *Craspedocarpus humilis* sp. nov. (Cystocloniaceae, Gigartinales) from North Carolina, and a reappraisal of the genus. *Phycologia* 27:1–9.

———. 1989. Two new species of *Dasysiphonia* (Dasyaceae, Rhodophyta) from the southeastern United States (Carolinas). *Bot. Mar.* 32:521–26.

Schneider, C. W., and N. J. Eiseman. 1979. *Searlesia*, a new genus from the western Atlantic based on *Membranoptera subtropica* (Rhodophyta, Delesseriaceae). *Phycologia* 18:319–24.

Schneider, C. W., and R. P. Reading. 1987. Revision of the genus *Peyssonnelia* (Rhodophyta, Cryptonemiales) from North Carolina, including *P. atlantica* new species. *Bull. Mar. Sci. Gulf Caribbean* 40:175–92.

Schneider, C. W., and R. B. Searles. 1973. North Carolina marine algae. II. New records and observations of the benthic offshore flora. *Phycologia* 12:201–11.

———. 1975. North Carolina marine algae. IV. Further contributions from the continental shelf, including two new species of Rhodophyta. *Nova Hedwigia* 24:83–103.

———. 1976. North Carolina marine algae. VII. New species of *Hypnea* and *Petroglossum* (Gigartinales) and additional records of other Rhodophyta. *Phycologia* 15:51–60.

———. 1979. Standing crop of benthic seaweeds on the Carolina continental shelf. Pp. 293–301. In A. Jensen and J. R. Stein, eds., *Proceedings of the Ninth International Seaweed Symposium*, 293–301. Science Press, Princeton, N.J.

Schnetter, R., I. Hörnig, and G. Weber-Peukert. 1987. Taxonomy of some North Atlantic *Dictyota* species (Phaeophyta). *Hydrobiologia* 151/152:193–97.

Schramm, A., and H. Mazé. 1865. *Essai de classification des algues de la Guadeloupe.* i–ii + 1–52 pp. Basse-Terre.

Seagrief, S. C. 1984. A catalogue of South African green, brown and red marine algae. *Mem. Bot. Sur. S. Africa.* 47:i–vi, 1–72.

Searles, R. B. 1972. North Carolina marine algae. I. Three new species from the continental shelf. *Phycologia* 11:19–24.

———. 1981. Seaweeds from Gray's Reef, Georgia. *Northeast Gulf Sci.* 5:45–48.

———. 1983. Vegetative and reproductive morphology of *Dudresnaya georgiana* sp. nov. (Rhodophyta, Dumontiaceae). *Phycologia* 22:309–16.

———. 1984a. Seaweed biogeography of the mid-Atlantic coast of the United States. *Helgoländer Wiss. Meeresuntersuch.* 38:259–71.

———. 1984b. North Carolina marine algae. XII. *Gloioderma rubrisporum* sp. nov. (Rhodophyta, Rhodymeniales). *Bull. Torrey Bot. Club* 111:217–21.

———. 1987. Phenology and floristics of seaweeds from the offshore waters of Georgia. *Northeast Gulf Sci.* 9:99–108.

———. 1988. An illustrated field and laboratory guide to the seaweeds of Gray's Reef National Marine Sanctuary. NOAA Technical Memorandum Series NOS MEMD 22. Pp. i–xv, 1–122. Washington, D.C.

Searles, R. B., and D. L. Ballantine. 1986. *Dudresnaya puertoricensis* sp. nov. (Dumontiaceae, Gigartinales, Rhodophyta). *J. Phycol.* 22:389–94.

Searles, R. B., M. H. Hommersand, and C. D. Amsler. 1984. The occurrence of *Codium fragile* subsp. *tomentosoides* and *C. taylorii* (Chlorophyta) in North Carolina. *Bot. Mar.* 27:185–87.

Searles, R. B., and G. L. Leister. 1980. North Carolina marine algae. IX. *Onslowia endophytica* gen. et sp. nov. (Phaeophyta, Sphacelariales) and notes on other new records for North Carolina. *J. Phycol.* 16:35–40.

Searles, R. B., and S. M. Lewis. 1982. North Carolina marine algae. XI. A new species of *Helminthocladia* (Liagoraceae, Nemaliales, Rhodophyta) with a reappraisal of the generic limits of *Helminthocladia* and *Helminthora*. *J. Phycol.* 19:164–72.

Searles, R. B., and C. W. Schneider. 1978. A checklist and bibliography of North Carolina seaweeds. *Bot. Mar.* 21:99–108.

———. 1980. Biogeographic affinities of the shallow and deep-water benthic marine algae of North Carolina. *Bull. Mar. Sci. Gulf Caribbean* 30:732–36.

———. 1989. New genera and species of Ceramiaceae (Rhodophyta) from the southeastern United States. *J. Phycol.* 25:731–40.

Sears, J. R., and R. T. Wilce. 1970. Reproduction and systematics of the marine alga *Derbesia* (Chlorophyceae) in New England. *J. Phycol.* 6:381–92.

Setchell, W. A. 1925. Notes on *Microdictyon*. *Univ. Calif. Publ. Bot.* 13:101–7.

Setchell, W. A., and N. L. Gardner. 1930. Marine algae of the Revillagigedo Islands Expedition in 1925. *Proc. Calif. Acad. Sci.*, 4th ser. 19:109–215.

Setchell, W. A., and L. R. Mason. 1943. *Goniolithon* and *Neogoniolithon*: Two genera of crustaceous coralline algae. *Proc. Natl. Acad. U.S.A.* 29:87–92.

Silva, P. C. 1952. A review of nomenclatural conservation in the algae from the point of view of the type method. *Univ. Calif. Publ. Bot.* 25:241–323.

———. 1955. The dichotomous species of *Codium* in Britain. *J. Mar. Biol. Assoc. U.K.* 34:565–77, 1 pl.

———. 1960. *Codium* (Chlorophyta) in the tropical western Atlantic. *Nova Hedwigia* 1:497–536, pls 107–23.

Silva, P. C., E. G. Menez, and R. L. Moe. 1987. Catalog of the benthic marine algae of the Philippines. *Smithsonian Contrib. Mar. Sci.* 27:i–iv, 1–179.

Sluiman, H. J., K. R. Roberts, K. D. Stewart, and K. R. Mattox. 1983. Comparative cytology and taxonomy of the Ulvophyceae. IV. Mitosis and cytokinesis in *Ulothrix* (Chlorophyta). *Acta Bot. Neerl.* 32: 257–69.

Smith, J. E. 1790–1814. *English Botany.* 36 vols. + indices. London.

Solier, A. J. J. 1846. Sur deux algues zoosporées formant le nouveau genre *Derbesia. Rev. Bot. Recueil Mens.* 3:452–54.

Sommerfeld, M. R., and H. W. Nichols. 1973. The life cycle of *Bangia fuscopurpurea* in culture. I: Effects of temperature and photoperiod on the morphology and reproduction of the *Bangia* phase. *J. Phycol.* 9:205–10.

Sommerfelt, S. C. 1826. *Supplementum florae lapponicae.* Pp. i–xii, 1–331, 3 pls. Oslo.

Sonder, O. G. 1845. Nova algarum genera et species, quas in itinere ad oras occidentales Nova Hollandiae, collegit C. Preiss, Ph. Dr. *Bot. Zeitung (Berlin)* 3:49–57.

———. 1871. Die Algen des tropischen Australiens. *Abh. Naturwiss. Naturwiss. Verein. Hamburg* 5:33–74, pls 1–6.

South, G. R. 1974. Contributions to the flora of marine algae of eastern Canada. II. Family Chaetophoraceae. *Naturaliste Canad.* 101:905–23.

South, G. R., and N. M. Adams. 1976. *Erythrotrichia foliiformis* sp. nov. (Rhodophyta, Erythropeltidaceae) from New Zealand. *J. Roy. Soc. New Zealand* 6:399–405.

South, G. R., and I. T. Tittley. 1986. *A checklist and distributional index of the benthic marine algae of the North Atlantic Ocean.* Huntsman Marine Laboratory and British Museum (Natural History), St. Andrews and London. Newfoundland. 1–76 pp.

Spencer K. G., M.-H. Yu, J. A. West, and A. N. Glazer. 1981. Phycoerythrin and interfertility patterns in *Callithamnion* (Rhodophyta) isolates. *Brit. Phycol. J.* 16:331–43.

Sprengel, C. 1827. *Caroli Linnaei... Systema vegetabilium.* Sixteenth ed. 4(1):[i–iv], 1–592. Göttingen.

Stackhouse, J. 1795. *Nereis britannica.* Fasc. 1. Pp. i–viii, 1–30, pls 1–8. Bath.

———. 1797. *Nereis britannica.* Fasc. 2. Pp. ix–xxiv, 31–70, pls 9–13.

———. 1802 ["1801"]. *Nereis britannica.* Fasc. 3. Pp. xxv–xl, 71–112, 1–4, 1–3, pls 13–17, pls A–G.

———. 1809. Tentamen marino-cryptogamicum. *Mém. Soc. Imp. Naturalistes Moscou* 2:50–97.

———. 1816. *Nereis britannica.* Second ed. Pp. i–xii, 1–68 pp., 20 pls. Oxford.

Stefansson, U., and L. P. Atkinson. 1967. Physical and chemical properties of the shelf and slope waters off North Carolina. Tech. Rep., Duke Univ. Mar. Lab., Beaufort, N.C., pp. 1–230.

Stefansson, U., L. P. Atkinson, and D. F. Bumpus. 1971. Hydrographic properties and circulation of the North Carolina shelf and slope waters. *Deep-Sea Res.* 18:383–420.

Stegenga, H., and W. J. Borsje. 1977. The morphology and life history of *Acrochaetium polyblastum* (Rosenv.) Børg. and *Acrochaetium hallandicum* (Kylin) Hamel (Rhodophyta, Nemaliales). *Acta Bot. Neerl.* 26:451–70.

Stegenga, H., and M. Vroman. 1976. The morphology and life history of *Acrochaetium densum* (Drew) Papenfuss (Rhodophyta, Nemaliales). *Acta Bot. Neerl.* 25:257–80.

Stephenson, T. A., and A. Stephenson. 1952. Life between the tide-marks in North America. II. Northern Florida and the Carolinas. *J. Ecol.* 40:1–49, 6 pls.

Stewart, J. 1968. Morphological variation in *Pterocladia pyramidale. J. Phycol.* 4:76–84.

Stewart, J., and J. N. Norris. 1981. Gelidiaceae (Rhodophyta) from the northern Gulf of California, Mexico. *Phycologia* 20:273–84.

Strömfelt, H. F. G. 1884. Om algvegetationen i Finlands sydvestra skärgörd. *Bidrag Kännedom Finlands Natur Folk* 39:112–40.

———. 1888. Algae novae quas ad litora Scandinavica. *Notarisia* 3:381–84.

Sutherland, J. P., and R. H. Karlson. 1977. Development and stability of the fouling community at Beaufort, North Carolina. *Ecol. Monogr.* 47:425–46.

Suyemoto, M. M. 1980. *A Contribution to the Non-Articulated Corallinaceae (Rhodophyta, Cryptonemiales) of Onslow Bay, North Carolina.* M.S. thesis. Pp. i–ix, 1–88. Duke University, Durham, N.C.

Tanaka, J., and M. Chihara. 1982. Morphology and taxonomy of *Mesospora schmidtii* Weber van Bosse, Mesosporaceae fam. nov. (Ralfsiaceae, Phaeophyceae). *Phycologia* 21:382–89.

Tatewaki, M. 1969. Culture studies on the life history of some species of the genus *Monostroma. Sci. Pap. Inst. Algol. Res. Fac. Sci. Hokkaido Imp. Univ.* 6:1–56.

Taylor, W. R. 1928. The marine algae of Florida with special reference to the Dry Tortugas. *Publ. Carnegie Inst. Wash.* 379:[i–v], 1–219, 37 pls.

———. 1937. *Marine Algae of the Northeastern Coast of North America.* Univ. Mich. Stud. Sci. Ser. XIII. Pp. i–vii, [1–3], 1–427. University of Michigan Press, Ann Arbor.

———. 1942. Caribbean marine algae of the Allan Hancock Expedition, 1939. *Allan Hancock Atlantic Exped. Rep.* 2:1–193.

———. 1943. Marine algae from Haiti collected by H. H. Bartlett in 1941. *Pap. Michigan Acad. Sci.* 28:143–63, 4 pls. (1942)

———. 1945. Pacific marine algae of the Allan Hancock expeditions to the Galapagos Islands. *Allan Hancock Pacific Exped.* 12:[3], i–iv, 1–528.

———. 1950. Reproduction of *Dudresnaya crassa* Howe. *Biol. Bull. Mar. Biol. Lab. Woods Hole* 99:272–84.

———. 1955. Marine algal flora of the Caribbean and its extension into neighboring seas. In *Essays in the Natural Sciences in Honor of Captain Allan Hancock,* 259–77. Univ. of Southern California Press, Los Angeles.

———. 1957. *Marine Algae of the Northeastern Coast of North America.* Rev. ed. Univ. of Mich. Stud. Sci. Ser. XIII. Pp. i–ix, 1–509. University of Michigan Press, Ann Arbor.

———. 1960. *Marine Algae of the Eastern Tropical and Subtropical Coasts of the Americas.* Pp. i–xi, 1–879. University of Michigan Press, Ann Arbor.

———. 1961. Notes on three Bermudian marine algae. *Hydrobiologia* 18:277–83.

Taylor, W. R., and A. J. Bernatowicz. 1952. Marine species of *Vaucheria* at Bermuda. *Bull. Mar. Sci. Gulf Caribbean* 2:405–13.

Taylor, W. R., A. B. Joly, and A. J. Bernatowicz. 1953. The relation of *Dichotomosiphon pusillus* to the algal genus *Boodleopsis. Pap. Michigan Acad. Sci.* 38:97–108, pls 1–3. (1952)

Thuret, G. 1854. Note sur la synonomie des *Ulva lactuca* et *latissima,* L.... Ulvacées. *Mém. Soc. Sci. Nat. Cherbourg.* 2:17–32.

Thuret, G., and E. Bornet. 1878. *Etudes phycologiques. Analyses d'algues marines.* Pp. i–iii, 1–105, 51 pls. Paris.

Tokida, J., and T. Masaki. 1956. Studies on the reproductive organs of red algae. II. On *Erythrophyllum gmelini* (Grun.) Yendo. *Bull. Fac. Fish. Hokkaido Univ.* 7:63–71.

Trevisan, V. B. A. 1845. *Nomenclator algarum.* 80 pp. Padoue [Padua].

Turner, D. 1802. *A synopsis of the British Fuci.* Vols. 1 and 2. Pp. i–xlvi, 1–400. Yarmouth.

———. 1808. *Fuci.* Vol. 1, 164 + [2] pp., pls 1–71. London.

———. 1809. *Fuci.* Vol. 2, 162 pp., pls 72–134. London.

———. 1811. *Fuci.* Vol. 3, 148 pp., pls 135–96. London.

———. 1815–19. *Fuci.* Vol. 4, [1] + 153 + [2] + 7 pp., pls 197–258. London.

U.S. Department of Commerce. 1987. *Tide Tables 1987: East Coast of North and South America.* U.S. Department of Commerce, National Oceanic and Atmospheric Administration, National Ocean Service. 289 pp.

Vahl, M. 1802. Endeel kryptogamiske planter fra St.-Croix. *Skr. Naturhist.-Selsk.* 5:29–47.

Vickers, A. 1905. Liste des algues marines de la Barbade. *Ann. Sci. Nat. Bot.,* ser. 9, 1:45–66.

Vinogradova, K. L., and E. Sosa. 1977. Additamenta ad floram Rhodophycearum insulae Cuba. *Nov. Sist. Nizshikh Rast.* 14:8–19.

Weber, F., and D. M. H. Mohr. 1804. *Naturhistorische Reise durch einen Theil Schwedens.* Pp. i–xii, 13–207, 3 pls. Göttingen.

Weber-van Bosse, A. 1898. Monographie des Caulerpes. *Ann. Jard. Bot. Buitenzorg* 15:243–401, pls 20–34.

West, J. A., A. R. Polanshek, and M. D. Guiry. 1977. The life history in culture of *Petrocelis cruenta* J. Ag. (Rhodophyta) from Ireland. *Brit. Phycol. J.* 12:45–53.

Whittick, A., and R. G. Hooper. 1977. The reproduction and phenology of *Antithamnion cruciatum* (Rhodophyta: Ceramiaceae) in insular Newfoundland. *Canad. J. Bot.* 55:520–24.

Whitford, L. A., and G. J. Schumacher. 1969. *A Manual of the Fresh-water Algae of North Carolina.* Tech. Bull. No. 188. Pp. [4] + 1–313. North Carolina Agricultural Experiment Station, Raleigh.

Wilkinson, M., and E. M. Burrows. 1972. An experimental taxonomic study of the algae confused under the name *Gomontia polyrhiza. J. Mar. Biol. Assoc. U.K.* 52:49–57.

Wille, N. 1901. Studien über Chlorophyceen. I–VIII. *Skr. Vidensk.-Selsk. Christiana, Math.-Naturvidensk. Kl.* 1900(6):3–46, 3 pls.

Williams, L. G. 1948a. Seasonal alternation of marine floras at Cape Lookout, North Carolina. *Amer. J. Bot.* 35:682–95.

———. 1948b. The genus *Codium* in North Carolina. *J. Elisha Mitchell Sci. Soc.* 64:107–16.

———. 1949. Marine algal ecology at Cape Lookout, North Carolina. *Bull. Furman Univ.* 31:1–21.

———. 1951. Algae of the Black Rocks. In A. S. Pearse and L. G. Williams, The biota of the reefs off the Carolinas, 149–59. *J. Elisha Mitchell Sci. Soc.* 67:133–63.

Wiseman, D. R. 1966. *A Preliminary Survey of the Rhodophyta of South Carolina.* M.A. thesis, pp. i–v, 1–190. Duke University, Durham, N.C.

———. 1978. Benthic marine algae. In R. G. Zingmark, ed., *An Annotated Checklist of the Biota of the Coastal Zone of South Carolina,* 23–36. Univ. of South Carolina Press, Columbia.

Wiseman, D. R., and C. W. Schneider. 1976. Investigations of the marine algae of South Carolina. I. New records of Rhodophyta. *Rhodora* 78:516–24.

Withering, W. 1796. *An arrangement of British plants.* Third ed., 4:1–418, pls 17, 18, 31. Birmingham.

Woelkerling, W. J. 1971. Morphology and taxonomy of the *Audouinella* complex (Rhodophyta) in southern Australia. *Austral. J. Bot.,* Suppl. 1:1–91.

———. 1972. Studies on the *Audouinella microscopica* (Naeg.) Woelk. complex (Rhodophyta). *Rhodora* 74:85–96.

————. 1973a. The *Audouinella* complex (Rhodophyta) in the western Sargasso Sea. *Rhodora* 75:78–101.

————. 1973b. The morphology and systematics of the *Audouinella* complex (Acrochaetiaceae, Rhodophyta) in northeastern United States. *Rhodora* 75:529–621.

————. 1983. The *Audouinella* (*Acrochaetium-Rhodochorton*) complex (Rhodophyta): present perspectives. *Phycologia* 22:59–92.

————. 1988. *The Coralline Red Algae: An Analysis of the Genera and Subfamilies of Nongeniculate Corallinaceae.* xi + 268 pp. Brit. Mus. (Nat. Hist.), Oxford Univ. Press, London and Oxford.

Woelkerling, W. J., Y. M. Chamberlain, and P. C. Silva. 1985. A taxonomic and nomenclatural reassessment of *Tenarea, Titanoderma* and *Dermatolithon* (Corallinaceae, Rhodophyta) based on studies of type and other critical specimens. *Phycologia* 24:317–37.

Wolfe, J. J. 1918. Alternation and parthenogenesis in *Padina*. *J. Elisha Mitchell Sci. Soc.* 31:12–26.

Wollaston, E. M. 1968. Morphology and taxonomy of southern Australian genera of Crouanieae Schmitz (Ceramiaceae, Rhodophyta). *Austral. J. Bot.* 16:217–417.

Womersley, H. B. S. 1967. A critical survey of the marine algae of southern Australia. II. Phaeophyta. *Austral. J. Bot.* 15:189–270.

————. 1978. Southern Australian species of *Ceramium* Roth (Rhodophyta). *Austral. J. Mar. Freshwater Res.* 29:205–57.

————. 1979. Southern Australian species of *Polysiphonia* Greville (Rhodomelaceae). *Austral. J. Bot.* 27:459–528.

————. 1984. *The Marine Benthic Flora of Southern Australia.* Part I. 329 pp. Adelaide.

————. 1987. *The Marine Benthic Flora of Southern Australia.* Part II. 484 pp. Adelaide.

Womersley, H. B. S., and A. Bailey. 1970. Marine algae of the Solomon Islands. *Philos. Trans.*, ser. B, 259:257–352, 4 pls.

Woodward, T. J. 1794. Description of *Fucus dasyphyllus*. *Trans. Linn. Soc. London* 2:239–41, pl. 23, figs 1–3.

————. 1797. Observations upon the generic character of *Ulva*, with descriptions of some new species. *Trans. Linn. Soc. London* 3:46–58.

Wulfen, F. X. 1786. In N. J. Jacquin, ed., *Collectanea ad botanicam, chemiam, et historiam naturalem spectantia,...* Vol. 1, 386 pp., 22 pls. Vienna.

————. 1803. Cryptogamia aquatica. *Arch. Bot. (Leipzig)* 3:1–64, pl. 1.

Wynne, M. J. 1984. The correct name for the type of *Hypoglossum* Kützing (Delesseriaceae, Rhodophyta). *Taxon* 33:85–87.

————. 1985a ["1984"]. Notes on *Herposiphonia* (Rhodomelaceae, Rhodophyta) in South Africa, with a description of a new species. *Cryptog. Algol.* 5:167–77.

————. 1985b. Concerning the names *Scagelia corallina* and *Heterosiphonia wurdemannii* (Ceramiales, Rhodophyta). *Cryptog. Algol.* 6:81–90.

————. 1985c. Nomenclatural assessment of *Goniotrichum* Kützing, *Erythrotrichia* Areschoug, *Diconia* Harvey, and *Stylonema* Reinsch (Rhodophyta). *Taxon* 34:502–5.

————. 1986. A checklist of benthic marine algae of the tropical and subtropical western Atlantic. *Canad. J. Bot.* 64:2239–81.

————. 1989. Towards the resolution of taxonomic and nomenclatural problems concerning the typification of *Acrosorium uncinatum* (Delesseriaceae, Rhodophyta). *Brit. Phycol. J.* 24:245–52.

————. 1990. Observations on *Haraldia* and *Calloseris*, two rare Delesseriaceae (Rhodophyta) from the western Atlantic. *Contr. Univ. Michigan Herb.* 17:327–34.

Wynne, M. J., and D. L. Ballantine. 1985. Notes on the marine algae of Puerto Rico. IV.

The taxonomic placement of *Grallatoria* (Ceramiaceae, Rhodophyta). *Cryptog. Algol.* 6:219–29.

———. 1986. The genus *Hypoglossum* Kützing (Delesseriaceae, Rhodophyta) in the tropical western Atlantic, including *H. anomalum* sp. nov. *J. Phycol.* 22:185–93.

Yamamoto, H. 1975. The relationship between *Gracilariopsis* and *Gracilaria* from Japan. *Bull. Fac. Fish. Hokkaido Univ.* 26:217–22.

———. 1978. Systematic and anatomical study of the genus *Gracilaria* in Japan. *Mem. Fac. Fish. Hokkaido Univ.* 25:97–152.

Yarish, C. 1975. A cultural assessment of the taxonomic criteria of selected marine Chaetophoraceae (Chlorophyta). *Nova Hedwigia* 26:385–430.

Zanardini, G. 1840. Sopra le alghe del mare Adriatico. Lettera seconda... alla Direzione della Biblioteca Italiana. *Biblioteca Italiana* (Milano), 99:195–229.

———. 1841. Synopsis algarum in mari Adriatico hucusque collectarum. *Mem. Reale Accad. Sci. Torino,* ser. 2, 4:105–255, pls 1–8.

———. 1843. *Saggio di classificazione naturale delle Ficee.* 1–64 pp., [1] pl. Venezia.

———. 1844. Corallinee. (Polipae calciferi di Lamouroux). In *Enciclopedia italiana* 6:1013–36. Venezia.

———. 1866. Scelta di Ficee nuove o piú rare dei mari Mediterraneo ed Adriatico. *Mem. Reale Ist. Veneto Sci. Lett. Arti* 13:143–76, pls 2–9.

———. 1869. Scelta di Ficee nuove o più rare dei mari Mediterraneo ed Adriatico. Part 10. *Mem. Reale Inst. Veneto Sci. Lett. Arti* 14:439–76, pls 73–80.

Zaneveld, J. S., and W. M. Willis. 1976. The marine algae of the American coast between Cape May, N.J., and Cape Hatteras, N.C. III. The Phaeophyceae. *Bot. Mar.* 19:33–46.

INDEX TO SCIENTIFIC NAMES

Note: Accepted names are in Roman type, synonyms are in italics. Primary listings are indicated in boldface type.

About the Authors

Craig W. Schneider is Professor of Biology at Trinity College, Hartford, Connecticut. Richard B. Searles is Professor of Botany at Duke University. Both scholars are veterans of numerous oceanographic expeditions and offshore dives.